CONTRIBUTORS

Terence Bennett
M. Berger
Walter J. Bock
William A. Calder
J. S. Hart
James R. King
M. Menaker
A. Oksche

AVIAN BIOLOGY
Volume IV

AVIAN BIOLOGY
Volume IV

EDITED BY

DONALD S. FARNER
Department of Zoology
University of Washington
Seattle, Washington

JAMES R. KING
Department of Zoology
Washington State University
Pullman, Washington

TAXONOMIC EDITOR

KENNETH C. PARKES
Curator of Birds
Carnegie Museum
Pittsburgh, Pennsylvania

1974

ACADEMIC PRESS   New York and London
A Subsidiary of Harcourt Brace Jovanovich, Publishers

COPYRIGHT © 1974, BY ACADEMIC PRESS, INC.
ALL RIGHTS RESERVED.
NO PART OF THIS PUBLICATION MAY BE REPRODUCED OR
TRANSMITTED IN ANY FORM OR BY ANY MEANS, ELECTRONIC
OR MECHANICAL, INCLUDING PHOTOCOPY, RECORDING, OR ANY
INFORMATION STORAGE AND RETRIEVAL SYSTEM, WITHOUT
PERMISSION IN WRITING FROM THE PUBLISHER.

ACADEMIC PRESS, INC.
111 Fifth Avenue, New York, New York 10003

United Kingdom Edition published by
ACADEMIC PRESS, INC. (LONDON) LTD.
24/28 Oval Road, London NW1

Library of Congress Cataloging in Publication Data
Main entry under title:

Avian biology.

Includes bibliographies.
1. Ornithology.   2. Zoology–Ecology.   I. Farner,
Donald Stanley, Date   ed.   II. King, James Roger,
ed.   [DNLM: 1. Birds.   QL673 F235a]
QL673.A9   598.2   79-178216
ISBN 0-12-249404-0 (v. 4)

PRINTED IN THE UNITED STATES OF AMERICA

These volumes are dedicated to the memory of
A. J. "JOCK" MARSHALL
(1911–1967)

whose journey among men was too short by half

CONTENTS

LIST OF CONTRIBUTORS	ix
PREFACE	xi
NOTE ON TAXONOMY	xv
CONTENTS OF OTHER VOLUMES	xix
OBITUARY	xxi

Chapter 1. The Peripheral and Autonomic Nervous Systems
Terence Bennett

I.	Introduction	1
II.	Peripheral Nervous System	2
III.	Autonomic Nervous System	15
	References	46

Chapter 2. The Avian Pineal Organ
M. Menaker and A. Oksche

I.	Function	80
II.	Structure	89
III.	General Conclusions	109
	References	114

Chapter 3. The Avian Skeletomuscular System
Walter J. Bock

I.	Introduction	120
II.	Descriptive Morphology	124
III.	Functional Morphology	137

IV.	Physiological Adaptation	223
V.	Ecological Morphology and Adaptation	226
VI.	Comparative Morphology and Systematics	228
VII.	Epilogue	250
	References	250

Chapter 4. Thermal and Caloric Relations of Birds

William A. Calder and James R. King

I.	Introduction	260
II.	A Simplified Heat Exchange Model	263
III.	Basic Principles of Heat Transfer	287
IV.	Physiological Responses to Heat and Cold	309
V.	Hypothermia	343
VI.	Integration of Thermal and Caloric Responses	354
VII.	Evolutionary Aspects of Thermoregulation and Energy Metabolism	386
	References	393

Chapter 5. Physiology and Energetics of Flight

M. Berger and J. S. Hart

I.	Introduction	416
II.	Respiratory Mechanics	416
III.	Respiratory Gas Exchange	426
IV.	Circulation	434
V.	Temperature Regulation	443
VI.	Water Loss	455
VII.	Energy Turnover in Migratory Flights	459
	References	467

AUTHOR INDEX	479
INDEX TO BIRD NAMES	494
SUBJECT INDEX	499

LIST OF CONTRIBUTORS

Numbers in parentheses indicate the pages on which the authors' contributions begin.

TERENCE BENNETT (1), Department of Physiology, The Medical School, The University of Nottingham, Nottingham, England

M. BERGER (415), Landesmuseum für Naturkunde, Münster, Germany

WALTER J. BOCK (119), Department of Biological Sciences, Columbia University, New York, New York

WILLIAM A. CALDER, JR. (259), Department of Biological Sciences, University of Arizona, Tucson, Arizona

J. S. HART* (415), Division of Biological Sciences, National Research Council of Canada, Ottawa, Ontario, Canada

JAMES R. KING (259), Department of Zoology, Washington State University, Pullman, Washington

M. MENAKER (79), Department of Zoology, The University of Texas at Austin, Austin, Texas

A. OKSCHE (79), Anatomisches Institut der Justus Liebig–Universität, Giessen, Federal Republic of Germany

*Deceased.

PREFACE

The birds are the best-known of the large and adaptively diversified classes of animals. About 8600 living species are currently recognized, and it is unlikely that more than a handful of additional species will be discovered. Although much remains to be learned, the available knowledge of the distribution of living species is much more nearly complete than that for any other class of animals. It is noteworthy that the relatively advanced status of our knowledge of birds is attributable to a very substantial degree to a large group of dedicated and skilled amateur ornithologists.

Because of the abundance of basic empirical information on distribution, habitat requirements, life cycles, breeding habits, etc., it has been relatively easier to use birds instead of other animals in the study of the general aspects of ethology, ecology, population biology, evolutionary biology, physiological ecology, and other fields of biology of contemporary interest. Model systems based on birds have played a prominent role in the development of these fields. The function of this multivolume treatise in relation to the place of birds in biological science is therefore envisioned as twofold. We intend to present a reasonable assessment of selected aspects of avian biology for those for whom this field is their primary interest. But we view as equally important the contribution of these volumes to the broader fields of biology in which investigations using birds are of substantial significance.

Only slightly more than a decade has passed since the publication of A. J. Marshall's "Biology and Comparative Physiology of Birds," but progress in most of the fields included in this treatise has made most of the older chapters obsolete. Avian biology has shared in the so-called information explosion. The number of serial publications devoted mainly to avian biology has increased by about 20% per decade since

1940, and the spiral has been amplified by the parallel increase in page production and by the spread of publication into ancillary journals. By 1964, there were about 215 exclusively ornithological journals and about 245 additional serials publishing appreciable amounts of information on avian biology (P. A. Baldwin and D. E. Oehlerts, *Studies in Biological Literature and Communications, No. 4. The Status of Ornithological Literature, 1964.* Biological Abstracts, Inc., Philadelphia, 1964).

These stark numbers reflect only the quantitative acceleration in the output of information in recent times. The qualitative changes have been much more impressive. Avifaunas that were scarcely known except as lists of species a decade ago have become accessible to scientific inquiry as a consequence of improved transportation and facilities in many parts of the world. Improved or new instrumentation has allowed the development of new fields of study and has extended the scope of old ones. Examples that come readily to mind include the use of radar in visualizing migration, of telemetry in studying the physiology of flying birds, and of spectrography in analyzing bird sounds. The development of mathematical modeling, for instance in evolutionary biology and population ecology, has supplied new perspectives for old problems and has created a new arena for the examination of empirical data. All of these developments — social, practical, and theoretical — have profoundly affected many aspects of avian biology in the last decade. It is now time for another inventory of information, hypotheses, and new questions.

Marshall's "Biology and Comparative Physiology of Birds" was the first treatise in the English language that regarded ornithology as consisting of more than anatomy, taxonomy, oology, and life history. This viewpoint was in part a product of the times, but it also reflected Marshall's own holistic philosophy and his understanding that "life history" had come to include the whole spectrum of physiological, demographic, and behavioral adaptation. This treatise is a direct descendent of Marshall's initiative. We have attempted to preserve the view that ornithology belongs to anyone who studies birds, whether it be on the level of molecules, individuals, or populations. To emphasize our intentions we have called the work "Avian Biology."

It has been proclaimed by various oracles that sciences based on taxonomic units (such as insects, birds, or mammals) are obsolete, and that the forefront of biology is process-oriented rather than taxon-oriented. This narrow vision of biology derives from the hyperspecialization that characterizes so much of science today. It fails to notice that lateral synthesis as well as vertical analysis are inseparable partners in the search for biological principles. Avian biologists of both stripes have together contributed a disproportionately large share of the information and thought that have produced contemporary principles in zoogeography, systematics, ethology, demography, comparative physiology, and other fields too numerous to mention.

In part, this progress results from the attributes of birds themselves. They are active and visible during the daytime; they have diversified into virtually all major habitats and modes of life; they are small enough to be studied in useful numbers but not so small that observation is difficult; and, not least, they are esthetically attractive. In short, they are relatively easy to study. For this reason we find gathered beneath the rubric of avian biology an alliance of specialists and generalists who regard birds as the best natural vehicle for the exploration of process and pattern in the biological realm. It is an alliance that seems still to be increasing in vigor and scope.

In the early planning stages we established certain working rules that we have been able to follow with rather uneven success.

1. "Avian Biology" is the conceptual descendent of Marshall's earlier treatise, but is more than simply a revision of it. We have deleted some topics and added or extended others. Conspicuous among the deletions is avian embryology, a field that has expanded and specialized to the extent that a significant review of recent advances would be a treatise in itself.

2. Since we expect the volumes to be useful for reference purposes as well as for instruction of advanced students, we have asked authors to summarize established facts and principles as well as to review recent advances.

3. We have attempted to arrange a balanced account of avian biology as its exists at the beginning of the 1970's. We have not only retained chapters outlining modern concepts of structure and function in birds, as is traditional, but have also encouraged contributions representing a multidisciplinary approach and synthesis of new points of view. Several such chapters appear in this volume.

4. We have attempted to avoid a parochial view of avian biology by seeking diversity among authors with respect to nationality, age, and ornithological heritage. In this search we have benefited by advice from many colleagues to whom we are grateful.

5. As a corollary of the preceding point, we have not intentionally emphasized any single school of thought, nor have we sought to dictate the treatment given to controversial subjects. Our single concession to conceptual uniformity is in taxonomic usage, as explained by Kenneth Parkes in the Note on Taxonomy.

We began our work with a careful plan for a logical topical sequence through all volumes. Only its dim vestiges remain. For a number of reasons we have been obliged to sacrifice logical sequence and have given first priority to the maintenance of general quality, trusting that each reader would supply logical cohesion by selecting chapters that are germane to his individual interests.

DONALD S. FARNER
JAMES R. KING

NOTE ON TAXONOMY

Early in the planning stages of "Avian Biology" it became apparent to the editors that it would be desirable to have the manuscript read by a taxonomist, whose responsibility it would be to monitor uniformity of usage in classification and nomenclature. Other multiauthored compendia have been criticized by reviewers for use of obsolete scientific names and for lack of concordance from chapter to chapter. As neither of the editors is a taxonomist, they invited me to perform this service.

A brief discussion of the ground rules that we have tried to follow is in order. Insofar as possible, the classification of birds down to the family level follows that presented by Dr. Storer in Chapter 1, Volume I.

Within each chapter, the first mention of a species of wild bird includes both the scientific name and an English name, or the scientific name alone. If the same species is mentioned by English name later in the same chapter, the scientific name is usually omitted. Scientific names are also usually omitted for domesticated or laboratory birds. The reader may make the assumption throughout the treatise that, unless otherwise indicated, the following statements apply:

1. "The duck" or "domestic duck" refers to domesticated forms of *Anas platyrhynchos.*

2. "The goose" or "domestic goose" refers to domesticated forms of *Anser anser.*

3. "The pigeon" or "domesticated pigeon" or "homing pigeon" refers to domesticated forms of *Columba livia.*

4. "The turkey" or "domestic turkey" refers to domesticated forms of *Meleagris gallopavo.*

5. "The chicken" or "domestic fowl" refers to domesticated forms of *Gallus gallus;* these are often collectively called *"Gallus domesticus"* in biological literature.

6. "Japanese Quail" refers to laboratory strains of the genus *Coturnix,* the exact taxonomic status of which is uncertain. See Moreau and Wayre, *Ardea* **56**, 209–227, 1968.

7. "Canary" or "domesticated canary" refers to domesticated forms of *Serinus canarius.*

8. "Guinea Fowl" or "Guinea Hen" refers to domesticated forms of *Numida meleagris.*

9. "Ring Dove" refers to domesticated and laboratory strains of the genus *Streptopelia,* often and incorrectly given specific status as *S. "risoria."* Now thought to have descended from the African Collared Dove (*S. roseogrisea*), the Ring Dove of today *may* possibly be derived in part from *S. decaocto* of Eurasia; at the time of publication of Volume 3 of Peters' "Check-list of Birds of the World" (p. 92, 1937), *S. decaocto* was thought to be the direct ancestor of *"risoria."* See Goodwin, *Pigeons and Doves of the World* **129**, 1967.

As mentioned above, an effort has been made to achieve uniformity of usage, both of scientific and English names. In general, the scientific names are those used by the Peters "Check-list"; exceptions include those orders and families covered in the earliest volumes for which more recent classifications have become widely accepted (principally Anatidae, Falconiformes, and Scolopacidae). For those families not yet covered by the Peters' list, I have relied on several standard references. For the New World I have used principally Meyer de Schauensee's "The Species of Birds of South America and Their Distribution" (1966), supplemented by Eisenmann's "The Species of Middle American Birds" (*Trans. Linnaean Soc. New York* **7**, 1955). For Eurasia I have used principally Vaurie's "The Birds of the Palaearctic Fauna" (1959, 1965) and Ripley's "A Synopsis of the Birds of India and Pakistan" (1961). There is so much disagreement as to classification and nomenclature in recent checklists and handbooks of African birds that I have sometimes had to use my best judgment and to make an arbitrary choice. For names of birds confined to Australia, New Zealand, and other areas not covered by references cited above, I have been guided by recent regional checklists and by general usage in recent literature. English names have been standardized in the same way, using many of the same reference works. In both the United States and Great Britain, the limited size of the avifauna has given rise to some rather provincial English names; I have added appropriate (and often previously used) adjectives to these. Thus *Sturnus vulgaris* is "European Starling," not simply "Starling"; *Cardinalis cardinalis* is "North American Cardinal," not simply "Cardinal"; and *Ardea cinerea* is "Gray Heron," not simply "Heron."

Reliance on a standard reference, in this case Peters, has meant that certain species appear under scientific names quite different from those used in most of the ornithological literature. For example, the Zebra Finch, widely used as a laboratory species, was long known as *Taeniopygia castanotis*. In Volume 14 of the Peters' "Check-list" (pp. 357–358, 1968), *Taeniopygia* is considered a subgenus of *Poephila*, and *castanotis* a subspecies of *P. guttata*. Thus the species name of the Zebra Finch becomes *Poephila guttata*. In such cases, the more familiar name will usually be given parenthetically.

For the sake of consistency, scientific names used in Volume I will be used throughout "Avian Biology," even though these may differ from names used in standard reference works that would normally be followed, but which were published after the editing of Volume I had been completed.

Strict adherence to standard references also means that some birds will appear under names that, for either taxonomic or nomenclatorial reasons, would *not* be those chosen by either the chapter author or the taxonomic editor. As a taxonomist, I naturally hold some opinions that differ from those of the authors of the Peters' list and the other reference works used. I feel strongly, however, that a general text such as "Avian Biology" should not be used as a vehicle for taxonomic or nomenclatorial innovation, or for the furtherance of my personal opinions. I therefore apologize to those authors in whose chapters names have been altered for the sake of uniformity, and offer as solace the fact that I have had my objectivity strained several times by having to use names that do not reflect my own taxonomic judgment.

KENNETH C. PARKES

CONTENTS OF OTHER VOLUMES

Volume I

Classification of Birds
 Robert W. Storer

Origin and Evolution of Birds
 Pierce Brodkorb

Systematics and Speciation in Birds
 Robert K. Selander

Adaptive Radiation of Birds
 Robert W. Storer

Patterns of Terrestrial Bird Communities
 Robert MacArthur

Sea Bird Ecology and the Marine
  Environment
 N. Philip Ashmole

Biology of Desert Birds
 D. L. Serventy

Ecological Aspects of Periodic
  Reproduction
 Klaus Immelmann

Population Dynamics
 Lars von Haartman

Ecological Aspects of Reproduction
 Martin L. Cody

Ecological Aspects of Behavior
 Gordon Orians

AUTHOR INDEX – INDEX TO BIRD NAMES
 – SUBJECT INDEX

Volume II

The Integument of Birds
 Peter Stettenheim

Patterns of Molting
 Ralph S. Palmer

Mechanisms and Control of Molt
 Robert B. Payne

The Blood-Vascular System of Birds
 David R. Jones and Kjell Johansen

Respiratory Function in Birds
 Robert C. Laswieski

Digestion and the Digestive System
 Vinzenz Ziswiler and Donald S. Farner

The Nutrition of Birds
 Hans Fisher

The Intermediary Metabolism of Birds
 Robert L. Hazelwood

Osmoregulation and Excretion in Birds
 Vaughan H. Shoemaker

AUTHOR INDEX – INDEX TO BIRD NAMES
 – SUBJECT INDEX

Volume III

Reproduction in Birds
 B. Lofts and R. K. Murton

The Adenohypophysis
  A. Tixier-Vidal and B. K. Follett
The Peripheral Endocrine Glands
  Ivan Assenmacher
Neuroendocrinology in Birds
  Hideshi Kobayashi and Masaru Wada
Avian Vison
  Arnold J. Sillman
Chemoreception
  Bernice M. Wenzel

Mechanoreception
  J. Schwartzkopff
Behavior
  Robert A. Hinde

AUTHOR INDEX – INDEX TO BIRD NAMES
  – SUBJECT INDEX

OBITUARY
J. Sanford Hart (1916–1973)

J. Sanford Hart, coauthor of Chapter 5, died on May 6, 1973. For one like myself who has spent much time comparing metabolic processes and thermoregulation in mammals, it appeared that Hart's venture into studies of metabolic energetics of birds in flight would depart from the long and elaborate experiences, and sometimes dull conventions, of mammalian physiology. Mammalian physiologists have provided only meager comparative views of such apparently intense and prolonged metabolic activities as those which transport birds in their long, swift flights. As Hart outlined his intentions, I looked forward with excitement to the new physiological insights that he would develop, if measurements could be made, to show the essential processes involved in expenditure of energy and regulation of temperature by birds in flight.

Within a few years, in a series of brilliantly designed and beautifully executed experiments, Hart had measured the metabolic energetics in flight and the essential partition of dissipation of metabolic heat between conduction from surfaces and evaporation. The measurements involved consideration for the special respiratory system of avian lungs and air sacs that are so different from the lungs of mammals. He utilized the rather few, and sometimes good reports of other studies in avian physiology, restricted as they often were by severe experimental limitations. He combined their useful information with the more comprehensive views derived from his own experimentation and the careful logic for which he has been noted.

There are still many kinds of birds and modes of flight, with specific characteristics in performances in various circumstances. But I venture that Hart has opened a new road to understanding the essential physiology of the metabolic energetics of flight. In doing this he has enlarged our general biological comprehension of the metabolic provision for the motion of animals. He has made this special contribution for our information and edification in his last years. I think that it will especially satisfy us to see the exposition of this thoroughly original addition to knowledge by the colleague and friend whom we so long regarded with admiration and deep affection.

LAURENCE IRVING
Institute of Arctic Biology
University of Alaska
College, Alaska

Chapter 1

PERIPHERAL AND AUTONOMIC NERVOUS SYSTEMS

Terence Bennett

I.	Introduction	1
II.	Peripheral Nervous System	2
	A. Definition	2
	B. General Anatomy	2
	C. Somatic Efferent Innervation	9
	D. Somatic Afferent Innervation	11
	E. Visceral Afferent Innervation	13
III.	Autonomic Nervous System	15
	A. Definition	15
	B. General Anatomy	16
	C. Peripheral Neuroeffector Systems	22
	D. Autonomic Ganglia	25
	E. Autonomic Innervation of Effector Tissues	31
	F. Autonomic Denervation	45
References		46

I. Introduction

It is, of course, not possible in a single chapter to present a detailed account of the peripheral and autonomic nervous systems. The material dealt with has therefore been deliberately selected to give an impression of the more recent studies. In many cases individual investigations are not discussed in detail, but they are cited in the list of references.

II. Peripheral Nervous System

A. DEFINITION

The peripheral nervous system is here defined as that portion of the nervous system that lies outside the main confines of the brain and spinal cord. In the form of the cranial and spinal nerves, the peripheral nervous system is concerned with the efferent and afferent innervation of skeletal muscle and, strictly, with the afferent innervation of all tissues (but see Section II,E below).

Although peripheral efferent fibers innervate skeletal muscle, it does not follow that the innervation of all striated muscle is provided by the peripheral nervous system. In birds, the intrinsic musculature of the eye is composed largely of striated muscle fibers, yet it is quite clear that it is innervated by autonomic nerve fibers (see Sections III,D,1 and III,E,8 below).

B. GENERAL ANATOMY

1. Cranial Nerves

The disposition of the cranial nerves in birds has received the attention of many investigators. Among the useful contributions are those of Bonsdorff (1852a,b), Marshall (1878), Laffont (1885), Magnien (1885a,b), Rochas (1885a,b,c, 1886, 1887), Marage (1889), Staderini (1889), Gadow (1891), Couvreur (1892), Thébault (1898), Jaquet (1901), Cords (1904), Carpenter (1906, 1911), Kaupp (1918), Terni (1924), Simonetta (1933), Nolf (1934a), Stiemens (1934), Coulouma (1935, 1939c), Kappers *et al.* (1936), Smith, (1941), Tixier-Durivault (1942), Hsieh (1951), Grewe (1951), Barnikol (1954), Webb (1957), Watanabe (1960, 1964, 1968), Malinovský (1962), Fedde *et al.* (1963), Chang (1964), Lucas and Stettenheim (1965), Watanabe *et al.* (1967), Karten and Hodos (1967), Watanabe and Yasuda (1968, 1970), Bubien-Waluszewska (1968), Evans (1969), Jungherr (1969), Yasuda and Lepkovsky (1969), Cohen and Schnall (1970), and Cohen *et al.* (1970).

The terminal distributions of the fibers of various cranial nerves are considered in the sections dealing with the innervation of specific effector tissues. The following generalized account and diagrammatic summary (Fig. 1) are included for completeness.

a. Olfactory Nerve (I). The central projections of the afferent fibers of the olfactory nerve are associated with the fila olfactoria of the olfactory bulbs. The nerve arises from the latter, through numerous fine rootlets, and runs into the ocular cavity through the olfactory foramen. It passes forward along the dorsal edge of the orbital septum and, anteriorly, divides into internal and external branches that supply the specialized olfactory mucosa of the dorsal turbinate. The nervus terminalis is reportedly absent in the adult fowl (Watanabe and Yasuda, 1968; see Simonetta, 1933). The majority of fibers in the olfactory nerves are sensory, but there are indications that efferent fibers may join the olfactory nerves from the ethmoidal ganglion (see Hsieh, 1951).

b. Optic Nerve (II). The optic nerve is, strictly speaking, not a cranial nerve, but an extracranial fiber tract. The major optic centers are represented by the optic tectum, the lateral geniculate nucleus, and the nucleus of the basal optic root. The nerve runs from the optic chiasma, below the thalamus, through the optic foramen, and laterally, to the eye. Apart from the numerous afferent fibers in the nerve, there is good evidence for the presence of efferent fibers.

c. Oculomotor Nerve (III). The oculomotor nucleus may be divided into an accessory (Edinger–Westphal) nucleus and the chief oculomotor nucleus with dorsomedial, dorsolateral, and ventromedial components. The accessory nucleus forms a linear mass in the periventricular gray at the rostral level of the nucleus ruber, comes to lie dorsal to the chief oculomotor nucleus, and terminates at isthmo-optic levels. The accessory nucleus contains the preganglionic cell bodies whose axons synapse on neurons in the ciliary ganglion. The efferent fibers of the oculomotor nerve appear rostrally in sections through the accessory nucleus and are present through most of the chief nucleus. They emerge close to the midline of the mesencephalon on either side of the mammillary body. The oculomotor nerve runs through the olfactory foramen; apart from the fibers to the ciliary ganglion (see Fig. 1), the nerve gives rise to a superior branch to the superior rectus muscle of the eye and an inferior branch that supplies the inferior rectus, medial rectus, and ventral oblique muscles of the eye. Branches to the superior and inferior palpebral muscles and the Harderian gland have also been reported (see Slonaker, 1918). The majority of efferent fibers in the oculomotor nerve presumably derive from the oculomotor nuclei, but due to the association of the ventral branch of the oculomotor nerve with the superior branch of the nerve of the pterygoid canal (Vidian nerve) (See Fig. 1; Hsieh, 1951), fibers from other sources may run with the oculomotor nerve.

d. Trochlear Nerve (IV). The trochlear nucleus is continuous with the dorsomedial and dorsolateral regions of the chief oculomotor nucleus. The trochlear nerve emerges from the dorsolateral surface of the mesencephalon and runs through the trochlear foramen to supply the dorsal oblique muscle of the eye with its motor innervation.

e. Trigeminal Nerve (V). The trigeminal nerve has both sensory and motor centers. The principal sensory nucleus lies in the region of the lateral pole of the fourth ventricle. The ventral, secondary center is represented by the nucleus of the descending root, which lies caudal to the principal nucleus. The nucleus of the mesencephalic root is represented by large neurons that lie in a discontinuous row in the region of the optic tectum. The sensory root fibers of the trigeminal run directly to the caudal pole of principal sensory nucleus. They are the central projections of cell bodies lying in the Gasserian (semilunar) ganglion; the peripheral extensions of these neurons run in the ophthalmic, maxillary, and mandibular branches of the nerve. The ophthalmic branch passes to the dorsal part of the nasal septum; the maxillary branch to the nasal, maxillary, and palatine regions; the mandibular branch supplies the muscles and glands of the jaws and face. The sensory fibers are concerned with the sensations of temperature and pain and with the proprioceptor innervation of the masticatory muscles (Vegetti and Palmieri, 1965, Bortolami and Vegetti, 1967). The trigeminal motor nucleus is divided into dorsal and ventral portions; the ventral motor nucleus appears medial to the tail of the principal sensory nucleus, while the dorsal motor nucleus lies caudal and dorsomedial to the ventral nucleus in the subventricular area. The motor fibers innervate the masticatory muscles (see Fig. 1).

f. Abducent Nerve (VI). The nucleus of the abducent nerve is divided into a major and accessory portion. The major nucleus lies in the region of the dorsomedian aspect of the anterior medulla; the accessory nucleus is found in the same area, but does not extend so far caudally. The nerve emerges from the floor of the medulla and runs through the foramen lacerum orbitale. A dorsal branch supplies the lateral rectus muscles of the eye and the quadrate muscle of the nictitating membrane; a ventral branch supplies the pyramidalis muscle of the nictitating membrane.

g. Facial Nerve (VII). The nucleus of the facial nerve is divided into a poorly defined sensory area, lying medial and dorsal to the major abducent nucleus, and a facial motor nucleus. The central projections of sensory cell bodies lying in the geniculate ganglion are associated with the sensory area, while their peripheral extensions provide an afferent innervation for the tongue muscles, mucous membranes of the mouth, pharynx, salivary glands, lacrimal gland, and the skin in the facial region. The facial motor nucleus is divided into dorsal, ventral, and intermediate regions. The ventral motor nucleus occurs at levels through the major abducent nucleus and lies in the ventrolateral aspect of the medulla.

The intermediate and dorsal motor nuclei occur as a discontinuous band of cell bodies extending diagonally from the dorsal pole of the ventral nucleus toward the ventricular floor. The motor–root fibers of the facial nerve arise from the ventral nucleus and run dorsomedially to the intermediate and dorsal groups; they then swing laterally around the nucleus of the descending root of the trigeminal nerve and emerge on the lateral surface of the medulla. The motor fibers supply the muscles lying between the mandibular and basisphenoid bones, the lateral tympanic muscle, and the cutaneous muscles of the ventral aspect of the pharynx. There appears to be no specific description of a superior salivatory nucleus in birds, but it is clear that the facial nerve contributes to the efferent innervation of the salivary glands.

h. Acoustic Nerve (VIII). The acoustic nerve is divisible into anterior and posterior portions (the vestibular nerve and cochlear nerve, respectively), and each arise from separate nuclei. The vestibular centers are divisible into superior, dorsolateral, ventrolateral, dorsomedial, tangential, and descending root nuclei, and these lie dorsal and caudal to the principal sensory nucleus of the trigeminal nerve. The fibers of the vestibular nerve arise from cell bodies in the vestibular ganglion; the majority of the central projections are associated with the tangential vestibular nucleus. The peripheral extensions of these cells innervate the anterior and external ampulla and the utriculus, primarily. The cochlear centers are divided into angular, laminar, and magnocellular nuclei that lie close to the abducent nucleus in dorsolateral and dorsomedial regions of the medulla. The cell bodies of the cochlear nerve fibers lie in the cochlear ganglion. The centrally directed fibers enter the dorsolateral corner of the medulla, slightly caudal to the vestibular fibers. They are associated with the angular and magnocellular nuclei. The peripheral cochlear fibers innervate the posterior ampulla, sacculus, and the papillae of the cochlea and lagena. There is some evidence for the presence of efferent nerve fibers in the acoustic nerve.

i. Glossopharyngeal Nerve (IX). The glossopharyngeal nerve has both sensory and motor nuclei. The sensory nucleus lies in a subventricular area above the dorsal motor nucleus of the vagus. Sensory cell bodies lie in the ganglion of the root of the glossopharyngeal (jugular ganglion) and the petrosal ganglion. These cells provide sensory innervation of part of the tongue, pharynx, larynx, and esophagus, and also of such specialized tissues as the carotid body. The motor nucleus of the glossopharyngeal is probably divisible into dorsal and ventral portions; the ventral portion lies in the ventrolateral aspect of the medulla at anterior levels of the dorsal motor nucleus of the vagus. The motor fibers emerge on the dorsolateral surface of the medulla and innervate the ceratomandibular muscle, the submandibular salivary glands, and the cranial part of the esophagus.

j. Vagus (X). The sensory nucleus of the vagus is combined with that of the glossopharyngeal nerve. Sensory cell bodies lie in the ganglion of the root of the vagus and in the nodose (thoracic) ganglion. Their central processes are associated with the combined sensory area of the glossopharyngeal and vagal nerves and with the fasciculus solitarius. The peripheral extensions of the sensory cell bodies are associated mainly with the thoracic and abdominal viscera. The motor centers of the vagus are divided into dorsal, intermediate, and ventral nuclei, and the fibers derived therefrom innervate the pharynx, larynx, trachea, bronchi, lungs, thyroid, thymus, heart, esophagus, proventriculus, gizzard, duodenum, pancreas, liver, and oblique septum. However, nerve fibers find their way indirectly to various other effector tissues.

k. Accessory Nerve (XI). The nuclear center of the spinal accessory nerve is represented by a small group of cells in the anterior cervical cord; the cells lie in the lateral aspect of the gray matter, approximately at the level of the central canal. The fibers leave the dorsolateral cervical cord as a series of fine rootlets that course through the foramen magnum, join the cranial roots, and run in intimate connection with the vagus through the accessory nerve foramen. It is variously reported that the accessory nerve innervates the proximal end of the lateral cutaneous colli muscle (Watanabe, 1960), the trapezius muscle (Watanabe, 1964), or the cucullaris muscle (Gadow, 1891).

l. Hypoglossal Nerve (XII). There is some controversy about the nuclei of the hypoglossal nerve, but a ventral nucleus is generally recognized. The cells of this nucleus occur in the posterior regions of the vagal motor nuclei. The hypoglossal nerve runs through the hypoglossal foramen and is associated with the superior cervical ganglion and the first cervical spinal nerve. The hypoglossal nerve innervates lingual, hyoid, laryngeal, and syringeal muscles.

2. *Spinal Nerves*

Although the spinal nerves have received less attention than the cranial nerves, their anatomy is reasonably well documented (e.g., Peck, 1889; Ramón y Cajal, 1890a,b,c; von Lenhossék, 1890; Van Gehuchten, 1893; Streeter, 1904; Kaiser, 1924; Huber, 1936; Kappers *et al.*, 1936; Yasuda, 1960, 1961, 1964; Brinkman and Martin, 1969; van den Akker, 1970; Martin and Brinkman, 1970; and others).

The spinal nerves arise from ventral (motor) and dorsal (sensory) roots. The ventral roots arise from cell bodies lying in the ventral cornua of the gray matter of the spinal cord, while the dorsal roots arise from cell bodies in the spinal ganglia. The first pair of cervical spinal nerves passes between the skull and the atlas; the succeeding pairs are given the same number as the following vertebra. The number of cervical vertebrae in birds is variable and thus the number of pairs of cervical spinal nerves also varies. There are, for example, fourteen pairs in the pigeon (Huber, 1936), twelve in the Budgerigar *Melopsittacus undulatus* (Evans, 1969), and fifteen in the domestic fowl (Jungherr, 1969). Each spinal nerve divides into a dorsal branch that supplies the dorsal spinal muscles and a ventral branch to the ventral spinal muscles and the body wall. In the Budgerigar, the first three pairs of cervical nerves give branches to the hypoglossal nerve and distribute to hyoid and cervical muscles (Evans, 1969). In the fowl, the first eleven pairs of cervical nerves give lateral branches to the vagus and hypoglossal nerves, and the dorsal and ventral branches also supply the cutaneous muscles and skin of the neck.

There are six pairs of thoracic spinal nerves in the pigeon, eight in the Budgerigar, and seven or more in the fowl. The dorsal branches supply the muscles of the back, while the ventral branches supply the intercostal and abdominal muscles. In the fowl, the brachial plexus is formed by the convergence of the ventral or lateral branches of the last four cervical spinal nerves and the first one or two thoracic nerves. The following nerves arise from the plexus: the dorsal thoracic nerve, from the caudal part of the plexus, supplying the latissimus dorsi muscle; the short thoracic nerve, from the caudal part of the plexus, supplying the serratus muscles; the external thoracic nerve, from the same region, supplying the skin of the lateral thoracic region; the cranial pectoral nerve, from the cranial part of the plexus, running forward and supplying the ventral supracoracoid and the deep pectoral muscles; the axillary (cranial and caudal) nerve, supplying the teres major, teres minor, and scapularis muscles; the radial nerve, from the cranial part of the plexus, innervating the triceps brachii, deltoideus, spinatus, coracobrachialis dorsalis, and the muscles that extend the elbow, tense the prepatagium, and extend the carpus and digits; the median nerve, from the caudal part of the plexus, supplying branches to the superficial pectoral and the two ventral coracobrachial muscles and, beyond these, branches to the extensors of the elbow and flexors of the carpus and digits; the ulnar nerve, from the caudal part of the plexus with the median, supplying the extensors of the elbow and the flexors of the carpus and digits.

In the fowl there are fourteen pairs of lumbosacral spinal nerves. The lumbosacral plexus arises from the first to the eighth lumbosacral nerves inclusive; the following account is based mainly on the work of Yasuda (1961). The first, second, and third femoral nerves arise mainly from the first, second, and third lumbosacral spinal nerves. There is a cutaneous branch to the skin over the thigh, and branches to the sartorius, tensor fasciae latae, rectus femoris, lateral vastus, medial vastus, pectineus, and iliac muscles. The obturator nerve arises from the roots of the three femoral nerves and innervates the external obturator, internal obturator, gemellus, and the adductor longus and adductor magnus muscles. The lumbal plexus arises from the common trunk of the second and third femoral nerves and supplies the anterior part of the superficial gluteal, the medial gluteal, and the deep gluteal muscles. The ischiadic plexus arises from the third to the eighth lumbosacral nerves, and gives rise to various branches which supply the biceps femoris, semitendinosus, semimembranosus, the posterior part of the superficial gluteal, the quadratus femoris and the caudofemoralis muscles. The ischiadic plexus gives rise to three main nerves – the ischiadic nerve (innervating the accessory semimembranosus and intermediate vastus muscles) and peroneal and tibial nerves; the latter innervates the gastrocnemius muscle, and they both supply the flexors and extensors of the digits. The internal pudic (arising from lumbosacral nerves ten and eleven) and coccygeal (from the twelfth to fourteenth) nerves supply the cloacal and coccygeal regions.

Within the spinal nerves run afferent fibers from muscle spindles, tendon organs, encapsulated receptors, and receptors associated with the skin and viscera. The efferent fibers present are concerned with the innervation of extrafusal and intrafusal muscle fibers and peripheral effectors such as pennamotor muscles and the cutaneous and skeletal muscle vasculature. The nerve fibers to the latter tissues arise from paravertebral ganglia.

C. Somatic Efferent Innervation

1. Characteristics of Striated Muscle

The work of Krüger (1950, 1952), and Krüger and Gunther (1956, 1958) indicated there were two types of striated skeletal muscle fibers —*Fibrillenstruktur*, showing an orderly structure, and *Felderstruktur*, having no organized structural pattern. Although there have been suggestions that these fiber types are basically the same (Yeh *et al.*, 1963), ultrastructural observations show clear-cut differences between them. Thus, Hess (1961, 1967, 1970) observed that *Fibrillenstruktur* muscle fibers contain regular fibrils, each surrounded by sarcoplasm and granules, with the Z disk running in a straight course across the fibrils. However, in *Felderstruktur* fibers, the fibrils are more randomly disposed and are not regularly surrounded by sarcoplasm or granules; moreover, the Z disk runs in a zig-zag course across the fibrils. In a more detailed study, Page (1969) showed that *Fibrillenstruktur* fibers are characterized by possession of a regular T system and sarcoplasmic reticulum that frequently came into contact to form triads; two series of triads occur in each sarcomere near the A–I band boundary. In *Felderstruktur* fibers, the T system and sarcoplasmic reticulum are less regular and have only small contact areas in which dyads are usually formed. Other published accounts of the ultrastructure of avian skeletal muscle (e.g., Bennett and Porter, 1953; Walker and Schrodt, 1966; Ashhurst, 1969; Grinyer and George, 1969a,b; and others) are in general agreement with the above account.

The rather simple description given above is complicated by a consideration of the contractile properties and pattern of innervation of various types of muscle fiber; no account will be given here of the histochemical characteristics of avian muscle fibers, since these have been dealt with at length in a recent book (George and Berger, 1966).

Various investigators have shown that avian muscles may be divided into slow or fast types, depending on their contractile properties (Ginsborg, 1959, 1960a; Ginsborg and Mackay, 1960; Page and Slater, 1965; Page, 1969; Cambier, 1969). For various reasons it has been considered that slow muscle fibers are of the *Felderstruktur* type, whereas fast muscle fibers are of the *Fibrillenstruktur* type, but due to recent findings, this supposition must be revised. Since the recent findings are concerned with the motor innervation of the muscle, they are considered below.

2. Characteristics of Motor Innervation

Morphological observations on the motor innervation of avian skeletal muscle have been made over a period of years (e.g., Rouget,

1862; Fischer, 1887; Kuhne, 1887; Botezat, 1909; Tiegs, 1953; Silver, 1963; Chinoy and George, 1965; Boesiger, 1965, 1968; Barker, 1968), and various descriptions have been given of the appearance of the motor terminals. However, only recently did it become clear that some muscle fibers receive only one motor ending (usually an *en plaque* ending), whereas others have several endings on them (usually *en grappe* endings) (Ginsborg, 1959, 1960a,b; Ginsborg and Mackay, 1960, 1961; Hess, 1961). From his morphological and functional observations, Ginsborg (1960a) suggested that multiple innervation, *Felderstruktur*, and the ability to develop contractures are probably properties of the same muscle fiber.

This view was tenable until the publication of some ultrastructural observations by Hess (1966) and Zenker and Krammer (1967) on the striated muscle of the avian iris. Hess (1966) found that the muscle fibers of the chicken iris contained an extensive, branching, sarcoplasmic reticulum together with a well developed T system characteristic of *Fibrillenstruktur* fibers. However, multiple motor-nerve endings are present on each muscle fiber, and form elongated terminals below which there are no foldings of the postjunctional sarcolemma. Thus, Hess pointed out, the internal structure of the muscle fibers was characteristic of twitch fibers, whereas the pattern and form of innervation was of the sort assumed to be associated with slow muscle fibers. Similar observations were made by Zenker and Krammer (1967), but they considered that each muscle fiber received only a single motor ending.

These ultrastructural peculiarities were investigated recently by physiological means (Pilar and Vaughan, 1969a,b). Pilar and Vaughan found that single muscle fibers respond to adequate neuronal stimuli by firing action potentials, but they also observed that end plate potentials often show facilitation when the motor nerves are stimulated repetitively, indicating multiple innervation. In 30% of the muscle cells examined, facilitation was not seen, thus indicating focal innervation of those muscle fibers. Pilar and Vaughan showed that a single muscle fiber could respond to an appropriate stimulus with a twitch or a contracture. As these workers pointed out, the observation that *Fibrillenstruktur* fibers show prolonged sensitivity to acetylcholine indicates that a well organized sarcoplasmic reticulum does not prevent a muscle fiber from behaving tonically. On the basis of their findings, Pilar and Vaughan (1969b) suggested that the innervation controlling the muscle membrane is important in determining whether the muscle is capable of a tonic response. This suggestion is supported, to some extent, by the findings on cross-innervated fast and slow muscles described in the next section.

3. Cross Innervation of Striated Muscle

As originally described by Ginsborg and Mackay (1960, 1961), the anterior latissimus dorsi of the chick consists of multiple-innervated slow-muscle fibers, whereas the posterior latissimus dorsi consists mainly of focally innervated twitch fibers. The effects of cross innervating these muscles has been examined by several workers (Feng *et al.*, 1965; Zelená *et al.*, 1967; Hník *et al.*, 1967). Zelená *et al.* (1967) found that when the posterior latissimus dorsi was reinnervated by fibers usually supplying the anterior latissimus dorsi, the muscle action potentials were only elicited following tetanic stimulation of the nerve. Furthermore, the reinnervated muscle fibers possessed multiple *en grappe* motor endings. When the anterior latissimus dorsi was reinnervated by a nerve that usually supplies a muscle that consists of twitch fibers only (anconeus scapularis), then the newly established motor endings were of the focal *en plaque* type, and the reinnervated muscle fibers fired action potentials in response to single shocks applied to the motor nerve. Zelená *et al.* (1967) observed no changes in the contraction times of the reinnervated muscles, and in a subsequent study (Hník *et al.*, 1967), it was found that there were no ultrastructural changes in the extrajunctional regions of the reinnervated muscle fibers. From these findings, it seems clear that the ultrastructure of the muscle fiber is determined independently of the innervation and that the form of innervation is not affected by the muscle fiber innervated.

D. Somatic Afferent Innervation

1. Muscle Spindles

There have only been scattered observations on avian muscle spindles (Huber and DeWitt, 1898; Regaud and Favre, 1904; Tello, 1922; Barker, 1960, 1968; Stammer, 1962; Germino and D'Albora, 1965; Palmieri, 1965; de Anda and Rebollo, 1967; Rebollo and de Anda, 1967; Saglam, 1968; Maier and Eldred, 1969). De Anda and Rebollo (1967) have examined muscle spindles in the gastrocnemius of the chicken; they found that the sensory fibers associated with the thick (nuclear bag) and thin (nuclear chain) intrafusal muscle fibers are similar, forming wide areas of apposition and ending in the region of the nuclei in hammer- or basket-shaped formations. The nuclear-bag fibers have *en plaque* motor endings, whereas the nuclear-chain fibers are supplied with simple endings. Saglam (1968), from an examination of the nuchal musculature of two species of woodpecker (*Dendrocopos major, D. syriacus*), has identified thick and thin intrafusal muscle fibers, but claimed that the thick fibers showed no nuclear bag or myotube regions, and that nuclear chain regions were absent from the thin intrafusal fibers. Saglam has described the primary sensory endings as simple, and observed that the motor endings were usually of the *en grappe* type.

Barker (1968) has stated that the muscle spindles of the anterior and posterior latissimus dorsi of the chicken possess one to eight intrafusal muscle fibers, each with an equatorial nuclear sac. Intrafusal muscle fibers with both primary and secondary sensory endings were about twice as numerous as those with only primary endings. In the anterior latissimus dorsi the fusimotor endings were only of the *en grappe* type, the motor fibers penetrating the poles of the spindles as the collaterals of extrafusal motor fibers, and also running indirectly through the capsule with the sensory fibers. In the spindles of the posterior latissimus dorsi, the intrafusal muscle fibers received *en plaque* motor terminals at the poles from collaterals of the extrafusal innervation, whereas the muscle fibers in the equatorial region received *en grappe* endings from motor fibers running with the sensory axons. As yet there have been no studies on muscle spindles and their innervation in cross-innervated muscles.

Although there have been several investigations of proprioceptor physiology in birds (Manni *et al.*, 1965; Bortolami and Vegetti, 1967; Azzena and Palmieri, 1967; Azzena *et al.*, 1970), the only investigation that deals specifically with muscle-spindle physiology is that of Dorward (1966, 1970a). Dorward (1970a) found avian muscle spindles comparable in behavior to those of mammals. Contrary to Barker's (1968) description, Dorward could find no indication of a fusimotor innervation deriving from the alpha motor supply to the extrafusal musculature; however, she did not exclude this possibility. The spindles showed typical in-parallel behavior. There was no indication, from measuring conduction velocities, of a dual afferent innervation of spindles as described morphologically by Barker (1968).

2. *Tendon Organs*

Tendon organs in birds have received only passing attention (e.g., Tello, 1922; Barker, 1960), but it appears that they take the form of discrete spray endings (Barker, 1960) and are functionally very similar to those in mammals (Dorward, 1970a).

3. *Sensory Corpuscles*

Various types of corpuscular receptors have been described in birds

(Krogis, 1931, 1932; Kappers *et al.*, 1936; Pellegrino de Iraldi and Rodriguez-Perez, 1961; Rodriguez-Perez and Pellegrino de Iraldi, 1961; Quilliam, 1962; Stammer, 1962; Wang and Ho, 1966; Saxod, 1967, 1968, 1970; Malinovský, 1967a,b, 1968; Andersen and Nafstad, 1968; Nafstad and Andersen, 1969, 1970; Andres, 1969; Malinovský and Zemanek, 1970; Cobb and Bennett, 1970b; and others), but once again physiological data are scarce. Dorward (1970b), in a study of the response patterns of cutaneous mechanoreceptors in the domestic duck, identified vibration-sensitive units that followed a stimulus 1:1 up to frequencies of 400–800 cps, and gave a steady response at 600–1000 cps; Dorward found that single Herbst corpuscles gave a similar response. It has been suggested (Dorward, 1970b; Cobb and Bennett, 1970b) that Herbst corpuscles associated with the wing feathers may be important in the regulation of flight.

4. Cutaneous Receptors

Apart from the numerous sensory corpuscles found in association with the skin and feathers, free-lying nerve endings have also been described (e.g., Schartau, 1938; Stammer, 1961b; Capanna and Civitelli, 1963; Ostmann *et al.*, 1963; Tetzlaff *et al.*, 1965). The only physiological indication of the existence of such cutaneous receptors in birds is, once again, the work of Dorward (1970b). In the wing of *Anas platyrhynchos,* Dorward identified sensory units, associated with the down feathers, that were quickly adapting, had large receptive fields and low mechanical thresholds; touch receptors that had localized receptive fields, medium to high mechanical thresholds and showed a range of adaptation rates; and, finally, pressure, or nociceptive receptors, that had diffuse receptive fields, high thresholds, and were slow to adapt.

There have been some morphological descriptions of presumed receptor endings associated with the joints in birds (e.g., Polacek *et al.*, 1966; Palmieri, 1969; Sklenska, 1969), but there are no physiological studies of this problem.

E. VISCERAL AFFERENT INNERVATION

As Kuntz (1953) has pointed out, there is nothing distinctly autonomic about visceral afferent fibers, since, in mammals, visceral and somatic afferent neurons are associated with preganglionic autonomic cell bodies and with somatic efferent neurons. Assuming the same may be true in birds, then visceral afferent fibers are included in this section on the peripheral nervous system.

1. Respiratory Afferents

Morphological studies of afferent endings associated with the respiratory tract are scarce. Van Campenhout's (1955, 1956) identification of sensory structures in the avian lung is somewhat doubtful, but the recent ultrastructural observations of Cook and King (1969a,b) are more convincing. These workers observed specialized cells in the epithelium of the primary bronchus; the cells contained dense-cored vesicles and were in close association with axon profiles. The apparent similarity between these cells and those of the carotid body (Cook and King, 1969b) is misleading, however, since no cells localized in the primary bronchus of the avian lung contain catecholamines (Bennett, 1971b), whereas carotid body cells do contain catecholamines (Kobayashi, 1969a; Bennett, 1971a).

In spite of the paucity of morphological evidence, physiological observations on the involvement of afferent fibers in avian respiration are numerous (e.g., Graham, 1940; Hiestand and Randall, 1941, 1943; Sinha, 1958; Fedde et al., 1963; Johansen and Reite, 1964; King, 1966; McCrady et al., 1966; Butler, 1967, 1970; Burger, 1968; Butler and Jones, 1968; King et al., 1968a,b; Peterson and Fedde, 1968; Richards, 1968, 1969, 1970a,b; Ray and Fedde, 1969; Jones, 1969; Fedde, 1970; Fedde and Peterson, 1970; Cohen and Schnall, 1970; Jones and Purves, 1970b). A discussion of the afferent control of avian respiration is included in Chapter 5 of Volume II of this treatise.

2. Cardiovascular Afferents (see also Chapter 4, Volume II)

There have been several descriptions of presumed receptor nerve endings in the heart (e.g., Ábrahám and Stammer, 1957; Ábrahám, 1962, 1969) and blood vessels (Takino, 1932; Nonidez, 1935; Tcheng, 1963; Tcheng et al., 1965a,b; Ábrahám, 1969) of birds, but most effort has been expended in describing the carotid body and supracardial paraganglia (Kose, 1902, 1904, 1907a,b; Mackenzie, 1910; Mackenzie and Robertson, 1910; Muratori, 1931, 1932a,b,c, 1933, 1934, 1937, 1962; Palme, 1934; Watzka, 1934; Nonidez, 1935; de Meyer, 1952; Chowdhary, 1953; de Kock, 1958, 1959; Fu et al., 1962; Tcheng and Fu, 1962; Tcheng, 1963; Tcheng et al., 1963a,b, 1965a,b; Wirtz, 1968; Kobayashi, 1969a,b; Morozov, 1969, 1970; Bennett, 1971a,b). Research on cardiovascular reflexes is concerned also with respiratory reflexes, and as in that field, there is no shortage of demonstrations of the involvement of afferent mechanisms (Eliassen, 1960; Andersen, 1963, 1966; Feigl and Folkow, 1963; Hollenberg and Uvnäs, 1963; Johansen and Aakhus, 1963; Reite et al., 1963; Aakhus and Johansen, 1964; Durfee, 1964; Johansen, 1964; Johansen and Reite, 1964; Folkow et al., 1966, 1967; Richards and Sykes, 1967a,b; Butler, 1967, 1970; Butler and Jones, 1968; Cohn et al., 1968; Djojosugito et al., 1968, 1969; Jones, 1969; Jones and Purves, 1969, 1970a).

It is of considerable interest that recent findings (Hollenberg and Uvnäs, 1963; Jones and Purves, 1969, 1970a) indicate a marked functional involvement of the carotid bodies of the duck in the cardiovascular response to immersion, since in chickens it has been claimed that this region is not involved in cardiovascular reflexes (e.g., van der Linden, 1934; Durfee, 1964). The neural control of the avian cardiovascular system is considered fully in Chapter 4, Volume II.

3. Gastrointestinal Afferents

The morphology and innervation of taste buds in birds are much neglected fields (see Kappers et al., 1936), but there are several physiological studies available on avian gustation (Kitchell et al., 1959; Halpern, 1963; Kadono et al., 1966).

There are no physiological observations on afferent fibers associated with the avian gut other than those that demonstrate vagovagal reflexes (e.g., Mangold, 1911). However, there are a few morphological studies that describe presumed sensory endings (e.g., Milokhin, 1962a,b; Kolossow, 1963; Solovieva, 1965; Ábrahám, 1966, 1967).

In various other systems, e.g., the urogenital system (Sykes, 1955b; Gilbert, 1965b) and autonomic ganglia (Milokhin, 1960, 1962b, 1964; Kolossow, 1963, 1965, 1970; Lukashin, 1969), the presence of afferent fibers has been assumed on the basis of physiological or morphological observations.

III. Autonomic Nervous System

A. DEFINITION

Following Campbell's (1970) suggestion, the autonomic nervous system may be most concisely defined as comprising all those efferent pathways that have ganglionic synapses outside the confines of the central nervous system. Thus, the avian autonomic nervous system innervates all smooth muscle, some types of secretory cells, cardiac striated muscle, and the striated intrinsic muscle of the eye. In this definition, the innervation of chromaffin cells (see Section III,E,6,a below) is regarded as ganglionic rather than terminal.

Langley (1921) divided the autonomic nervous system into parasympathetic (craniosacral), sympathetic (thoracolumbar), and enteric components. This anatomical differentiation was supported by the observation that parasympathetic fibers apparently ran to the effector tissues without passing through the paravertebral chains but formed synapses in peripheral ganglia closely associated with the effector tissues. Sympathetic fibers, however, passed through the paravertebral chains and formed synapses there or in prevertebral ganglia remote from the effector tissue. Furthermore, parasympathetic fibers appeared to act on effector tissues by releasing acetylcholine, whereas sympathetic fibers did so by releasing an adrenaline-like substance (Langley, 1921; Dale, 1933).

Since this scheme is so delightfully simple, there has come about the widely held belief that there is a rigid delineation between the anatomical disposition and physiological characteristics of the so-called sympathetic and parasympathetic divisions of the autonomic nervous system. This belief persists in spite of the numerous known exceptions to the scheme (see Mitchell, 1956; Campbell, 1970).

The situation in nonmammalian vertebrates, particularly birds, is even more confused, since there is often no clear-cut anatomical basis on which to define the components of the autonomic nervous system. Furthermore, the practice of using "sympathetic" and "adrenergic" interchangeably and "parasympathetic" and "cholinergic" as synonymous is particularly misleading in this context, since there is a marked deviation in the anatomical and physiological characteristics of the nerve fibers (see Burn, 1968; Bennett, 1970b). A further complication arises from the possible existence of fibers that act by releasing substances other than acetylcholine or noradrenaline (see Campbell, 1970; and Section III,C,3 below). For these various reasons, the terms sympathetic and parasympathetic will not be used in this review; autonomic nerves will be considered on the basis of the transmitter released by them and, where possible, the location of their cell bodies.

B. General Anatomy

1. Cephalic Autonomic System

The cephalic autonomic system is composed of fibers that arise from the brain and cranial autonomic ganglia and fibers deriving from the superior cervical ganglion and cervical paravertebral chain ganglia (see Fig. 1). Autonomic fibers run with the cranial nerves and with blood vessels; there are numerous anastomoses between the various pathways, and thus the picture is confusing.

Preganglionic fibers running with the oculomotor nerve arise from cell bodies in the accessory nucleus of the oculomotor nerve (see

Kappers *et al.*, 1936; Kappers, 1947; Goller, 1969) and synapse with cell bodies in the ciliary ganglion (see Sections III,D,2 and III,E,8 below). Postganglionic fibers from the ciliary ganglion supply the eye (see Section III,E,8 below). Although little work has been done on the origin of preganglionic fibers running with the facial and glossopharyngeal nerves, it is well established that ethmoidal, sphenopalatine, and submandibular ganglia are present and are concerned with the innervation of various glandular structures in the cephalic region (Laffont, 1885; Magnien, 1885a,b, 1887; Rochas, 1885a,b,c, 1886, 1887; Hsieh, 1951) (see Section II,B,1 and Fig. 1).

More is known of the origin of the autonomic fibers running with the vagus (see Kappers *et al.*, 1936), and recent studies have clarified this point. Watanabe (1968) observed in the chicken that cell bodies in the ventral and medial areas of the medial and rostral regions of the dorsal vagal nucleus on both sides gave rise to fibers that were associated with the proventriculus and gizzard. Cells of the middle and rostral regions of the dorsal vagal nucleus sent fibers to the thoracic organs, while cells in the dorsal and lateral areas of the middle and rostral regions of the dorsal vagal nucleus sent fibers to the pharynx and larynx.

Cohen *et al.* (1970) found in the pigeon that the representation of the abdominal vagus within the dorsal motor nucleus was widespread and occupied the rostral two-thirds of the nucleus, thus corresponding with Watanabe's (1968) findings in the chicken. Cohen *et al.* (1970) found that vagal cardioinhibitory fibers were represented throughout the rostral half of the dorsal motor nucleus, but were most concentrated in the ventral portion. These morphological observations were confirmed by physiological findings (Cohen and Schnall, 1970). The distributions of the vagus and glossopharyngeal nerves have been dealt with in most investigations concerned with the cranial nerves (see Section II,B,1 above); the vagal and glossopharyngeal supply to various effector tissues will be dealt with in the relevant sections below.

Cephalic extensions from the superior cervical ganglion, and their associations with various cranial nerves have long been known (Cuvier, 1802; Tiedemann, 1810; Emmert, 1811; Weber, 1817; Laffont, 1885; Magnien, 1885a,b; Rochas, 1885a,b,c, 1886, 1887; Marage, 1889; Thébault, 1898; Gadow, 1891; Cords, 1904; Tixier-Durivault, 1942; Hsieh, 1951; Hammond and Yntema, 1958; and others). A detailed account will not be given here, but it should be remembered that the superior cervical ganglion is connected with the first, third, fourth, fifth, sixth, seventh, and ninth cranial nerves by the

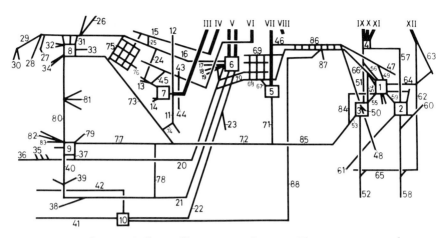

FIG. 1. Distribution of efferent fibers in cranial nerves. This is a composite diagram made up from various accounts and illustrations. There is inconsistency in the literature with regard to certain relationships and terminologies, but the present scheme is in line with most accounts. III. oculomotor nerve; IV. trochlear nerve; V. trigeminal nerve; VI. abducent nerve; VII. facial nerve; VIII. acoustic nerve; IX. glossopharyngeal nerve; X. vagus; XI. accessory nerve; XII. hypoglossal nerve. 1. Superior cervical ganglion; 2. ganglion of the root of the vagus; 3. petrosal ganglion; 4. ganglion of the root of the glossopharyngeal (jugular ganglion); 5. geniculate ganglion; 6. gasserian (semilunar) ganglion; 7. ciliary ganglion; 8. ethmoidal (orbitonasal) ganglion; 9. sphenopalatine ganglion; 10. submandibular ganglion; 11. ventral branch of III; 12. dorsal branch of III; 13. long ciliary nerves; 14. short ciliary nerves; 15. peripheral extensions of IV; 16. deep ophthalmic branch of V; 17. superficial ophthalmic branch of V; 18. lacrimal and palpebral branch of V; 19. external cutaneous branch of V; 20. infraorbital branch of V; 21. branch of V to the angle of the mouth; 22. mandibular branch of V; 23. masseter branch of V; 24. anastomosis between deep ophthalmic branch of V and the long ciliary nerves; 25. anastomosis between deep ophthalmic branch of V and the trochlear nerve; 26. frontal rami of the deep ophthalmic branch of V; 27. superior palpebral rami of the deep ophthalmic branch of V; 28. lateral nasal rami of the deep ophthalmic branch of V; 29. medial nasal rami of the deep ophthalmic branch of V; 30. internal nasal rami of the deep ophthalmic branch of V; 31. anastomoses between the ethmoidal ganglion and the deep ophthalmic branch of V; 32. rami from the ethmoidal ganglion to the nasal gland; 33. rami from the ethmoidal ganglion to the periorbita and Harderian gland; 34. rami from the ethmoidal ganglion to the olfactory nerves; 35. inferior internal nasal rami of the infraorbital branch of V; 36. external nasal rami of the infraorbital branch of V; 37. anastomosis between the sphenopalatine ganglion and the infraorbital branch of V; 38. anterior paleal rami of the branch of V to the angle of the mouth; 39. palatine rami of the branch of V to the angle of the mouth; 40. anastomosis between the sphenopalatine ganglion and the palatine rami of the branch of V to the angle of the mouth; 41. alveolar mandibular rami of the mandibular branch of V; 42. lingual rami of the mandibular branch of V; 43. dorsal branch of VI; 44. ventral branch of VI; 45. anastomosis between dorsal branch of VI and the long ciliary nerves; 46. hyomandibular branch of VII; 47. lateral branch of VII; 48. medial branch of VII; 49. anastomosis between the superior cervical ganglion and the lateral branch of VII; 50. anastomosis between the petrosal ganglion and the medial branch of VII; 51. glossopharyngeal nerve; 52. peripheral branch of glossopharyngeal nerve;

temporolacrimal and cephalic carotid nerves and their extensions and receives communicating branches from the glossopharyngeal and vagus directly and from the hindbrain and the hypoglossal nerves by the interganglionic cord (Hsieh, 1951) (see Fig. 1 and Section II,B,1). Furthermore, there are demonstrable autonomic nerve trunks along the cephalic blood vessels (see Hsieh, 1951), and thus there are pathways whereby, directly or indirectly, fibers from the superior cervical ganglion may reach all cephalic effector tissues.

2. Spinal Preganglionic System

The studies available on the spinal preganglionic system are in general agreement (e.g., Terni, 1923; Huber, 1936; Kappers *et al.,* 1936; Staudacher, 1940; Levi-Montalcini, 1950; Pera, 1953; Matsushita, 1968; Macdonald and Cohen, 1970). It appears that the thoracolumbar preganglionic efferent column lies between the first thoracic and second lumbar segments in the form of two mediodorsal tracts. Axons arise from these cells and leave the spinal cord in the ventral roots to synapse in the paravertebral chain or the prevertebral ganglia. In the cervical region, an organized preganglionic column is absent; thus preganglionic fibers must pass rostrad along the paravertebral

---

53. inferior branch of tympanic nerve (from petrosal ganglion); 54. superior branch of tympanic nerve (from petrosal ganglion); 55. anastomosis between petrosal ganglion and superior cervical ganglion; 56. anastomosis between glossopharyngeal nerve and superior cervical ganglion; 57. vagus nerve; 58. peripheral branch of vagus nerve; 59. anastomosis between the ganglion of the root of the vagus and the superior cervical ganglion; 60. dorsal branch of XI; 61. ventral branch of XI; 62. anterior branch of XII; 63. posterior branch of XII; 64. anastomosis between anterior branch of XII and the superior cervical ganglion; 65. anastomosis between XII, X and IX; 66. temporolacrimal (internal maxillary) nerve (arising from superior cervical ganglion); 67. anastomosis between geniculate ganglion and terminal extensions of temporolacrimal nerve; 68. lesser superficial petrosal nerve; 69. infratemporal (external ophthalmic) plexus; 70. anastomoses between infratemporal plexus and branch of V; 71. greater superficial petrosal nerve; 72. nerve of the pterygoid canal (vidian nerve); 73. nasopalatine nerve (superior branch of vidian nerve); 74. anastomosis between nasopalatine nerve and ventral branch of III; 75. plexus arising from nasopalatine nerve anastomosing with deep ophthalmic branch of V and the trochlear nerve; 76. branch from nasopalatine plexus to periorbita and harderian gland; 77. pterygopalatine nerve (inferior branch of vidian nerve); 78. posterior palatine nerve, giving branches to the palatine glands; 79. branch from sphenopalatine ganglion to periorbita and harderian gland; 80. anastomosis between sphenopalatine and ethmoidal ganglia; 81. branches to periorbita and Harderian gland from anastomosis between sphenopalatine and ethmoidal ganglia; 82. branches from sphenopalatine ganglion to palatine and maxillary glands, palate, and nasal septum; 83. salivary branch from sphenopalatine ganglion; 84. cephalic carotid nerve (arising from superior cervical ganglion); 85. deep petrosal nerve; 86. anastomosis between temporolacrimal nerve and hyomandibular branch of VII; 87. auricular nerve; 88. chorda tympani.

chains to synapse in the superior cervical ganglion. Langley (1904) claimed that the last cervical spinal nerve was the first through which fibers pass from the spinal cord into the cervical paravertebral chain, but Terni (1924, 1929a, 1931) and Yntema and Hammond (1945) found that the highest obvious preganglionic contribution to the paravertebral chain is through the first thoracic spinal nerve.

One point of interest that has not been resolved is the occurrence of efferent fibers in the dorsal roots of the cervical spinal nerves. Von Lenhossék (1890) observed fibers running from the lateral anterior horn, through the dorsal roots, and thus to the periphery, without connection with the cells of the spinal ganglia. These findings were confirmed by Ramón y Cajal (1890c) and Van Gehuchten (1893). Although the observations of Zorzoli and Maggi (1950) also tended to support this claim, Matsushita (1968) has stated that such fibers, if they exist, must be few. In mammals, the present evidence favors the existence of dorsal root efferents (see Mitchell, 1956).

Besides the thoracolumbar preganglionic system described above, there is a similar column extending from the eighth to the fourteenth lumbosacral segment, or thereabouts (Terni, 1923; Huber, 1936; Levi-Montalcini, 1950; Browne, 1953; Pera, 1953; Yntema and Hammond, 1955). It occupies the same paraependymal position as the thoracolumbar column, and judging from the distribution of the peripheral extensions of the cell bodies (Browne, 1953; Yntema and Hammond, 1955), it represents a sacral preganglionic visceral system.

3. *Paravertebral Trunk and Prevertebral Ganglia*

There have been numerous accounts of the paravertebral trunk and prevertebral ganglia, and, with a few exceptions, the accounts are surprisingly consistent (e.g., Cuvier, 1802; Tiedemann, 1810; Emmert, 1811; Weber, 1817; Swan, 1835; Marage, 1889; Gadow, 1891; Couvreur, 1892; Thébault, 1898; Langley, 1904; Goormaghtigh, 1921; Terni, 1924, 1929a, 1931; Tello, 1925; Bremer, 1926; Uchida, 1927; Popa and Popa 1931; van Campenhout, 1930a,b,c, 1931, 1933; Stiemens, 1934; Goloube, 1936; Kostinowitsch, 1936; Poustilnik, 1940; Weber, 1940; Yntema and Hammond, 1945, 1954, 1955; Hammond and Yntema, 1947; Hammond, 1949; Pera, 1950; Hsieh, 1951; Malinovský, 1962; Baumel, 1964; Pastěa, 1965; Pastěa and Pastča, 1967; Freedman, 1968; and others). The inconsistencies that do exist appear to result from variations in terminology. It is generally agreed that nerve trunks running from the paravertebral ganglia of the second to sixth thoracic segments are associated to form the celiac plexus on the aorta, at the root of the celiac artery. Branches from the vagi join the ganglionated celiac plexus; thus, the nerve trunks that distribute to the effector tissues probably contain nerve fibers from paravertebral and prevertebral ganglia and from vagal sources. Nerve trunks from the celiac plexus follow the branches of the celiac artery and are associated with splenic, hepatic, pancreaticoduodenal, gastric, and various other subsidiary plexuses (see Hsieh, 1951).

The aortic plexus lies on the dorsal aorta and is associated with the nerve trunks arising from the paravertebral ganglia of the fifth thoracic to the sixth lumbosacral segments; this plexus thus includes the adrenal, anterior mesenteric, genital, and renal plexuses. The right and left adrenal plexuses are joined by branches around the aorta, each plexus being formed by nerve trunks from the last three thoracic and the first two or three lumbosacral paravertebral ganglia (Hsieh, 1951; Freedman, 1968). There are numerous large ganglia in each plexus, particularly in association with the adrenal capsule. The anterior mesenteric plexus invests the root of the anterior mesenteric artery and is distributed with the arterial branches; this plexus contributes fibers to Remak's nerve (see Section III,D,4 below). The plexuses associated with the genital and urinary system arise from the main aortic plexus and also receive fibers from the ganglia of the adrenal glands and from the pelvic nerves (Hsieh, 1951).

In some accounts, the prevertebral plexus arising from the caudal lumbosacral paravertebral ganglia is termed the inferior mesenteric plexus, while in others it is referred to as the hypogastric plexus. Hsieh (1951) refers to it as the hypogastric plexus and finds that it derives from the nerve trunks of the sixth to the twelfth lumbosacral ganglia of both paravertebral chains. The hypogastric plexus contributes to the posterior mesenteric and pelvic plexuses. The posterior mesenteric plexus runs with the posterior mesenteric artery and thus to the rectum and the caudal branches of Remak's nerve; these caudal branches, together with the associated nerve trunks from the hypogastric plexus and pelvic splanchnic nerves, form the pelvic plexuses associated with the rectum. The pelvic plexuses are continuous with the cloacal plexus, which derives from the pelvic splanchnic nerves; the cloacal plexus distributes fibers to the genital and urinary tracts and the distal regions of the rectum and cloaca (Hsieh, 1951). Neuronal relationships in the paravertebral trunks and prevertebral ganglia are discussed in Sections III,D,2 and 3 below. There are two marked peculiarities of the autonomic nervous system in birds—the disposition of the paravertebral chain in the cervical region (see Bennett, 1971a) and the presence of the ganglionated nerve of Remak (see Section III,D,4 below).

C. PERIPHERAL NEUROEFFECTOR SYSTEMS

The organization of avian effector systems has not been extensively studied. Sommer and Johnson (1969) described cell types and relationships in avian cardiac muscle and referred to earlier publications on this topic. The ultrastructural organization of visceral smooth muscle in birds has been described in several papers (Imaizumi, 1968; Imaizumi and Hama, 1969; Bennett and Cobb, 1969a,b; Cobb and Bennett, 1969a,b, 1970a). The findings indicate that in this effector system cell relationships are clearly defined, and appear to be important in the development and functioning of the system (Bennett and Cobb, 1969b; Cobb and Bennett, 1969a,b, 1970a). In the following sections little reference is made to the effect of drugs on tissues as a means of analyzing the patterns of autonomic innervation. This is deliberate, since it is quite clear that some tissues will respond to supposed transmitter substances even though they lack an innervation (e.g., Baur, 1928; Moynahan *et al.*, 1970).

1. Cholinergic Innervation

Although true cholinesterase in birds appears to be peculiar in some ways (Blaber and Cuthbert, 1962), various histochemical techniques for the localization of cholinesterases have been used to detect supposed cholinergic innervation of autonomic effectors (see relevant sections below). The main drawback with this method is the possibility that the cholinesterase localized may be associated with nerves that are not functionally cholinergic (see, e.g., Eränkö *et al.*, 1970). Thus, the histochemical method should always be used in conjunction with physiological techniques. Recently, a histochemical method for the localization of choline acetyltransferase has been successfully applied to avian tissue (Kasa *et al.*, 1970), and it is hoped that this technique will provide a less dubious means of identifying nerves that are functionally cholinergic.

With the electron microscope it is possible, in various effector tissues, to demonstrate axon profiles containing agranular vesicles. These vesicles are 400–600 Å in diameter and are comparable to those seen at the motor end plates associated with skeletal muscle. Since such axons are frequently found in situations where cholinergic nerves would be expected to occur on the basis of physiological findings, then it is assumed that they are cholinergic. The correlation between the presence of small agranular vesicles and cholinergic transmission is now so widely accepted that the demonstration of axons containing such vesicles is often assumed to be a demonstration of cholinergic innervation; but this assumption is not strictly justifiable.

Cholinergic junctional transmission has been studied only in the gut (Ohashi and Ohga, 1967; Bennett, 1969b,c,d, 1970a). In this situation, it is found that stimulation of cholinergic nerves elicits, in the smooth-muscle cells, excitatory junction potentials that show facilitation and summation and thereby provoke action potentials. The excitatory junction potentials are blocked by atropine (Ohashi and Ohga, 1967) or hyoscine (Bennett, 1969b,d) and enhanced by anticholinesterases (Bennett, 1969d).

2. Adrenergic Innervation

Recently, a sensitive and specific fluorescence histochemical method for the localization of biogenic amines was developed by Hillarp and his co-workers (see Falck and Owman, 1965). Under controlled conditions it is possible to use this technique to identify catecholamine-containing nerve fibers and cell bodies, and the method has been applied with success to various avian systems (e.g., Govyrin and Leontieva, 1965a,b; Enemar et al., 1965; Leontieva, 1966a,b; Folkow et al., 1966; Costa, 1966a,b; Sano et al., 1967; Everett and Mann, 1967; Baumgarten and Holstein, 1968; Bennett, 1969a, 1971a,b,c; Bennett and Cobb, 1969b; Doležel and Žlábeck, 1969; Akester and Mann, 1969a,b; Akester et al., 1969; Otsuka and Tomisawa, 1969; Bell, 1969; Bennett and Malmfors, 1970; Bennett et al., 1971a,b, 1973a,b,c, 1974a,c).

The color of the fluorescence and the conditions under which it develops indicate that the main catecholamine in avian adrenergic nerves is noradrenaline (see Bennett and Malmfors, 1970). However, there are some quantitative observations on the catecholamine content of the chick heart that indicate that adrenaline is present in higher concentrations than noradrenaline (Callingham and Cass, 1965, 1966; Ignarro and Shideman, 1968; Lin and Sturkie, 1968; Lin et al., 1970; Sturkie et al., 1970), although other findings do not support this contention (Govyrin and Leontieva, 1962; Govyrin, 1965; Enemar et al., 1965). It was recently suggested that the high levels of adrenaline in chick hearts are probably associated with nonneuronal tissue (Bennett and Malmfors, 1970), and this is supported by the finding that other tissues innervated by adrenergic nerves contain high levels of noradrenaline but little adrenaline (Sjöstrand, 1965; Baumgarten and Holstein, 1968). Furthermore, it has recently been observed that on stimulation of the cardiac nerves, only noradrenaline is released (Sturkie et al., 1970; Tummons and Sturkie, 1970).

Under the electron microscope, axon profiles in various peripheral effector tissues are seen to contain two types of granular vesicles—

small granular vesicles ranging from 400 to 600 Å in diameter, and large granular vesicles measuring 800 to 1600 Å in diameter. In the same axon profiles, both large and small vesicles are usually present. The larger vesicles are similar to those seen in the catecholamine-containing chromaffin cells of the adrenal gland (e.g., Kano, 1959; Coupland and Hopwood, 1966; Grignon et al., 1966; Hatier et al., 1969). It is interesting to note that some axon profiles contain mainly small granular vesicles (Bennett et al., 1970), whereas others contain a preponderance of large granular vesicles (Bennett and Cobb, 1969c; Akester, 1970). Until recently, it was assumed that axon profiles containing granular vesicles were adrenergic by analogy with the adrenal medulla and on the basis of such axons being present in systems in which an adrenergic innervation had been demonstrated by histochemical and physiological means. Recently, however, it was found that injections of 6-hydroxydopamine into chicks caused an intense granulation of vesicles in some axons (Bennett et al., 1970), and it was clearly shown by various techniques that the vesicles were contained within axons that were functionally adrenergic; 6-hydroxydopamine can thus be used as a marker for adrenergic nerves.

As yet there are few electrophysiological studies of adrenergic transmission (von Golenhofen, 1967; Bennett, 1969b,c, 1970a), and discrete adrenergic junction potentials have not been observed. In the case of the gut (Bennett, 1969b,c, 1970a), this seems to result from the lack of intimate contact between the nerves and the smooth muscle cells.

3. Noncholinergic, Nonadrenergic Innervation

Excitation resulting from stimulation of noncholinergic, nonadrenergic nerves has been suggested to occur in the chicken esophagus (Hassan, 1969); no comment on this possibility can be made at present, except to say that postinhibitory rebound excitation was not excluded (see below).

It was recently suggested that extrinsic nerves of the avian gut were associated with intramural inhibitory neurons that were neither cholinergic nor adrenergic (Everett, 1968; Bennett, 1969b,c, 1970a; Sato et al., 1970; Nakazato et al., 1970). Stimulation of these neurons gave rise to inhibition that was frequently followed by excitation (Everett, 1968; Nakazato et al., 1970; Sato et al., 1970). The electrophysiological basis of the inhibition in the gizzard was shown to be a transient hyperpolarization of the smooth-muscle cells, resulting from the summation of individual inhibitory junction potentials (Bennett, 1969b,c, 1970a). Following nerve stimulation, the membrane potential of the smooth muscle showed a rebound depolarization that could give rise to action potentials (Bennett, 1969b). It was suggested that the inhibitory junction potentials observed were not adrenergic, since stimulation of adrenergic nerves gave rise to excitatory effects (Bennett, 1969b). However, this is no longer a valid argument, since Sato et al. (1970) have demonstrated that adrenergic nerves can have inhibitory effects on the gizzard. Thus, while the observations described above remain acceptable, it is necessary to question whether they result from stimulation of nonadrenergic inhibitory neurons.

In mammals, it has been assumed that some of the inhibitory neurons innervating the gut are nonadrenergic, since it can be shown pharmacologically that the cell bodies of these inhibitory neurons lie in the enteric plexuses, but histochemical examination has not revealed the presence of adrenergic cell bodies in the enteric plexuses (see Campbell, 1970). The other arguments supporting the existence of nonadrenergic neurons can readily be countered, as Campbell (1970) has pointed out. Thus, if adrenergic cell bodies can be demonstrated in the enteric plexuses, there is no convincing evidence for the nonadrenergic nature of the inhibitory neurons of the gut. The same reasoning holds true for the avian system.

It has become clear that there is present in Auerbach's plexus of the avian gizzard a system of neurons that contain catecholamines (Bennett, 1969a, 1971c; Bennett and Malmfors, 1970; Bennett et al., 1971a, 1973c). These neurons are susceptible to 6-hydroxydopamine and other drugs that are known to affect adrenergic neurons (Bennett, 1971c; Bennett et al., 1971a, 1973c) and would thus appear to be, by definition, adrenergic. Whether or not these neurons are identical with the inhibitory neurons demonstrated physiologically remains to be determined.

D. AUTONOMIC GANGLIA

1. Ciliary Ganglion

The anatomical relationships of the ciliary ganglion are considered in Section III,E,8 below, but it should be noted that there are pathways to the ciliary ganglion whereby fibers from the oculomotor, trigeminal, and abducent nerves and from the superior cervical ganglion could reach it. The histology of the avian ciliary ganglion was first described in detail by von Lenhossék (1911) and Carpenter (1911). Carpenter (1911) observed large, medullated oculomotor fibers entering the ganglion and terminating in calyciform or arborescent endings on about three-quarters of the cells. Fine, weakly medullated fibers forming the radix longa from the deep ophthalmic nerve entered the distal portion of the ganglion and terminated in delicate end nets on the cells. This account differs from several subsequent descriptions (e.g., Seto, 1931; Terzuolo, 1951; Taxi, 1965; Oehme, 1968) in that it has been found that trigeminal fibers do not synapse in the ciliary ganglion. Thus, the calyx endings in the proximal regions and the pericellular endings in the distal regions belong to oculomotor fibers. However, it seems clear that trigeminal fibers do run through the ganglion (Oehme, 1968), and there is a possibility that fibers from the superior cervical ganglion are also present (Oehme, 1968). This last point is of interest, since histochemical studies (Ehinger, 1967) have shown that there are adrenergic pericellular endings around the cell bodies in the distal portion of the ganglion. Stimulation of extrinsic nerves does not elicit adrenergic effects, however (Marwitt et al., 1971; G. Pilar, personal communication, 1970), and thus, the pericellular endings may derive from cell bodies within the ganglion (Ehinger, 1967).

The early light microscopical findings on calyciform endings in the ciliary ganglion have been confirmed by ultrastructural observations (De Lorenzo, 1960, 1966; Hamori and Dyachkova, 1964; Szentágothai, 1964; Hess, 1965; Taxi, 1965; Takahashi and Hama, 1965a,b; Takahashi, 1967; Koenig, 1967; and others). However, it is clear that there are marked post-hatching developmental changes in synaptic relationships. Hamori and Dyachkova (1964) found, in the newly hatched chick, that each ganglion cell complex was surrounded by a loose capsule of Schwann cells that became more tightly packed during development and eventually formed a myelin-like sheath. These workers found that the calyciform endings were solid only during the first few days of post-hatching development; subsequently the calyces broke up into many branches of the same preganglionic fiber, and processes of the soma grew out and invaginated the presynaptic profile.

In newly hatched chicks, Hamori and Dyachkova (1964) observed some synaptic thickenings between pre- and postsynaptic elements, but accumulations of synaptic vesicles were usually lacking. On, or about, the fifth day after hatching, synaptic vesicles appeared in the synaptic regions, and by the seventh day, the adult type of polarized synapses was seen. Hess (1965) described similar developmental modifications, but the time courses of the changes he observed are radically different from those given by Hamori and Dyachkova (1964).

Martin and Pilar (1963a,b, 1964a,b,c), using electrophysiological techniques, demonstrated both electrical and chemical transmission from pre- to postsynaptic elements in the avian ciliary ganglion, but found that electrical transmission was established at the time when Hamori and Dyachkova (1964) observed that the calyx endings were disrupted. Thus, the assumption that the calyces were responsible for the electrical coupling was difficult to justify. This problem has been resolved by the demonstration that the myelin sheath surrounding the ganglion cells having calyx endings is responsible for electrical coupling between pre- and postganglionic elements (Hess *et al.*, 1969; Marwitt *et al.*, 1971). However, there is a possibility that specialized junctional regions between pre- and postganglionic membranes also contribute to the electrical coupling (De Lorenzo, 1966; De Lorenzo and Barnett, 1966). Recent results (Landmesser and Pilar, 1970; Pilar *et al.*, 1970) indicate that the relationships between pre- and postganglionic elements in the avian ciliary ganglion are more specific than is usually the case in autonomic ganglia.

2. Paravertebral Ganglia

The organization of the paravertebral ganglia has been considered by several investigators (e.g., Bidder and Volkmann, 1842; Marage, 1889; Huber, 1900; Langley, 1902, 1904; Rossi, 1922; Terni, 1924, 1929a, 1931; Kolossow and Sabussow, 1929; Bennett and Malmfors, 1970; Bennett, 1971a). Huber (1900), in summarizing earlier work, stated that axons of paravertebral chain ganglia rarely ran in peripheral branches to the viscera, but more frequently ran up or down the chain to adjacent ganglia, or peripherally in a ventral branch of a spinal nerve. Langley (1902) found that rami communicantes contained medullated postganglionic fibers and that such fibers also ran along the paravertebral chain in company with afferent fibers. A more detailed study by Langley (1904), however, indicated that between the ganglia in the upper cervical paravertebral chain there are no medullated fibers present, and this was accounted for by the observation that the postganglionic fibers pass into the spinal nerves of the corresponding segments. Langley denied the presence of commissural fibers (i.e., fibers arising from cell bodies in one ganglion and ending in another ganglion of the chain), but found that some preganglionic fibers sent branches to two or three successive paravertebral ganglia. Terni (1929a) found that the cervical paravertebral ganglia are traversed by fibers, some of which arise in the same or adjacent ganglia, while others are preganglionic fibers running in a cranial direction. Terni (1929a) found branches between the paravertebral ganglia and the ganglia of the cervical retrocarotid nerve trunk and claimed that the branches conveyed preganglionic fibers to the retrocarotid ganglia. Histochemical observations (Bennett, 1971a) have shown that adrenergic fibers arising from cervical paravertebral ganglia run out along the corresponding spinal nerves rather than in the paravertebral chain, an observation that agrees with Langley's (1904) findings.

Microscopic examination of paravertebral ganglia has shown the presence of various types of neurons, many of which appear to be adrenergic (Wechsler and Schmekel, 1966a,b, 1967; Bennett and Malmfors, 1970; Bennett, 1971a). Ganglionic synapses have been observed in the primary paravertebral trunk of the chick embryo (Wechsler and Schmekel, 1966a,b, 1967), but no systematic ultrastructural analysis of the paravertebral ganglia has been made, and physiological studies are entirely lacking.

3. Prevertebral and Terminal Ganglia

In spite of Weber's (1817) claim that ganglion cells are absent from the celiac and mesenteric prevertebral plexuses, subsequent studies (see references in Section III,B,3 above) have shown that there are extensive ganglia associated with the plexuses of the celiac, anterior, and posterior mesenteric arteries. Particularly in association with the adrenal glands, ovaries, and testes, the ganglia are widespread and frequently extend into the tissues innervated (see relevant sections below); thus there is no marked delineation between prevertebral and terminal ganglia. Recent observations (Bennett and Malmfors, 1970; Bennett, 1971c; Cobb and Bennett, 1971; Bennett *et al.*, 1974a) have shown that the terminal extensions of the prevertebral ganglia associated with the inferior vena cava are composed of adrenergic and cholinesterase-positive neurons. There are adrenergic and cholinergic pericellular endings associated with the cell bodies, but only cholinergic fibers form specialized synapses; the adrenergic fibers form only *en passage* relationships. These systems remain to be investigated further with ultrastructural and physiological techniques.

4. Remak's Nerve

The ganglionated nerve of Remak has received much attention (Swan, 1835; Remak, 1847; Marage, 1889; His, 1897; Thébault, 1898; Kuntz, 1910; Tello, 1924; Szantroch, 1927; van Campenhout, 1931, 1933; Nolf, 1934c; Coulouma, 1939a,b; Coulouma and Herrath, 1939; Hsieh, 1951; Pera, 1952a,b, 1962a,b, 1963a,b; Browne, 1953; Yntema and Hammond, 1955; Enemar *et al.*, 1965; Costa, 1966a,b; Cantino, 1968; Bennett and Malmfors, 1970; and others). Remak's nerve extends from the caudal end of the rectum to the upper end of the duodenum; its distance from the gut varies at different points. At the rostral end, the nerve is continuous with the celiac and mesenteric plexuses, and vagal fibers join the nerve at this level (Pera, 1962a,b, 1963a,b). At its caudal end, Remak's nerve is continuous with the lower extensions of the paravertebral trunks and with the ganglionated pelvic plexuses that derive, in part, from preganglionic fibers running through the sacral nerves. There are numerous connections between the paravertebral chains and Remak's nerve and between the gut and Remak's nerve at all points along its length.

Pera (1962a,b, 1963a,b) claimed that vagal fibers synapsed on cell bodies in the nerve as far down as the rectum. More recent histochemical observations (Costa, 1966a,b; Bennett and Malmfors, 1970) indicated that there were a large number of ascending adrenergic fibers in the rectal portion of the nerve; many ended around the numerous nonfluorescent cell bodies in this region, while others ran through side branches to the rectum or ascended the main nerve trunk. Toward the duodenum, the numbers of adrenergic cell bodies in the nerve increased, as did the numbers of descending adrenergic fibers, while pericellular adrenergic endings became fewer. At all points along the nerve, adrenergic fibers joined it from the prevertebral and paravertebral ganglia, and adrenergic fibers ran from the nerve into the gut. As yet, no investigation has been made of the origin or distribution of other types of axons in Remak's nerve. Ultrastructural observations (Cantino, 1968; Cobb and Bennett, 1971) showed that some cell bodies in Remak's nerve had the characteristics of adrenergic neurons. In spite of the histochemical demonstration of numerous adrenergic pericellular endings in the rectal portions of the nerve, no specialized adrenergic synapses were identified with the electron microscope (Cobb and Bennett, 1971), a finding similar to that made in prevertebral ganglia (Cobb and Bennett, 1971) and Auerbach's plexus (Bennett and Cobb, 1969c; Bennett et al., 1971a, 1973c).

The general opinion is that Remak's nerve is comparable to the hypogastric or pelvic nerve plexus of mammals. Kuntz (1910, 1911) considered that Remak's nerve is associated with the oviparous habit of birds. However, histochemical and physiological findings (see Section III,E,3,d below) indicate that Remak's nerve is mainly concerned with the innervation of the gut, although this does not preclude the possibility that it contributes to the innervation of the genital, or other, systems, as mentioned in the relevant sections.

5. Enteric Plexuses

Studies of the enteric plexuses indicate that their anatomical disposition varies in different regions of the gut (e.g., Mangold, 1906; Iwanow, 1930; Kolossow et al., 1932; Iwanow and Radostina, 1933; Okamura, 1934; Nolf, 1934a,b; Ábrahám, 1935, 1936; Coulouma and Herrath, 1939; Kolossow, 1959; Malinovský, 1963, 1964; Enemar et al., 1965; Everett and Mann, 1967; Csoknya, 1968; Bennett, 1969a; Bennett and Cobb, 1969b,c; Cantino, 1970), while the types of neurons present in the plexuses also vary in different regions.

Experimental studies by Kolossow et al. (1932) showed that vagal fibers synapse on cells in the enteric plexuses as far down as the duodenum. Subsequent studies by Iwanow and Radostina (1933) demonstrated that there are anastomoses between Auerbach's and Meissner's plexuses in the small intestine of the pigeon. Cells of Meissner's plexus were observed to send processes into Auerbach's plexus and also into other ganglia of Meissner's plexus. Cauterization of a portion of Meissner's plexus in the pigeon duodenum caused degeneration of fibers in the circular muscle coat and of pericellular fibers in Auerbach's plexus. Following cauterization of regions of Auerbach's plexus, however, degenerating fibers were only rarely seen in Meissner's plexus, but many pericellular endings in Auerbach's plexus and fibers in the muscle coats were affected.

Recent histochemical observations (Enemar et al., 1965; Everett and Mann, 1967; Bennett, 1969a; Bennett and Cobb, 1969c; Bennett and Malmfors, 1970; Bennett et al., 1971a, 1973c) have shown that many adrenergic fibers supplying the avian gut terminate in the enteric plexuses. However, there are indications that in some regions of the gut there may be a moderate adrenergic innervation of the musculature (see Bennett and Malmfors, 1970). In Auerbach's plexus of the gizzard, fluorescent cell bodies have been seen (Bennett, 1969a, 1971c; Bennett and Malmfors, 1970; Bennett et al., 1971a, 1973c), but whether these cells are adrenergic or not has yet to be resolved (see Section III,C,3 above). Thus, the finding of cholinesterase-positive and cholinesterase-negative cell bodies in the gizzard plexus (Bennett, 1969a; Bennett and Cobb, 1969b) is difficult to interpret at present.

From ultrastructural findings (Bennett and Cobb, 1969b,c; Bennett et al., 1971a), it seems that there are usually numerous cholinergic synapses on cell bodies in Auerbach's plexus of the gizzard, but adrenergic fibers, although they may be closely associated with ganglion cell bodies, do not appear to form specialized synapses.

Physiological studies have done little to resolve the problem of the organization of the enteric plexuses. The work of Nolf (1928, 1930,

1938a,b,c, 1939) is detailed but difficult to interpret (see Bennett, 1970b); his results indicate the presence of intrinsic excitatory and inhibitory neurons in Auerbach's plexus that are responsible for the spontaneous activity of the gut. More recently, electrophysiological studies (Bennett, 1969b,c, 1970a) have confirmed that there are intrinsic excitatory and inhibitory neurons in Auerbach's plexus of the gizzard, but attempts to demonstrate adrenergic modulation of ganglionic transmission in the plexus failed, although it was found that there was a complex neuronal control of individual smooth muscle cells (Bennett, 1969c, 1970a).

E. AUTONOMIC INNERVATION OF EFFECTOR TISSUES

1. Cardiovascular System (see also Chapter 4, Volume II)

a. Heart. The gross anatomy of cardiac innervation is dealt with in most works on autonomic anatomy (see Sections II,B,1 and 3 above), and there is general agreement that the heart is supplied by the vagi and by the cardiac nerves that arise from the paravertebral chain. A recent detailed account by Cohen et al. (1970) indicates that in the pigeon the vagi send several branches to the heart, forming a particularly marked plexus in the region of the sinoatrial and atrioventricular nodes. Macdonald and Cohen (1970) showed that preganglionic cardioaccelerator fibers always arise from the upper two thoracic segments of the spinal cord, and sometimes from the last cervical and midthoracic segments, in agreement with Paton (1912). The postganglionic cardiac nerve fibers arise from the three right caudal cervical paravertebral ganglia, and thus the postganglionic outflow is displaced with regard to the location of the preganglionic cell bodies. The left cardiac nerve has an inconsistent effect on heart rate. Tummons and Sturkie (1968, 1969) found, as did several earlier workers, that the cardiac nerves in the chicken arise from the first thoracic paravertebral ganglion.

There have been various observations on the microscopic anatomy of cardiac innervation (Vignal, 1881; Pisskunoff, 1911; Paton, 1912; Külbs, 1912; Kondratjew, 1926, 1933; Drennan, 1927; Ssinelnikow, 1928a,b; Szantroch, 1929; Szepsenwol and Bron, 1935a,b, 1936; Tixier-Durivault, 1942; Hsieh, 1951; de Meyer, 1952; Ábrahám and Stammer, 1957; Ábrahám, 1962, 1969; Hirsch, 1963; Yousuf, 1965; Jain, 1965), but these earlier studies did not indicate the nature of the cell bodies and nerve fibers associated with the heart. More recent studies have shown that both adrenergic (Govyrin and Leontieva, 1965a,b; Enemar et al., 1965; Otsuka and Tomisawa, 1969; Akester et al., 1969; Yamauchi, 1969; Bennett and Malmfors, 1970; Bennett,

1971b,c; Bennett *et al.*, 1971b, 1973a,b, 1974c) and cholinergic (Hirakow, 1966; Gossrau, 1968; Yamauchi, 1969) nerve fibers supply the cardiac muscle, but apart from those in cardiac paraganglia (Enemar *et al.*, 1965; Bennett, 1971b), catecholamine-containing cells are absent from the heart.

These findings are supported by physiological observations (Marage, 1889; Couvreur, 1892; Jürgens, 1909; Stübel, 1910; Paton, 1912; Johansen and Reite, 1964; Bolton and Raper, 1966; Bolton, 1967; Cohen and Pitts, 1967, 1968; Folkow and Yonce, 1967; Tummons and Sturkie, 1968, 1969, 1970; Bolton and Bowman, 1969; Macdonald and Cohen, 1970; Cohen and Schnall, 1970; Cohen *et al.*, 1970; Bennett, 1971c; Bennett *et al.*, 1974c; and see references in Section II,E,2 above), but the effects of stimulating extrinsic nerves do not always appear to be straightforward. Marage (1889) observed that vagal stimulation had excitatory and inhibitory effects on the heart; similarly, Couvreur (1892) found that positive inotropic and negative chronotropic effects resulted from vagal stimulation. The bradycardia was abolished by atropine, but no tachycardia was seen under those conditions. Paton (1912) found in the duck that vagal stimulation could arrest the auricles and ventricles, but after atropine, ventricular contractions were sometimes enhanced by vagal stimulation. More recently, Tummons and Sturkie (1970) demonstrated that stimulation of the cardiac nerves in chickens elicited excitatory adrenergic and inhibitory cholinergic effects on the heart. It seems that some of the conflicting effects of extrinsic nerve stimulation on the bird heart may result from excitation of mixed nerve trunks. The origins of the different types of nerve fibers remain to be determined.

b. Blood Vessels. Various investigators have made sundry observations on the innervation of avian blood vessels (e.g., Geberg, 1884; Botezat, 1906; Arimoto and Miyagawa, 1930; Takino, 1932; Laruelle *et al.*, 1951; de Meyer, 1952; Gilbert, 1961; Ábrahám, 1969; and others). Recent findings indicate that blood vessels in birds may be innervated by adrenergic (Enemar *et al.*, 1965; Leontieva, 1966a,b; Folkow *et al.*, 1966; Bennett, 1969a, 1971a,b,c; Akester and Mann, 1969a,b; Doležel and Žlábek, 1969; Bell, 1969; Bennett and Malmfors, 1970; Cobb and Bennett, 1971; Bennett *et al.*, 1971b, 1973a,b, 1974a,b,c) and cholinergic nerves (Bennett, 1969a; Bennett and Cobb, 1969a; Akester and Mann, 1969a,b; Bell, 1969; Bennett *et al.*, 1974a).

Functional studies on vascular innervation have been carried out in investigations of avian cardiovascular reflexes (see Section II,E,2) and in works dealing with autonomic physiology generally. Thus, Laffont (1885) observed that stimulation of the peripheral extensions of the trigeminal nerves or the cephalic branches of the superior cervical ganglion had vasomotor effects on cranial vessels. Jegorow (1890) found that stimulation of the cervical paravertebral chain caused constriction of the vessels of the integument, conjunctiva, eyeball, and eyelids; these observations were extended by Langley (1904) to include the vessels of the mucous membranes of the beak, and Laruelle *et al.* (1951) confirmed the findings on the vasomotor innervation of integumentary vessels. The only positive study dealing with vasomotion resulting from vagal stimulation appears to be that of Couvreur (1892), who demonstrated constriction of the vessels of the esophagus and crop. More recent investigations of isolated tissues indicate that vessels may be affected by cholinergic and adrenergic nerves (Bolton, 1968a,b, 1969; Bell, 1969; Bennett, 1971c; Bennett *et al.*, 1974b,c).

2. Respiratory System (see also Chapter 5, Volume II)

The gross anatomy of the innervation of the avian lung is well-documented (e.g., Marage, 1889; Gadow, 1891; Couvreur, 1892; Thébault, 1898; Hsieh, 1951; Watanabe, 1960; Malinovský, 1962; Fedde *et al.*, 1963). Hsieh (1951) stated that the pulmonary plexus lies on the lateral and anterior aspect of the root of the lung and was formed by the pulmonary branches of the vagal and cardiac nerves; the majority of the united nerve branches entered the lung parenchyma through the hilus of the lung. Hsieh found that the pulmonary plexus extended over the whole ventral surface of the lung beneath the visceral pleura; many nerve fibers pass into the substance of the lung, but some run to the pulmonary aponeurosis. Hsieh observed that the latter was also supplied by fibers arising from the esophageal plexus, as was the oblique septum (Watanabe, 1960).

The intrinsic innervation of the lung has been described by various workers (Eberth, 1863; Arimoto and Miyagawa, 1930; Okamura, 1930; Takino, 1932; Muratori, 1935; Hsieh, 1951; van Campenhout, 1955, 1956; McLelland, 1969). From histochemical observations (Akester and Mann, 1969b; Bennett and Malmfors, 1970; Bennett, 1971b), it seems that there is only a sparse adrenergic innervation of the bronchial musculature. Thus, the numerous pulmonary ganglion cells and nerve fibers demonstrated histologically (see references above) are probably cholinergic, as suggested by the findings of Akester and Mann (1969b). Ultrastructural observations on the innervation of pulmonary smooth muscle (Cook and King, 1969a, 1970) indicate that both cholinergic and adrenergic fibers are present. Cook and King (1969a, 1970) found that the primary bronchi are more densely innervated than the secondary or tertiary bronchi, but in the last, the neuromuscular relationships were more intimate.

Although Couveur (1892) claimed that stimulation of the peripheral end of the vagus had no effect on the lung, it has recently been suggested that the vagus supplies cholinergic excitatory fibers to the lung (Cowie and King, 1969; King and Cowie, 1969). As yet there is no convincing demonstration of an adrenergic influence on the avian lung.

3. Digestive System (see also Volume II, Chapter 6)

a. Salivary Glands. Hsieh (1951) has given an extensive account of his own and other work dealing with the innervation of avian salivary glands (see Section II,B,1), but physiological studies on this topic are rare. Graham (1938), who appears to be the only investigator to have dealt with the subject, found that stimulation of the chorda tympani in the duck causes salivary secretion that is blocked by atropine, indicating that the nerve fibers involved are cholinergic.

b. Esophagus and Crop. The gross anatomy of the nerve supply of the gastrointestinal tract is dealt with in those papers concerned with the distribution of cranial nerves and autonomic nerve trunks (see references in Sections II,B,1 and III,B,3). Hsieh (1951) has described the esophageal plexus lying on the ventral surface of the esophagus, dorsal and posterior to the base of the heart. The plexus is continuous with the ground plexus of the vagus anteriorly, and is connected with the pulmonary plexuses laterally and with the superior and inferior cardiac plexuses ventrally. Branches from the plexus enter the esophagus on the ventral and lateral aspects while the descending branch of the recurrent nerve is distributed to the dorsolateral aspect. The glossopharyngeal nerve also contributes to the innervation of the esophagus (Thébault, 1898; Watanabe, 1964; Bubien-Waluszewska, 1968; and see references in Section II,B,1 above).

The intramural innervation of the esophagus and crop has been dealt with by various investigators (Kolossow *et al.*, 1932; Csoknya, 1968; Bartlett and Hassan, 1968), and there are several physiological studies of this region of the gastrointestinal tract (Doyon, 1894b; Ihnen, 1928; Hanzlik and Butt, 1928; Ashcraft, 1930; Bowman and Everett, 1964; Everett, 1966; Hassan, 1967, 1969; Ohashi and Ohga, 1967; Nakayama, 1968; Sato, 1969; Sato *et al.*, 1970). Stimulation of the vagus produces cholinergic effects that are usually excitatory, but may, in some instances, be inhibitory (Hanzlik and Butt, 1928); part of the vagal cholinergic innervation is preganglionic since it is suppressed by hexamethonium and mecamylamine (Bowman and Everett, 1964; Everett, 1966). The glossopharyngeal innervation of the esophagus and crop is cholinergic and excitatory (Hassan, 1967). More recent findings complicate the picture; Hassan (1969) has described a hyoscine-resistant contraction of the esophagus in response to stimulation of the vagal and descending esophageal nerves. However, since Sato et al. (1970) have observed relaxation of the crop in response to vagal stimulation, it is possible that the contractions observed by Hassan (1969) were rebound contractions following stimulation of inhibitory nerves (see Section III,C,3 above).

Hanzlik and Butt (1928) claimed that stimulation of "sympathetic" nerves to the crop of the pigeon caused cholinergic contractions of the longitudinal muscle, but from their anatomical description it seems likely that they were stimulating the descending esophageal nerves of the glossopharyngeal. The more reliable study of Sato et al. (1970) indicates that perivascular stimulation has no effect on the crop.

Various workers (e.g., Bowman and Everett, 1964; Everett, 1966; Ohashi and Ohga, 1967) have assumed that the cholinergic innervation of the esophagus and crop is vagal in origin (at least by implication, since they refer to the innervation as "parasympathetic"). However, the point is not at all clear and serves to demonstrate the need for careful anatomical and physiological studies of the origin and distribution of the innervation of this part, and indeed all parts, of the avian gut.

c. Proventriculus and Gizzard. All the general references given in Sections II,B,1 and III,B,3 deal in varying extent with the innervation of the proventriculus and gizzard, but this topic is considered in more detail by Nolf (1934a,b), Coulouma (1935, 1939c), Hsieh (1951), Malinovský (1963), and Bennett (1969b). There is general agreement that vagal and perivascular nerves supply the proventriculus and gizzard; the perivascular nerves run from the celiac and mesenteric plexuses, but their exact site of origin is not known.

The microscopic anatomy of the innervation of the proventriculus and gizzard has received intermittent attention (Mangold, 1906; Iwanow and Radostina, 1933; Okamura, 1934. Ábrahám, 1936; Coulouma and Herrath, 1939; Kolossow, 1959; Malinovský, 1964; Csoknya, 1968; Milokhin and Reshetnikov, 1968; Bennett, 1969a; Bennett and Cobb, 1969a,b,c; Cantino, 1970; Bennett and Malmfors, 1970; Bennett et al., 1971a, 1973c). Neuronal relationships in the enteric plexuses are considered in Section III,D,5, but here it should be noted that there is evidence for a cholinergic innervation of the smooth muscle of this region of the gut. Adrenergic nerves are localized mainly in the enteric plexuses, but appear to affect the smooth muscle directly (see below).

Physiological studies of the innervation of the proventriculus and gizzard (Doyon, 1894a,b, 1925; Elliot, 1905; Mangold, 1906, 1911; Stübel, 1911; Nolf, 1925a,b,c, 1927, 1935; Bogdanov and Kibyakov, 1955; Hassan, 1967; Bennett, 1969b,c,d, 1970a; Kazumoto, 1969; Sato, 1969; Sato *et al.*, 1970; Nakazato *et al.*, 1970) generally indicate that vagal and perivascular pathways may contain both excitatory and inhibitory fibers for the proventriculus and gizzard. The excitatory fibers in the vagus are cholinergic and may be pre- or postganglionic; the excitatory fibers in the perivascular nerve trunks may be pre- or postganglionic cholinergic or postganglionic adrenergic (see Bennett, 1969b). The picture is complicated by the observation that the adrenergic fibers in the perivascular nerve trunks may also have inhibitory effects on the proventriculus and gizzard (Kazumoto, 1969; Sato *et al.*, 1970). Furthermore, there may be present, in vagal and perivascular nerves, pre- and postganglionic inhibitory nerve fibers that are resistant to the effects of various antiadrenergic drugs (Bennett, 1969b,c, 1970a; Sato, 1969; Sato *et al.*, 1970). There is also a suggestion (Nakazato *et al.*, 1970) that the proventriculus is innervated by preganglionic fibers that synapse with atropine-resistant, intramural, excitatory neurons. Electrophysiological studies (Bennett, 1969b,c,d, 1970a) have shown that single smooth-muscle cells of the gizzard may be affected by excitatory cholinergic fibers, excitatory adrenergic fibers, and inhibitory fibers whose nature is obscure at present (see Section III,C,3 above). The origin and relationships of these various nerve types remain to be determined, as does the problem of whether the adrenergic fibers act directly on the smooth-muscle cells *in vivo*.

d. Intestine, Rectum, and Cloaca. Apart from the general references on autonomic anatomy (see Sections II,B,1 and III,B,3) many studies dealing with Remak's nerve have considered its contribution to the innervation of the distal portions of the gastrointestinal tract (see references in Section III,D,4). There is some confusion among them; for example, Thébault (1898) maintained that the vagus supplies the entire intestine, whereas Stiemens (1934) denied this and claimed that the vagus supplies only the proximal duodenum. While it may be true that no macroscopic vagal branches run the entire length of the intestine, there is every possibility that vagal fibers reach the distal intestine following other pathways. Indeed, Kolossow *et al.* (1932) have demonstrated that vagal fibers are present in the enteric plexuses of the intestine, and Pera (1962b, 1963a,b) has shown that vagal fibers run in Remak's nerve as far as the rectum. Since Remak's nerve is continuous proximally with the celiac and mesenteric prevertebral plexuses and distally with the caudal paravertebral chains and pelvic plexuses (see Section III,D,4 above), then there are numerous routes whereby fibers from the vagi, the paravertebral and prevertebral ganglia, and the pelvic plexuses could reach the intestine, rectum, and cloaca.

The intramural innervation of these regions of the gut has been described in developing (Kuntz, 1910; Abel, 1910, 1912; Tello, 1924; van Campenhout, 1931, 1932; Enemar et al., 1965; Cordier, 1969) and adult birds (Iwanow, 1930; Kolossow et al., 1932; Iwanow and Radostina, 1933; Ábrahám, 1935, 1936; Kolossow, 1959; Pintea et al., 1967; Everett and Mann, 1967; Everett, 1968; Batekhina, 1968; Csoknya, 1968). But this region of the avian gut has been particularly neglected from a physiological point of view. Elliot (1905) observed that vagal stimulation has slight excitatory effects on the duodenum of the chicken. Furthermore, stimulation of the nerve trunks arising from the third and fourth thoracic and first lumbar segments causes the duodenum to blanch and contract slowly. Nolf (1934a) found that stimulation of the vagus below the level of the oblique septum causes excitatory effects throughout the whole small intestine. He considered the effect to be transmitted to the intestine via Auerbach's plexus of the gizzard and Remak's nerve. Nolf (1934b) observed that the celiac and mesenteric plexuses give rise to fibers that innervate the entire small intestine; stimulation of the nerves accompanying the celiac artery caused an augmentation of the contractions of the duodenum, whereas the middle and lower ileum showed an initial inhibition followed by a period of hyperactivity. Stimulation of the nerves accompanying the anterior mesenteric artery caused an initial inhibition and subsequent excitation of the whole intestine. Nolf found the excitatory effects of vagal and celiac nerve stimulation were more marked after section of Remak's nerve, an observation that it is difficult to reconcile with his earlier suggestion that vagal fibers reached the intestine via Remak's nerve. Stimulation of Remak's nerve (Nolf, 1934c) produced a reaction in the rectum and in the ileum as far as the duodenum. In the upper and midileum the effect was excitatory, whereas in the lower ileum the effect was diphasic, being an inhibition followed by an excitation. Nolf (1934c) found that when Remak's nerve was stimulated at different points along its course, there was an initial inhibition of the ileum for some distance above the point of stimulation. Nolf considered that the inhibitory effect was mediated by ascending fibers derived from the mesenteric nerves in the ileal course of Remak's nerve. Excitation of the caudal end of Remak's nerve evoked a tetanic contraction of the rectum (Nolf, 1934c).

More recently, Everett (1968) found that the duodenum, ileum, and rectal ceca of very young chicks responded to perivascular or transmural stimulation in a biphasic fashion. The excitatory component of the response was blocked by anticholinergic drugs and was suppressed by ganglion blocking drugs. The inhibitory response was blocked by antiadrenergic drugs, although under certain conditions inhibitory responses resistant to antiadrenergic drugs were seen. These observations are compatible with the earlier findings, and support the histochemical evidence for an adrenergic (Enemar et al., 1965; Everett and Mann, 1967; Bennett and Malmfors, 1970) and cholinergic (Everett, 1968) innervation of the avian intestine, as do more recent pharmacological investigations (Kagawa et al., 1969; Shimada, 1969).

4. Excretory System

a. Kidney and Ureter. Since the organs in the lower abdominal region are often closely associated with extensive nerve plexuses (see Hsieh, 1951), it is difficult to give a complete anatomical description of their innervation; the kidneys and ureters are supplied, at least, by fibers from the aortic and pelvic plexuses. Although there are some descriptions of the anatomy of the urinary system (e.g., Bortolami and Palmieri, 1962; Liu, 1962), the microscopic innervation of this region has received scant attention. Shvalev (1965) described small ganglia associated with the kidneys and ureters lying along blood vessels and the ureters and between the channels of the mesonephros and along the larger nerve trunks. The ganglia consist of small neurons and larger neurons with extensive processes; Shvalev found that interganglionic fibers persisted after vagotomy.

Histochemical observations (Akester and Mann, 1969a; Bennett and Malmfors, 1970) indicate that the adrenergic innervation of the kidney is sparse, the nerve fibers being associated with blood vessels. Bennett and Malmfors (1970) found that the ureters receive a moderate adrenergic innervation, the fibers being most numerous in the inner circular muscle layer.

Gibbs (1929) observed that the ureter of the fowl undergoes peristalses, but no antiperistalses were seen. Adrenaline increased the frequency but decreased the amplitude of spontaneous contractions. Gibbs (1929) suggested that the excitatory effects of vagal stimulation on the ureter were due to reflex release of catecholamines from the adrenal medulla, following the vagally evoked fall in blood pressure.

b. Nasal Gland (see also Chapter 9, Volume II). The innervation of the nasal gland is dealt with in many papers concerning autonomic anatomy (see, especially, Grewe, 1951; Hsieh, 1951). More recent observations (Fänge *et al.*, 1958, 1963; Schmidt-Nielsen, 1960; Håkansson and Malcus, 1969; Ash *et al.*, 1969; Cottle and Pearce, 1970) have confirmed the earlier studies and indicate that the preganglionic innervation of the nasal gland derives from the facial nerve, with no contribution from the ophthalmic branch of the trigeminal. There is some variation in the accounts with regard to the source of the postganglionic innervation of the gland, it being either from the ethmoidal ganglion (see Hsieh, 1951; Schmidt-Nielsen, 1960) or from the sphenopalatine ganglion via the secretory nerve ganglion (Ash *et al.*, 1969; Cottle and Pearce, 1970). Ash *et al.* (1969) suggested that the secretory nerve (i.e., the nerve running between the sphenopalatine and secretory nerve ganglia) contained somatic afferent and postganglionic adrenergic fibers, since it was devoid of cholinesterase activity. Cottle and Pearce (1970) pointed out that the preganglionic innervation could be noncholinergic, although the synapses in the secretory nerve ganglion did not differ from classic cholinergic synapses. Histochemical studies have shown that the nasal gland receives a fairly dense cholinesterase-positive innervation (Fänge *et al.*, 1963; Ballantyne and Fourman, 1967; Fourman, 1969; Ash *et al.*, 1969) that may, in part, arise from cell bodies within the gland (Cottle and Pearce, 1970). Fourman (1969) found that the adrenergic innervation of the nasal gland was sparse.

Physiological studies indicate that the cholinergic innervation is excitatory (Fänge *et al.*, 1958, 1963; Schmidt-Nielsen, 1960; Håkansson and Malcus 1969; Ash *et al.*, 1969), although there has been some discussion whether the fibers act on the gland cells or on blood vessels (see Schmidt-Nielsen, 1960; Bonting *et al.*, 1964). Stimulation of the cervical paravertebral chain was found to have no effect on glandular secretion (Schmidt-Nielsen, 1960; Fänge *et al.*, 1963) but did cause a small increase in glandular blood pressure. Fänge *et al.* (1963) suggested that adrenergic vasoconstrictor fibers reached the gland via facial nerve pathways.

5. *Genital System*

a. Ovary and Oviduct. From the general references on autonomic anatomy (see Section III,B,3) and the specific descriptions of the innervation of the female reproductive tract (de Lisi, 1924; Johnson, 1925; Mauger, 1941; Freedman, 1962; Freedman and Sturkie, 1963; Oribe, 1963, 1965, 1969), it is clear that this system may be supplied by fibers from the paravertebral chain, the prevertebral plexuses, the pelvic plexus, and Remak's nerve. Observations on the microscopic innervation of the ovary (Ganfini, 1907; Biswal, 1954; van Campenhout, 1962, 1964, 1967; Oribe, 1963, 1965, 1969; Gilbert, 1965a,b, 1967, 1968, 1969; Bennett and Malmfors, 1970) have frequently indicated that the density of innervation increases with age. Since the terminal innervation of the ovary derives, in part, from intramural ganglion cells, then this increase is possibly correlated with the observed increase in numbers of intramural ganglion cells. Ovarian interstitial tissue has been shown to receive a dense innervation (van Campenhout, 1962, 1964, 1967; Dahl, 1970). Gilbert (1969) and Bennett and Malmfors (1970) found that the adrenergic innervation of the ovary, and particularly of the mature follicle, is also very dense. This is a particularly interesting point, since functional studies have indicated that an adrenergic mechanism may be involved in ovulation (Ferrando and Nalbandov, 1969).

The intramural innervation of the oviduct has received little attention (e.g., Biswal, 1954; Gilbert and Lake, 1963; Bennet and Malmfors, 1970), but recent observations indicate an apparent age-dependent change in the pattern of adrenergic innervation (Bennett and Malmfors, 1970). Pharmacological studies (McKenney et al., 1932; Sykes, 1955a; Prasad et al., 1964) have shown that different regions of the oviduct are variably responsive to adrenergic and cholinergic drugs; it seems that the response may be determined by the hormonal state of the animal (Prasad et al., 1964). However, physiological investigations (Sturkie and Freedman, 1962) have produced little information on the involvement of the autonomic nervous system in oviduct function.

b. Testis and Vas Deferens. The gross anatomy of the innervation of the male genital system is comparable to that of the female (see Hseih, 1951). Although there are scattered accounts of the anatomy of the genital system itself (e.g., MacDonald and Taylor, 1933; Gray, 1937; Lake, 1957; Höhn, 1960; Mehrotra, 1964; Hassa and Calislar, 1964; Komarek and Marvan, 1969; Marvan, 1969; Traciuc, 1969), descriptions of the microscopic anatomy of the innervation of the testis and vas deferens are extremely rare. Baumgarten and Holstein (1968) found adrenergic nerves associated with the epididymal ducts and testicular interstitial cells in the Mute Swan (*Cygnus olor*). In mature domestic cocks, Bennett and Malmfors (1970) found that adrenergic fibers are sparse in the testis and are usually associated with blood vessels or the smooth muscle of the testicular capsule. However, the epididymal ducts and vasa deferentia are densely innervated, and in the latter, the adrenergic nerve fibers form an extremely dense submucosal plexus. The density of innervation of the vas deferens corresponds with its very high catecholamine content (Sjöstrand, 1965). There appear to have been no physiological observations on the autonomic control of the male genitalia.

6. Endocrine Organs

a. Adrenal Gland. There have been several detailed descriptions of the innervation of the adrenal glands (Fusari, 1892, 1893; Kuntz, 1910; Ganfini, 1916; Goormaghtigh, 1921; Rau and Johnston, 1923; Müller, 1929; Brauer, 1932; Goloube, 1936; Hammond and Yntema, 1947; Hsieh, 1951; Yntema and Hammond, 1955; Ábrahám, 1963; Freedman, 1968). It is generally agreed that the adrenal innervation passes through splanchnic nerve trunks associated with the fourth thoracic to the third lumbosacral segment.

Studies on the minute anatomy of adrenal innervation (Rabl, 1891; Giacomini, 1898; Hoshi, 1926; Kura, 1927; Knouff and Hartman, 1951; Biswal, 1954; Kano, 1959; Ábrahám and Stammer, 1959; Ábrahám, 1963; Shioda and Nishida, 1967; Bennett and Malmfors, 1970) have generally indicated that there is a dense innervation of the chromaffin cells; this innervation appears to be cholinergic (Ábrahám and Stammer, 1959). The nerve fibers form classic synapses on the chromaffin cells (Kano, 1959), and thus the evidence indicates that the innervation is preganglionic, as in mammals. Although there are various reports on the occurrence of adrenaline and noradrenaline-containing cells in the chromaffin tissue (Eränkö, 1957; Benedeczky *et al.*, 1964; Coupland and Hopwood, 1966; Sivaram, 1968; Cuello, 1970), there have been no physiological observations on nerve-mediated release of adrenaline or noradrenaline from the avian adrenal gland.

The innervation of the cortical (interrenal) cells of the adrenal glands has been considered to be absent (Ábrahám and Stammer, 1959), sparse (Giacomini, 1898; Kura, 1927), or moderate (Shioda and Nishida, 1967). Shioda and Nishida (1967) concluded that fine nerve fibers are distributed to nests of cortical cells as well as to blood vessels. Supposed afferent endings were observed between the cortical cells and on the larger blood vessels.

b. Pineal Gland. It is generally agreed that the pineal gland is innervated mainly by fibers that reach it running with blood vessels (Stammer, 1961a; Quay, 1965; Kappers, 1965; Fujie, 1968; Bischoff, 1969; Oksche and Kirschstein, 1969; Oksche *et al.*, 1969). Although there have been descriptions of ganglion cells within the pineal parenchyma (e.g., Ueck, 1970), this is denied by the majority of investigators. Adrenergic (Oksche *et al.*, 1969; Wight and MacKenzie, 1970) and cholinergic (Wight and MacKenzie, 1970) nerve fibers have been observed in association with the pineal gland. It seems that some of the fibers arise from, or pass through, the superior cervical ganglion (McFarland *et al.*, 1966; Lauber *et al.*, 1968).

c. Thymus Gland. Terni (1929a,b) observed that the thymus receives a dense innervation from perivascular and vagal fibers, but that the majority of nerves end in association with myoid and epthelioid cells. Terni also noted the presence of small ganglia along the nerve trunks that accompany blood vessels. Hsieh (1951) claimed that branches of the recurrent and retrovertebral nerves innervate the thymus. In a recent histochemical study (Bennett, 1971a), it was found that there are only a few adrenergic nerves within the thymus and that they are associated with blood vessels; no adrenergic cell bodies were seen within the thymus. Thus, it seems that the dense innervation of the thymus, reported by Terni (1929a,b), is not adrenergic.

d. Thyroid Gland. Terni (1929a) reported that only few perivascular nerve fibers innervate the thyroid gland, but Hsieh (1951) found that collateral branches from the vagus and fibers from the recurrent, retrocarotid, and retrovertebral nerves supply the gland. Similar observations were made by Pastěa and Pastěa (1966). The microscopic anatomy of thyroid innervation has been described by Legait and Legait (1952), who found large numbers of sensory cell bodies associated with the nerve trunks supplying the gland. However, a recent reinvestigation of this problem (Makita *et al.*, 1966) indicated that, although the thyroid is supplied by fibers from the nodose ganglion, there were few cell bodies associated with the gland. Makita *et al.* (1966) observed nerve plexuses surrounding the thyroid blood vessels, but only few nerve fibers are associated with thyroid follicles. Bennett (1971a), using a fluorescence histochemical technique, found that the thyroid vessels were densely innervated by adrenergic fibers, but no perifollicular nerve fibers were found. It seems that the follicular innervation described by Makita *et al.* (1966) is nonadrenergic and probably afferent.

e. Parathyroid Gland. Although some observations indicate that the parathyroid gland is poorly innervated (Terni, 1929a), other findings suggest the gland is well innervated (Hsieh, 1951; Legait and Legait, 1952). This latter view has been supported by recent findings (Bennett, 1971a) that showed a dense adrenergic innervation of the parathyroid vasculature and parenchyma. It remains to be demonstrated physiologically that such an innervation is involved in the control of the parathyroid gland.

f. Ultimobranchial Body. It is generally agreed that the ultimobranchial body is supplied by vagal and perivascular nerve trunks (Terni, 1927, 1929a; Watzka, 1933; Nonidez, 1935; Dudley, 1942). Ulstrastructural observations have shown that some C cells of the ultimobranchial body are closely associated with nerve fibers (Stoeckel and Porte, 1967, 1969a,b; Hodges and Gould, 1969), and it has recently been found that vagal stimulation causes a fall in plasma calcium (Hodges and Gould, 1969), presumably due to nerve-mediated release of calcitonin. The vagal effect was blocked by atropine; there is no indication of an adrenergic innervation of the ultimobranchial body (Bennett, 1971a). However, it is of interest to note that the C cells of the ultimobranchial are able to take up, metabolize, and store various amines and their precursors (Hachmeister *et al.,* 1967; Bennett, 1971a).

g. Pancreas. The islet cells of the pancreas have been shown to contain catecholamines (Falck and Hellman, 1963; Bennett and Malmfors, 1970), but the significance of this finding is obscure. The innervation of islet cells has been reported to be absent or scarce (Kobayashi and Fujita, 1969; Bennett and Malmfors, 1970) or moderate (Falck and Hellman, 1963). These variations are likely to be due to species differences. No physiological data are available on the nervous control of pancreatic islets.

7. Integumentary System

a. Integumentary Smooth Muscle. While the gross anatomy of the innervation of integumentary smooth muscle has received little attention, available studies indicate that pennamotor fibers run from paravertebral chain ganglia to the effector tissue in company with spinal or cranial nerves (Jegorow, 1887, 1890; Langley, 1902, 1904; Fänge *et al.,* 1963; McFarland *et al.,* 1966; Lauber *et al.,* 1968; Bennett, 1971a). All observers agree that integumentary smooth muscle is densely innervated (De Lucchi, 1936; Schartau, 1938; Stammer, 1961b; Ostmann *et al.,* 1963; Capanna and Civitelli, 1963; Tetzlaff *et al.,* 1965; Bennett and Malmfors, 1970; Bennett *et al.,* 1970); histochemical and physiological findings indicate that the innervation is mainly or completely adrenergic (von Golenhofen, 1967, 1968; von Golenhofen and Petry, 1968; Bennett and Malmfors, 1970; Bennett *et al.,* 1970; Buckley and Lwin, 1970). (See also Chapter 1, Volume II.)

b. Harderian Gland. The innervation of the Harderian gland has been described as arising from the oculomotor nerve (e.g., Slonaker, 1918), the Vidian nerve (e.g., Hsieh, 1951) or the sphenopalatine ganglion (e.g., Webb, 1957). Stammer (1964) could find no macroscopically distinct nerve branches supplying the gland, but observed a dense innervation of the parenchyma from ganglion cells lying just beneath the capsule of the gland and from fibers accompanying blood vessels; the majority of ganglion cells and nerve fibers were cholinesterase negative. However, Fourman and Ballantyne (1967) claimed the majority of nerve fibers supplying the gland are cholinesterase positive. There are no physiological studies available that would serve to resolve this discrepancy.

8. The Eye

There have been numerous studies on the gross anatomy of the innervation of the eye (Rochas, 1885b; Schwalbe, 1879; D'Erchia, 1894, 1895; Langendorff, 1894, 1900; Holtzmann, 1896; Consiglio, 1900; Carpenter, 1906, 1911; von Lenhossék, 1911; Slonaker, 1918; Seto, 1931; Terzuolo, 1951; Ábrahám and Stammer, 1966; Watanabe *et al.,* 1967; Oehme, 1968), and while there is certainly not complete agreement between these acounts, there are various points of concurrence. Thus, it is agreed that the ciliary ganglion lying on the oculomotor trunk receives preganglionic fibers running with the oculomotor nerve. Some of these fibers synapse on cells whose axons run through the long ciliary nerves to the ciliary muscles and iris, while others are associated with neurons whose axons pass through the short ciliary nerves to innervate the smooth muscle of the choroidea (see Landmesser and Pilar, 1970; Marwitt *et al.*, 1971). Afferent fibers from trigeminal sources reach the long and short ciliary nerves through rami between them and the deep ophthalmic branch of the trigeminal; fibers from the superior cervical ganglion may follow the same final pathway (Oehme, 1968; but see Hsieh, 1951).

The structure and innervation of the striated musculature of the iris and ciliary body are well documented (Krohn, 1837; Dogiel, 1870, 1886; Geberg, 1884; Melkich, 1895; Zietzschmann, 1910; Urra, 1923; Boeke, 1933; Yamasaki, 1939; Stammer, 1962; Ábrahám and Stammer, 1966; Hess, 1966; Zenker and Krammer, 1967; Zenker, 1968; Mayr and Zenker, 1969; Oehme, 1969). It is generally accepted that the striated muscle fibers of the iris are arranged both radially and circularly, thus forming a dilator and sphincter pupillae; the majority of the fibers receive multiple *en grappe* motor terminals. Physiological studies indicate that the sphincter pupillae is controlled by oculomotor fibers (Zeglinski, 1885; Jegorow, 1887), whereas the dilator pupillae responds to stimulation of the trigeminal nerve (Zeglinski, 1885; Jegorow, 1887). However, caution should be employed in interpreting the latter findings, since it is known in mammals that pupillary responses to trigeminal stimulation are due to antidromic firing of afferent fibers (see Ambache, 1955).

Recently, it has become clear that the earlier reports of a smooth-muscle dilator system in the avian iris were correct (Oehme, 1969; Nishida and Sears, 1970), the stratum pigmentum iridis (anterior epithelial layer) being formed of radially oriented epitheliomuscle cells. From histochemical (Ehinger, 1967) and ultrastructural (Nishida and Sears, 1970) observations, it seems that adrenergic and cholinergic nerves supply this system. These findings are of particular interest, since the effect on the iris of stimulating the cervical paravertebral chain has never been adequately resolved. Zeglinski (1885), Jegorow (1887, 1890), Langley, (1904), and Koppanyi and Sun (1926) denied that the iris is affected by stimulation of the cervical paravertebral chain, but Gruenhagen (1887) found that such stimulation caused pupillodilatation. Furthermore, some studies have indicated that the avian iris is affected only by cholinergic drugs (Zeglinski, 1885; Koppanyi and Sun, 1926; Campbell and Smith, 1962; Zenker and Krammer, 1967), whereas other investigators have found pupillodilatation in response to adrenergic drugs (Clouse and Rigdon, 1959). It is hoped that future work will throw new light on the avian iris.

F. Autonomic Denervation

1. Surgical Denervation

Due to the peripheral location of many of the ganglia that give rise to the terminal innervation of effector tissues and to the varied routes the fibers take in reaching their destination, surgical denervation poses a particularly difficult problem. Thus, although it has proved possible to section the cardiac nerves (Tummons and Sturkie, 1968, 1969), it is unlikely that this leads to complete adrenergic denervation of the heart, since there are many adrenergic ganglia closely associated with the heart, and numerous adrenergic fibers follow routes other than through the cardiac nerves (Bennett and Malmfors, 1970; Bennett, 1971a). Obviously, other methods of denervation are desirable.

2. Chemical Denervation

Following the demonstration that intravenous injection of 6-hydroxydopamine into mammals causes degeneration of terminal adrenergic nerves, similar observations were made in birds (Bennett *et al.*, 1970). Extensive studies on various effector tissues have shown that the effect of the drug is variable, and the reasons for this have been discussed (Bennett, 1971c; Cobb and Bennett, 1971; Bennett *et al.*, 1971c, 1973a, 1974c). In certain systems, 6-hydroxydopamine causes widespread adrenergic denervation, and it seems likely that the denervation so achieved is more extensive than would be attained by surgical means. It should be remembered, however, that following treatment with 6-hydroxydopamine, adrenergic nerve terminals regenerate rapidly (Bennett, 1971c; Bennett *et al.*, 1973a, 1974c), and thus chemical denervation is less persistent than surgical denervation.

Recently, it was shown that intravenous injections of vinblastine sulfate causes damage of terminal adrenergic nerves in chicks (Bennett *et al.*, 1971b, 1973b). However, the damage did not appear to be confined to adrenergic nerves, since the animals showed marked locomotor disturbances. It seems likely that such treatment caused widespread damage of the whole nervous system, and for that reason vinblastine is a less useful tool than 6-hydroxydopamine.

ACKNOWLEDGMENTS

I am indebted to my colleague Mr. G. Mucznik for his unstinting assistance in the translation of Ukranian, Russian, and German papers. My thanks are also due to Dr. Y. Uehara and Dr. M. Costa for translations of Japanese and Italian papers, respectively.

REFERENCES

Aakhus, T., and Johansen, K. (1964). Angiocardiography of the duck during diving. *Acta Physiol. Scand.* **62,** 10–17.

Abel, W. (1910). The development of the autonomic nervous mechanism of the alimentary canal of the bird. *Proc. Roy. Soc. Edinburgh* **30,** 327–347.

Abel, W. (1912). Further observations on the development of the sympathetic nervous system in the chick. *J. Anat.* **47,** 37–45.

Ábrahám, A. (1935). Über die Nerven der Vogelkloake. *Arb. Ung. Biol. Forschungsinst.* **8,** 1–8.

Ábrahám, A. (1936). Beiträge zur Kenntnis der Innervation des Vogeldarmes. *Z. Zellforsch. Mikrosk. Anat.* **23,** 737–745.

Ábrahám, A. (1962). Die intramurale Innervation des Vogelherzens. *Z. Mikro.-Anat. Forsch.* **69,** 195–216.

Ábrahám, A. (1963). The nerve supply of the adrenal gland in birds. *Gen. Comp. Endocrinol.* **3,** 680.

Ábrahám, A. (1966). Baroreceptors in the gastrointestinal system of birds. *Acta Biol. (Budapest)* **17,** 402.

Ábrahám, A. (1967). The structure of baroreceptors in pathological conditions in man. *In* "Baroreceptors and Hypertension" (P. Kezdi, ed.), pp. 273–292. Pergamon, Oxford.

Ábrahám, A. (1969). "Microscopic Innervation of the Heart and Blood Vessels in Vertebrates Including Man." Pergamon, Oxford.

Ábrahám, A., and Stammer, A. (1957). Die mikroskopische Innervation des Vogelherzens. Acta Univ. Szeged, Acta Biol. 3, 247–273.
Ábrahám, A., and Stammer, A. (1959). Untersuchungen über die Struktur, die mikroskopische Innervation und die Cholinesteraseaktivität der Nebennieren von Vögeln. Acta Univ. Szeged, Acta Biol. 5, 85–95.
Ábrahám, A., and Stammer, A. (1966). Über die Struktur und die Innervierung der Augenmuskeln der Vögel unter Berücksichtigung des Ganglion ciliare. Acta Univ. Szeged, Acta Biol. 12, 87–118.
Akester, A. R. (1970). Dense core vesiculated axons in the renal portal valve of the domestic fowl. J. Anat. 106, 185–186.
Akester, A. R., and Mann, S. P. (1969a). Adrenergic and cholinergic innervation of the renal portal valve in the domestic fowl. J. Anat. 104, 241–252.
Akester, A. R., and Mann, S. P. (1969b). Ultrastructure and innervation of the tertiary bronchial unit in the lung of Gallus domesticus. J. Anat. 105, 202–204.
Akester, A. R., Akester, B., and Mann, S. P. (1969). Catecholamines in the avian heart. J. Anat. 104, 591.
Ambache, N. (1955). The use and limitations of atropine for pharmacological studies on autonomic effectors Pharmacol. Rev. 7, 467–494.
Andersen, A. E., and Nafstad, P. H. J. (1968). An electron microscopic investigation of the sensory organs in the hard palate of the hen (Gallus domesticus). Z. Zellforsch. Mikrosk. Anat. 91, 391–401.
Andersen, H. T. (1963). The reflex nature of the physiological adjustments to diving and their afferent pathway. Acta Physiol. Scand. 59, 12–23.
Andersen, H. T. (1966). Physiological adaptations in diving vertebrates. Physiol. Rev. 46, 212–243.
Andres, K. H. (1969). Zur Ultrastruktur verschiedener Mechanorezeptoren von höheren Wirbeltieren. Anat. Anz. 124, 551–565.
Arimoto, K., and Miyagawa, K. (1930). Histological studies on the innervation of the lung. (In Japanese.) Mitt. Med. Akad. Kioto 4, 736–748.
Ash, R. W., Pearce, J. W., and Silver, A. (1969). An investigation of the nerve supply to the salt gland of the duck. Quart. J. Exp. Physiol., Cog. Med. Sci. 54, 281–295.
Ashcraft, D. W. (1930). Correlative activities of the alimentary canal of fowl. Amer. J. Physiol. 93, 105–110.
Ashhurst, D. E. (1969). The fine structure of pigeon breast muscle. Tissue & Cell 1, 485–496.
Azzena, G. B., and Palmieri, G. (1967). A trigeminal monosynaptic reflex in birds. Exp. Neurol. 18, 184–193.
Azzena, G. B., Desole, C., and Palmieri, G. (1970). Cerebellar projections of the masticatory and extraocular muscle proprioception. Exp. Neurol. 27, 151–161.
Ballantyne, B., and Fourman, J. (1967). Cholinesterases and the secretory activity of the duck supraorbital gland. J. Physiol. (London) 188, 32P–33P.
Barker, D. (1960). The evolution of stretch receptors in vertebrates. Proc. Centen. Bicenten. Congr. Biol., 1958 pp. 53–57.
Barker, D. (1968). L'innervation motrice du muscle strié des vertébrés. Actual. Neurophysiol. 8, 23–72.
Barnikol, A. (1954). Zur Morphologie des Nervus Trigeminus der Vögel unter besonderer Berücksichtigung der Acciptres, Cathartidae, Striges und Anseriformes. Z. Wiss. Zool. 157, 285–332.
Bartlett, A. L., and Hassan, T. (1968). The action of physostigmine and the distribution of cholinesterases in the chicken oesophagus. Brit. J. Pharmacol. 33, 531–536.

Batekhina, N. K. (1968). Innervatsiya pryamoi kishki rekotorykh zhivotnykh. *Sb. Nauch. Rab. Vologograd. Med. Inst.* **21**, 26–29.

Baumel, J. J. (1964). Vertebral dorsal carotid artery interrelationships in the pigeon and other birds. *Anat. Anz.* **114**, 113–130.

Baumgarten, H. G., and Holstein, A. F. (1968). Adrenerge Innervation in Hoden und Nebenhoden vom Schwan *(Cygnus olor)*. *Z. Zellforsch. Mikrosk. Anat.* **91**, 402–410.

Baur, M. (1928). Versuche am Amnion von Huhn und Gans. *Naunyn-Schmiedebergs Arch. Exp. Pathol. Pharmakol.* **134**, 49–65.

Bell, C. (1969). Indirect cholinergic vasomotor control of intestinal blood flow in the domestic chicken. *J. Physiol. (London)* **205**, 317–328.

Benedeczky, I., Puppi, A., Tigyi, A., and Lissak, K. (1964). Various cell types of the adrenal medulla. *Nature (London)* **204**, 591–592.

Bennett, H. S., and Porter, K. R. (1953). An electron microscope study of sectioned breast muscle of the domestic fowl. *Amer. J. Anat.* **93**, 61–105.

Bennett, T. (1969a). Studies on the avian gizzard: Histochemical analysis of the extrinsic and intrinsic innervation. *Z. Zellforsch. Mikrosk. Anat.* **98**, 188–201.

Bennett, T. (1969b). Nerve-mediated excitation and inhibition of the smooth muscle cells of the avian gizzard. *J. Physiol. (London)* **204**, 669–686.

Bennett, T. (1969c). Interaction of nerve-mediated excitation and inhibition of single smooth muscle cells of the avian gizzard. *Aust. J. Exp. Biol. Med. Sci.* **47**, P. 2.

Bennett, T. (1969d). The effects of hyoscine and anticholinesterases on cholinergic transmission to the smooth muscle cells of the avian gizzard. *Brit. J. Pharmacol.* **37**, 585–594.

Bennett, T. (1970a). Interaction of nerve-mediated excitation and inhibition of single smooth muscle cells of the avian gizzard. *Comp. Biochem. Physiol.* **32**, 669–680.

Bennett, T. (1970b). Autonomic neuroeffector systems with particular reference to birds. Ph.D. Thesis, Melbourne University.

Bennett, T. (1971a). The neuronal and extra-neuronal localizations of biogenic amines in the cervical region of the domestic fowl. *(Gallus gallus domesticus* L.). *Z. Zellforsch. Mikrosk. Anat.* **112**, 443–464.

Bennett, T. (1971b). The adrenergic innervation of the pulmonary vasculature, the lung and the thoracic aorta, and on the presence of aortic bodies in the domestic fowl. *(Gallus gallus domesticus* L.). *Z. Zellforsch. Mikrosk. Anat.* **114**, 117–134.

Bennett, T. (1971c). Fluorescence histochemical and functional studies on adrenergic nerves following treatment with 6-hydroxydopamine. *In* "6-Hydroxydopamine and Catecholamine Neurons" (T. Malmfors and H. Thoenen, eds.), pp. 303–314. North-Holland Publ., Amsterdam.

Bennett, T., and Cobb, J. L. S. (1969a). Studies on the avian gizzard: Morphology and innervation of the smooth muscle. *Z. Zellforsch. Mikrosk. Anat.* **96**, 173–185.

Bennett, T., and Cobb, J. L. S. (1969b). Studies on the avian gizzard: Development of the gizzard and its innervation. *Z. Zellforsch. Mikrosk. Anat.* **98**, 599–621.

Bennett, T., and Cobb, J. L. S. (1969c). Studies on the avian gizzard: Auerbach's plexus. *Z. Zellforsch. Mikrosk. Anat.* **99**, 109–120.

Bennett, T., and Malmfors, T. (1970). The adrenergic nervous system of the domestic fowl *(Gallus domesticus* L.). *Z. Zellforsch. Mikrosk. Anat.* **106**, 22–50.

Bennett, T., Burnstock, G., Cobb, J. L. S., and Malmfors, T. (1970). An ultrastructural and histochemical study of the short-term effects of 6-hydroxydopamine on adrenergic nerves in the domestic fowl. *Brit. J. Pharmacol.* **38**, 802–809.

Bennett, T., Cobb, J. L. S., and Malmfors, T. (1971a). Fluorescence histochemical observations on Auerbach's plexus and the problem of the inhibitory innervation of the gut. *J. Physiol. (London)* **218**, 77–78P.

Bennett, T., Cobb, J. L. S., and Malmfors, T. (1971b). The effects of intravenous injections of vinblastine, on adrenergic nerves. *J. Physiol. (London)* **217**, 26–27P.

Bennett, T., Cobb, J. L. S., and Malmfors, T. (1971c). Differential effects of 6-hydroxydopamine on the terminals and non-terminal axons of adrenergic neurones. *Brit. J. Pharmacol.* **43**, 445–446P.

Bennett, T., Malmfors, T., and Cobb, J. L. S. (1973a). Fluorescence histochemical observations on degeneration and regeneration of noradrenergic nerves in the chick following treatment with 6-hydroxydopamine. *Z. Zellforsch. Mikroskop. Anat.* **142**, 103–130.

Bennett, T., Cobb, J. L. S., and Malmfors, T. (1973b). Fluorescence histochemical and ultrastructural observations on the effects of intravenous injections of vinblastine on noradrenergic nerves. *Z. Zellforsch. Mikroskop. Anat.* **141**, 517–527.

Bennett, T., Malmfors, T., and Cobb, J. L. S. (1973c). Fluorescence histochemical observations on catecholamine-containing cell bodies in Auerbach's plexus. *Z. Zellforsch. Mikroskop. Anat.* **139**, 69–81.

Bennett, T., Cobb, J. L. S., and Malmfors, T. (1974a). The vasomotor innervation of the inferior vena cava of the domestic fowl. I. Structure. (In preparation.)

Bennett, T., Malmfors, T., and Cobb, J. L. S. (1974b). The vasomotor innervation of the inferior vena cava of the domestic fowl. II. Function. (In preparation.)

Bennett, T., Malmfors, T., and Cobb, J. L. S. (1974c). Fluoresence histochemical and functional studies on regeneration of the noradrenergic innervation of the cardiovascular system following treatment with 6-hydroxydopamine. (In preparation.)

Bidder, and Volkmann. (1842). "Die Selbständigkeit des sympathischen Nervensystems." Leipzig (original not seen, see Langley, 1921).

Bischoff, M. B. (1969). Photoreceptoral and secretory structures in the avian pineal organ. *J. Ultrastruct. Res.* **28**, 16–26.

Biswal, G. (1954). Additional histological findings in the chicken reproductive tract. *Poultry Sci.* **33**, 843–851.

Blaber, L. C., and Cuthbert, A. W. (1962). Cholinesterases in the domestic fowl and the specificity of some reversible inhibitors. *Biochem. Pharmacol.* **11**, 113–124.

Boeke, J. (1933). Innervationsstudien. III. Die Nervenversorgung des M. ciliaris und des M. sphincter Iridis bei Säugern und Vögeln. *Z. Mikrosk. Anat. Forsch.* **33**, 233–259.

Boesiger, B. (1965). Comparaison des terminaisons motrices du muscle pectoral majeur et latissimus dorsi anterieur chez *Coturnix coturnix japonica*. *Arch. Anat. Microsc. Morphol. Exp.* **54**, 823–846.

Boesiger, B. (1968). Comparaison morphologique, biométrique et histochimique des fibres musculaires et de leurs terminaisons neurveuses motrices de différents muscles de la caille japonaise, *Coturnix coturnix japonica*. *Acta Anat.* **71**, 274–310.

Bogdanov, R. Z., and Kibyakov, A. V. (1955). K mechanizmu innervatsii zheludka ptits. *Fiziol. Zh. SSSR im. I. M. Sechenova* **41**, 239–242.

Bolton, T. B. (1967). Intramural nerves in the ventricular myocardium of the domestic fowl and other animals. *Brit. J. Pharmacol.* **31**, 253–268.

Bolton, T. B. (1968a). Studies on the longitudinal muscle of the anterior mesenteric artery of the domestic fowl. *J. Physiol. (London)* **196**, 273–281.

Bolton, T. B. (1968b). Electrical and mechanical activity of the longitudinal muscle of the anterior mesenteric artery of the domestic fowl. *J. Physiol. (London)* **196**, 283–292.
Bolton, T. B. (1969). Spontaneous and evoked release of neurotransmitter substances in the longitudinal muscle of the anterior mesenteric artery of the fowl. *Brit. J. Pharmacol.* **35**, 112–120.
Bolton, T. B., and Bowman, W. C. (1969). Adrenoreceptors in the cardiovascular system of the domestic fowl. *Eur. J. Pharmacol.* **5**, 121–132.
Bolton, T. B., and Raper, C. (1966). Innervation of domestic fowl and guinea-pig ventricles. *J. Pharm. Pharmacol.* **18**, 192–193.
Bonsdorff, E. J. (1852a). Symbolae ad anatomiam comparatam nervorum animalium vertebratorum. I. Nervi cerebrales *Corvi cornicis* (Linn). *Acta Soc. Sci. Fenn.* **3**, 505–569.
Bonsdorff, E. J. (1852b). Symbolae ad anatomiam comparatam nervorum animalium vertebratorum. II. Nervi cerebrales *Gruis cinereae* (Linn). *Acta Soc. Sic. Fenn.* **3**, 591–624.
Bonting, S. L., Caravaggio, L. L., Canady, M. R., and Hawkins, N. M. (1964). Studies on sodium-potassium activated adenosine triphosphatase. XI. The salt gland of the Herring Gull. *Arch. Biochem. Biophys.* **106**, 49–56.
Bortolami, R., and Palmieri, G. (1962). Macroscopic and microscopic anatomy of the ureters of some birds. *Riv. Biol.* **55**, 95–146.
Bortolami, R., and Vegetti, A. (1967). Function of the trigeminal mesencephalic nucleus in the duck. *Riv. Biol.* **60**, 325–334.
Botezat, E. (1906). Die Nervenendapparate in den Mundteilen der Vögel und die einheitliche Endigungsweise der peripheren Nerven bei den Wirbeltieren. *Z. Wiss. Zool.* **84**, 205–360.
Botezat, E. (1909). Fasern und Endplatten von Nerven zweiter Art an den quergestreiften Muskeln der Vögel, *Anat. Anz.* **35**, 396–398.
Bowman, W. C., and Everett, S. D. (1964). An isolated parasympathetically-innervated oesophagus preparation from the chick. *J. Pharm. Pharmacol.* **16**, Suppl., 72T–79T.
Brauer, A. (1932). A topographical and cytological study of the sympathetic nervous components of the suprarenal of the chick embryo. *J. Morphol.* **53**, 277–326.
Bremer, J. L. (1926). The influence of nerves on the position of the coeliac artery in the chick. *Anat. Rec.* **33**, 299–310.
Brinkman, R., and Martin, A. H. (1969). Large neurons in the substantia gelatinosa of Rolando. *Experientia* **25**, 962–963.
Browne, M. J. (1953). A study of the sacral autonomic nerves in a chick and a human embryo. *Anat. Rec.* **116**, 189–203.
Bubien-Waluszewska, A. (1968). Le groupe caudale des nerfs crâniens de la poule domestique *(Gallus domesticus)*. *Acta Anat.* **69**, 445–457.
Buckley, G. A., and Lwin, S. (1970). The expansor secundariorum of the domestic fowl: A smooth muscle-nerve preparation without cholinergic receptors. *Brit. J. Pharmacol.* **39**, 245P.
Burger, R. E. (1968). Pulmonary chemosensitivity in the domestic fowl. *Fed. Proc., Fed. Amer. Soc. Exp. Biol.* **27**, 328.
Burn, J. H. (1968). The development of the adrenergic fibre. *Brit. J. Pharmacol.* **32**, 575–582.
Butler, P. J. (1967). The effect of progressive hypoxia on the respiratory and cardiovascular systems of the chicken. *J. Physiol. (London)* **191**, 309–324.

Butler, P. J. (1970). The effect of progressive hypoxia on the respiratory and cardiovascular systems of the pigeon and duck. *J. Physiol. (London)* **211**, 527–538.

Butler, P. J., and Jones, D. R. (1968). Onset of and recovery from diving bradycardia in ducks. *J. Physiol. (London)* **196**, 255–272.

Callingham, B. A., and Cass, R. (1965). Catecholamine levels in the chick. *J. Physiol. (London)* **176**, 32P–33P.

Callingham, B. A., and Cass, R. (1966). Catecholamines in the chick. *In* "Physiology of the Domestic Fowl" (C. Horton-Smith and E. C. Amoroso, eds.), pp. 279–285. Oliver & Boyd, Edinburgh.

Cambier, E. (1969). Propriétés mécaniques et fonctions d'un muscle tonique d'oiseau: Le latissimus dorsi anterior. *Arch. Int. Physiol. Biochim.* **77**, 416–426.

Campbell, G. (1970). Autonomic nervous supply to effector tissues. *In* "Smooth Muscle" (E. Bülbring, *et al.*, eds.), pp. 418–450. Arnold, London.

Campbell, H. S., and Smith, J. L. (1962). The pharmacology of the pigeon pupil. *Arch. Ophthalmol.* [N.S.] **67**, 501–504.

Cantino, D. (1968). Données préliminaires sur l'ultrastructure de nerf ganglionné de Remak. *C. R. Ass. Anat.* **63**, 640–644.

Cantino, D. (1970). An histochemical study of the nerve supply to the developing alimentary tract. *Experientia* **26**, 766–767.

Capanna, E., and Civitelli, M. V. (1963). Innervation in relation to the feathers of the chicken. *Rend. Ist. Sci. Univ. Camerino* **4**, 18–32.

Carpenter, F. W. (1906). The development of the oculomotor nerve, the ciliary ganglion and the abducent nerve in the chick. *Bull. Mus. Comp. Zool., Harvard Univ.* **48**, 139–230.

Carpenter, F. W. (1911). The ciliary ganglion of birds. *Folia Neurobiol., Leipzig* **5**, 738–754.

Chang, H-Y. (1964). Nervus facialis and nervus acusticus in domestic fowl. *Acta Zool. Sinica* **16**, 539–554. (In Chinese.)

Chinoy, N. J., and George, J. C. (1965). Cholinesterases in the pectoral muscle of some vertebrates. *J. Physiol. (London)* **177**, 346–354.

Chowdhary, D. S. (1953). The carotid body and carotid sinus of the fowl. Ph.D. Thesis, Edinburgh University.

Clouse, M. E., and Rigdon, R. H. (1959). The pharmacological action of autonomic drugs on the eye of fowls. *Tex. Rep. Biol. Med.* **17**, 305–311.

Cobb, J. L. S., and Bennett, T. (1969a). A study of nexuses in visceral smooth muscle. *J. Cell Biol.* **42**, 287–297.

Cobb, J. L. S., and Bennett, T. (1969b). A study of intercellular relationships in developing and mature visceral smooth muscle. *Z. Zellforsch. Mikrosk. Anat.* **100**, 516–526.

Cobb, J. L. S., and Bennett, T. (1970a). An ultrastructural study of mitotic division in differentiated gastric smooth muscle cells. *Z. Zellforsch. Mikrosk. Anat.* **108**, 177–189.

Cobb, J. L. S., and Bennett, T. (1970b). Herbst corpuscles in the smooth muscles in the wings of chicks. *Experientia* **26**, 768–769.

Cobb, J. L. S., and Bennett, T. (1971). An electron microscopic examination of the short term effects of 6-hydroxydopamine on the peripheral adrenergic nervous system. *In* "6-Hydroxydopamine and Catecholamine Neurons" (T. Malmfors and H. Thoenen, eds.), pp. 33–46. North-Holland Publ., Amsterdam.

Cohen, D. H., and Pitts, L. H. (1967). Vagal and sympathetic contributions to classically conditioned tachycardia in the pigeon. *Physiologist* **10**, 145.

Cohen, D. H., and Pitts, L. H. (1968). Vagal and sympathetic components of conditioned cardioacceleration in the pigeon. *Brain Res.* **9**, 15–31.

Cohen, D. H., and Schnall, A. M. (1970). Medullary cells of origin of vagal cardioinhibitory fibers in the pigeon. II. Electrical stimulation of the dorsal motor nucleus. *J. Comp. Neurol.* **140**, 321–342.

Cohen, D. H., Schnall, A. M., Macdonald, R. L., and Pitts, L. H. (1970). Medullary cells of origin of vagal cardioinhibitory fibers in the pigeon. I. Anatomical studies of peripheral vagus nerve and the dorsal motor nucleus. *J. Comp. Neurol.* **140**, 299–320.

Cohn, J. E., Krog, J., and Shannon, S. (1968). Cardiopulmonary responses to head immersion in domestic geese. *J. Appl. Physiol.* **25**, 36–41.

Consiglio, M. (1900). Sul decorso delle fibre irido-constrittrici negli ucceli; nota sperimentale. *Arch. Farmacol. Ter.* **8**, 269–275.

Cook, R. D., and King, A. S. (1969a). Nerves of the avian lung: Electron microscopy. *J. Anat.* **105**, 202.

Cook, R. D., and King, A. S. (1969b). A neurite-receptor complex in the avian lung: Electron microscopical observations. *Experientia* **25**, 1162–1164.

Cook, R. D., and King, A. S. (1970). Observations on the ultrastructure of the smooth muscle and its innervation in the avian lung. *J. Anat.* **106**, 273–284.

Cordier, A. (1969). L'innervation de la bourse de Fabricius durant l'embryogenèse et la vie adulte. *Acta Anat.* **73**, 38–47.

Cords, E. (1904). Beiträge zur Lehre vom Kopfnervensystem der Vögel. *Anat. Hefte* **26**, 49–100.

Costa, M. (1966a). Osservazioni preliminari sulla presenza di cellule e fibre nervose adrenergiche nel nervo gangliare di Remak nel pollo. *Boll. Soc. Ital. Biol. Sper.* **42**, 976–977.

Costa, M. (1966b). Contribution a l'étude de l'innervation adrenergique du canal alimentaire: Le nerf ganglionné de Remak. *C. R. Ass. Anat.* **51**, 249–252.

Cottle, M. K. W., and Pearce, J. W. (1970). Some observations on the nerve supply to the salt gland of the duck. *Quart. J. Exp. Physiol. Cog. Med. Sci.* **55**, 207–212.

Cŏulouma, P. (1935). La terminaison des nerfs pneumogastrique chez quelques vertébrés. *C. R. Ass. Anat.* **30**, 120–150.

Coulouma, P. (1939a). L'anatomie comparée du nerf intestinal chez les vertébrés. *C. R. Ass. Anat.* **34**, 88–99.

Coulouma, P. (1939b). Le nerf intestinal chez la poule et le canard. Son équivalent chez les Mammifères. Sa signification. *C. R. Ass. Anat.* **34**, 100–114.

Coulouma, P. (1939c). L'anastomose vago-sympathique abdominale chez les oiseaux. *C. R. Ass. Anat.* **34**, 115–126.

Coulouma, P., and Herrath, E. (1939). Recherches histologiques sur le plexus gastrique sous-sereux et les ganglions du nerf intestinal chez le poulet. *C. R. Ass. Anat.* **34**, 127–132.

Coupland, R. E., and Hopwood, D. (1966). Mechanism of a histochemical reaction differentiating between adrenaline- and noradrenaline-storing cells in the electron microscope. *Nature (London)* **209**, 590–591.

Couvreur, E. (1892). Sur la pneumogastrique des oiseaux. *Ann. Univ. Lyon* **2**, 1–107.

Cowie, A. F., and King, A. S. (1969). Further observations on the bronchial muscle of birds. *J. Anat.* **104**, 177–178.

Csoknya, M. (1968). Data to the knowledge of innervation of the bird's digestive tract. *Acta Univ. Szeged, Acta Biol.* **14**, 57–63.

Cuello, A. C. (1970). Occurrence of adrenaline and noradrenaline cells in the adrenal gland of the Gentoo Penguin (*Pygoscelis papua*). *Experientia* **26**, 416–418.

Cuvier, G. (1802). "Lectures on Comparative Anatomy" (translated by W. Ross), Vol. 2. Longman & Rees, Paternoster Row.
Dahl, E. (1970). Studies of the fine structure of ovarian interstitial tissue. 3. The innervation of the thecal gland of the domestic fowl. Z. Zellforsch. Mikrosk. Anat. 109, 212-226.
Dale, H. H. (1933). Nomenclature of fibres in the autonomic nervous system and their effects. J. Physiol. (London) 80, 10P-11P.
de Anda, G., and Rebollo, M. A. (1967). The neuromuscular spindles in the adult chicken. I. Morphology Acta Anat. 67, 437-451.
de Kock, L. L. (1958). On the carotid body of certain birds. Acta Anat. 35, 161-178.
de Kock, L. L. (1959). The carotid body system of higher vertebrates. Acta Anat. 37, 265-279.
de Lisi, L. (1924). Caratteri sessuali dei gangli simpatici perisurrenali degli uccelli. Monit. Zool. Ital. 35, 62-68.
De Lorenzo, A. J. (1960). The fine structure of synapses in the ciliary ganglion of the chick. J. Biophys. Biochem. Cytol. 7, 31-36.
De Lorenzo, A. J. (1966). Electron microscopy: Tight junctions in synapses of the chick ciliary ganglion. Science 152, 76-78.
De Lorenzo, A. J., and Barnett, G. R. (1966). Fine-structural basis for chemical and electrotonic transmission in a parasympathetic ganglion. Biol. Bull. 131, 380-381.
De Lucchi, G. (1936). Contributo istologico alla inervazione della muscolatura liscia delle cute. II. Ricerche sui muscoli delle penne. Atti Soc. Med.-Chir. Padova Fac. Med. Chir. Univ. Padova 15, 554-565.
de Meyer, R. (1952). Contribution à l'étude du tissu nodal du coeur des oiseaux. Arch. Biol. 63, 455-514.
D'Erchia, F. (1894). Contributo allo studio della struttura e delle connessioni del ganglio ciliare. I. Sulla struttura del ganglio ciliare. Monit. Zool. Ital. 5, 235-238.
D'Erchia, F. (1895). Contributo allo studio della struttura e delle connessioni del ganglio ciliare. II. Connessioni del ganglio ciliare. Monit. Zool. Ital. 6, 157-164.
Djojosugito, A. M., Folkow, B., and Kovach, A. G. B. (1968). The mechanisms behind the rapid blood volume restoration after haemorrhage in birds. Acta Physiol. Scand. 74, 114-122.
Djojosugito, A. M., Folkow, B., and Yonce, L. R. (1969). Neurogenic adjustments of muscle blood flow, cutaneous A-V shunt flow, and of venous tone during "diving" in ducks. Acta Physiol. Scand. 75, 377-386.
Dogiel, J. (1870). Ueber den Musculus dilator pupillae bei Säugethieren, Menschen und Vögeln. Arch. Mikrosk. Anat. 7, 89-99.
Dogiel, J. (1886). Neue Untersuchungen über den pupillenerweiternden Muskel der Säugethiere und Vögel. Arch. Mikrosk. Anat. 27, 403-409.
Doležel, S., and Žlábek, K. (1969). Über einen monoaminergen Mechanismus im Nierenpfortadersystem der Vögel. Z. Zellforsch. Mikrosk. Anat. 100, 527-535.
Dorward, P. K. (1966). Responses of mechanoreceptors and intraspinal projection of their afferent fibres in the domestic duck. Ph.D. Thesis, Monash University.
Dorward, P. K. (1970a). Response characteristics of muscle afferents in the domestic duck. J. Physiol. (London) 211, 1-17.
Dorward, P. K. (1970b). Response patterns of cutaneous mechanoreceptors in the domestic duck. Comp. Biochem. Physiol. 35, 729-735.
Doyon, M. (1894a). Contribution à l'étude des phénomènes mécaniques de la digestion gastrique chez les oiseaux. Arch. Physiol. Norm. Pathol. 6, 869-878.
Doyon, M. (1894b). Recherches expérimentales sur l'innervation gastrique des oiseaux. Arch. Physiol. Norm. Pathol. 6, 887-898.

Doyon, M. (1925). Influence des nerfs sur la motricité de l'estomac chez l'oiseau. *C. R. Soc. Biol.* **93**, 578–580.

Drennan, M. R. (1927). The auriculo-ventricular bundle in the bird's heart. *Brit. Med. J.* **1**, 321–322.

Dudley, J. (1942). The development of the ultimo-branchial body of the fowl, *Gallus domesticus*. *Amer. J. Anat.* **71**, 65–98.

Durfee, W. K. (1964). Cardiovascular reflex mechanisms in the fowl. *Diss. Abstr.* **24**, 2966.

Eberth, C. J. (1863). Über den feineren Bau der Lunge. *Z. Wiss. Zool.* **12**, 427–454.

Ehinger, B. (1967). Adrenergic nerves in the avian eye and ciliary ganglion. *Z. Zellforsch. Mikrosk. Anat.* **82**, 577–588.

Eliassen, E. (1960). Cardiovascular responses to submersion aphyxia in avian divers. *Arb. Univ. Bergen, Mat.-Natur. Ser.* **2**, 1–100.

Elliot, T. R. (1905). The action of adrenalin. *J. Physiol. (London)* **32**, 401–467.

Emmert, A. F. (1811). Beobachtungen über einige anatomische Eigenheiten der Vögel. *Arch. Physiol. (Reil's Arch.)* **10**, 377–392.

Enemar, A., Falck, B., and Håkanson, R. (1965). Observations on the appearance of norepinephrine in the sympathetic nervous system of the chick embryo. *Develop. Biol.* **11**, 268–283.

Eränkö, O. (1957). Distribution of adrenaline and noradrenaline in hen adrenal gland. *Nature (London)* **179**, 417.

Eränko, O., Rechardt, L., Eränkö, L., and Cunningham, A. (1970). Light and electron microscopic histochemical observations on cholinesterase-containing sympathetic nerve fibres in the pineal body of the rat. *Histochem. J.* **2**, 479–490.

Evans, H. E. (1969). Anatomy of the Budgerigar. *In* "Diseases of Cage and Aviary Birds" (M. L. Petrak, ed.), pp. 45–112. Lea & Febiger, Philadelphia, Pennsylvania.

Everett, S. D. (1966). Pharmacological responses of the isolated oesophagus and crop of the chick. *In* "Physiology of the Domestic Fowl" (C. Horton-Smith and E. C. Amoroso, eds.), pp. 261–273. Oliver & Boyd, Edinburgh.

Everett, S. D. (1968). Pharmacological responses of the isolated innervated intestine and rectal caecum of the chick. *Brit. J. Pharmacol.* **33**, 342–356.

Everett, S. D., and Mann, S. P. (1967). Catecholamine release by histamine from the isolated intestine of the chick. *Eur. J. Pharmacol.* **1**, 310–320.

Falck, B., and Hellman, B. (1963). Evidence for the presence of biogenic amines in pancreatic islets. *Experientia* **19**, 139–140.

Falck, B., and Owman, C. (1965). A detailed methodological description of the fluorescence method for the cellular demonstration of biogenic monoamines. *Acta Univ. Lund., Sect. 2* **7**, 1–23.

Fänge, R., Schmidt-Nielsen, K., and Robinson, M. (1958). Control of secretion from the avian salt gland. *Amer. J. Physiol.* **195**, 321–326.

Fänge, R., Krog, J., and Reite, O. B. (1963). Blood flow in the avian salt gland studied by polarographic oxygen electrodes. *Acta Physiol. Scand.* **58**, 40–47.

Fedde, M. R. (1970). Peripheral control of avian respiration. *Fed. Proc., Fed. Amer. Soc. Exp. Biol.* **29**, 1664–1673.

Fedde, M. R., and Peterson, D. F. (1970). Intrapulmonary receptor response to changes in airway–gas composition in *Gallus domesticus*. *J. Physiol. (London)* **209**, 609–626.

Fedde, M. R., Burger, R. E., and Kitchell, R. L. (1963). Localization of vagal afferents involved in the maintenance of normal avian respiration. *Poultry Sci.* **42**, 1224–1236.

Feigl, E., and Folkow, B. (1963). Cardiovascular responses in "diving" and during brain stimulation in the duck. *Acta Physiol. Scand.* **57**, 99–110.
Feng, T. P., Wu, W. Y., and Yang, F. Y. (1965). Selective reinnervation of a "slow" or "fast" muscle by its original motor supply during regeneration of mixed nerve. *Sci. Sinica* **14**, 1717–1720.
Ferrando, G., and Nalbandov, A. V. (1969). Direct effect on the ovary of the adrenergic blocking drug dibenzyline *Endocrinology* **85**, 38–42.
Fisher, E. (1887). Über die Endigung der Nerven im quergestreiften Muskel der Wirbeltiere. *Arch. Mikrosk. Anat.* **13**, 365–390.
Folkow, B., and Yonce, L. R. (1967). The negative inotropic effect of vagal stimulation on the heart ventricles of the duck. *Acta Physiol. Scand.* **71**, 77–84.
Folkow, B., Fuxe, K., and Sonnenschein, R. R. (1966). Responses of skeletal musculature and its vasculature during "diving" in the duck. Peculiarities of the adrenergic vasoconstrictor innervation. *Acta Physiol. Scand.* **67**, 327–342.
Folkow, B., Nilsson, N. J., and Yonce, L. R. (1967). Effects of "diving" on cardiac output in ducks. *Acta Physiol. Scand.* **70**, 347–361.
Fourman, J. (1969). Cholinesterase activity in the supra-orbital salt secreting gland of the duck. *J. Anat.* **104**, 233–239.
Fourman, J., and Ballantyne, B. (1967). Cholinesterase activity in the Harderian gland of *Anas domesticus*. *Anat. Rec.* **159**, 17–28.
Freedman, S. L. (1962). Innervation and blood vessels to the uterus of the chicken. Ph.D. Thesis, Rutgers University, New Brunswick, New Jersey.
Freedman, S. L. (1968). The innervation of the suprarenal gland of the fowl. *Acta Anat.* **69**, 18–25.
Freedman, S. L., and Sturkie, P. D. (1963). Extrinsic nerves of the chicken's uterus. *Anat. Rec.* **147**, 431–439.
Fu, S. K., Chen, T. Y., and Tcheng, K. T. (1962). Studies on the glomera aortica of the Great Reed Warbler and von Schrenk's Little Bittern. *Acta Zool. Sinica* **14**, 297–299.
Fujie, E. (1968). Ultrastructure of the pineal body of the domestic chicken with special reference to the changes induced by altered photoperiods. *Arch. Histol.* **29**, 271–289.
Fusari, R. (1892). Contributo allo studio dello sviluppo delle capsule surrenali e del simpatico nel pollo e nei mammiferi. *Arch. Sci. Med.* **16**, 248–301.
Fusari, R. (1893). Contribution à l'étude du développement des capsules surrénales et du sympathique chez le poulet et chez les mammifères. *Arch. Ital. Biol.* **18**, 161–182.
Gadow, H. (1891). Vögel. *Bronn's Klassen* **6**, No. 4. Winter, Leipzig.
Ganfini, C. (1907). Sulla presenza di cellule gangliari nell'ovario di *Gallus domesticus*. *Bibliogr. Anat.* **16**, 128–132.
Ganfini, C. (1916). Lo sviluppo del sistema nervoso simpatico negli Uccelli. *Arch. Ital. Anat. Embriol.* **15**, 91–138.
Geberg, A. (1884). Ueber die Nerven der Iris und des Ciliarkörpers bei Vögeln. *Int. Monatsschr. Anat. Histol.* **1**, 7–52.
George, J. C., and Berger, A. J. (1966). "Avian Myology." Academic Press, New York.
Germino, N. I., and D'Albora, H. (1965). Succinic dehydrogenase activity in the neuromuscular spindles of the chick. *Experientia* **21**, 45–46.
Giacomini, E. (1898). Sur les terminaisons nerveuses dans les capsules surrénales des oiseaux. *Arch. Ital. Biol.* **29**, 482–483.
Gibbs, O. S. (1929). Functions of the fowl's ureter. *Amer. J. Physiol.* **87**, 594–601.

Gilbert, A. B. (1961). The innervation of the renal portal valve of the domestic fowl. *J. Anat.* **95**, 594–598.

Gilbert, A. B. (1965a). Triple silver impregnation for selective staining of avian nerves. *Stain Technol.* **40**, 301–304.

Gilbert, A. B. (1965b). Innervation of the ovarian follicle of the domestic hen. *Quart. J. Exp. Physiol. Cog. Med. Sci.* **50**, 437–445.

Gilbert, A. B. (1967). Formation of the egg in the domestic chicken. *Advan. Reprod. Physiol.* **2**, 111–180.

Gilbert, A. B. (1968). Innervation of the avian ovary. *J. Physiol. (London)* **196**, 4P–5P.

Gilbert, A. B. (1969). Innervation of the ovary of the domestic hen. *Quart. J. Exp. Physiol. Cog. Med. Sci.* **54**, 404–411.

Gilbert, A. B., and Lake, P. E. (1963). Terminal innervation of the uterus and vagina of the domestic hem. *J. Reprod. Fert.* **5**, 41–48.

Ginsborg, B. L. (1959). Multiple innervation of chick muscle fibres. *J. Physiol. (London)* **148**, 50P–51P.

Ginsborg, B. L. (1960a). Spontaneous activity in muscle fibres of the chick. *J. Physiol. (London)* **150**, 707–717.

Ginsborg, B. L. (1960b). Some properties of avian skeletal muscle fibres with multiple neuromuscular junctions. *J. Physiol. (London)* **154**, 581–598.

Ginsborg, B. L., and Mackay, B. (1960). The latissimus dorsi muscles of the chick. *J. Physiol. (London)* **153**, 19P–20P.

Ginsborg, B. L., and Mackay, B. (1961). A histochemical demonstration of two types of motor innervation in avian skeletal muscle. *Bibl. Anat.* **2**, 174–181.

Goller, H. (1969). Topographical demonstration of nuclei in the medulla oblongata of the fowl. *(Gallus domesticus). Zentralbl. Veterinaermed. Reihe A* **16**, 257–270.

Goloube, D. M. (1936). Sur le développement de la glande surrénale et de ses nerfs chez le poulet. *Ann. Anat. Pathol. Anat. Norm. Med.-Chir.* **13**, 1055–1066.

Goormaghtigh, N. (1921). Organogenèse et histogenèse de la capsule surrénale et du plexus coeliaque. *Arch. Biol.* **31**, 83–165.

Gossrau, R. (1968). Über das Reizleitungssystem der Vögel. Histochemische und elektronenmikroskopische Untersuchungen. *Histochemie* **13**, 111–159.

Govyrin, V. A. (1965). O roli adrenalina i noradrenalina v adaptatsionnotroficheskoi funktsii nervnoi sistemy u teplokrovnykh i kholodnokrovnykh zhivotnykh. *In* Funktsional'naya evolutsiya nervoni sistemy," pp. 115–123. "Nauka," Leningrad.

Govyrin, V. A., and Leontieva, G. R. (1962). Catecholamines of the heart of birds during ontogenesis. *Dokl. Biol. Sci.* **147**, 1407–1409.

Govyrin, V. A., and Leontieva, G. R. (1965a). Raspredylenie katekholovykh aminov v miokarde pozvonochnykh. *Zh. Evol. Biokhim. Fiziol.* **1**, 38–44.

Govyrin, V. A., and Leontieva, G. R. (1965b). K voprosu o khromaffinoi tkani i istochnikakh katyekhinovykh zhivotnykh. *Byull. Eksp. Biol. Med.* **59**, 98–100.

Graham, J. D. P. (1938). A note on the action of atropine in the bird. *J. Physiol. (London)* **93**, 56P–58P.

Graham, J. D. P. (1940). Respiratory reflexes in the fowl. *J. Physiol. (London)* **97**, 525–532.

Gray, J. C. (1937). The anatomy of the male genital ducts in the fowl. *J. Morphol.* **60**, 393–401.

Grewe, F. J. (1951). Nuwe gegewens aangaande die ontogenese van die neuskliere, die orgaan van Jacobson en die Dekbene van die skedel by die genus *Anas*. *Ann. Univ. Stellenbosch, Ser. A* **3**, 69–99.

Feigl, E., and Folkow, B. (1963). Cardiovascular responses in "diving" and during brain stimulation in the duck. *Acta Physiol. Scand.* **57**, 99–110.

Feng, T. P., Wu, W. Y., and Yang, F. Y. (1965). Selective reinnervation of a "slow" or "fast" muscle by its original motor supply during regeneration of mixed nerve. *Sci. Sinica* **14**, 1717–1720.

Ferrando, G., and Nalbandov, A. V. (1969). Direct effect on the ovary of the adrenergic blocking drug dibenzyline *Endocrinology* **85**, 38–42.

Fisher, E. (1887). Über die Endigung der Nerven im quergestreiften Muskel der Wirbeltiere. *Arch. Mikrosk. Anat.* **13**, 365–390.

Folkow, B., and Yonce, L. R. (1967). The negative inotropic effect of vagal stimulation on the heart ventricles of the duck. *Acta Physiol. Scand.* **71**, 77–84.

Folkow, B., Fuxe, K., and Sonnenschein, R. R. (1966). Responses of skeletal musculature and its vasculature during "diving" in the duck. Peculiarities of the adrenergic vasoconstrictor innervation. *Acta Physiol. Scand.* **67**, 327–342.

Folkow, B., Nilsson, N. J., and Yonce, L. R. (1967). Effects of "diving" on cardiac output in ducks. *Acta Physiol. Scand.* **70**, 347–361.

Fourman, J. (1969). Cholinesterase activity in the supra-orbital salt secreting gland of the duck. *J. Anat.* **104**, 233–239.

Fourman, J., and Ballantyne, B. (1967). Cholinesterase activity in the Harderian gland of *Anas domesticus*. *Anat. Rec.* **159**, 17–28.

Freedman, S. L. (1962). Innervation and blood vessels to the uterus of the chicken. Ph.D. Thesis, Rutgers University, New Brunswick, New Jersey.

Freedman, S. L. (1968). The innervation of the suprarenal gland of the fowl. *Acta Anat.* **69**, 18–25.

Freedman, S. L., and Sturkie, P. D. (1963). Extrinsic nerves of the chicken's uterus. *Anat. Rec.* **147**, 431–439.

Fu, S. K., Chen, T. Y., and Tcheng, K. T. (1962). Studies on the glomera aortica of the Great Reed Warbler and von Schrenk's Little Bittern. *Acta Zool. Sinica* **14**, 297–299.

Fujie, E. (1968). Ultrastructure of the pineal body of the domestic chicken with special reference to the changes induced by altered photoperiods. *Arch. Histol.* **29**, 271–289.

Fusari, R. (1892). Contributo allo studio dello sviluppo delle capsule surrenali e del simpatico nel pollo e nei mammiferi. *Arch. Sci. Med.* **16**, 248–301.

Fusari, R. (1893). Contribution à l'étude du développement des capsules surrénales et du sympathique chez le poulet et chez les mammifères. *Arch. Ital. Biol.* **18**, 161–182.

Gadow, H. (1891). Vögel. *Bronn's Klassen* **6**, No. 4. Winter, Leipzig.

Ganfini, C. (1907). Sulla presenza di cellule gangliari nell'ovario di *Gallus domesticus*. *Bibliogr. Anat.* **16**, 128–132.

Ganfini, C. (1916). Lo sviluppo del sistema nervoso simpatico negli Uccelli. *Arch. Ital. Anat. Embriol.* **15**, 91–138.

Geberg, A. (1884). Ueber die Nerven der Iris und des Ciliarkörpers bei Vögeln. *Int. Monatsschr. Anat. Histol.* **1**, 7–52.

George, J. C., and Berger, A. J. (1966). "Avian Myology." Academic Press, New York.

Germino, N. I., and D'Albora, H. (1965). Succinic dehydrogenase activity in the neuromuscular spindles of the chick. *Experientia* **21**, 45–46.

Giacomini, E. (1898). Sur les terminaisons nerveuses dans les capsules surrénales des oiseaux. *Arch. Ital. Biol.* **29**, 482–483.

Gibbs, O. S. (1929). Functions of the fowl's ureter. *Amer. J. Physiol.* **87**, 594–601.

Gilbert, A. B. (1961). The innervation of the renal portal valve of the domestic fowl. *J. Anat.* **95**, 594–598.

Gilbert, A. B. (1965a). Triple silver impregnation for selective staining of avian nerves. *Stain Technol.* **40**, 301–304.

Gilbert, A. B. (1965b). Innervation of the ovarian follicle of the domestic hen. *Quart. J. Exp. Physiol. Cog. Med. Sci.* **50**, 437–445.

Gilbert, A. B. (1967). Formation of the egg in the domestic chicken. *Advan. Reprod. Physiol.* **2**, 111–180.

Gilbert, A. B. (1968). Innervation of the avian ovary. *J. Physiol. (London)* **196**, 4P–5P.

Gilbert, A. B. (1969). Innervation of the ovary of the domestic hen. *Quart. J. Exp. Physiol. Cog. Med. Sci.* **54**, 404–411.

Gilbert, A. B., and Lake, P. E. (1963). Terminal innervation of the uterus and vagina of the domestic hem. *J. Reprod. Fert.* **5**, 41–48.

Ginsborg, B. L. (1959). Multiple innervation of chick muscle fibres. *J. Physiol. (London)* **148**, 50P–51P.

Ginsborg, B. L. (1960a). Spontaneous activity in muscle fibres of the chick. *J. Physiol. (London)* **150**, 707–717.

Ginsborg, B. L. (1960b). Some properties of avian skeletal muscle fibres with multiple neuromuscular junctions. *J. Physiol. (London)* **154**, 581–598.

Ginsborg, B. L., and Mackay, B. (1960). The latissimus dorsi muscles of the chick. *J. Physiol. (London)* **153**, 19P–20P.

Ginsborg, B. L., and Mackay, B. (1961). A histochemical demonstration of two types of motor innervation in avian skeletal muscle. *Bibl. Anat.* **2**, 174–181.

Goller, H. (1969). Topographical demonstration of nuclei in the medulla oblongata of the fowl. *(Gallus domesticus). Zentralbl. Veterinaermed. Reihe A* **16**, 257–270.

Goloube, D. M. (1936). Sur le développement de la glande surrénale et de ses nerfs chez le poulet. *Ann. Anat. Pathol. Anat. Norm. Med.-Chir.* **13**, 1055–1066.

Goormaghtigh, N. (1921). Organogenèse et histogenèse de la capsule surrénale et du plexus coeliaque. *Arch. Biol.* **31**, 83–165.

Gossrau, R. (1968). Über das Reizleitungssystem der Vögel. Histochemische und elektronenmikroskopische Untersuchungen. *Histochemie* **13**, 111–159.

Govyrin, V. A. (1965). O roli adrenalina i noradrenalina v adaptatsionnotroficheskoi funktsii nervnoi sistemy u teplokrovnykh i kholodnokrovnykh zhivotnykh. *In* Funktsional'naya evolutsiya nervoni sistemy," pp. 115–123. "Nauka," Leningrad.

Govyrin, V. A., and Leontieva, G. R. (1962). Catecholamines of the heart of birds during ontogenesis. *Dokl. Biol. Sci.* **147**, 1407–1409.

Govyrin, V. A., and Leontieva, G. R. (1965a). Raspredylenie katekholovykh aminov v miokarde pozvonochnykh. *Zh. Evol. Biokhim. Fiziol.* **1**, 38–44.

Govyrin, V. A., and Leontieva, G. R. (1965b). K voprosu o khromaffinoi tkani i istochnikakh katyekhinovykh zhivotnykh. *Byull. Eksp. Biol. Med.* **59**, 98–100.

Graham, J. D. P. (1938). A note on the action of atropine in the bird. *J. Physiol. (London)* **93**, 56P–58P.

Graham, J. D. P. (1940). Respiratory reflexes in the fowl. *J. Physiol. (London)* **97**, 525–532.

Gray, J. C. (1937). The anatomy of the male genital ducts in the fowl. *J. Morphol.* **60**, 393–401.

Grewe, F. J. (1951). Nuwe gegewens aangaande die ontogenese van die neuskliere, die orgaan van Jacobson en die Dekbene van die skedel by die genus *Anas. Ann. Univ. Stellenbosch, Ser. A* **3**, 69–99.

Grignon, G., Hatier, R., Guedenet, J. C., and Dollander, A. (1966). Aspects ultrastructuraux de la glande surrénale au cours du développement chez l'embryon de poulet. *C. R. Soc. Biol.* **160**, 1654–1657.

Grinyer, I., and George, J. C. (1969a). An electron microscopic study of the pigeon breast muscle. *Can. J. Zool.* **47**, 517–524.

Grinyer, I., and George, J. C. (1969b). Some observations on the ultrastructure of the humming bird pectoral muscles. *Can. J. Zool.* **47**, 771–774.

Gruenhagen, A. (1887). Ueber den Einfluss des Sympathicus auf die Vogelpupille. *Pflueger's Arch. Gesamte Physiol. Menschen Tiere* **40**, 65–67.

Hachmeister, U., Kracht, J., Kruse, H., and Lenke, M. (1967). Lokalisation von C-Zellen im Ultimobranchialkörper des Haushuhns. *Naturwissenschaften* **54**, 619.

Håkansson, C. H., and Malcus, B. (1969). Secretive response of the electrically stimulated salt gland in *Larus argentatus* (Herring Gull). *Acta Physiol. Scand.* **76**, 385–392.

Halpern, B. P. (1963). Gustatory nerve responses in the chicken. *Amer. J. Physiol.* **203**, 541–544.

Hammond, W. S. (1949). Formation of the sympathetic nervous system in the trunk of the chick embryo following removal of the thoracic neural tube. *J. Comp. Neurol.* **91**, 67–85.

Hammond, W. S., and Yntema, C. L. (1947). Depletions in the thoraco-lumbar sympathetic system following removal of neural crest in the chick. *J. Comp. Neurol.* **86**, 237–265.

Hammond, W. S., and Yntema, C. L. (1958). Origin of ciliary ganglia in the chick. *J. Comp. Neurol.* **110**, 367–390.

Hamori, J., and Dyachkova, L. N. (1964). Electron microscope studies on developmental differentiation of ciliary ganglion synapses in the chick. *Acta Biol. (Budapest)* **15**, 213–230.

Hanzlik, P. J., and Butt, E. M. (1928). Reactions of the crop (oesophageal) muscles under tension, with a consideration of the anatomical arrangement, innervation and other factors. *Amer. J. Physiol.* **85**, 271–282.

Hassa, O., and Calislar, T. (1964). Anatomical and histological investigations on the penis of domestic male ducks. *Vet. Fak. Dergisi* **11**, 8–27.

Hassan, T. (1967). Effects of stimulation of the cervical vagus and descending oesophageal nerves on the alimentary tract of the domestic fowl. *Zentralbl. Veterinaermed. Reihe A* **14**, 854–861.

Hassan, T. (1969). A hyoscine-resistant contraction of isolated chicken oesophagus in response to stimulation of parasympathetic nerves. *Brit. J. Pharmacol.* **36**, 268–275.

Hatier, R., Grignon, G., and Guedenet, J. C. (1969). Etude ultrastructurale des cellules adrenales chez le poussin après traitement par la résérpine. *C. R. Soc. Biol.* **163**, 491–494.

Hess, A. (1961). Structural differences of fast and slow extrafusal muscle fibres and their nerve endings in chickens. *J. Physiol. (London)* **157**, 221–231.

Hess, A. (1965). Developmental changes in the structure of the synapse on the myelinated cell bodies of the chicken ciliary ganglion. *J. Cell Biol.* **25**, 1–19.

Hess, A. (1966). The fine structure of striated muscle fibres and their nerve terminals in the avian iris: Morphological "twitch-slow" fibres. *Anat. Rec.* **154**, 357.

Hess, A. (1967). The structure of vertebrate slow and twitch muscle fibres. *Invest. Ophthalmol.* **6**, 217–225.

Hess, A. (1970). Vertebrate slow muscle fibres. *Physiol. Rev.* **50**, 40–62.

Hess, A., Pilar, G., and Weakly, J. N. (1969). Correlation between transmission and structure in avian ciliary ganglion synapses. *J. Physiol. (London)* **202**, 339–354.

Hiestand, W. A., and Randall, W. C. (1941). Species differentiation in the respiration of birds following carbon dioxide administration and the location of inhibitory receptors in the upper respiratory tract. *J. Cell. Comp. Physiol.* **17**, 333–340.

Hiestand, W. A., and Randall, W. C. (1943). Influence of proprioceptive vagal afferents on panting and accessory panting movements in mammals and birds. *Amer. J. Physiol.* **138**, 12–15.

Hirakow, R. (1966). Fine structure of Purkinje fibres in the chick heart. *Arch. Histol.* **27**, 485–500.

Hirsch, E. F. (1963). The innervation of the human heart. V. A comparative study of the intrinsic innervation of the heart in vertebrates. *Exp. Mol. Pathol.* **2**, 384–401.

His, W., Jr. (1897). Über die Entwicklung des Bauchsympathicus beim Hühnchen und Menschen. *Arch. Anat. Physiol., Physiol. Abt.* **21**, Suppl., 137–170.

Hnĭk, P., Jirmanová, I., Vyklický, L., and Zelena, J. (1967). Fast and slow muscles of the chick after nerve cross-union. *J. Physiol. (London)* **193**, 309–325.

Hodges, R. D., and Gould, R. P. (1969). Partial nervous control of the avian ultimobranchial body. *Experientia* **25**, 1317–1319.

Höhn, E. O. (1960). Seasonal changes in the mallard's penis and their hormonal control. *Proc. Zool. Soc. London* **134**, 547–555.

Hollenberg, N. K., and Uvnäs, B. (1963). The role of the cardiovascular response in the resistance to asphyxia of avian divers. *Acta Physiol. Scand.* **58**, 150–161.

Holtzmann, H. (1896). Untersuchungen über Ciliarganglion und Ciliarnerven. *Morphol. Arb.* **6**, 114–142.

Hoshi, T. (1926). Histology of nerves in the suprarenals. (In Japanese.) *Tohoku Med. J.* **9**, 443–468.

Hsieh, T. M. (1951). The sympathetic and parasympathetic nervous systems of the domestic fowl. Ph.D. Thesis, Edinburgh University.

Huber, G. C. (1900). A contribution on the minute anatomy of the sympathetic ganglia of the different classes of vertebrates. *J. Morphol.* **16**, 27–90.

Huber, G. C., and DeWitt, L. (1898). A contribution to the motor nerve endings and on the nerve endings in muscle spindles. *J. Comp. Neurol.* **7**, 169–230.

Huber, J. F. (1936). Nerve roots and nuclear groups in the spinal cord of the pigeon. *J. Comp. Neurol.* **65**, 43–91.

Ignarro, L. J., and Shideman, F. E. (1968). Appearance and concentrations of catecholamines and their biosynthesis in the embryonic and developing chick. *J. Pharmacol. Exp. Ther.* **159**, 38–48.

Ihnen, K. (1928). Beiträge zur Physiologie des Kropfes bei Huhn und Taube. I. Bewegung und Innervation des Kropfes. *Pfluegers Arch. Gesamte Physiol. Menschen Tiere* **218**, 767–782.

Imaizumi, M. (1968). A study of the fine structure of the smooth muscle cell. (In Japanese.) *Osaka Med. J.* **20**, 467–474.

Imaizumi, M., and Hama, K. (1969). An electron microscope study of the interstitial cells of the gizzard of the love bird (*Uroloncha domestica*). *Z. Zellforsch. Mikrosk. Anat.* **97**, 351–357.

Iwanow, J. F. (1930). Die sympathische Innervation der Verdauungstraktes einiger Vogelarten (*Columbia livia* L., *Anser cinereus* L., und *Gallus domesticus*). *Z. Mikrosk.-Anat. Forsch.* **22**, 469–492.

Iwanow, J. F., and Radostina, T. N. (1933). Sur la morphologie du système nerveux autonome du tube digestif chez certains mammifères et quelques oiseaux. *Trab. Inst. Cajal Invest. Biol.* **28**, 303–321.

Jain, P. D. (1965). Studies on the structure development, innervation and physiology of the heart and its conducting tissue in bat, pigeon and rabbit. *Agra Univ. J. Res., Sci.* **14,** 207-213.

Jaquet, M. (1901). Anatomie comparée due système nerveux sympathique cervical dans la serie des vertébrés. *Bul. Soc. Sci. (Bucharest)* **10,** 240-315.

Jegorow, J. (1887). Ueber den Einfluss des Sympathicus auf die Vogelpupille. *Pflueger's Arch. Gesamte Physiol. Menschen Tiere* **41,** 326-348.

Jegorow, J. (1890). Ueber das Verhältniss des Sympathicus zur Kopfverzierung einiger Vögel. *Arch. Anat. Physiol., Physiol. Abt.* **14,** Suppl., 33-56.

Johansen, K. (1964). Regional distribution of circulating blood during submersion asphyxia in the duck. *Acta Physiol. Scand.* **62,** 1-9.

Johansen, K., and Aakhus, T. (1963). Central cardiovascular responses to submersion asphyxia in the duck. *Amer. J. Physiol.* **205,** 1167-1171.

Johansen, K., and Reite, O. B. (1964). Cardiovascular responses to vagal stimulation and cardioaccelerator nerve blockade in birds. *Comp. Biochem. Physiol.* **12,** 477-489.

Johnson, J. S. (1925). The innervation of the female genitalia of the domestic fowl. *Anat. Rec.* **29,** 387.

Jones, D. R. (1969). Avian afferent vagal activity related to respiratory and cardiac cycles. *Comp. Biochem. Physiol.* **28,** 961-966.

Jones, D. R., and Purves, M. J. (1969). Arterial chemoreceptors and the diving response in the duck. *J. Physiol. (London)* **203,** 41P.

Jones, D. R., and Purves, M. J. (1970a). The carotid body in the duck and the consequences of its denervation upon the cardiac responses to immersion. *J. Physiol. (London)* **211,** 279-294.

Jones, D. R., and Purves, M. J. (1970b). The effect of carotid body denervation upon the respiratory response to hypoxia and hypercapnia in the duck. *J. Physiol. (London)* **211,** 295-309.

Jungherr, E. L. (1969). The neuroanatomy of the domestic fowl (*Gallus domesticus*). (Completed by C. F. Helmboldt and P. Timmins.) "Avian Diseases," Spec. Issue.

Jürgens, H. (1909). Über die Wirkung des Nervus vagus auf das Herz der Vögel. *Pflueger's Arch. Gesamte Physiol. Menschen Tiere* **129,** 506-524.

Kadono, H., Okada, T., and Ohno, K. (1966). Neurophysiological studies on the sense of taste in the chicken. *Res. Bull. Fac. Agr., Gifu Univ.* **22,** 149-160.

Kagawa, K., Yagasaki, O., Takewaki, T., and Yanagiya, I. (1969). Changes of the inner plexus control in the chick intestinal motility through the development. *Jap. J. Vet. Sci.* **31,** 23-28.

Kaiser, L. (1924). L'innervation segmentale de la peau chez le pigeon (*Columba livia* var. *domestica*). *Arch. Neer. Sci. Soc.* **9,** 299-379.

Kano, M. (1959). Electron microscopic study on the adrenal medulla of the domestic fowl. (In Japanese.) *Arch. Histol.* **18,** 25-56.

Kappers, C. U. A. (1947). "Anatomie comparée du système nerveux, particulièrement de celui des mammifères et de l'homme." Masson, Paris.

Kappers, C. U. A., Huber, G. C., and Crosby, E. C. (1936). "The Comparative Anatomy of the Nervous System of Vertebrates Including Man," Vol. I. Macmillan, New York.

Kappers, J. A. (1965). Survey of the innervation of the epiphysis cerebri and the accessory pineal organs of vertebrates. *Progr. Brain Res.* **10,** 87-153.

Karten, H. J., and Hodos, W. (1967). "A Stereotaxic Atlas of the Brain of the Pigeon *Columba livia*." John Hopkins Press, Baltimore, Maryland.

Kasa, P., Mann, S. P., and Hebb, C. (1970). Localization of choline acetyltransferase. *Nature (London)* **226**, 812–816.

Kaupp, B. F. (1918). "The Anatomy of the Domestic Fowl." Saunders, Philadelphia, Pennsylvania.

Kazumoto, F. (1969). On the responses of chicken gizzard to stimulation of vagus and splanchnic nerves and medulla oblongata. (In Japanese.) *Jap. J. Smooth Muscle Res.* **5**, 84–92.

King, A. S. (1966). Afferent pathways in the vagus and their influence on avian breathing: A review. *In* "Physiology of the Domestic Fowl" (C. Horton Smith and E. C. Amoroso, eds.), pp. 302–310. Oliver & Boyd, Edinburgh.

King, A. S., and Cowie, A. F. (1969). The functional anatomy of the bronchial muscle of the bird. *J. Anat.* **105**, 323–336.

King, A. S., McLelland, J., Molony, V., and Mortimer, M. F. (1968a). Respiratory afferent activity in the avian vagus: Eupnoea, inflation and deflation. *J. Physiol. (London)* **201**, 35P–36P.

King, A. S., Molony, V., McLelland, J., Bowsher, D. R., and Mortimer, M. F. (1968b). Afferent respiratory pathways in the avian vagus. *Experientia* **24**, 1017–1018.

Kitchell, R. L., Strom, L., and Zotterman, Y. (1959). Electrophysiological studies of thermal and taste reception in chickens and pigeons. *Acta Physiol. Scand.* **46**, 133–151.

Knouff, R. A., and Hartman, F. A. (1951). A microscopic study of the adrenal of the Brown Pelican. *Anat. Rec.* **109**, 161–187.

Kobayashi, S. (1969a). Catecholamines in the avian carotid body. *Experientia* **25**, 1075–1076.

Kobayashi, S. (1969b). On the fine structure of the carotid body of the bird, *Uroloncha domestica*. *Arch. Histol.* **31**, 9–19.

Kobayashi, S., and Fujita, T. (1969). Fine structure of mammalian and avian pancreatic islets, with special reference to D cells and nervous elements. *Z. Zellforsch. Mikrosk. Anat.* **100**, 340–363.

Koenig, H.-L. (1967). Quelques particularités ultrastructurales des zones synaptiques dans le ganglion ciliaire du poulet. *C. R. Ass. Anat.* **52**, 711–719.

Kolossow, N. G. (1959). Weitere Beobachtungen am Nervensystem des Darmes. *Z. Mikrosk.-Anat. Forsch.* **65**, 557–573.

Kolossow, N. G. (1963). Afferente Elemente in den Ganglien des vegetativen Nervensystems. *Acta Univ. Szeged, Acta Biol.* **9**, 175–190.

Kolossow, N. G. (1965). Concerning the afferent neurons of the sympathetic nervous system. *Dokl. Biol. Sci.* **161**, 490–492.

Kolossow, N. G. (1970). Chuvstvitel 'nye elementy v avtonomnoi nervnoi sisteme. *Arkh. Anat., Gistol. Embriol.* **58**, 3–14.

Kolossow, N. G., and Sabussow, G. H. (1929). Beiträge zum Studium der sympathischen und spinalen Ganglien einiger Reptilien und Vögel. Die Ganglien von *Emys europaea* L., *Anser cinereus* L. und *Columba livia* L. *Z. Mikrosk.-Anat. Forsch.* **18**, 5–36.

Kolossow, N. G., Sabussow, G. H., and Iwanow, J. F. (1932). Zur Innervation des Verdauungskanales der Vögel: Eine experimentell-morphologische Untersuchung. *Z. Mikrosk.-Anat. Forsch.* **30**, 257–294.

Komarek, V., and Marvan, F. (1969). Beitrag zur mikroskopischen Anatomie des Kopulationsorganes der Entenvögel. *Anat. Anz.* **124**, 467–476.

Kondratjew, N. S. (1926). Zur Frage über die intrakardiale Innervation der Vögel. I Mitteilung. *Z. Anat. Entwicklungsgesch.* **79**, 753–761.

Kondratjew, N. (1933). Ueber Formbildungstypen der kardialen nervösen Grundgeflechts bei verschiedenen Gruppen der Wirbeltiere. III Mitteilung. Waran (*Varanus griseus*), Schelto pusik (*Ophisarus apus*) und Eule (*Asio accipitrinus* Pall.). *Z. Anat. Entwicklungsgesch.* **100**, 712–734.

Koppanyi, T., and Sun, K. H. (1926). Comparative studies on pupillary reaction in tetrapods. III. The reactions of the avian iris. *Amer. J. Physiol.* **78**, 364–367.

Kose, W. (1902). Über das Vorkommen einer "Carotisdrüse" und der "chromaffinen Zellen" bei Vögeln. *Anat. Anz.* **22**, 162–170.

Kose, W. (1904). Ueber die "Carotisdrüse" und das "chromaffine Gewebe" der Vögel. *Anat. Anz.* **25**, 609–617.

Kose, W. (1907a). Die Paraganglien bei den Vögeln. Erster Teil. *Arch. Mikrosk. Anat.* **69**, 563–664.

Kose, W. (1907b). Die Paraganglien bei den Vögeln. Zweiter Teil. *Arch. Mikrosk. Anat.* **69**, 665–790.

Kostinowitsch, L. I. (1936). Über Truncus Collateralis thoracis bei Tauben. *Anat. Anz.* **82**, 191–199.

Krogis, A. (1931). La disposition des corpuscles de Herbst et de Grandry dans le bec des oiseaux adultes. *C. R. Soc. Biol.* **108**, 742–744.

Krogis, A. (1932). On the topography of Herbst's and Grandry's corpuscles in the adult and embryo duck-bill. *Acta Zool. (Stockholm)* **12**, 241–263.

Krohn, A. (1837). Üeber die Structur der Iris der Vögel in ihren Bewegungsmechanismus. *Arch. Anat., Physiol. Wiss. Med.* pp. 357–380.

Krüger, P. (1950). Untersuchungen am Vögelflügel. *Zool. Anz.* **145**, Ergänzungsbd. 445–460.

Krüger, P. (1952). "Tetanus und Tonus der quergesteiften Skelettmuskeln der Wirbeltiere und des Menschen." Geest & Portig, Leipzig.

Krüger, P., and Gunther, P. G. (1956). Das sarkoplasmische Reticulum in der quergestreiften muskelfasern der Wirbeltiere und des Menschen. *Acta Anat.* **28**, 135–149.

Krüger, P., and Gunther, P. G. (1958). Innervation und pharmakologisches Verhalten des M. gastrocnemius und M. pectoralis major der Vögel. *Acta Anat.* **33**, 325–338.

Kuhne, W. (1887). Neue Untersuchungen über motorische Nervenendigungen. *Z. Biol. (Munich)* **23**, 1–148.

Külbs, F. (1912). Ueber das Reizleitungssystem bei Amphibien, Reptilien und Vögeln. *Z. Exp. Pathol. Ther.* **11**, 51–68.

Kuntz, A. (1910). The development of the sympathetic nervous system in birds. *J. Comp. Neurol.* **20**, 283–308.

Kuntz, A. (1911). The evolution of the sympathetic nervous system in vertebrates. *J. Comp. Neurol.* **21**, 215–236.

Kuntz, A. (1953). "The Autonomic Nervous System." Lea & Febiger, Philadelphia.

Kura, N. (1927). The innervation of the adrenal gland. (In Japanese.) *Mitt. Med. Akad. Kioto* **1**, 107–124.

Laffont, M. (1885). Recherches sur l'anatomie et la physiologie comparée du nerfs trijumeau, facial et sympathique céphalique des oiseaux. *C. R. Acad. Sci.* **101**, 1286–1289.

Lake, P. E. (1957). The male reproductive tract of the fowl. *J. Anat.* **91**, 116–129.

Landmesser, L., and Pilar, G. (1970). Selective reinnervation of the cell populations in the adult pigeon ciliary ganglion. *J. Physiol. (London)* **211**, 203–216.

Langendorff, O. (1894). Ciliarganglion und Oculomotorius. *Arch. Gesamte Physiol. Menschen Tiere* **56**, 522–527.

Langendorff, O. (1900). Zur Verständigung über die Natur des Ciliarganglions. *Klin. Monatsbl. Augenheilk.* **38**, 307–314.

Langley, J. N. (1902). Preliminary note on the sympathetic system of the bird. *J. Physiol. (London)* **27**, 35P–36P.

Langley, J. N. (1904). On the sympathetic system of birds and on the muscles which move the feathers. *J. Physiol. (London)* **30**, 221–252.

Langley, J. N. (1921). "The Autonomic Nervous System," Part I. Heffer & Sons, Cambridge.

Laruelle, L., Reumont, M., and Legait, E. (1951). Recherches sur le mécanisme des changements de couleur des caroncles vascularies du Dindon (*Meleagris gallopavo* L.). *Arch. Anat. Microsc. Morphol. Exp.* **40**, 91–113.

Lauber, J. K., Boyd, J. E., and Axelrod, J. (1968). Enzymatic synthesis of melatonin in avian pineal body: Extra-retinal response to light. *Science* **161**, 489–490.

Legait, E., and Legait, H. (1952). Quelle est l'importance de l'équipement ganglionaire du corps thyroïde? *Arch. Anat., Histol. Embryol.* **34**, 261–270.

Leontieva, G. R. (1966a). Raspredylenie katekholovykh aminov v stienke krovosynkh sosudov tyeplokrovnykh zhivotnykh. *Arkh. Anat., Gistol. Embriol.* **50**, 36–41.

Leontieva, G. R. (1966b). K voprosu ob adrenergycheskoi innertsii kapillarov. *Zh. Evol. Biokhim. Fiziol.* **2**, 457–461.

Levi-Montalcini, R. (1950). The origin and development of the visceral system in the spinal cord of the chick embryo. *J. Morphol.* **86**, 253–284.

Lin, Y. C., and Sturkie, P. D. (1968). Effect of environmental temperature on the catecholamines of chickens. *Amer. J. Physiol.* **214**, 237–240.

Lin, Y. C., Sturkie, P. D., and Tummons, J. L. (1970). Effect of cardiac sympathectomy, reserpine and environmental temperatures on the catecholamine levels in the chicken heart. *Can. J. Physiol. Pharmacol.* **48**, 182–184.

Liu, H. C. (1962). The comparative structure of the ureter. *Amer. J. Anat.* **111**, 1–16.

Lucas, A. M., and Stettenheim, P. R. (1965). Avian anatomy. *In* "Disease of Poultry" (H. E. Biester and L. H. Schwarte, eds.), pp. 1–59. Iowa State Univ. Press, Ames.

Lukashin, V. G. (1969). On sensitive innervation of vegetative ganglia. *Dokl. Biol. Sci.* **189**, 909–912.

McCrady, J. D., Vallbona, C., and Hoff, H. E. (1966). Neural origin of the respiratory heart rate response. *Amer. J. Physiol.* **211**, 323–328.

Macdonald, E., and Taylor, L. W. (1933). The rudimentary copulatory organ of the domestic fowl. *J. Morphol.* **54**, 429–449.

Macdonald, R. L., and Cohen, D. H. (1970). Cells of origin of sympathetic pre- and post-ganglionic cardioacceleratory fibers in the pigeon. *J. Comp. Neurol.* **140**, 343–358.

McFarland, L. Z., Homma, K., and Wilson, W. O. (1966). Superior cervical ganglionectomy and its effect on reproduction in the Japanese Quail (*Coturnix c. japonica*.). *Anat. Rec.* **154**, 385.

McKenney, F. D., Essex, H. E., and Mann, F. C. (1932). The action of certain drugs on the oviduct of the domestic fowl. *J. Pharmacol. Exp. Ther.* **45**, 113–119.

Mackenzie, I. (1910). Zur Frage eines Koordinationssystems im Herzen. *Verh. Deut. Ges. Pathol.* **14**, 90–97.

Mackenzie, I., and Robertson, J. I. (1910). Recent researches on the anatomy of the bird's heart. *Brit, Med. J.* **2**, 1161–1164.

McLelland, J. (1969). Observations with the light microscope on the ganglia and nerve plexuses of the intrapulmonary bronchi of the bird. *J. Anat.* **105**, 202.

Magnien, L. (1885a). Sur le ganglion géniculé des oiseaux. *C. R. Acad. Sci.* **100**, 1507–1509.

Magnien, L. (1885b). Recherches sur l'anatomie comparée de la corde du tympan des oiseaux. *C. R. Acad. Sci.* **101**, 1013-1016.

Magnien, L. (1887). Etude des rapport qui existent entre les nerfs craniens et la sympathique céphalique chez les oiseaux. *C. R. Acad. Sci.* **104**, 77-79.

Maier, A., and Eldred, E. (1969). The structure of avian intrafusal fibers. *Physiologist* **12**, 291.

Makita, T., Shioda, T., and Nishida, S. (1966). A histological study on the innervation of the avian thyroid gland. *Arch. Histol.* **26**, 203-214.

Malinovský, L. (1962). Contribution to the anatomy of the vegetative nervous system in the neck and thorax of the domestic pigeon. *Acta Anat.* **50**, 326-338.

Malinovský, L. (1963). The nerve supply of the stomach in the domestic pigeon. *Cesk. Morfol.* **11**, 16-27.

Malinovský, L. (1964). Mikroskopicka struktura nervovych pleteni svalnatého žaludku holuba domaciho (*Columba domestica*). *Cesk. Morfol.* **12**, 30-39.

Malinovský, L. (1967a). Some problems connected with the evaluation of skin receptors and their classification. *Folia Morphol. (Prague)* **15**, 18-25.

Malinovský, L. (1967b). Die Nervenendkörperchen in der Haut von Vögeln und ihre Variabilität. *Z. Mikrosk.-Anat. Forsch.* **77**, 279-303.

Malinovský, L. (1968). Types of sensory corpuscles common to mammals and birds. *Folia Morphol. (Prague)* **16**, 61-73.

Malinovský, L., and Zemanek, R. (1970). Sensory corpuscles in the beak skin of the domestic pigeon. *Folia Morphol. (Prague)* **17**, 241-250.

Mangold, E. (1906). Der Muskelmagen der körnerfressenden Vögel, seine motorischen Funktionen und ihre Abhängigkeit vom Nervensystem. *Pflueger's Arch. Gesamte Physiol. Menschen Tiere* **111**, 163-240.

Mangold, E. (1911). Die Magenbewegungen der Krähe und Dohle und ihre Beeinflussung vom Vagus. *Pflueger's Arch. Gesamte Physiol. Menschen Tiere* **138**, 1-13.

Manni, E., Bortolami, R., and Azzena, G. B. (1965). Jaw muscle proprioception and mesencephalic trigeminal cells in birds. *Exp. Neurol.* **12**, 320-328.

Marage, R. (1889). Anatomie descriptive du sympathique chez les oiseaux. *Ann. Sci. Natur.: Zool.* [7] **7**, 1-72.

Marshall, M. (1878). The development of the cranial nerves in the chick. *Quart. J. Microsc. Sci.* **18**, 10-40.

Martin, A. H., and Brinkman, R. (1970). The dorsal horn of the avian spinal cord, a reexamination. *Experientia* **26**, 887-889.

Martin, A. R., and Pilar, G. (1963a). Dual mode of synaptic transmission in the avian ciliary ganglion. *J. Physiol. (London)* **168**, 443-463.

Martin, A. R., and Pilar, G. (1963b). Transmission through the ciliary ganglion of the chick. *J. Physiol. (London)* **168**, 464-475.

Martin, A. R., and Pilar, G. (1964a). An analysis of electrical coupling at synapses in the avian ciliary ganglion. *J. Physiol. (London)* **171**, 454-475.

Martin, A. R., and Pilar, G. (1964b). Quantal components of the synaptic potential in the ciliary ganglion of the chick. *J. Physiol. (London)* **175**, 1-16.

Martin, A. R., and Pilar, G. (1964c). Pre-synaptic and post-synaptic events during post-tetanic potentiation and facilitation in the avian ciliary ganglion. *J. Physiol. (London)* **175**, 17-30.

Marvan, F. (1969). Postnatal development of the male genital tract of the *Gallus domesticus*. *Anat. Anz.* **124**, 443-462.

Marwitt, R., Pilar, G., and Weakly, J. N. (1971). Characterization of two ganglion cell populations in avian ciliary ganglia. *Brain Res.* **25**, 317-334.

Matsushita, M. (1968). Zur Zytoarchitektonik des Hühnerrückenmarkes nach Silberimprägnation. *Acta Anat.* **70**, 238-259.

Mauger, H. M., Jr. (1941). The autonomic innervation of the female genitalia in the domestic fowl and its correlation with the aortic branchings. *Amer. J. Vet. Res.* **2**, 447-453.

Mayr, R., and Zenker, W. (1969). Narrowings of the synaptic cleft in myoneural junctions. *Experientia* **25**, 1319-1321.

Mehrotra, P. N. (1964). On the microscopic anatomy of the epididymis of *Anser melanotus*, L. *Trans. Amer. Microsc. Soc.* **83**, 456-460.

Melkich, A. (1895). Zur Kenntnis des Cilarkörpers und der Iris bei Vögeln. *Anat. Anz.* **10**, 28-35.

Milokhin, A. A. (1960). Über die afferente Innervation der peripheren vegetativen Neuronen. *Z. Mikrosk.-Anat. Forsch.* **66**, 483-488.

Milokhin, A. A. (1962a). Über eine besondere Form der Interorezeptoren des Verdauungstraktes. *Z. Mikrosk.-Anat. Forsch.* **69**, 145-152.

Milokhin, A. A. (1962b). New facts concerning the afferent innervation of peripheral autonomic neurones. *Dokl. Biol. Sci.* **141**, 726-729.

Milokhin, A. A. (1964). Afferentnaya innervatsiya nervnykh kletok vegetativnykh gangliev. *Arkh. Anat., Gistol. Embriol.* **47**, 27-34.

Milokhin, A. A., and Reshetnikov, A. B. (1968). Akso-aksonal'nye sinapticheskie svyazi v gangliyakh auerbakhova spleteniya. *Arkh. Anat., Gistol. Embriol.* **54**, 25-30.

Mitchell, G. A. G. (1956). "Cardiovascular Innervation." Livingstone, Edinburgh.

Morozov, E. K. (1969). K voprosu ob innervatsii sinokarotidinoi refleksogennoi zony u ptits. *Arkh. Anat., Gistol. Embriol.* **56**, 35-38.

Morozov, E. K. (1970). Otnoshenie yazykoglotochnogo nerva k karotidnomu glomusu u ptits. *Arkh. Anat., Gistol. Embriol.* **58**, 79-83.

Moynahan, E. J., Sethi, N., and Jowett, P. (1970). Reactions of the small vessels in the extra-embryonic membranes and the limb bud of the developing chick to vasoactive agents. *Brit. J. Dermatol.* **82**, Suppl. 5, 77-85.

Müller, J. (1929). Die Nebennieren von *Gallus domesticus* und *Columba livia domestica*. *Z. Mikrosk.-Anat. Forsch.* **17**, 303-352.

Muratori, G. (1931). Connessioni tra sistema del vago e sistema del paraganglio carotico. *Boll. Soc. Ital. Biol. Sper.* **6**, 861-863.

Muratori, G. (1932a). Ricerche istologiche e sperimentali sull'innerverzione del tessuto paragangliare annesso al sistema del vago (Paraganglio carotico: Paragangli iustavagali et intravagali). *Boll. Soc. Ital. Biol. Sper.* **7**, 137-142.

Muratori, G. (1932b). Contributo all'innervazione del tessuto paragangliare annesso al sistema del vago (glomo carotico, paragangli estravagali ed intravagali) e all innervazione del seno carotideo. *Anat. Anz.* **75**, 115-123.

Muratori, G. (1932c). Recherches histologiques et experiméntales sur l'innervation du tissu paraganglionnaire (=phèochrome) annexé au système du nerf vague des amniotes (Glomus carotidien; paragangdons extravagaux et intravagaux). *C. R. Ass. Anat.* **27**, 409-415.

Muratori, G. (1933). Ricerche istologiche sull'innervazione del glomo carotico. *Arch. Ital. Anat. Embriol.* **30**, 573-602.

Muratori, G. (1934). Contributo istologico all'innervazione della zona arteriosa glomocarotidea. *Arch. Ital. Anat. Embriol.* **33**, 421-442.

Muratori, G. (1935). Contributo istologico alla conoscenza dell'innervazione polmonare. *Arch. Ital. Anat. Embriol.* **34**, 45-71.

Muratori, G. (1937). Osservazione istologiche e considerazioni embriologiche sui recettori aortici degli amnioti. *Anat. Anz.* **83**, 367-379.

Muratori, G. (1962). Histological observations on the cervicothoracic paraganglia of amniotes. *Arch. Int. Pharmacodyn. Ther.* **140,** 217–226.
Nafstad, P. H. J., and Andersen, A. E. (1969). Further observations on the Herbst corpuscle in the domestic fowl. *J. Ultrastruct. Res.* **29,** 573.
Nafstad, P. H. J., and Andersen, A. E. (1970). Ultrastructural investigation of the innervation of the Herbst corpuscle. *Z. Zellforsch. Mikrosk. Anat.* **103,** 109–114.
Nakayama, S. (1968). Intrinsic and extrinsic reflexes of the oesophagus in hen. (In Japanese.) *Jap. J. Smooth Muscle Res.* **4,** 177–182.
Nakazato, Y., Sato, M., and Ohga, A. (1970). Evidence for a neurogenic "rebound" contraction of the smooth muscle of the chicken proventriculus. *Experientia* **26,** 50–51.
Nishida, S., and Sears, M. (1970). Fine structure of the anterior epithelial cell layer of the iris of the hen. *Exp. Eye Res.* **9,** 241–245.
Nolf, P. (1925a). Influence du vague sur la motricité de l'estomac de l'oiseau. *C. R. Soc. Biol.* **93,** 454–455.
Nolf, P. (1925b). Influence de l'hypercapnée et de l'anoxèmie sur la motricité de l'estomac musculaire de l'oiseau. *C. R. Soc. Biol.* **93,** 455–456.
Nolf, P. (1925c). Influence des nerfs sympathiques sur la motricité de l'estomac de l'oiseau. *C. R. Soc. Biol.* **93,** 839.
Nolf, P. (1927). Du rôle des nerfs vague et sympathique dans l'innervation motrices de l'estomac de l'oiseau. *Arch. Int. Physiol.* **28,** 309–428.
Nolf, P. (1928). Le système nerveux entérique. Essai d'analyse par la methode à la nicotine de Langley. *Arch. Int. Physiol.* **30,** 317–492.
Nolf, P. (1930). Action de l'atropine sur les éléments nerveux du plexus entérique chez l'oiseau. *Arch. Int. Pharmacodyn. Ther.* **38,** 591–617.
Nolf, P. (1934a). Les nerfs extrinsèques de l'intestin chez l'oiseau. I. Les nerfs vagues. *Arch. Int. Physiol.* **39,** 113–163.
Nolf, P. (1934b). Les nerfs extrinsèques de l'intestin chez l'oiseau. II. Les nerfs coeliaques et mésentèriques. *Arch. Int. Physiol.* **39,** 165–226.
Nolf, P. (1934c). Les nerfs extrinsèques de l'intestin de l'oiseau. III. Le nerf de Remak. *Arch. Int. Physiol.* **39,** 227–256.
Nolf, P. (1935). Influence de l'anoxemie sur les effets de la stimulation des nerfs extrinsèques de l'éstomac de l'oiseau. *Arch. Int. Physiol.* **41,** 340–375.
Nolf, P. (1938a). L'appareil nerveux de l'automatisme gastrique de l'oiseau. I. Essai d'analyse par la nicotine. *Arch. Int. Physiol.* **46,** 1–85.
Nolf, P. (1938b). L'appareil nerveux de l'automatisme gastrique de l'oiseau. II. Étude des effets causés par une ou plusieurs sections de l'anneau nerveux du gesier. *Arch. Int. Physiol.* **46,** 441–559.
Nolf, P. (1938c). Les éléments intrinsèques de l'anneau nerveux du gesier de l'oiseau granivore. I. *Arch. Int. Physiol.* **47,** 453–518.
Nolf, P. (1939). Les éléments intrinsèques de l'anneau nerveux du gesier de l'oiseau granivore. II. *Arch. Int. Physiol.* **48,** 451–542.
Nonidez, J. F. (1935). The presence of depressor nerves in the aorta and carotid of birds. *Anat. Rec.* **62,** 47–74.
Oehme, H. (1968). Das Ganglion ciliare der Rabenvögel (*Corvidae*). *Anat. Anz.* **123,** 261–277.
Oehme, H. (1969). Der Bewegungsapparat der Vogeliris. (Eine vergleichende morphologischfunktionelle Untersuchung.) *Zool. Jahrb., Abt. Anat. Ontog. Tiere* **86,** 96–128.
Ohashi, H., and Ohga, A. (1967). Transmission of excitation from parasympathetic nerve to smooth muscle. *Nature (London)* **216,** 291–292.

Okamura, C. (1930). Über den Nervenapparat der Respirationsorgane. *Z. Anat. Entwicklungsgesch.* **92,** 20–26.

Okamura, C. (1934). Über die Darstellung des Nervenapparates in der Magendarmwand mittels der Vergoldungsmethode. *Z. Mikrosk.-Anat. Forsch.* **35,** 218–253.

Oksche, A., and Kirschstein, H. (1969). Elektronenmikroskopische Untersuchungen am Pinealorgan von *Passer domesticus. Z. Zellforsch. Microsk. Anat.* **102,** 214–241.

Oksche, A., Morita, Y., and Vaupel-von Harnack, M. (1969). Zur Feinstruktur und Funktion des Pinealorgans der Taube. *Z. Zellforsch. Mikrosk. Anat.* **102,** 1–30.

Oribe, T. (1963). Studies on the innervation of the female gonads of domestic fowl. (In Japanese.) *J. Tokyo Soc. Vet. Zootech. Sci.* **13,** 5–13.

Oribe, T. (1965). Studies on the innervation of the ovary of adult domestic fowls. (In Japanese.) *Bull. Coll. Agr. Vet. Med., Nihon. Univ.* **21,** 20–26.

Oribe, T. (1969). On the morphological structure of the ovary of the Japanese Hawkeagle. (In Japanese.) *Bull. Hiroshima Agr. Coll.* **3,** 211–216.

Ostmann, O. W., Ringer, R. K., and Tetzlaff, M. J. (1963). The anatomy of the feather follicle and its immediate surroundings. *Poultry Sci.* **42,** 958–969.

Otsuka, N., and Tomisawa, M. (1969). Fluorescence microscopy of the catecholamine containing nerve fibres in the vertebrate hearts. *Acta Anat. Nippon (Tokyo)* **44,** 1–6.

Page, S. G. (1969). Structure and some contractile properties of fast and slow muscles of the chicken. *J. Physiol. (London)* **205,** 131–146.

Page, S. G., and Slater, C. R. (1965). Observations on fine structure and rate of contraction of some muscles from the chicken. *J. Physiol. (London)* **179,** 58P.

Palme, F. (1934). Die Paraganglien über dem Herzen und im Endigungsgebiet des Nervus depressor. *Z. Mikrosk.-Anat. Forsch.* **36,** 391–420.

Palmieri, G. (1965). Sulla presenza di fusi neuromuscolari nei muscoli oculari estrinseci di alcuni mammiferi ed uccelli. *Biochim. Biol. Sper.* **4,** 65–72.

Palmieri, G. (1969). Sensory nervous expansions in duck and chicken joints. *Riv. Biol.* **62,** 345–350.

Pastěa, E. (1965). Plexurile, cardio-bronho-gastrice la păsările domestice. *Morfol. Norm. Patol.* **10,** 537–543.

Pastěa, E., and Pastěa, Z. (1966). Vascularizatia si inervatia macroscopica tiroparatiroidiana la gaina. *Lucr. Stiint., Ser. C* **9,** 627–632.

Pastěa, E., and Pastěa, Z. (1967). Plexurile hepato-gastro-splenice la ratá. *Lucr. Stiint., Ser. C* **10,** 171–179.

Paton, D. N. (1912). On the extrinsic nerves of the heart of the bird. *J. Physiol. (London)* **45,** 106–114.

Peck, J. I. (1889). Variation of the spinal nerves in the caudal region of the domestic pigeon. *J. Morphol.* **3,** 127–136.

Pellegrino de Iraldi, A., and Rodriguez-Perez, A. P. (1961). The ultrastructure of Grandry's corpuscles in the duck's beak. *Trab. Inst. Cajal Invest. Biol.* **53,** 185–189.

Pera, L. (1950). Il sistema dei NN. Splanchnici negli Uccelli (*Passer italiae*). *Proc. Verb. Soc. Tosc. Sci. Natur., Mem. Ser. B* **57,** 29–77.

Pera, L. (1952a). Contributo allo studio del nervo intestinale degli uccelli (*Passer italiae*). (Nota preliminare). *Proc. Verb. Soc. Tosc. Sci. Natur., Mem. Ser. B* **59,** 119–121.

Pera, L. (1952b). La morfologia del nervo intestinale di Remak in *Passer italiae. Proc. Verb. Soc. Tosc. Sci. Natur., Mem. Ser. B* **59,** 222–261.

Pera, L. (1953). Contributo allo studio dei centri pregangliari spinali degli uccelli. *Proc. Verb. Soc. Tosc. Sci. Natur., Mem. Ser. B* **60,** 51–87.

Pera, L. (1962a). Contributo allo studio della distribuzione della porzione abdominale del nervo vago negli uccelli (studio sperimentale). *Atti Soc. Ital. Anat.* **21**, 382–383.

Pera, L. (1962b). Sui rapporti tra fibre vagali e gangli del nervo intestinale di Remak in *Gallus domesticus*. *Atti Soc. Ital. Anat.* **22**, 82.

Pera, L. (1963a). Sul decorso delle fibre vagali contenute nel nerve intestinale del polla. *Proc. Verb. Soc. Tosc. Sci. Natur., Mem. Ser. B* **70**, 26–39.

Pera, L. (1963b). Sùlla presenza di sinapsi tra neuroni pregangliari vagali e gangli del nervo intestinale nel pollo. *Proc. Verb. Soc. Tosc. Sci. Natur., Mem. Ser. B* **70**, 40–52.

Peterson, D. F., and Fedde, M. R. (1968). Receptors sensitive to carbon dioxide in lungs of chicken. *Science* **162**, 1499–1501.

Pilar, G., and Vaughan, P. C. (1969a). Electrophysiological investigations of the pigeon iris neuromuscular junctions. *Comp. Biochem. Physiol.* **29**, 51–72.

Pilar, G., and Vaughan, P. C. (1969b). Mechanical responses of the pigeon iris muscle fibres. *Comp. Biochem. Physiol.* **29**, 73–87.

Pilar, G., Jenden, D., and Campbell, B. (1970). Change in acetylcholine content in postganglionic cells of adult pigeon ciliary ganglion after denervation. *Physiologist* **13**, 284.

Pintea, V., Constantinescu, G., and Radu, C. M. (1967). Vascular and nervous supply of the bursa of Fabricius in the hen. *Acta Vet. (Budapest)* **17**, 263–268.

Pisskunoff, N. N. (1911). Zur Frage nach den Ganglien in den Herzkammern von Vögeln. *Anat. Anz.* **38**, 394–395.

Polacek, P., Sklenska, A., and Malinovskỹ, L. (1966). Contribution to the problem of joint receptors in birds. *Folia Morphol. (Prague)* **14**, 33–40.

Popa, G., and Popa, F. (1931). The influence of the sympathetic on the pigeon's wing. *J. Anat.* **65**, 407–410.

Poustilnik, E. (1940). Do pitannia pro innervacin tazovykh i cherevnikh organiv u ptakhiv. *Med. Zh. (Kiev)* **10**, 1211–1214.

Prasad, A., Davis, L. E., and Dale, H. E. (1964). Adrenergic response of the avian uterus. *Poultry Sci.* **43**, 192–196.

Quay, W. B. (1965). Histological structure and cytology of the pineal organ in birds and mammals. *Progr. Brain Res.* **10**, 49–84.

Quilliam, T. A. (1962). Growth degrowth and regrowth in the Herbst corpuscle. *Anat. Rec.* **142**, 322.

Rabl, H. (1891). Die Entwicklung und Struktur der Nebennieren bei den Vögeln. *Arch. Mikrosk. Anat.* **38**, 492–523.

Ramón y Cajal, S. (1890a). Sur l'origine et les ramifications des fibres nerveuses de la moelle embryonnaire. *Anat. Anz.* **5**, 85–95.

Ramón y Cajal, S. (1890b). Sur l'origine et les ramifications des fibres nerveuses de la moelle embryonnaire. *Anat. Anz.* **5**, 111–119.

Ramón y Cajal, S. (1890c). A quelle époque aparaissent les expansions des cellules nerveuses de la moelle epinière du poulet? *Anat. Anz.* **5**, 609–613.

Rau, S. A., and Johnston, P. H. (1923). Observations on the development of the sympathetic system and the suprarenal glands in the sparrow. *Proc. Zool. Soc. London* **93**, 741–768.

Ray, P. J., and Fedde, M. R. (1969). Responses to alterations in respiratory PO_2 and PCO_2 in the chicken. *Resp. Physiol.* **6**, 135–143.

Rebollo, M. A., and de Anda, G. (1967). The neuromuscular spindles in the adult chicken. II. Histochemistry. *Acta Anat.* **67**, 595–608.

Regaud, C., and Favre, M. (1904). Les terminaisons nerveuses et les organes nerveux sensitifs de l'appareil locomoteur. *Rev. Gen. Histol.* **1,** 1–140.

Reite, O. B., Krog, J., and Johansen, K. (1963). Development of bradycardia during submersion of the duck. *Nature (London)* **200,** 684–685.

Remak, R. (1847). "Ueber ein selbständiges Darmnervensystem." Berlin (original not seen, see Marage, 1889.)

Richards, S. A. (1968). Vagal control of thermal panting in mammals and birds. *J. Physiol. (London)* **199,** 89–101.

Richards, S. A. (1969). Vagal function during respiration and the effects of vagotomy in the domestic fowl (*Gallus domesticus*). *Comp. Biochem. Physiol.* **29,** 955–964.

Richards, S. A. (1970a). The role of hypothalamic temperature in the control of panting in the chicken exposed to heat. *J. Physiol. (London)* **211,** 341–358.

Richards, S. A. (1970b). The biology and comparative physiology of thermal panting. *Biol. Rev. Cambridge Phil. Soc.* **45,** 223–264.

Richards, S. A., and Sykes, A. H. (1967a). Response of the domestic fowl (*Gallus domesticus*) to occlusion of the cervical arteries and veins. *Comp. Biochem. Physiol.* **21,** 39–50.

Richards, S. A., and Sykes, A. H. (1967b). The effects of hypoxia, hypercapnia and asphyxia in the domestic fowl (*Gallus domesticus*). *Comp. Biochem. Physiol.* **21,** 691–701.

Rochas, F. (1885a). Sur quelques particularités relatives aux connexions des ganglions cervicaux du grande sympathique et à la distribution de leurs rameaux, afférents et efférents chez l'*Anas boschas*. *C. R. Acad. Sci.* **100,** 649–651.

Rochas, F. (1885b). Des nerfs qui ont été appelés Vidiens chez les oiseaux. *C. R. Acad. Sci.* **101,** 573–575.

Rochas, F. (1885c). Du mode de distribution de quelques filets sympathiques intracraniens et de l'existence d'une racine sympathique du ganglion ciliare chez l'oie. *C. R. Acad. Sci.* **101,** 829–831.

Rochas, F. (1886). De l'existence chez les oiseaux d'une série de ganglions cephalique de nature sympathique, correspondent aux nerfs craniens segmentaires. *C. R. Acad. Sci.* **102,** 1028–1031.

Rochas, F. (1887). De la signification morphologique du ganglion cervical supérieur et de la nature de quelques-uns filets qui y aboutissent ou en émanent chez divers vertébrés. *C. R. Acad. Sci.* **104,** 865–868.

Rodriquez-Perez, A. P., and Pellegrino de Iraldi, A. (1961). Some data on the electron microscopy of the Herbst corpuscles in the duck's beak. *Trab. Inst. Cajal Invest. Biol.* **53,** 191–195.

Rossi, O. (1922). On the afferent paths of the sympathetic system, with special reference to nerve cells of spinal ganglia sending their peripheral processes into the rami communicantes. *J. Comp. Neurol.* **34,** 493–505.

Rouget, C. (1862). Note sur la terminaison des nerfs moteurs, dans les muscles chez les reptiles, les oiseaux et les mammifères. *C. R. Acad. Sci.* **55,** 548.

Saglam, M. (1968). Morphologische und quantitative Untersuchungen über die Muskelspindeln in der Nackenmuskulatur (M. biventer cervicis, M. rectus capitis dorsalis und M. rectus capitis lateralis) des Bunt- und Blutspechtes. *Acta Anat.* **69,** 87–104.

Sano, Y., Odake, G., and Yonezawa, T. (1967). Fluorescence microscopic observations of catecholamines in cultures of the sympathetic chains. *Z. Zellforsch. Mikrosk.-Anat.* **80,** 345–352.

Sato, H. (1969). The role of the autonomic nerves on the motility of the stomach of the domestic fowl. *Jap. J. Vet. Res.* **17,** 88–89.

Sato, H., Ohga, A., and Nakazato, Y. (1970). The excitatory and inhibitory innervation of the stomachs of the domestic fowl. *Jap. J. Pharmacol.* **20**, 382–397.
Saxod, R. (1967). Histogenèse des corpuscules sensoriels cutanés chez le poulet et le canard. *Arch. Anat. Microsc. Morphol. Exp.* **56**, 153–166.
Saxod, R. (1968). Ultrastructure des corpuscules sensoriels cutanés de Herbst et de Grandry chez le canard. *Arch. Anat. Microsc. Morphol. Exp.* **57**, 379–400.
Saxod, R. (1970). Etude au microscope électronique de l'histogenèse du corpuscule sensoriel cutané de Grandry chez le canard. *J. Ultrastruct. Res.* **32**, 477–496.
Schartau, O. (1938). Die periphere Innervation der Vogelhaut. *Zoologica, Stuttgart* **95**, 1–17.
Schmidt-Nielsen, K. (1960). The salt-secreting gland of marine birds. *Circulation* **21**, 955–967.
Schwalbe, G. (1879). Das Ganglion oculomotorii, ein Beitrag zur vergleichenden Anatomie der Kopfnerven. *Jena. Z. Naturwiss.* **13**, 173–268.
Seto, H. (1931). Anatomisch-histologische Studien über das Ganglion ciliare der Vögel nebst seinen sinn- und austretenden Nerven. I. Mitt. Bei den erwachsenen Hühnern. *J. Orient. Med.* **15**, 123–129.
Shimada, M. (1969). Physiological significance of intestinal inner plexus. (In Japanese.) *Bull. Univ. Osaka Prefect., Ser. B* **21**, 221–277.
Shioda, T., and Nishida, S. (1967). The innervation of the adrenal cortex. *Arch. Histol.* **28**, 23–44.
Shvalev, V. N. (1965). Aktual'nye problemy morfologii i prirody nervnogo apparata pochek. *Arkh. Anat., Gistol. Embriol.* **49**, 54–64.
Silver, A. (1963). A histochemical investigation of cholinesterases at neuromuscular junctions in mammalian and avian muscle. *J. Physiol. (London)* **169**, 386–393.
Simonetta, B. (1933). Ricerche sull'origine e sullo sviluppo del nervo olfattivo negli uccelli. Esistono negli uccelli il nervo terminale e l'organo di Jacobson? *Arch. Ital. Anat. Embriol.* **31**, 396–424.
Sinha, M. P. (1958). Vagal control of respiration as studied in the pigeon, *Helv. Physiol. Pharmacol. Acta* **16**, 58–78.
Sivaram, S. (1968). Histochemical studies on the developing adrenal gland of *Gallus domesticus*. *Histochemie* **12**, 316–325.
Sjöstrand, N. O. (1965). High noradrenaline content in the vas deferens of the cock and the tortoise. *Experientia* **21**, 96.
Sklenska, A. (1969). Über Unterschiede in Häufigkeit, Verteilung and Struktur der Rezeptoren in einigen grossen Gelenken der Extremitäten von Wirbeltieren. *Z. Mikrosk.-Anat. Forsch.* **80**, 249–259.
Slonaker, J. R. (1918). A physiological study of the anatomy of the eye and its accessory parts in the English Sparrow *Passer domesticus*. *J. Morphol.* **31**, 351–460.
Smith, M. L. (1941). Anatomy of the brain and cranial nerves of the turkey. *Univ. Colo. Stud., Ser. A* **26**, 135.
Solovieva, I. A. (1965). Razvitie afferentnoi innervatsii pishchevoda tsyplyat. *Arkh. Anat., Gistol. Embriol.* **49**, 64–70.
Sommer, J. R., and Johnson, E. A. (1969). Cardiac muscle. A comparative ultrastructural study with special reference to frog and chicken hearts. *Z. Zellforsch. Mikrosk. Anat.* **98**, 437–468.
Ssinelnikow, R. (1928a). Die Herznerven der Vögel. *Z. Anat. Entwicklungsgesch.* **86**, 540–562.
Ssinelnikow, R. (1928b). Die intramurale Nervensystem des Vogelherzens. *Z. Anat. Entwicklungsgesch.* **86**, 563–578.

Staderini, R. (1889). Sopra la distribuzione dei nervi glosso-faringeo vago e ipoglosso in alcuni rettili ed uccelli. *Atti Accad. Fisiocrit. Siena* [4] **1**, 585–599.

Stammer A. (1961a). Untersuchungen über die Struktur und die Innervation der Epiphyse bei Vögeln. *Acta Univ. Szeged, Acta Biol.* **7**, 65–76.

Stammer, A. (1961b). Die Nervenendorgane der Vogelhaut. *Acta Univ. Szeged, Acta Biol.* **7**, 115–131.

Stammer, A. (1962). Nervenverbindungen in der Tunica Vasculosa. *Acta Univ. Szeged, Acta Biol.* **8**, 143–159.

Stammer, A. (1964). Ein Beitrag zur Struktur und mikroskopischen Innervation der Harderschen Drüse der Vögel. *Acta Univ. Szeged, Acta Biol.* **10**, 99–107.

Staudacher, E. V. (1940). Contributo sperimentale alla conoscenza dell'origine della catena simpatica laterovertebrale, con particolare riguardo al sistema pregangliare. *Arch Ital. Anat. Embriol.* **43**, 99–118.

Stiemens, M. J. (1934). Anatomische Untersuchungen über die vago-sympathische Innervation der Baucheingeweide bei den Vertebraten. *Verh. Kon. Ned. Akad. Wetensch., Afd. Natuurk., Sect. II* **33**, 1–356.

Stoeckel, M. E., and Porte, A. (1967). Sur l'ultrastructure des corps ultimobranchiaux du poussin. *C. R. Acad. Sci., Ser. D* **265**, 2051–2053.

Stoeckel, M. E., and Porte, A. (1969a). Etude ultrastructurale des corps ultimobranchiaux du poulet. I. Aspect normal et développement embryonnaire. *Z. Zellforsch. Mikrosk. Anat.* **94**, 495–512.

Stoeckel, M. E., and Porte, A. (1969b). Localisation ultimobranchiale et thyroïdienne des cellules C (cellules à calcitonine) chez deux Columbidae: Le pigeon et le tourtereau. Etude au microscope électronique. *Z. Zellforsch. Mikrosk. Anat.* **102**, 376–386.

Streeter, G. L. (1904). The structure of the spinal cord of the Ostrich. *Amer. J. Anat.* **3**, 1–27.

Stübel, H. (1910). Beiträge zur Kenntnis der Physiologie des Blutkreislaufes bei verschiedenen Vogelarten. *Pflueger's Arch. Gesamte Physiol. Menschen Tiere* **135**, 249–365.

Stübel, H. (1911). Der Erregungsvorgang in der Magenmuskulatur nach Versuchen am Frosch- und am Vogelmagen. *Pflueger's Arch. Gesamte Physiol. Menschen Tiere* **143**, 381–394.

Sturkie, P. D., and Freedman, S. L. (1962). Effects of transection of pelvic and lumbosacral nerves on ovulation and oviposition in the fowl. *J. Reprod. Fert.* **4**, 81–85.

Sturkie, P. D., Poorvin, D., and Ossorio, N. (1970). Levels of epinephrine and norepinephrine in blood and tissues of duck, pigeon, turkey, and chicken. *Proc. Soc. Exp. Biol. Med.* **135**, 267–270.

Swan, J. (1835). "Illustrations of the Comparative Anatomy of the Nervous System." London (original not seen, see Marage, 1889.)

Sykes, A. H. (1955a). The effect of adrenaline on oviduct motility and egg production in the fowl. *Poultry Sci.* **34**, 662–668.

Sykes, A. H. (1955b). Further observations on reflex bearing-down in the fowl. *J. Physiol. (London)* **128**, 249–257.

Szantroch, Z. (1927). Morphologie des Darmnerven beim Hühnchen. *Bull. Int. Acad. Pol. Sci. Lett., Cl. Sci. Math. Natur., Ser. B* **3**, 211–282.

Szantroch, Z. (1929). L'histogenèse des ganglions nerveux du coeur. *Bull. Int. Acad. Pol. Sci. Lett., Cl. Sci. Math. Natur., Ser. B* **5**, 417–431.

Szentágothai, J. (1964). The structure of the autonomic interneuronal synapse. *Acta Neuroveg.* **26**, 338–359.

Szepsenwol, J., and Bron, A. (1935a). Le premier contact du système nerveux vagosympathique avec l'appareil cardio-vasculaire chez les embryons d'oiseaux (canard et poulet). *C. R. Soc. Biol.* **118**, 946-948.

Szepsenwol, J., and Bron, A. (1935b). L'origine des cellules nerveuses sympathiques dans le coeur des oiseaux. *C. R. Soc. Biol.* **118**, 1030-1031.

Szepsenwol, J., and Bron, A. (1936). Origine et nature de l'innervation primitive du coeur chez les embryons d'oiseaux. *Rev. Suisse Zool.* **43**, 1-23.

Takahashi, K. (1967). Special somatic spine synapses in the ciliary ganglion of the chick. *Z. Zellforsch. Mikrosk. Anat.* **83**, 70-75.

Takahashi, K., and Hama, K. (1965a). Some observations on the fine structure of the synaptic area in the ciliary ganglion of the chick. *Z. Zellforsch. Mikrosk. Anat.* **67**, 174-184.

Takahashi, K., and Hama, K. (1965b). Some observations on the fine structure of nerve cell bodies and their satellite cells in the ciliary ganglion of the chick. *Z. Zellforsch. Mikrosk. Anat.* **67**, 835-843.

Takino, M. (1932). Über der Innervation der Blutgefässe der Lunge beim Vögel (Taube und Haushuhn), besonder über das Vorkommen der Ganglienzellen in oder der Wand der Venae und Arteriae pulmonales und über Verbreitung der Blutgefässenerven daselbst. II. Mitteilung. *Acta Sch. Med. Univ. Imp. Kioto* **15**, 308-320.

Taxi, J. (1965). Contribution à l'étude des connexions des neurones moteurs du système nerveux autonome. *Ann. Sci. Natur., Zool. Biol. Anim.* [12] **7**, 413-674.

Tcheng, K.-T. (1963). Receptors of the pulmonary artery in birds. *Acta Univ. Szeged, Acta Biol.* **9**, 281-289.

Tcheng, K.-T., and Fu, S.-K. (1962). The structure and innervation of the aortic body of the Yellow-breasted Bunting. *Sci. Sinica* **11**, 221-232.

Tcheng, K. T., Fu, S. K., and Chen, T. Y. (1963a). Supracardial encapsulated receptors of the aorta and the pulmonary artery in birds. *Sci. Sinica* **12**, 73-82.

Tcheng, K.-T., Fu, S.-K., and Chen, T.-Y. (1963b). On the vasculature of the aortic bodies in birds. *Sci. Sinica* **12**, 339-346.

Tcheng, K.-T., Chen, T.-Y., and Fu, S.-K. (1965a). Intramural baroreceptors of the pulmonary arteries in birds. *Sci. Sinica* **14**, 938-940.

Tcheng, K.-T., Fu, S.-K., and Chen, T.-Y. (1965b). Intravagal receptors and sensory neurons in birds. *Sci. Sinica* **14**, 1323-1331.

Tello, J. F. (1922). Die Enstehung der motorischen und sensiblen Nervenendigungen. I. In dem lokomotorischen System der höheren Wirbeltiere. Muskuläre Histogenese. *Z. Anat. Entwicklungsgesch.* **64**, 348-440.

Tello, J. F. (1924). La précocité embryonnaire du plexus d'Auerbach et ses différences dans les intestins antérieur et postérieur. *Trab. Inst. Cajal Invest. Biol.* **22**, 317-328.

Tello, J. F. (1925). Sur la formation des chaines primaires et secondaires du grand sympathique dans l'embryon de poulet. *Trab. Inst. Cajal Invest. Biol.* **23**, 1-28.

Terni, T. (1923). Ricerche anatomiche sul sistema nervoso autonomo degli uccelli. I. Il sistema pregangliare spinale. *Arch. Ital. Anat. Embriol.* **20**, 433-510.

Terni, T. (1924). Il ganglio toracico e la porzione cervicale del vago negli uccelli. *Arch. Ital. Anat. Embriol.* **21**, 404-434.

Terni, T. (1927). Il corpo ultimobranchiale degli uccelli. Ricerche embriologiche anatomiche e istologiche su *Gallus domesticus. Arch. Ital. Anat. Embriol.* **24**, 407-531.

Terni, T. (1929a). Recherches morphologiques sur le sympathique cervicale des oiseaux et sur l'innervation autonomes de quelques organes glandulaires de cou. *C. R. Ass. Anat.* **24**, 473-480.

Terni, T. (1929b). Ricerche istologiche sull'innervazione del timo dei sauropsidii. *Z. Zellforsch. Mikrosk. Anat.* **9**, 377–424.

Terni, T. (1931). Il simpatico cervicale degli amnioti (richerche di morfologia comparata). *Z. Anat. Entwicklungsgesch.* **96**, 289–426.

Terzuolo, C. (1951). Richerche sul ganglio ciliare degli uccelli. Connessioni, mutamenti in relazione all'eta e dopo recisione delle fibre pregangliari. *Z. Zellforsch. Mikrosk. Anat.* **36**, 255–267.

Tetzlaff, M. J., Peterson, R. A., and Ringer, R. K. (1965). A phenylhydrazine-leucofuchsin sequence for staining nerves and nerve endings in the integument of poultry. *Stain Technol.* **40**, 313–316.

Thébault, V. (1898). Etude des rapports qui existent entre les systèmes pneumogastrique et sympathique chez les oiseaux. *Ann. Sci. Natur., Zool.* [8] **6**, 1–243.

Tiedemann, F. (1810). "Zoologie," Vol. 2. Anatomie und Naturgeschichte der Vögel. Landshut, Heidelberg.

Tiegs, O. W. (1953). Innervation of voluntary muscle. *Physiol. Rev.* **33**, 90–144.

Tixier-Durivault, A. (1942). Contribution à l'étude de l'innervation du coeur chez les oiseaux. *Oiseau Rev. Fr., Ornithol.* **12**, 81–97.

Traciuc, E. (1969). La structure de l'epididyme de *Coloeus monedula* (Corvidae). *Anat. Anz.* **125**, 49–67.

Tummons, J. L., and Sturkie, P. D. (1968). Cardio-accelerator nerve stimulation in chickens. *Life Sci.* **7**, 377–380.

Tummons, J. L., and Sturkie, P. D. (1969). Nervous control of heart rate during excitement in the adult white leghorn cock. *Amer. J. Physiol.* **216**, 1437–1440.

Tummons, J. L., and Sturkie, P. D. (1970). Beta adrenergic and cholinergic stimulants from the cardio-accelerator nerve in the domestic fowl. *Z. Vergl. Physiol.* **68**, 268–271.

Uchida, S. (1927). Ueber die Entwicklung des sympathischen Nervensystems. I Mitteilung. Ueber die Entwicklung des Sympathicus bei Vögeln. *Acta Sch. Med. Univ. Imp. Kioto* **10**, 63–94.

Ueck, M. (1970). Weitere Untersuchungen zur Feinstruktur und Innervation des Pinealorgans von *Passer domesticus* L. *Z. Zellforsch. Mikrosk. Anat.* **105**, 276–302.

Urra, M. (1923). Estructura y detalles fisiológicos de las placas motrices del ciliar en el hombre, mamíferos y aves de lento y rápido vuelo. *Arch. Espan. Oftalmol.* **23**, 33–42.

van Campenhout, E. (1930a). Contribution to the problem of the development of the sympathetic nervous system. *J. Exp. Zool.* **56**, 295–320.

van Campenhout, E. (1930b). Historical survey of the development of the sympathetic nervous system. *Quart. Rev. Biol.* **5**, 23–50.

van Campenhout, E. (1930c). Historical survey of the development of the sympathetic nervous system (concluded). *Quart. Rev. Biol.* **5**, 217–234.

van Campenhout, E. (1931). Le développement du système nerveux sympathique chez le poulet. *Arch. Biol.* **42**, 479–506.

van Campenhout, E. (1932). Further experiments on the origin of the enteric nervous system in the chick. *Physiol. Zool.* **5**, 333–353.

van Campenhout, E. (1933). The innervation of the digestive tract in the 6-day chick embryo. *Anat. Rec.* **56**, 111–118.

van Campenhout, E. (1955). Innervation du poumon de la poule adulte. *C. R. Ass. Anat.* **42**, 1338–1341.

van Campenhout, E. (1956). Contribution à l'étude de l'innervation du poumon chez la poule adulte. *Arch. Biol.* **67**, 1–19.

1. PERIPHERAL AND AUTONOMIC NERVOUS SYSTEMS

van Campenhout, E. (1962). Innervation de l'ovaire de la poule. *C. R. Ass. Anat.* **48**, 1333-1337.
van Campenhout, E. (1964). L'innervation de l'ovaire de la poule. *Ann. Soc. Roy. Zool. Belg.* **95**, 9-28.
van Campenhout, E. (1967). L'innervation de l'ovaire des oiseaux. *C. R. Ass. Anat.* **52**, 1180-1183.
van den Akker, L. M. (1970). The termination of three long descending systems in the cord of the pigeon. *Psychiat. Neurol. Neurochir.* **72**, 11-16.
van der Linden, P. (1934). Recherches sur les réflexes sino-carotidiens chez les oiseaux. *Arch. Int. Physiol.* **40**, 59-82.
van Gehuchten, A. (1893). Les éléments nerveux moteurs des racines postérieures. *Anat. Anz.* **8**, 215-225.
Vegetti, A., and Palmieri, G. (1965). The mesencephalic nucleus of the trigeminus of birds. *Riv. Biol.* **58**, Suppl., 93-115.
Vignal, W. (1881). Appareil ganglionnaire du coeur des vertébrés. *Arch. Physiol. Norm. Pathol.* **6**, 947-962.
von Golenhofen, K. (1967). Elektrophysiologische Untersuchungen am elastischmuskulösen System der Vogelhaut. *Pfluegers Arch. Gesamte Physiol. Menschen Tiere* **297**, R5.
von Golenhofen, K. (1968). Die neuromuskuläre Erregungsübertragung an der glatten Federmuskulatur der Taube. *Pfluegers Arch. Gesamte Physiol. Menschen Tiere* **300**, 79.
von Golenhofen, K., and Petry, G. (1968). Physiologische Untersuchungen an der innervierten glatten Muskulatur der Vogelfedern (Mm. pennarum). *Experientia* **24**, 1137-1138.
von Lenhossék, M. (1890). Über Nervenfasern in den hinteren Wurzeln, welche aus dem Vorderhorn entspringen. *Anat. Anz.* **5**, 360-362.
von Lenhossék, M. (1911). Das ganglion ciliare der Vögel. *Arch. Mikrosk. Anat. Entwicklungsmech.* **76**, 745-769.
Walker, S. M., and Schrodt, G. R. (1966). T-System connexions with the sarcolemma and sarcoplasmic reticulum. *Nature (London)* **211**, 935-938.
Wang, C. C., and Ho, W. Y. (1966). The nerve endings and receptors of duck's bill and tongue. (In Chinese.) *Acta Anat. Sinica* **9**, 250-257.
Watanabe, T. (1960). Comparative and topographical anatomy of the fowl. VII. On the peripheral course of the vagus nerve in the fowl. (In Japanese.) *Jap. J. Vet. Sci.* **22**, 145-154.
Watanabe, T. (1964). Comparative and topographical anatomy of the fowl. XVII. Peripheral courses of the hypoglossal, accessory and pharyngeal nerves. (In Japanese.) *Jap. J. Vet. Sci.* **26**, 249-258.
Watanabe, T. (1968). A study of retrograde degeneration in the vagal nuclei of the fowl. *Jap. J. Vet. Sci.* **30**, 331-340.
Watanabe, T., and Yasuda, M. (1968). Comparative and topographical anatomy of the fowl. LI. Peripheral course of the olfactory nerve in the fowl. (In Japanese.) *Jap. J. Vet. Sci.* **30**, 275-279.
Watanabe, T., and Yasuda, M. (1970). Comparative and topographical anatomy of the fowl. XXVI. Peripheral course of the trigeminal nerve. (In Japanese.) *Jap. J. Vet. Sci.* **32**, 43-58.
Watanabe, T., Isomura, G., and Yasuda, M. (1967). Comparative and topographical anatomy of the fowl. XXX. Distribution of nerves in the oculomotor and ciliary muscles. (In Japanese.) *Jap. J. Vet. Sci.* **29**, 151-158.

Watzka, M. (1933). Vergleichende Untersuchungen über den ultimobranchialen Körper. *Z. Mikrosk.-Anat. Forsch.* **34**, 485–533.

Watzka, M. (1934). Von Paraganglion Caroticum. *Anat. Anz.* **78**, Ergänzungsheft, 108–120.

Webb, M. (1957). The ontology of the cranial bones, cranial peripheral and cranial parasympathetic nerves, together with a study of the visceral muscles of *Struthio*. *Acta Zool. (Stockholm)* **38**, 1–203.

Weber, A. (1940). Développement du plexus sympathique gastro-duodenal chez l'embryon du poulet. *C. R. Soc. Biol.* **133**, 537–538.

Weber, E. H. (1817). Beitrag zur vergleichenden Anatomie des sympathischen Nerven. *Arch. Anat. Physiol. (Meckel's Deut. Arch. Physiol.)* **3**, 396–417.

Wechsler, W., and Schmekel, L. (1966a). Elektronenmikroskopischer Nachweis spezifischer Grana in den Sympathicoblasten der Grenzstrangganglien von Hühnerembryonen. *Experientia* **22**, 296–297.

Wechsler, W., and Schmekel, L. (1966b). Elektronenmikroskopischer Beitrag zur Frühentwicklung der Sympatiloblasten des Grenzstranges bei verschiedenen Vogelarten. *Anat. Anz.* **120**, Erganzungsheft, 317–325.

Wechsler, W., and Schmekel, L. (1967). Elektronenmikroskopische Untersuchung der Entwicklung der Vegetativen (Grenzstrang-) und spinalen Ganglien bei *Gallus domesticus*. *Acta Neuroveg.* **30**, 427–433.

Wight, P. A. L., and Mackenzie, G. M. (1970). Dual innervation of the pineal of the fowl, *Gallus domesticus*. *Nature (London)* **228**, 474–475.

Wirtz, B. (1968). Ueber das Vorkommen von Glomera im Herzen des Menschen, der Taube und der Huhnes. *Arch. Kreisslaufforsch.* **56**, 89–105.

Yamasaki, I. (1939). Über den M. dilator pupillae beim Menschen und höheren Wirbeltieren. *Acta Soc. Ophthalmol. Jap.* **42**, 17–20.

Yamauchi, A. (1969). Innervation of the vertebrate heart as studied with the electron microscope. *Arch. Histol.* **31**, 83–117.

Yasuda, M. (1960). Comparative and topographical anatomy of the fowl. III. On the nervous supply of the thoracic limb in the fowl. (In Japanese.) *Jap. J. Vet. Sci.* **22**, 89–102.

Yasuda, M. (1961). Comparative and topographical anatomy of the fowl. XI. On the nervous supply of the hind limbs. (In Japanese.) *Jap. J. Vet. Sci.* **23**, 145–155.

Yasuda, M. (1964). Comparative and topographical anatomy of the fowl. XXXIV. Distribution of cutaneous nerves in the fowl. (In Japanese.) *Jap. J. Vet. Sci.* **26**, 241–254.

Yasuda, M., and Lepkovsky, S. (1969). The chicken diencephalon in stereotaxic coordinates. *Jap. J. Zootech. Sci.* **40**, 417–431.

Yeh, Y., Huang, S. K., and Feng, T. P. (1963). Apparent transformation from Felderstruktur to Fibrillenstruktur with stretch and the converse change with shortening in certain muscles of the chick. *Sci. Sinica* **12**, 1242–1244.

Yntema, C. L., and Hammond, W. S. (1945). Depletions and abnormalities in the cervical sympathetic system of the chick following extirpation of neural crest. *J. Exp. Zool.* **100**, 237–263.

Yntema, C. L., and Hammond, W. S. (1954). The origin of intrinsic ganglia of trunk viscera from vagal neural crest in the chick embryo. *J. Comp. Neurol.* **101**, 515–541.

Yntema, C. L., and Hammond, W. S. (1955). Experiments on the origin and development of the sacral autonomic nerves in the chick embryo. *J. Exp. Zool.* **129**, 375–414.

Yousuf, N. (1965). The conducting system of the heart of the House Sparrow, *Passer domesticus indicus*. *Anat. Rec.* **152**, 235–250.

Zeglinski, N. (1885). Experimentelle Untersuchungen über die Irisbewegung. *Arch. Anat. Physiol., Physiol. Abt.* **9**, 1–37.

Zelená, J., Vyklický, L., and Jirmanová, I. (1967). Motor end-plates in fast and slow muscles of the chick after cross-union of their nerves. *Nature (London)* **214**, 1010–1011.

Zenker, W. (1968). Zur Charakteristik vegetativ innervierter quergestreifter Muskulatur. Die innere Augenmuskulatur des Huhnes. *Anat. Anz.* **121**, Ergänzungsband, 19–23.

Zenker, W., and Krammer, E. (1967). Untersuchungen über Feinstruktur und Innervation der inneren Augenmuskulatur des Huhnes. *Z. Zellforsch. Mikrosk. Anat.* **83**, 147–168.

Zeitzschmann, O. (1910). Der Musculus dilator pupillae des Vögels. *Arch. Vergl. Ophthalmol.* **1**, 9–19.

Zorzoli, G. C., and Maggi, G. (1950). Contributo allo studio parasimpatico spinale dei palmipedi. *Monit. Zool. Ital.* **57**, 73–77.

After this review had been finished the following papers were published or discovered. They are not discussed here, but are included for the sake of completeness.

Akester, A. R., and Akester, B. (1971a). Some ultrastructural characteristics of conducting cells in the avian heart. *J. Anat.* **108**, 616–617.

Akester, A. R., and Akester, B. (1971b). Double innervation of the avian cardiovascular system. *J. Anat.* **108**, 618–619.

Asahiro, N. (1970). Developmental study of the iris and ciliary body of the man and bird. *Nippon Ika Daigaku Zasshi* **37**, 143–152.

Bartlett, A. L., and Hassan, T. (1971). Contraction of chicken rectum to nerve stimulation after blockade of sympathetic and parasympathetic transmission. *Quart. J. Exp. Physiol. Cog. Med. Sci.* **56**, 178–183.

Bock, W. J., and Hikida, R. S. (1968). An analysis of twitch and tonus fibers in the hatching muscle. *Condor* **70**, 211–222.

Bretz, W. C., and Schmidt-Nielsen, K. (1971). Bird respiration: flow patterns in the duck lung. *J. Exp. Biol.* **54**, 103–118.

Butler, P. J., and Jones, D. R. (1971). The effect of variations in heart rate and regional distribution of blood flow on the normal pressor response in diving ducks. *J. Physiol. (London)* **214**, 457–480.

Crain, S. M. (1971). Intracellular recordings suggesting synaptic functions in chick embryo spinal sensory ganglion cells isolated *in vitro*. *Brain Res.* **26**, 188–191.

Csoknya, M., Horvăth, I., and Halăsz, N. (1971). A contribution to the knowledge of receptors of the glandular stomach of birds. *Acta. Anat.* **79**, 126–137.

Dahl, E. (1971). Studies on the fine structure of ovarian interstitial tissue. I. A comparative study of the fine structure of the ovarian interstitial tissue in the rat and the domestic fowl. *J. Anat.* **108**, 275–290.

Dorward, P. K., and McIntyre, A. K. (1971). Responses of vibration sensitive receptors in the interosseous region of the duck's hind limb. *J. Physiol. (London)* **219**, 77–88.

Eaton, J. A., Jr., Fedde, M. R., and Burger, R. E. (1971). Sensitivity to inflation of the respiratory system of the chicken. *Resp. Physiol.* **11**, 167–177.

Geffen, L. B., and Hughes, C. C. (1972). Degeneration of sympathetic nerves *in vitro*, and development of smooth muscle supersensitivity to noradrenaline. *J. Physiol. (London)* **221**, 71–84.

Groth, H. P. (1972). Licht und fluoreszenzmikroskopische Untersuchungen zur Innervation des Luftsacksystems der Vögel. *Z. Zellforsch. Mikrosk. Anat.* **127,** 87–115.

Guttmann, E., Hájek, I., and Vitek, V. (1970). Compensatory hypertrophy of latissimus dorsi posterior muscle induced by elimination of the latissimus dorsi anterior muscle of the chicken. *Physiol. Bohemoslov.* **19,** 483–490.

Hanwell, A., Linzell, J. L., and Peaker, M. (1971a). Salt-gland secretion and blood flow in the goose. *J. Physiol. (London)* **213,** 373–388.

Hanwell, A., Linzell, J. L., and Peaker, M. (1971b). Cardiovascular responses to salt-loading in conscious domestic geese. *J. Physiol. (London)* **213,** 389–398.

Hedlund, L., Ralph, C. I., Chepko, J., and Lynch, H. J. (1971). A diurnal serotonin cycle in the pineal body of the Japanese Quail: Photoperiod phasing and the effect of superior cervical ganglionectomy. *Gen. Comp. Endocrinol.* **16,** 52–58.

Hikida, R. S., and Bock, W. J. (1970). The structure of pigeon muscle and its changes due to tenotomy. *J. Exp. Zool.* **175,** 343–356.

Hikida, R. S., and Bock, W. J. (1971). Innervation of the avian tonus latissimus dorsi anterior muscle. *Amer. J. Anat.* **180,** 269–280.

Jewett, P. H., Sommer, J. R., and Johnson, E. A. (1971). Cardiac muscle. Its ultrastructure in the finch and humming bird with special reference to the sarcoplasmic reticulum. *J. Cell Biol.* **49,** 50–65.

Jirmanova, I., and Zelena, J. (1970). Effect of denervation and tonotomy on slow and fast muscles of the chicken. *Z. Zellforsch. Mikrosk. Anat.* **106,** 333–347.

Jirmanová, I., Hník, P., and Zelená, J. (1971). Implantation of "fast" nerve into slow muscle in young chickens. *Physiol. Bohemoslov.* **20,** 199–204.

Kao, L. W. L., and Nalbandov, A. V. (1972). The effect of antiadrenergic drugs on ovulation in hens. *Endocrinology* **90,** 1343–1349.

Kano, M., and Shimada, Y. (1971). Innervation of skeletal muscle cells differentiated *in vitro* from chick embryo. *Brain Res.* **27,** 402–405.

Koenig, J. (1970). Contribution à l'étude de la morphologie des plaques motrices des glands dorsaux antérieur et postérieur du Poulet aprés innervation croisée. *Arch. Anat. Microsc. Morphol. Exp.* **59,** 403–426.

Lee, W. C., Lew, J. M., and Yoo, C. S. (1970). Studies on myocardial catecholamines related to species ages and sex. *Arch. Int. Pharmacodyn. Ther.* **184,** 259–268.

McGrath, J. J. (1971). An isolated avian cardiac muscle preparation. *J. Appl. Physiol.* **30,** 583–584.

McGregor, D. D. (1971). Vascular reactivity to noradrenaline in mesenteric arteries of young chickens. *Proc. Univ. Otago Med. Sch.* **49,** 49–51.

McLelland, J., and Abdala, A. B. (1972). The gross anatomy of the nerve supply to the lungs of *Gallus domesticus*. *Anat. Anz.* **131,** 448–453.

Maier, A., and Eldred, E. (1971). Comparisons in the structure of avian muscle spindles. *J. Comp. Neurol.* **143,** 25–40.

Malinovský, L., and Cĕch, S. (1970). Fluoreszenzmikroskopische Untersuchungen an sensiblen Nervenendkörperchen. *Z. Mikrosk.-Anat. Forsch.* **82,** 7–16.

Malinovský, L., and Zemanek, R. (1970). Sensory nerve endings in the joint capsules of the large limb joints in the domestic hen (*Gallus domesticus*) and the Rook (*Corvus frugilegus*). *Folia Morphol. (Prague)* **18,** 206–212.

Mayr, R. (1968). Morphologische und physiologische Untersuchungen über den aktiven Bewegungsapparat der Nickhaut des Huhnes. *Gegenbaurs Jahrb.* **112,** 113–122.

Melander, A., Owman, C., and Sundler, F. (1971). Concomittant depletion of dopamine and secretory granules from cells in the ultimobranchial gland of vitamin D_2-treated chicken. *Histochemie* **25,** 21–31.

Molony, V. (1971). Some characteristics of the CO_2-sensitive receptors present in the lower respiratory tract of the chicken. *J. Physiol. (London)* **219**, 35P–36P.

Morisset, J., and Webster, P. D. (1970). Effects of atropine on pigeon pancreas. *Amer. J. Physiol.* **219**, 1286–1291.

Ohashi, H. (1971). An electrophysiological study of transmission from intramural excitatory nerves to the smooth muscle cells of the chicken oesophagus. *Jap. J. Pharmacol.* **21**, 585–596.

Paul, E. (1971). Neurohistologische und fluoreszenzmikroskopische Untersuchungen über die Innervation des Glykogenkörpers der Vögel. *Z. Zellforsch. Mirkosk. Anat.* **112**, 516–525.

Pessacq, T. P. (1969). Quelques aspects de la vascularisation du corps glycogénique de la moelle épinière des oiseaux. *Acta Anat.* **72**, 33–37.

Pilar, G., and Vaughan, P. D. (1971). Ultrastructure and contractures of the pigeon iris striated muscle. *J. Physiol. (London)* **219**, 253–266.

Szantroch, Z. (1934). Über den feineren Bau der Ganglienknoten der Remakschen Darmnerven. *Z. Zellforsch. Mikrosk. Anat.* **20**, 417–422.

Takahashi, K. (1970). Fine structure of the iris muscles and their innervation in the pigeon eye. (In Japanese.) *Acta Soc. Ophthalmol. Jap.* **74**, 828–838.

Vyskočil, F., Vyklický, L., and Huston, R. (1971). Quantum content at the neuromuscular junction of fast muscle after cross-union with nerve of slow muscle in the chick. *Brain Res.* **25**, 443–445.

Wight, P. A. L., and MacKenzie, G. M. (1971). The histochemistry of the pineal gland of the domestic fowl. *J. Anat.* **108**, 261–274.

Chapter 2

THE AVIAN PINEAL ORGAN

M. Menaker and A. Oksche

I.	Function	80
	A. Control of Reproduction	82
	B. Photoreception	84
	C. Control of Circadian Rhythmicity	86
II.	Structure	89
	A. Cell Types	90
	B. Nervous Apparatus	101
	C. Secretory Apparatus	107
	D. Vasculature	109
III.	General Conclusions	109
References		114

The study of the vertebrate pineal organ is in an exciting, if temporarily frustrating, stage of development. Evidence has accumulated, slowly at first but much more rapidly during the past ten years, that indicates centrally important physiological functions for this organ which has historically been considered a functionally insignificant relic or, alternatively, the physical seat of the human soul. Unfortunately, we are still completely ignorant of many of the mechanisms involved in pineal function. The present tension between the soundly based intuition that the pineal's role is an important one and the very great ignorance of what that role actually is, is particularly apparent to the student of the avian pineal. The pineal of several mammalian species has been clearly shown to function in the control of the reproductive system. The pineals of at least some fish, amphibians, and reptiles are certainly photoreceptive; the avian pineal, on the other hand, may have both functions or neither and certainly has at least one other function, seemingly unique. In the birds, great variability in pineal structure among species is matched by an equal variability in the reported results of experimental manipulations, even in the same species. Because we have as yet no clear idea of the part the pineal plays in the overall physiological organization of birds, it is exceedingly difficult to separate significant variability from that which may be biologically trivial. The present situation is thus an unusually challenging one for the comparative physiologist and morphologist. In the search for clarification of the functions of the avian pineal, one is forced to "interpolate" from both evolutionary directions but must face the probability that each of the interpolations is likely to contain less truth than falsehood.

I. Function

We would like to begin this chapter in a somewhat unorthodox manner. We will consider, for the moment without regard to the quantity or quality of the evidence for each, the several functions for the avian pineal that have been suggested by modern biologists. They make an intriguing set.

1. *The avian pineal is involved in the control of reproductive function.* This hypothesis is of course suggested by the recent work which demonstrates such a role for the pineal of some mammals (see Wurtman *et al.*, 1968). This work, however, leaves, it should be stressed, a great many unanswered questions. In order to place this interpolation in proper perspective, one must note that the investigation in mammals has been directed almost exclusively at resolving the pineal's role in either the ontogeny of the reproductive system or in its maintenance in mature animals. Seasonal cycling of the state of the reproductive system, a phenomenon of much greater general significance in avian than in mammalian reproductive physiology, has only occasionally been considered by students of the mammalian pineal.

2. *The avian pineal is photoreceptive.* This hypothesis derives from interpolation from the reverse evolutionary direction. The

pineals of most of the lower vertebrates probably are photoreceptive structures. In a number of cases there is solid electrophysiological evidence that this is so (see Dodt et al., 1971). One must note, however, that their higher-order functions have not yet been discovered. Thus, we have very little idea what the lower vertebrates do with the information obtained by their photoreceptive pineals.

3. *The avian pineal is involved in the control of circadian rhythmicity.* Although this is the strongest of the three hypotheses, the information necessary to determine its generality is lacking. We know that pineal removal has profound effects on two overt circadian rhythms in one avian species and on at least one rhythm in two other species. As all three species are passeriforms (see p. 86), we do not know how the results apply to birds in general, nor do we know how much of the complex circadian system is affected in any one species. Comparable effects apparently do not occur in either reptiles or mammals.

The reader will recognize that these three hypothetical functions are not mutually exclusive. One, two, or all of them may in fact be

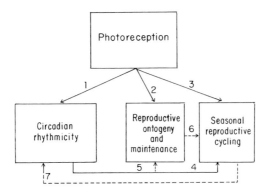

FIG. 1. Diagram of the known relationships (solid arrows) among photoreception, circadian rhythmicity, and two aspects of reproductive physiology in birds, as well as some of the more likely hypothetical relationships (dashed arrows). Arrow 1 refers primarily to the well-known effect of light cycles in synchronizing circadian rhythms; 2 refers to the effects of constant light and light cycles of various durations on ontogeny and maintenance of the reproductive system; 3 refers to the primary action of photoperiod in controlling seasonal reproductive rhythmicity in many species; 4 refers to the documented but still poorly understood role of the circadian system in the measurement of photoperiodic time; 5 refers to a possible role of the circadian system in ontogeny and a much more likely role in maintenance; 6 refers to the still uninvestigated possibility that events in reproductive ontogeny may influence later seasonal cycling; and 7 refers to the possibility of feedback from the seasonal rhythm on the circadian system.

correct. The possibility that they are all correct is what transforms them into an intriguing set, for there are already multiple well established relationships among rhythms, reproduction, and photoreception in birds. These relationships, along with others that are only hypothetical, are diagrammed in Fig. 1. At our present state of knowledge, we can think of no heuristically more valuable suggestion than that a good deal of what is shown in Fig. 1 is in fact taking place in the avian pineal. It must be admitted that the available evidence does not strongly support this view, but neither does it rule it out.

A. Control of Reproduction

1. Seasonal Cycles

The small number of published studies all report negative results. Surgical removal of the pineal or its electrolytic destruction does not appear to affect the course of testicular recrudescence in either the White-crowned Sparrow (*Zonotrichia leucophrys gambelii*) (Kobayashi, 1969; Oksche *et al.*, 1972), the House Finch (*Carpodacus mexicanus*) (Hamner and Barfield, 1970), or the House Sparrow (*Passer domesticus*) (Menaker *et al.*, 1970); testicular regression was also unaffected in *Zonotrichia leucophrys gambelii* (Kobayashi, 1969), in Harris's Sparrow (*Zonotrichia querula*) (Donham and Wilson, 1970), or the House Finch (Hamner and Barfield, 1970). In *Z. querula* (Donham and Wilson, 1970) it was shown that pinealectomy has no effect on either the testis weight of photorefractory birds held on short days or their recovery from photorefractoriness. In House Sparrows, Ralph and Lane (1969) could find no correlation between pineal cytology and the reproductive state of the birds, whereas Barfuss and Ellis (1971) have reported a seasonal cycle of HIOMT (hydroxyindole-*O*-methyltransferase) activity in the pineal; however, they feel this cycle is temporally related to the reproductive events in such a way as to render a causal relationship unlikely. In addition, there are unpublished reports by several investigators that confirm the general impression generated by the scanty literature of a lack of involvement of the pineal organ in the control of seasonal reproductive cycles. Nevertheless, it should be borne in mind that several other approaches will have to be tried before a firm conclusion can be drawn. There are at least four kinds of experiments that need to be done: (1) Other groups of birds must be examined. (2) Pinealectomized birds should be followed through an entire reproductive cycle; the possibility of very long-range effects cannot be excluded. (3) The effects of pinealectomy performed early in development on later reproductive cycling must be studied. (4) The pineal's possible role as a photoperiodic photoreceptor should be more thoroughly investigated. Despite these gaps in our knowledge, it seems less likely that the avian pineal is involved in the seasonal control of reproduction than in any of the other three functions diagrammed in Fig. 1.

2. Noncyclic Aspects of Reproduction

There is quite a large literature dealing with the influence of the avian pineal on the ontogeny of the reproductive system and on its maintenance, focused almost exclusively on four or five species of domestic birds. This literature is beset with difficulties due in large part, it seems fair to say, to a high proportion of poorly designed or anecdotally reported experiments. It contains very little that is not controversial. In effect, it is possible by the judicious choice of published data to support contentions that the avian pineal plays a progonadal role, an antigonadal role, no role at all, or the above three roles combined. As this literature has been very recently and competently reviewed by Ralph (1970) in a paper entitled with some wit "Structure and Alleged Functions of Avian Pineals," we will not attempt an inclusive review here. The reader desiring more direct access to this literature should consult Foá (1928), Shellabarger and Breneman (1950), Shellabarger (1953), Stalsberg (1965), and Sayler and Wolfson (1967).

In so far as it is possible to summarize the conflicting results obtained by means of surgical intervention, it seems that the pineal may have a progonadal role in young male chickens, an antigonadal role later in life, and no discernable role in females (Shellabarger, 1952, 1953). In contrast, the pineal of Japanese Quail appears to exert transitory progonadal effect in females (Sayler and Wolfson, 1967; see, also, Sayler and Wolfson, 1969). Recently, Cardinali et al. (1971) have reported a convincing effect of pinealectomy on the maintenance of testicular weight and in vitro biosynthesis of steroids in the domestic duck. Pinealectomy, performed during the breeding season, resulted, after a delay of 2 months, in pronounced testicular regression 4 months in advance of the sham-operated and intact controls. The changes produced in testicular steroidogenesis by pinealectomy were different from those that occurred during normal seasonal involution, and the annual cycle of testis functions of the pinealectomized birds appeared normal (at least until the middle of the following breeding period) subsequent to the abnormally early regression. It should be emphasized that published studies exist which appear to contradict all of the above statements with the exception of the Cardinali study.

The avian pineal contains at least some of the biochemical machinery familiar to students of the organ in mammals. Serotonin, melatonin, and several other biogenic amines have been detected, as have at least two associated enzymes, HIOMT and 5-hydroxytryptophan decarboxylase. Serotonin, melatonin, and HIOMT have been reported to vary diurnally in the presence of light cycles. Predictably, biologists have not failed to introduce some of the biogenic amines, notably melatonin, as well as whole or powdered pineal glands, into birds by one means or another. This is a scientifically hazardous undertaking, as not much is known of normal circulating levels, target organs, toxic side effects, nonspecific effects, or the influence of sex or stage of development on responsiveness, and the expected conflicting results have been obtained (Ralph, 1970).

Although the literature is confusing in the extreme, there are a large number of reports claiming some effect of pineal surgery or the introduction of exogeneous melatonin on the development of the reproductive system or its maintenance in domestic birds. Whether one wishes to interpret this literature *in toto* as experimental noise around a mean of zero (in terms of pineal involvement) or as the result of multiple inadequately controlled experiments which, because of great and unappreciated complexity of the system under study, have failed to create a clear and consistent picture of an important aspect of reproductive control, is a matter of personal preference. The authors are inclined to take the latter view.

B. PHOTORECEPTION

Evidence for the proposition that the avian pineal functions at least in part as a photoreceptor is also unfortunately conflicting. In this case, however, the conflict is not among variously derived experimental results but rather between the implications of the ultrastructural morphology, which at least in some species is strongly suggestive of such a function, and the small amount of physiological evidence, which is almost uniformly negative. Here we will deal only with the physiological evidence, reserving a discussion of the morphology for a later section.

The physiological experiments which concern us are of two kinds, which we will simply call direct and indirect. Direct experiments are those in which light is applied to the pineal and one or another parameter, in which a change might be expected, is measured. Indirect experiments involve surgically altering the pineal or associated structures and looking for effects that can be interpreted as being due to altered photoreceptivity of the organism.

Morita (1966) systematically explored the pineal of the pigeon with microelectrodes searching for an effect of light, applied directly to the pineal or to the lateral eyes, on electrical activity. He failed to find any effect, although he did detect spontaneous activity apparently arising deep within the organ. Ralph and Dawson (1968) obtained essentially the same results on illuminating the pineals of Japanese Quail and House Sparrows (see, also, D. I. Hamasaki, in Oksche and Kirschstein, 1969). Oishi and Kato (1968) used radioluminous paint to chronically illuminate the pineal region of intact Japanese Quail and reported a photostimulatory effect on gonadal growth which did not occur if the birds were previously pinealectomized. However, Homma and Sakakibara (1971) in a more carefully designed experiment were unable to repeat their results. Rosner *et al.* (1971) have reported a direct effect of light on HIOMT activity in organ cultured duck pineals. Glands cultured for 24 hours in constant light had almost double the HIOMT activity of those cultured for the same period in darkness.

The indirect experiments pose several difficulties in interpretation. In the most useful of them, Lauber *et al.* (1968) found that neither blinding nor removal of the superior cervical ganglion (which presumably denervates the pineal) blocked the rise in HIOMT activity which normally occurs in the pineal of chicks in response to constant light. One must conclude that either (1) the ganglionectomy was incomplete, (2) there is another unknown route by means of which extraretinally perceived light is able to influence pineal HIOMT levels, or (3) the chick pineal is itself photoreceptive. In the House Sparrow, Menaker (1968a,b), Menaker and Keatts (1968), Gaston and Menaker (1968), and Menaker *et al.* (1970) have demonstrated that neither the eyes nor the pineal organ are necessary for either entrainment of the circadian rhythm of locomotor activity to light cycles or photoperiodically stimulated testicular recrudescence. Harrison and Becker (1969) have found in chickens that the normal restriction of oviposition to the light portion of a light–dark cycle was unaffected by blinding, pinealectomy, or a combination of the two treatments. However, since the existence of as yet unidentified extraretinal photoreceptors is well documented in birds, these latter studies have no more bearing on the possible photosensitivity of the pineal than on that of the retina itself.

Because they are so few in number, it must be emphasized that the physiological experiments directed at discovering whether or not the avian pineal has a photoreceptive function suffer from the deficiencies of all experiments with negative results. Expecially needed are more experiments of the direct type in which a variety of photic stimuli are imposed and a variety of possible responses analyzed. The results obtained by Lauber et al. (1968) must be accounted for in terms of an input from an extraretinal, extrapineal photoreceptor, and a good deal more negative evidence will have to accumulate before we will be justified in concluding that the avian pineal is not photoreceptive.

C. CONTROL OF CIRCADIAN RHYTHMICITY

Gaston and Menaker (1968) have reported that pinealectomy of House Sparrows has the effect of abolishing the circadian rhythm of locomotor activity under constant environmental conditions. Whereas normal sparrows in constant darkness invariably express a classic circadian rhythm in which activity is restricted to an 8–12 hour portion of the cycle and systematically drifts (free runs) between ½ and 1 hour per day with respect to real time, the activity of pinealectomized birds is continuous and its distribution is approximately uniform in time (Fig. 2). Gaston (1971) has reported the same phenomenon in White-crowned Sparrows. Binkley (1970) and Binkley et al. (1972), by making quantitative measurements of activity and subjecting the records to power spectral analysis, have confirmed the validity of the original conclusions, which were based on semiquantitative data. Moreover, they have shown that the total amount of activity expressed by sparrows in constant darkness is unaffected by pinealectomy; only its distribution in time is modified. McMillan (1972) has shown that pinealectomy abolishes both the free running rhythm of activity in nonmigratory White-throated Sparrows (*Zonotrichia albicollis*) and the free running rhythm of nocturnal activity (*Zugunruhe*) which these birds express during the spring and fall migratory periods.

The activity of pinealectomized House Sparrows (Gaston and Menaker, 1968) and of White-throated Sparrows (McMillan, 1972), although arrhythmic in constant conditions, can still be synchronized by light cycles. Synchronization to a LD 12:12 cycle is essentially normal, but in House Sparrows is phase advanced with respect to that of intact birds, under LD 6:18 cycles (Gaston, 1971).

Binkley et al. (1971) have shown that pinealectomy also abolishes the free running rhythm of body temperature change in House Sparrows. Pinealectomized sparrows held in constant darkness elevate their temperatures to be between 40.5° and 42.5°C and, in marked contrast to the records of normal birds, no circadian periodicity is detectable (Fig. 3). When arrhythmic pinealectomized birds are transferred from constant darkness to the presence of a light cycle, locomotor activity is immediately synchronized whereas body temperature re-

2. THE AVIAN PINEAL ORGAN

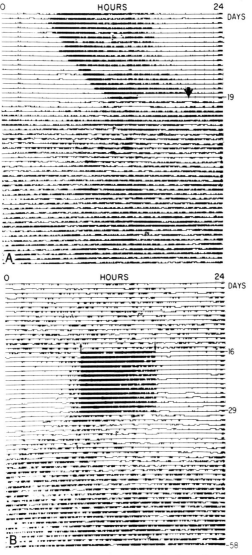

FIG. 2. Effect of pinealectomy on the circadian rhythm of locomotor activity of House Sparrows. In (A) a sparrow free running in constant darkness was pinealectomized on day 19 at the time indicated by the arrow. Note the almost immediate occurrence of arrhythmicity. In (B) a pinealectomized bird, arrhythmic in constant darkness, was exposed to a light cycle (↑ lights on; ↓ lights off) for 14 days beginning on day 16 and then returned to constant darkness on day 29. Note (1) that the pinealectomized bird entrains to the light cycle, (2) that during entrainment the onset of locomotor activity phase leads the onset of the light portion of the cycle, and (3) that the return to arrhythmicity following entrainment is gradual, requiring about 8 days. (From Gaston and Menaker, 1968, copyright 1968 by the American Association for the Advancement of Science.)

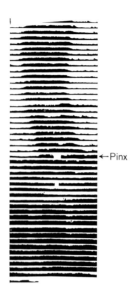

FIG. 3. Effect of pinealectomy on the circadian rhythm of body temperature in House Sparrows. The figure shows 46 days of continuously telemetered body temperature data from a single sparrow held in constant darkness. On the day indicated (← Pinx) the bird was pinealectomized. As with locomotor activity, rhythmicity disappears almost immediately. Several artifacts can be seen in the data during the 12 days following pinealectomy. (From Binkley et al., 1971, copyright 1971 by the American Association for the Advancement of Science.)

gains rhythmicity gradually over four or five cycles. Although the amplitude of the rhythm of locomotor activity of pinealectomized birds exposed to light cycles is the same as that of normal birds, the amplitude of the entrained rhythm of body temperature is always lower as a result of elevation of the temperature during the dark portion of the cycle.

The pineal of three species of passeriform birds is clearly involved in the regulation of circadian rhythmicity, but further interpretation is difficult at present. Pinealectomized birds behave in constant darkness very much as do normal birds in constant light. In fact, most of the effects of pinealectomy on circadian rhythmicity can be "explained" by the assumption that the pinealectomized bird "sees" darkness as light of moderate intensity. It should be stressed that this line of thought does not necessarily rest on the further assumption of pineal photoreceptivity. Alternatively, the observed effects of pinealectomy are equally compatible with the hypotheses that the pineal is a circadian driving oscillator or part of a mechanism coupling driving oscillations and overt driven rhythms (Gaston and Menaker, 1968).

In the chicken, Harrison and Becker (1969) have failed to find an effect of pinealectomy on the rhythm of oviposition in constant light. Whether or not this rhythm is in fact circadian is open to serious question, as it does not respond to light cycles as do all other known circadian rhythms. Nevertheless, their intriguing result opens the way to fruitful comparative work.

II. Structure

The pineal organ of birds shows a remarkable variety in size, form, and structure (Fig. 4). For a detailed review of the classic literature,

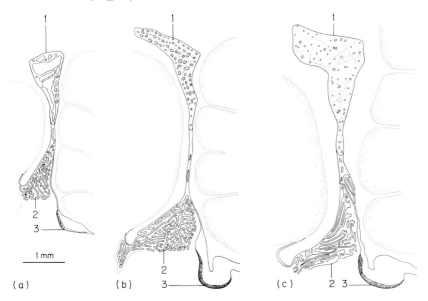

FIG. 4. Different types of the avian pineal organ; (a) saccular (e.g., *Passer domesticus*), (b) tubulofollicular (e.g., domestic pigeon, domestic duck), and (c) lobular (e.g., domestic fowl) patterns of structural organization. Note that a pineal lumen is very evident in type a, while type b is characterized by open tubules and follicles, and type c shows (in adults) only closed follicles; (1) pineal organ; (2) choroid plexus of the third ventricle; (3) subcommissural organ. (From Oksche and Vaupel-von Harnack, 1966; courtesy of Springer-Verlag.)

the monographs by Studnička (1905), Tilney and Warren (1919), or Bargmann (1943) and the comparative outline by Quay (1965) should be consulted. Three different structural types have been described for the avian pineal (Studnička, 1905): (1) sac-like hollow organ with folds and local proliferations of the wall (passerine birds); (2) system of tubules and follicles (domestic pigeon and duck); (3) nearly solid, lobular structure (adult fowl). The first type resembles the lacertilian pineal organ in some respects and may be regarded as more primitive in terms of comparative anatomy; in this type the primary connection of the pineal lumen with the third ventricle may persist. However, it is difficult to associate the three basic morphological types of avian pineal organ with differences in physiological activity. Electron microscopic and fluorescence microscopic characteristics may provide more precise comparative criteria than the mere description of pineal form.

Embryologically, the avian pineal is an evagination of the diencephalic roof. In embryos of several Columbiformes, Renzoni (1970) observed numbers of such evaginations. The parenchymal accessory structures that have been found in adult Columbidae are probably remnants of secondary evaginations. Krabbe (1955) described a "parietal corpuscle" detached from the pineal organ in the swan; he suggested that this corpuscle was a homolog of the parietal eye of lizards. However Renzoni's (1970) investigations show that this problem is somewhat more complex. Our further considerations will be limited to the avian pineal organ itself.

A. Cell Types

The pineal organ of vertebrates (Fig. 5) has undergone phylogenetic changes that have produced remarkable variations in cellular structures (see Oksche, 1965, 1971; Oksche *et al.*, 1971). The pineal systems of fishes, amphibians, and lizards, which are true photoreceptor organs (see Dodt *et al.*, 1971),* contain sensory cells [described by Eakin (1961)* in the frontal organ of a frog (*Hyla regilla*)] with regularly

---

Fig. 5. Comparative outline. Diagrammatic median sagittal sections of different pineal systems in relation to their characteristic type of pinealocyte. The submicroscopic diagrams illustrate the ultrastructure of sensory and secretory elements. (A) Teleostei: star = pineal organ; arrow = tractus pinealis. The outer segment of the receptor cell is relatively short and, in several species, domelike (overlapping the inner segment). (B) Anura: double star = frontal organ; double arrow = nervus pinealis; star = epiphysis cerebri; arrow = tractus pinealis. In Anura, outer segments of (1) regular (conelike), (2) irregular (vesiculated), and (3) overlapping (domelike) structures have been described. (C) Lacertilia: double star = parietal eye; double arrow = parietal nerve; star = epiphysis cerebri; arrow = tractus pinealis. (1) Long, regular outer segments which predominate in the parietal eye. In the epiphysis cerebri, in addition to short, regular outer segments (2), irregular, clublike, partly vacuolated forms (3) occur. The latter display dense-core vesicles. (D) Aves: arrow = pineal tract in the posterior wall of the organ. For the fine structure of the pinealocytes see Figs. 6–8. (E) Mammalia: Mammalian pinealocytes are exclusively secretory. (Drawing by Miss D. Vaihinger, modified after Oksche *et al.*, 1971; (E) based on findings by J. Ariëns Kappers.)

*See also additional references on p. 118.

2. THE AVIAN PINEAL ORGAN

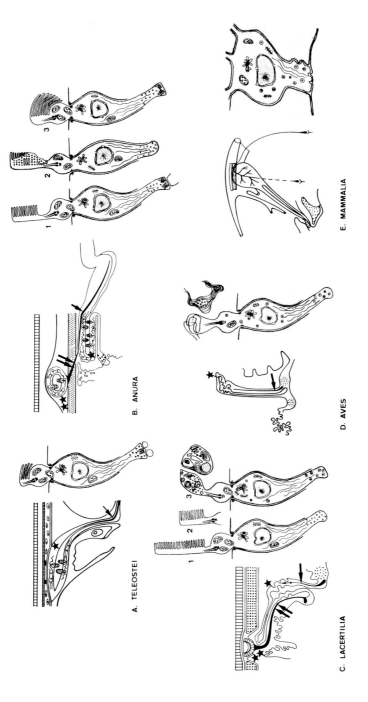

FIG. 5

organized conelike outer segments and distinct nervous pathways to the brain (see Oksche, 1971). The structure of the avian pineal organ differs in several important points from the unambiguous sensory structure of the pineal organs in lower vertebrates. Its structure appears to be predominantly secretory, although some morphological indications of sensory function remain.

The classic papers on the avian pineal organ distinguish between pineal "ependymocytes," "hypendymocytes," "pinealocytes," neuroglial elements, and occasional nerve cells (see Quay, 1965). Electron microscopy gives a more precise basis for classification of these cell types. In the current literature the general term "pinealocyte" has been often used to describe the various types of parenchymal cell of the pineal organ.

In terms of comparative ultrastructural morphology, the avian pineal organ displays the following cell types: (1) pineal cells of the "receptor" type; (2) "supportive" cells with ependymal characteristics; (3) pineal cells without connection with the lumen or its remnants; (4) nerve cells.

The first electron-microscopic description of an avian pineal organ was given by Oksche and Vaupel-von Harnack (1965a,b). Between 1966 and 1972 Oksche (1968, 1971), Oksche et al. (1966, 1969, 1971, 1972), and Collin (1966a,b,c, 1967a,b, 1968, 1969, 1971) have published a series of papers on the ultrastructure of avian pineals. Contributions to this subject have also been made by Bischoff (1967, 1969), Renzoni et al. (1968), and Mikami (1969).

1. Pinealocytes of the "Receptor" Type

Adjacent to the pineal lumina, slender cells occur that show a number of structures comparable to those found in the pineal receptor cells of lower vertebrates. These cells bear bulbous cilia of the 9-0 type which protrude into the lumen and form complex aggregates (Fig. 6). Although the cilia do not show lamellae comparable to the comblike arrangements of membranes in typical pineal receptors of lower vertebrates, they do form cytoplasmic laminae which envelop different numbers of ciliary extensions, thus producing concentric lamellar bodies (Figs. 7, 8). At least at the tip of single cilia, these lamellar bodies are continuous with the cytoplasm. On the other hand, there is evidence that the concentric lamellar bodies sometimes lose their cytoplasmic contacts and degenerate. Frequently the lamellar whorls seem to have a chaotic appearance.

Oksche and Vaupel-von Harnack (1965b) examined the developing

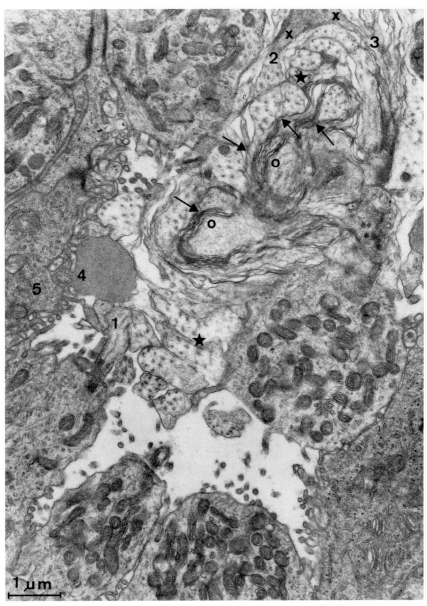

FIG. 6. Complex formed by bulbous cilia and lamellar whorls in the lumen of the pineal organ of *Passer domesticus* (O). Stars mark cross or oblique sections through cilia of the 9+0 type. Arrows indicate points where the cilia are continuous with single laminae of the lamellar body. (1) Axial section of a cilium that arises from a pinealocyte of the receptor type. Crosses indicate bilateral extensions of a receptorlike pinealocyte in close connection with a cilium (2) and an arrangement of lamellae (3). (4) Finely granular material adjacent to a supportive cell. Note that the apical surface on this supportive cell bears numerous microvilli. ×14,400. (From Oksche and Kirschstein, 1969; courtesy of Springer-Verlag.)

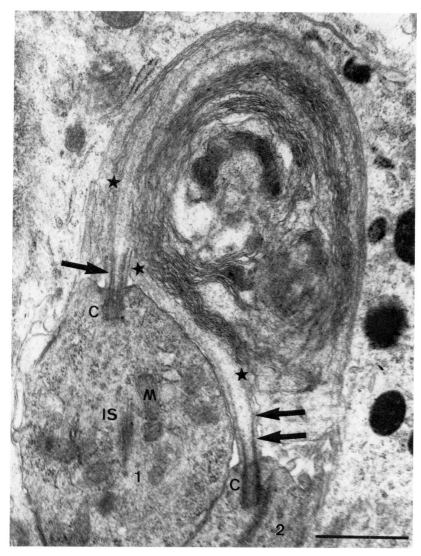

FIG. 7. Concentric lamellar body formed by the cilia (arrow and double arrow) of two receptorlike pinealocytes (1,2) of *Passer domesticus*. This structure only partly resembles the conelike outer segment of pineal receptors in lower vertebrates, although it shows a high degree of cytological organization. Frequently the avian lamellar whorls have a chaotic appearance. Stars indicate points where the lamellae are continuous with the cytoplasmic matrix of the cilia. C = centriole, IS = inner segment with numerous mitochondria (M). ×25,200; Bar = 1 μ. (Electron micrograph by H. Kirschstein; material A. Oksche and M. Menaker) (*Cell Tiss. Res.*, in preparation.)

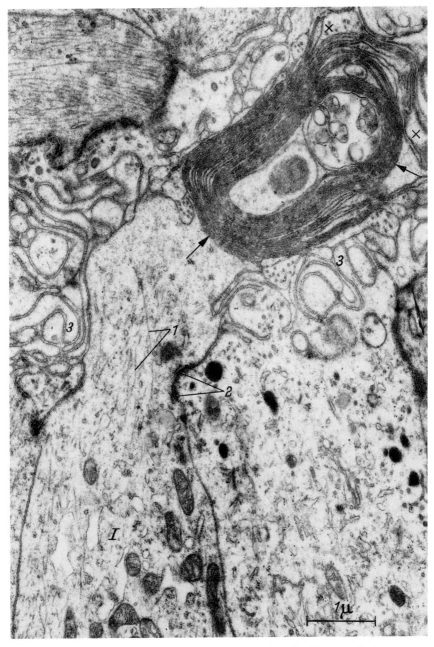

FIG. 8. Pinealocytes and lamellar structures (arrow) in the domestic duck; ×=cytoplasmic structures of bulbous cilia; I=inner segment of a pinealocyte containing mitochondria, endoplasmic reticulum and microtubules (1). Note the differentiated intercellular contacts (2) and interdigitations (3). ×18,500. (From Oksche and Vaupel-von Harnack, 1966; courtesy of Springer-Verlag.)

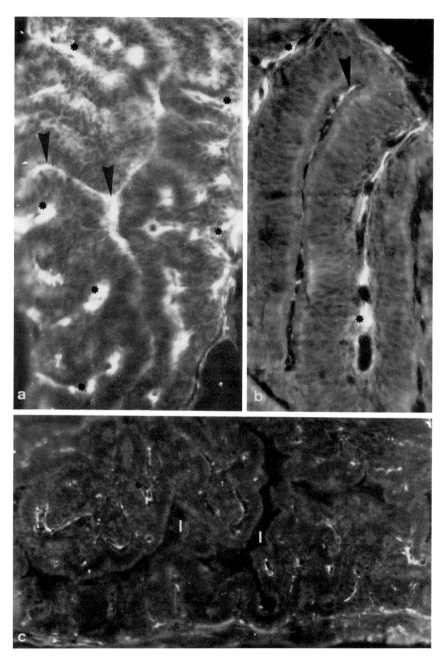

FIG. 9. *Passer domesticus.* (a) Yellow-fluorescent fluorophore in the pineal lumen (arrows) and in the autonomic nerve fibers (asterisks). Method by Falck-Hillarp; ×220. (b) After fading of the parenchymal fluorescence (arrow points to the lumen), the aminergic nerve fibers (asterisk) become more distinct. ×440. (c) No fluorescence in the lumen (I) and in the parenchyma is visible; the aminergic nerves show a green fluorescence. ×140. (Courtesy of M. Ueck, 1973, and Springer-Verlag.)

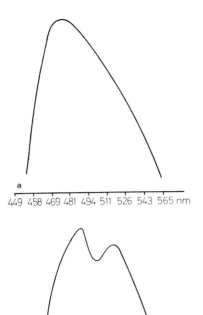

FIG. 10. Microspectrofluorometric estimation of emission maxima λ (relative values to the wavelength) in the pineal organ of *Passer domesticus*. (a) Green-fluorescent aminergic nerve fibers in the perilobular connective tissue. Maximum: 475 nm. (b) Pineal parenchyma. Two maxima: 470–485 nm and 515–530 nm. (Courtesy of M. Ueck, 1973, and Springer-Verlag.)

pineal organ of the domestic fowl with the electron microscope. Between the fourteenth and eighteenth day of embryonic development, when the first differentiation of retinal receptors can be observed (Meller and Breipohl, 1965), the columnar pineal cells form an inner segment and a cilium. At the same time a few lamellar whorls appear in the lumen, but in contrast to the retinal receptors, they do not show further development and regular differentiation. Two days after hatching, their substructural organization remains fundamentally unchanged.

The inner segment of the avian receptorlike pinealocytes resembles very much the inner segment of the pineal receptor cells in lower vertebrates. It is very rich in mitochondria and also contains Golgi zones and granular endoplasmic reticulum. Dense-core vesicles with

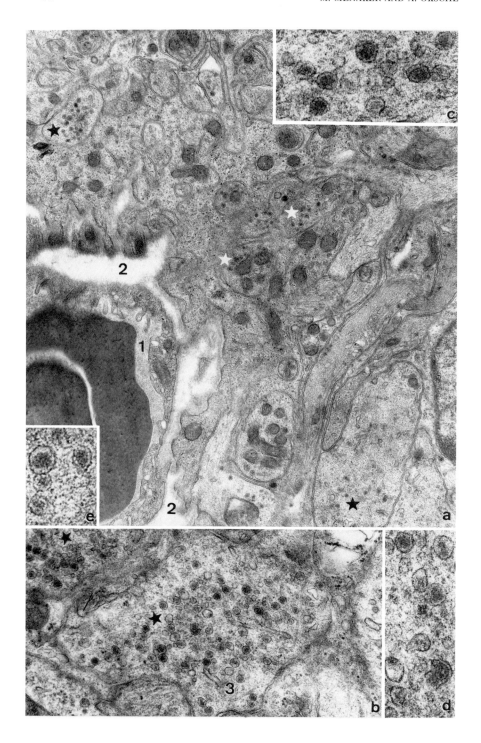

diameters of 800–1200 Å are elaborated in the Golgi complexes. The basal processes of the receptor cells usually contain filaments, microtubules, and dense-core vesicles. The end-foot of the basal process is often found in juxtaposition with a capillary. In some avian species (e.g., domestic pigeon) digitated end-feet containing accumulations of dense-core granules have been described (Oksche et al., 1969). The end-foot is always separated by a basal lamina (basement membrane) from the capillary. In Falck-Hillarp preparations such receptorlike pineal cells show a strong yellow fluorescence (Fig. 9). This fluorescence depends on a fluorophore of 5-HT (5-hydroxytryptophan) (Fig. 10), as has been shown in spectrographic studies by Ueck (1973).

In the House Sparrow, Ueck (1970) has observed "synaptic ribbons" in terminals of cells which display the other ultrastructural characteristics of receptor-like pinealocytes (see, also, Collin, 1969, 1971).

2. Supportive Cells

The supportive cells of the avian pineal organ belong to the group of ependymal elements. The apical surface of these cells bears numerous microvilli. Of particular interest are the circular tight junctions and desmosomal connections between the pineal receptors and the supportive cells. There is no ultrastructural indication that the avian pineal supportive cells contain "myeloid bodies" of the type described in the pineal supportive cells of Anura (see Oksche et al., 1963; Kelly and Smith, 1964; Ueck, 1971) which resemble the myeloid bodies in the pigment epithelium of the lateral eye.

3. Pinealocytes without Connections with the Lumen

In the thicker portions of the pineal organ of passerine birds, pineal cells occur that apparently do not have a connection with the pineal lumen. These cells produce dense-core vesicles in their Golgi zones and show a strong yellow fluorescence (Fig. 11). In the House Sparrow, groups of dark and light cells of this type have been observed (Fig. 12). M. Ueck, P. Zimmermann, and H. Kobayashi (unpublished) have analyzed the characteristics of these two groups of pineal cells. Morphometric measurements show that the nuclear area of the dark

FIG. 11., Granular materials in the pineal parenchyma of *Passer domesticus*. (a) Dense-core vesicles (560–780 Å in diameter) in circumscribed profiles (stars) within the pineal parenchyma. These profiles belong rather to extensions of secretory pinealocytes than to the autonomic nervous elements. 1 = capillary; 2 = perivascular space. ×20,700. (b) Dense-core vesicles (stars) in the perikarya of two adjacent pinealocytes. These granules measure 800–1000 Å in diameter. 3 = endoplasmic reticulum. ×24,000. (c, d, e) Dense-core vesicles under higher magnification; c, d, ×50,000; e, ×100,000. (From Oksche and Kirschstein, 1969; courtesy of Springer-Verlag.)

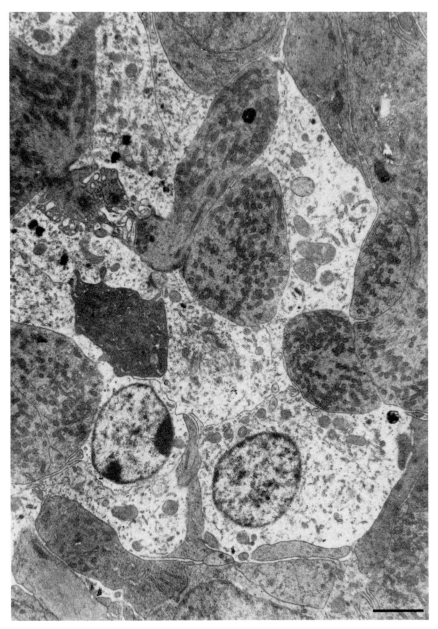

FIG. 12. Dark and light pinealocytes of *Passer domesticus*. ×6900; bar = 2 μ. (Electron micrograph by H. Kirschstein; material A. Oksche and M. Menaker.)

pinealocytes exceeds that of the light pinealocytes. Morphometric data obtained after different pharmacologic treatments did not lead to a clear decision as to whether these two cell groups should be considered as two different cell types or as two different functional stages of one basic cell type.

In the pineal follicles of the pigeon, all pineal cells seem to be columnar and to form one layer; they connect the follicular lumen with the basement membrane (see Oksche *et al.,* 1969). A diagram by Collin (1969) shows a similar arrangement in the pineal organ of the Black-billed Magpie (*Pica pica*).

4. Nerve Cells

The presence of nerve cells within the avian pineal organ has been frequently denied (for references, see Quay, 1965). Strong evidence for the existence of nerve cells in the pineal organ of the House Sparrow has been presented by Quay and Renzoni (1963) with silver impregnation methods (see, also, Oksche *et al.,* 1972). They distinguish between large and small intramural cells. Furthermore, Quay and Renzoni (1963) describe, at least in the House Sparrow, a small Gomori-positive neurosecretory nucleus and a commissuropineal tract. However, Oksche *et al.* (1972) have shown in the White-crowned Sparrow how the appearance of pineal neurosecretion might be produced by a particular type of active pinealocyte. Oksche *et al.* (1972) observed nerve cells in Nissl preparations from the pineal organ of the White-crowned Sparrow. These nerve cells can be localized much more specifically in acetylcholinesterase preparations. Ueck and Kobayashi (1972) have demonstrated a very abundant system of acetylcholinesterase-positive neurons in the pineal stalk of the House Sparrow and other passerine birds. Nerve cells have also been described in electron micrographs (Oksche *et al.,* 1969; Ueck, 1970).

B. NERVOUS APPARATUS

1. Intrapineal Neuronal Structures

The receptorlike pineal cells bearing bulbous cilia which are partly connected with lamellar whorls would, in terms of retinal morphology, have to be considered as degenerate. However, it is uncertain whether criteria defined for the retina can be used for pineal structures on the basis of simple analogy. Although the lamellar whorls seem to be disorganized, there is a degree of substructural order in that their mem-

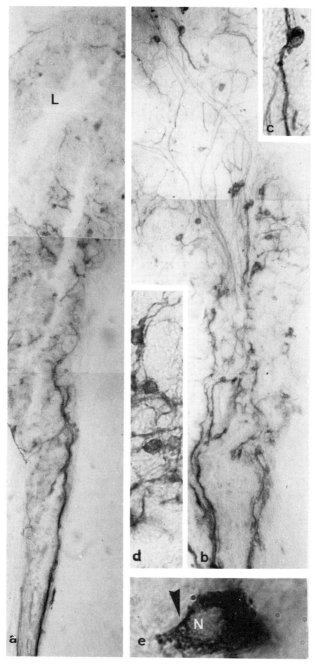

FIG. 13. Acetylcholinesterase-positive neurons in the pineal organ of *Passer domesticus*. (a) L = pineal lumen. Increasing number of acetylcholinesterase-positive neurons

brane distances measure approximately 200 Å (membrane distances or periods of approximately 200 Å are also characteristic of functional photoreceptors of the ciliary or microvillous type) (cf. Figs. 7 and 8). In addition the synaptic ribbons and membrane thickenings observed in the terminals of the inner segments of the receptorlike pineal cells of the House Sparrow (Ueck, 1970) speak in favor of some kind of photoreceptive activity. But interpretive caution is necessary; single synaptic ribbons have also been described by Wolfe (1965) in the wholly secretory pinealocytes of the rat.

While the functional significance of the pineal receptors has not been clarified by morphological studies, pineal nerve cells have been clearly identified. Even more convincing than the silver impregnation is the selective demonstration in the House Sparrow of their cellular images by the acetylcholinesterase method (Ueck and Kobayashi, 1972) (Fig. 13). Attempts to demonstrate these cells in the more "glandular" pineal organ of the domestic pigeon were practically negative. The axons of the acetylcholinesterase-positive neurons of the House Sparrow extend into the pineal stalk (Fig. 13). This fiber system is probably identical with the axon structures that have been shown by silver impregnation (Quay and Renzoni, 1963; Oksche *et al.*, 1972) (Fig. 14) and with the bundles of numerous unmyelinated and a few myelinated nerve fibers which have been observed with the electron microscope in the pineal stalk of passerine birds (Oksche *et al.*, 1969, 1972; Ueck, 1970) (Fig. 15). These fibers may be regarded as the homolog of the pineal tract of the lower vertebrates. In the pineal stalk they are surrounded by extensions of receptor pinealocytes. Occasionally a ribbon synapse or a synapse of the conventional type is observed in this neuropil (Ueck, 1970).

According to Renzoni (1970), the majority of the fibers of the avian pineal tract (see, also, Studnička, 1905) enter the posterior commissure and have an accessory connection directed toward the habenular commissure. Ueck (1970) has traced this fiber system to the intercommissural plate with the electron microscope. More precise information on this nervous pathway is very much needed.

No nerve cells could be demonstrated in avian pineals with the monoamine oxidase reaction, although the pinealocytes were strongly

---

toward the stalk portion of the organ. Note the distinct nervous pathway in its posterior wall (cf. Fig. 14B). (b–c) Images of acetylcholinesterase-positive neurons. (e) Reaction products (arrow) in a nerve cell perikaryon. N = cell nucleus. Magnifications: a, ×70; b, ×120; c, ×250; d, ×240; e, ×1000. (From Ueck and Kobayashi, 1972; courtesy of Springer-Verlag.)

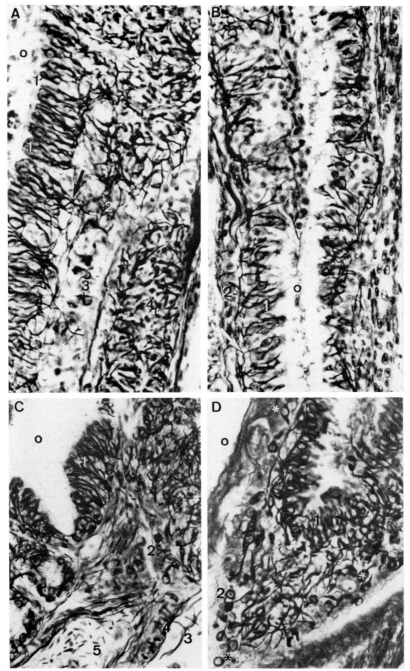

Fig. 14. Fiber structures in the pineal organ of the White-crowned Sparrow, *Zonotrichia leucophrys gambelii*; O = pineal lumen. (A) 1 = receptorlike pinealocytes showing looping fibers. Note the extension of these fibers into a basal process (arrow)

positive (M. Ueck and H. Kobayashi, unpublished). Granulated aminergic nerve cells have never been observed in avian pineal organs. Furthermore, only a very few granulated nerve fibers can be shown with the electron microscope in the intramural portion of the pineal stalk. The origin of these sparse elements is not known.

2. Autonomic Innervation

The avian pineal has an abundant autonomic (sympathetic) innervation. Hedlund (1970) and Hedlund and Nalbandov (1969) have shown that the pineal organs of Japanese Quail, White Leghorn cockerels, and Red Junglefowl (*Gallus gallus*) are innervated primarily, if not exclusively, from the superior cervical ganglia. The parenchymal cell complex is innervated extensively in chickens and junglefowl and only sparsely in quail. In the quail pineal, the green fluorescent (sympathetic) fibers are located primarily in the interlobular (septal) areas. In the pineal organ of the chicken, green fluorescent structures are scattered about the parenchymal cell mass; they disappear 2–3 days after ganglionectomy (Hedlund and Nalbandov, 1969).

The pineal organ of the House Sparrow receives a rich supply of sympathetic fibers (Ueck, 1970, 1973) that accompany the capsular blood vessels (Fig. 16). Microspectrographic measurements indicate that they contain noradrenalin or serotonin (cf. Fig. 9a). Ueck (1970) did not find any evidence in his material that sympathetic nerve fibers penetrate the pineal basement membrane. In electron micrographs, these fibers contain two different types of vesicular inclusions: (1) 300–500 Å in diameter and (2) 800–1200 Å in diameter (see Fig. 17). After administration of nialamide both types of vesicles show a dense core suggesting that they contain biogenic amines. In the pineal organ of the duck, Ueck (1973) could also trace fluorescent intraparenchymal fibers; in silver impregnations these nerve fibers are visible within the walls of the follicles (A. Oksche, unpublished results). In the pineal of the pigeon, fluorescent autonomic fibers are extremely sparse.

The autonomic nerve fibers in the connective tissue of the pineal capsule are arranged in small bundles (Fig. 17). In the White-crowned

---

and its relation to an outer reticular area (2). 3 = strongly impregnated small round cells; 4 = tangentially sectioned region of the pineal wall. (B) 1 = pinealocytes; arrow = pinealocyte basal process; 2 = longitudinal fiber system. (C) 1 = looping fibers and 2 = small round cells that lie at the base of this cell assemblage; 3 = perivascular nerve fibers; 4 = arteriole; 5 = small vein. (D) Pineal stalk. 1 = receptorlike pinealocytes, 2 and asterisks = round cells. Magnifications: A,B, ×270; C, ×380; D, ×410. (From Oksche *et al.*, 1972; courtesy of Springer-Verlag.)

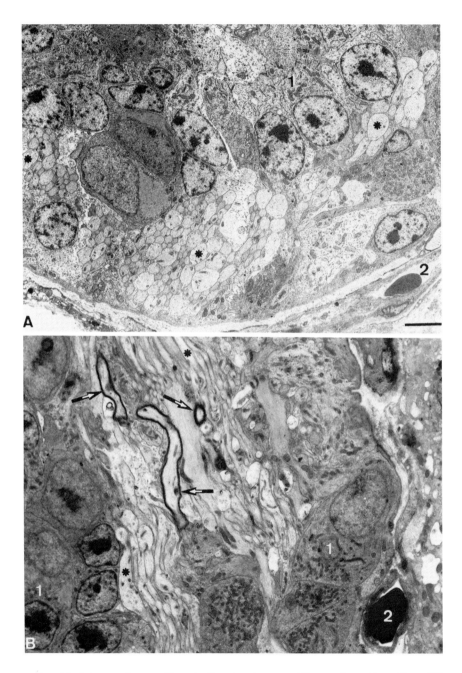

FIG. 15. Intramural nerve fibers in the pineal organ of *Passer domesticus*. Cross (A) and longitudinal (B) sections through the pineal stalk. Asterisks = unmyelinated fibers; arrows = myelinated fibers. The unmyelinated elements form bundles, the myelinated ones are sparse. 1 = pinealocytes; 2 = capillary. ×3450; bar = 3 μ. (Electron micrographs by H. Kirschstein; material, A. Oksche and M. Menaker *Cell Tissue Res.*, in press.)

Sparrow, they measure 0.1–0.4 μ in diameter (Oksche et al., 1972). These data are in good agreement with observations in the House Sparrow (Oksche and Kirschstein, 1969). In this species, there is new evidence that occasional myelinated elements also occur in the capsular nerve bundles (Material of M. Menaker, unpublished; A. Oksche unpublished) (Fig. 17).

With respect to the autonomic fibers that innervate the avian pineal organ, there are several problems remaining to be solved. In species with intramural sympathetic nerve fibers (e.g., duck), there is a lack of evidence concerning the exact structural type of the contacts that they make with the pineal cells. In the House Sparrow there is still the possibility that the penetration of the autonomic fibers through the basement membrane has been overlooked. The origin of some of the finest intrinsic nerve elements (approximately 0.1–0.2 μ in diameter) is still in doubt. In sparrows that have not received any pharmacological treatment, most of the larger capsular axons are free of dense-cored vesicles, but the absence of such dense-core material in any of the pineal fibers does not rule out the possiblity that these fibers may belong to the autonomic system.

C. Secretory Apparatus

Secretory granules that arise in the Golgi complex have been observed both in the receptorlike and glandular avian pinealocytes (Fig. 18). Ueck (1973) emphasizes the observation that dense-cored vesicles and yellow fluorescence are both found in the receptorlike pinealocytes (Fig. 19). This is in general agreement with the conclusions of Collin (1969, 1971) and Oksche (1971). The dense-cored vesicles belong to both the 1000 Å type (800–1200 Å, occasionally up to 1400 Å in diameter) and the 300–500 Å type. A considerable part of this material (both granular types) is accumulated in the perivascular end-feet of the pineal cells. However, as the granular processes of pinealocytes may in some cases be intermixed with intramural autonomic fibers (see Ueck, 1970) cautious interpretation, based on ultrastructural features other than granulation itself, is required. Part of the fluorescent and granular material located in avian pinealocytes is reserpine-resistant and can be augmented by administration of nialamide (for further details, see Ueck, 1973).

M. Ueck, P. Zimmermann, and H. Kobayashi (unpublished) have applied light- and electron-microscopic morphometric methods in their analyses of the secretory activity in the pineal organ of the House Sparrow. After injection of reserpine, the nuclear area of the dark pinealocytes increases, whereas there is a decrease

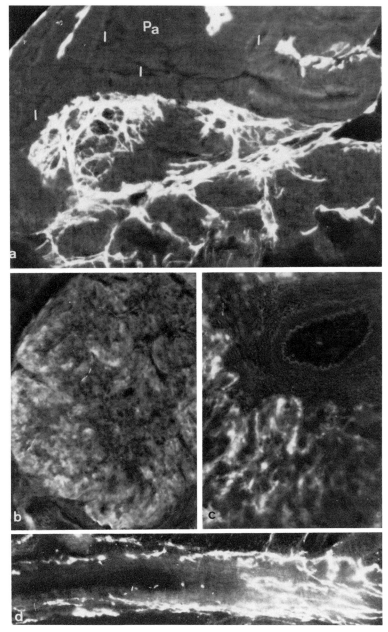

FIG. 16. Fluorescent autonomic nerve fibers in the avian pineal organ (method by Falck-Hillarp). (a) Greenfinch (*Carduelis chloris*). Yellow-green fluorescent aminergic fibers are seen in the interlobular connective tissue. No indication of fluorescent fibers

in the nuclear area of the light pinealocytes. Administration of nialamide does not alter the ratio of the nuclear area of dark pinealocytes to that of the light pinealocytes, but it does produce an increase in mitochondrial surface area in the inner segments of the receptor-like pinealocytes. The functional significance of these changes is not yet understood.

D. Vasculature

The vascular system of avian pineal organs has not yet been thoroughly investigated. (Consideration should be given to the detailed analysis of the blood vessels of the lacertilian pineal; see Lierse, 1965.) From our india ink-injected pineal organs of the House Sparrow (A. Oksche *et al.*, unpublished material), it can be concluded, that the capillary network of this organ is very dense if compared with other adjacent regions of the brain. M. Ueck, P. Zimmermann, and H. Kobayashi (unpublished) have shown with morphometric methods that after administration of nialamide the capillary surface area in the pineal region of the House Sparrow increased; at the same time, there was no such increase in the choroid plexus which was used as a control tissue.

III. General Conclusions

There are considerable interspecies differences among the pineal organs of passerine birds, domestic pigeons, and ducks. The pineal organ of the pigeon is very poorly supplied with autonomic nerves and there is no indication that it possesses acetylcholinesterase-positive neurons. The pineal organ of the duck differs from that of the pigeon (both belong to the follicular type) in that it has numerous intramural autonomic fibers. The functional significance of these differences is unknown. In the chicken, the autonomic innervation of the pineal organ is so abundant that it is difficult to understand the results of Lauber *et al.* (1968), which seem to indicate that the autonomic

---

between the parenchymal elements (Pa) or in the lumen (I). × 144. (b) *Passer domesticus*. After administration of reserpine the fluorescence of the aminergic fibers is abolished. The fluorescence of the parenchyma, however, persists. × 144. (c) Domesticated pigeon. Yellow fluorescence in the basal portion of the follicles; no evidence for aminergic fibers in the interlobular connective tissue or in perivascular position. × 360. (d) *Carduelis chloris*. Numerous perivascular aminergic fibers. × 342. (Courtesy of M. Ueck, 1973, and Springer-Verlag.)

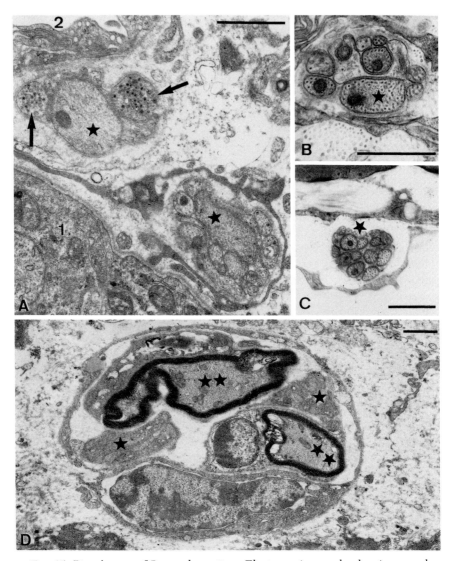

FIG. 17. Pineal organ of *Passer domesticus*. Electron micrographs showing capsular and perivascular nerve fibers. (A–C) Stars = unmyelinated elements. Arrows point to unmyelinated fibers bearing dense-core vesicles. 1 = pineal parenchyma; 2 = capillary. (D) Stars = bundle of unmyelinated fibers; double star = myelinated fibers. Magnifications: A, ×18,400; B, ×21,600; C, ×13,000; D, ×9200; bar = 1 μ. (A and D: Electron micrographs by H. Kirschstein; material, A. Oksche and M. Menaker, *Cell Tissue Res.*, in press.)

nerves are not necessary for light-dependent melatonin synthesis to occur.

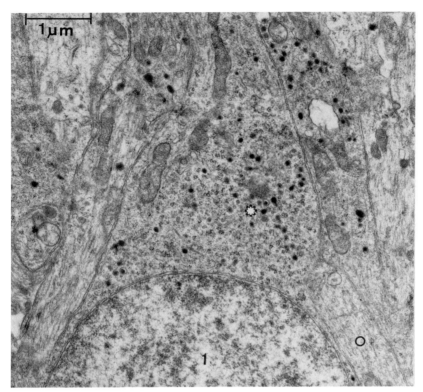

FIG. 18. Dense-core vesicles in a pinealocyte perikaryon (asterisk) of the pigeon; 1 = cell nucleus, O = microtubules. Adjacent granulated processes of other pinealocytes. × 17,600. (From Oksche et al., 1969, courtesy of Springer-Verlag.)

Among the passerine birds, the pineal organ of the House Sparrow (*Passer domesticus*) has been the most thoroughly studied. It contains pinealocytes with irregular outer segments and synaptic structures of ribbon and conventional type, as well as cholinergic nerve cells that form an intrinsic nervous pathway consisting of several hundred unmyelinated and a few myelinated fibers. This nerve tract can be traced to the commissural region (diencephalomesencephalic border area); its terminal central nervous projections are not known. In the frog (Paul et al., 1971), the apparent homolog of this tract makes a connection with the tegmentum of the brainstem. The function of the nervous apparatus of the pineal organ in the House Sparrow is not understood. It has not been possible to detect an electrical response to the onset or cessation of light (Ralph and Dawson, 1968; Hamasaki, in Oksche and

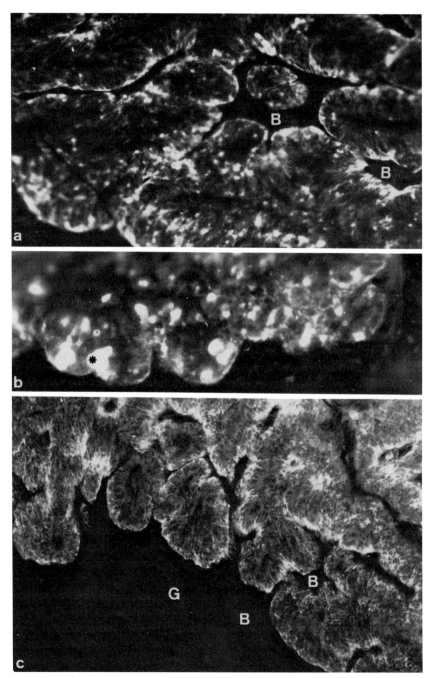

FIG. 19. Fluorescent material in the pineal organ of the pigeon (method of Falck-Hillarp). (a–b) Vesicular accumulations in the basal portion of the follicles (asterisk).

Kirschstein, 1969). In addition, the pineal organ of this sparrow receives a rich sympathetic innervation and it is highly vascularized.

For the pineal organ of the House Sparrow there is strong evidence that 5-HT is formed not only in glandular, but also in receptorlike pinealocytes. An agranular pool of 5-HT seems to exist in the functional pineal photoreceptors of fishes and frogs (Owman and Rüdeberg, 1970; Owman et al., 1970). One can suggest that in these sensory cells an input of photic information is converted into both electrical and neurohumoral outputs. Whether the avian receptorlike pinealocytes also have this capacity is an open question.

Despite its incompleteness (especially with respect to physiology), the presently available literature suggests to us that the avian pineal may be involved directly or indirectly in the processes of photoreception, reproduction, and circadian rhythmicity. To put this in a slightly different way, we feel that the pineal's role may well be one of integrating several aspects of the bird's response to its photic environment. It seems to us that it will not be possible adequately to test this admittedly very general hypothesis by an extension of the presently utilized techniques applied to the species that have been most extensively studied. If, as we suspect, the pineal's function is truly integrative and if each of the several processes involved in the integration is itself complex, it will be necessary to study them all simultaneously in a single species. The species used must have an ontogeny that is readily accessible to experimental manipulation; a photoperiodically controlled seasonal reproductive cycle; one, or better, several readily assayed overt circadian rhythms; and a surgically accessible pineal. Perhaps most important, it should not be a domesticated species, for the artificial selection associated with domestication is almost always directed at modifying or eliminating entirely the influence of the environment on reproduction. As it is very difficult to regulate the environmental influences to which the nestlings of altricial birds are exposed, the choice of experimental species is for practical purposes restricted to wild precocial birds. We do not mean to imply that much cannot be learned from studying passeriforms or domestic gallinaceous species, but simply that, in the end, the information derived from such studies will have to be understood against a background that is only obtainable elsewhere.

The ultra-structural equivalent of the fluorescent droplets is apparently a circumscribed accumulation of elementary granules. In the connective tissue, there is no indication of fluorescent nerve fibers. (c) Diffuse yellow fluorescence at the basis of pineal follicles. B, connective tissue; G, blood vessel. Magnifications: a, ×150; b, ×360; c, ×110. (Courtesy of M. Ueck, 1973, and Springer-Verlag.)

ACKNOWLEDGMENT

Some of the experimental work reported here was supported by NIH Grant HD03803-04 and a Career Development Award HD09327-04. The chapter was prepared while M. Menaker held a John Simon Guggenheim Fellowship. Acknowledgment is made to the Deutsche Forschungsgemeinschaft for support of the pertinent investigations of A. Oksche and colleagues cited herein. A. Oksche is greatly indebted to Miss I. Lyncker, Giessen, for her help in preparing this manuscript.

REFERENCES

Ariëns Kappers, J., and Schadé, J. P., eds. (1965). "Structure and Function of the Epiphysis Cerebri." Elsevier, Amsterdam.

Barfuss, D. W., and Ellis, L. C. (1971). Seasonal cycles in melatonin synthesis by the pineal gland as related to testicular function in the House Sparrow (*Passer domesticus*). *Gen. Comp. Endocrinol.* **17**, 183–193.

Bargmann, W. (1943). Die Epiphysis cerebri. *In* "Handbuch der mikroskopischen Anatomie des Menschen" (W. von Möllendorff, ed.), Vol. 6, Part 4, pp. 309–502. Springer-Verlag, Berlin and New York.

Binkley, S. (1970). The pineal organ and circadian organization in the House Sparrow. Doctoral dissertation, University of Texas, Austin.

Binkley, S., Kluth, E., and Menaker, M. (1971). Pineal function in sparrows: Circadian rhythms and body temperature. *Science* **174**, 311–314.

Binkley, S., Kluth, E., and Menaker, M. (1972). Pineal and locomotor activity levels and arrhythmia in sparrows. *J. Comp. Physiol.* **77**, 163–169.

Bischoff, M. B. (1967). Ultrastructural evidence for secretory and photoreceptor functions in the avian pineal organ. *J. Cell Biol.* **35**, 13A–14A.

Bischoff, M. B. (1969). Photoreceptoral and secretory structures in the avian pineal organ. *J. Ultrastruct. Res.* **28**, 16–26.

Cardinali, D. P., Cuello, A. E., Tramezzani, J. H., and Rosner, J. M. (1971). Effects of pinealectomy on the testicular function of the adult male duck. *Endocrinology* **89**, 301.

Collin, J.-P. (1966a). Contribution à l'étude des follicules de l'épiphyse embryonnaire d'Oiseau. *C. R. Acad. Sci., Ser. D* **262**, 2263–2266.

Collin, J.-P. (1966b). Etude préliminaire des photorécepteurs rudimentaires de l'épiphyse de *Pica pica* L. pendant la vie embryonnaire et postembryonnaire. *C. R. Acad. Sci., Ser. D* **263**, 660–663.

Collin, J.-P. (1966c). Sur l'évolution des photorécepteurs rudimentaires épiphysaires chez la Pie (*Pica pica* L.). *C. R. Soc. Biol.* **160**, 1876–1880.

Collin, J.-P. (1967a). Le photorécepteur rudimentaire de l'épiphyse d'Oiseau: Le prolongement basal chez la passereau *Pica pica* L. *C. R. Acad. Sci.* **265**, 48–51.

Collin, J.-P. (1967b). Nouvelles remarques sur l'épiphyse de quelques Lacertiliens et Oiseaux. *C. R. Acad. Sci.* **265**, 1725–1728.

Collin, J.-F. (1968). Rubans circonscrits par des vésicules dan les photorécepteurs rudimentaires épiphysaires de l'Oiseau: *Vanellus vanellus* (L.) et nouvelles considérations phylogénétiques relatives aux pinéalocytes (ou cellules principales) des Mammifères. *C. R. Acad. Sci.* **267**, 758–761.

Collin, J.-P. (1969). Contribution à l'étude de l'organe pinéal. De l'épiphyse sensorielle à la glande pinéale: Modalités de transformation et implications fonctionnelles. *Ann. Sta. Biol. Besse-en-Chandesse, Suppl.* **1**, 1–359.

Collin, J.-P. (1971). Differentiation and regression of the cells of the sensory line in the epiphysis cerebri. *Pineal Gland, Ciba Found. Symp., 1970* pp. 79–125.

Dodt, E., Ueck, M., and Oksche, A. (1971). Relation of structure and function. The pineal organ of lower vertebrates. *In* "J. E. Purkyně Centenary Symposium, Prag, 1969" (V. Kruta, ed.), pp. 253–278. Univ. Jana Evangelisty Purkyně, Brno.

Donham, R. S., and Wilson, F. E. (1970). Photorefractoriness in pinealectomized Harris Sparrows. *Condor* **72**, 101–102.

Eakin, R. M. (1961). Photoreceptors in the amphibian frontal organ. *Proc. Nat. Acad. Sci. U.S.* **47**, 1084–1088.

Foà, C. (1928). Nuovi esperimenti sulla fisiologia della ghiandola pineale (Physiology of the pineal body). *Arch. Sci. Biol. (Napoli)* **12**, 306–321.

Gaston, S. (1971). The influence of the pineal organ on the circadian activity rhythm in birds. *In* "Biochronometry" (M. Menaker, ed.), pp. 541–548. Nat. Acad. Sci., Washington, D.C.

Gaston, S., and Menaker, M. (1968). Pineal function: The biological clock in the sparrow? *Science* **160**, 1125–1127.

Hammer, W., and Barfield, R. J. (1970). Uneffectiveness of pineal lesions on the testis cycle of a Finch. *Condor* **72**, 99–101.

Harrison, P. C., and Becker, W. C. (1969). Extraretinal photocontrol of oviposition in pinealectomized domestic fowl. *Proc. Soc. Exp. Biol. Med.* **132**, 164–171.

Hedlund, L. (1970). Sympathetic innervation of the avian pineal body. *Anat. Rec.* **166**, 406.

Hedlund, L., and Nalbandov, A. V. (1969). Innervation of the avian pineal body. *Amer. Zool.* **9**, 1090.

Homma, K., and Sakakibara, Y. (1971). Encephalic photoreceptors and their significance in photoperiodic control of sexual activity in Japanese Quail. *In* "Biochronometry" (M. Menaker, ed.), pp. 333–341. Nat. Acad. Sci., Washington, D.C.

Kelly, D. E., and Smith, S. W. (1964). Fine structure of the pineal organs of the adult frog, *Rana pipiens. J. Cell Biol.* **22**, 653–674.

Kobayashi, H. (1969). Pineal and gonadal activity in birds. *In* "Seminar on Hypothalamic and Endocrine Functions in Birds" (H. Kobayashi and D. S. Farner, eds.), Abstracts, p. 72. Tokyo.

Krabbe, K. H. (1955). Development of the pineal organ and a rudimentary parietal eye in some birds. *J. Comp. Neurol.* **103**, 139–149.

Lauber, J. K., Boyd, J. E., and Axelrod, J. (1968). Enzymatic synthesis of melatonin in avian pineal body: Extraretinal response to light. *Science* **161**, 489–490.

Lierse, W. (1965). Die Gefässversorgung der Epiphyse und Paraphyse bei Reptilien. *In* "Structure and Function of the Epiphysis Cerebri" (J. Ariëns Kappers and J. P. Schadé, eds.), pp. 183–192. Elsevier, Amsterdam.

McMillan, J. (1972). Pinealectomy abolishes the circadian rhythm of migratory restlessness. *J. Comp. Physiol.* **79**, 105–112.

Meller, K., and Breipohl, W. (1965). Die Feinstruktur und Differenzierung des inneren Segments und des Paraboloids der Photorezeptoren in der Retina von Hühnerembryonen. *Z. Zellforsch. Mikrosk. Anat.* **66**, 673–684.

Menaker, M. (1968a). Extraretinal light perception in sparrows. I. Entrainment of the biological clock. *Proc. Nat. Acad. Sci. U.S.* **59**, 414–421.

Menaker, M. (1968b). Light perception by extraretinal receptors in the brain of the sparrow. *Proc. 76th Annu. Convention Amer. Pyschol. Ass.* pp. 299–300.

Menaker, M., and Keatts, H. (1968). Extraretinal light perception in the sparrow. II. Photoperiodic stimulation of testis growth. *Proc. Nat. Acad. Sci. U.S.* **60**, 146–151.

Menaker, M., Roberts, R., Elliott, J., and Underwood, H. (1970). Extraretinal light perception in the sparrow. III. The eyes do not participate in photoperiodic photoreception. *Proc. Nat. Acad. Sci. U.S.* **67**, 320-325.

Mikami, S. I. (1969). The morphology and fine structure of the pineal body of the chicken. *In* "Seminar on Hypothalamic and Endocrine Functions in Birds" (H. Kobayashi and D. S. Farner, eds.), Abstracts, pp. 69-70. Tokyo.

Morita, Y. (1966). Absence of electrical activity of the pigeon's pineal organ in response to light. *Experientia* **22**, 402.

Oishi, T., and Kato, M. (1968). Pineal organ as a possible photoperiodic testicular response in Japanese Quail. *Mem. Fac. Sci., Kyoto Univ., Ser. Biol.* **2**, 12-18.

Oksche, A. (1965). Survey of the development and comparative morphology of the pineal organ. *In* "Structure and Function of the Epiphysis Cerebri" (J. Ariëns Kappers and J. P. Schadé, eds.), pp. 3-29. Elsevier, Amsterdam.

Oksche, A. (1968). Zur Frage extraretinaler Photorezeptoren im Pinealorgan der Vögel. *Arch. Anat. Histol. Embryol.* **51**, 497-507.

Oksche, A. (1971). Sensory and glandular elements of the pineal organ. *Pineal Gland, Ciba Found. Symp., 1970* pp. 127-146.

Oksche, A., and Kirschstein, H. (1969). Elektronenmikroskopische Untersuchungen am Pinealorgan von *Passer domesticus*. *Z. Zellforsch. Mikrosk. Anat.* **102**, 214-241.

Oksche, A., and Vaupel-von Harnack, M. (1965a). Vergleichende elektronenmikroskopische Studien am Pinealorgan. *In* "Structure and Function of the Epiphysis Cerebri" (J. Ariëns Kappers and J. P. Schadé, eds.), pp. 237-258. Elsevier, Amsterdam.

Oksche, A., and Vaupel-von Harnack, M. (1965b). Über rudimentäre Sinneszellstrukturen im Pinealorgan des Hühnchens. *Naturwissenschaften* **52**, 662-663.

Oksche, A., and Vaupel-von Harnack, M. (1966). Elektronenmikroskopische Untersuchungen zur Frage der Sinneszellen im Pinealorgan der Vögel. *Z. Zellforsch. Mikrosk. Anat.* **69**, 41-60.

Oksche, A., and von Harnack, M. (1963). Elektronenmikroskopische Untersuchungen am Stirnorgan von Anuren. (Zur Frage der Lichtrezeptoren). *Z. Zellforsch. Mikrosk. Anat.* **59**, 239-288.

Oksche, A., Morita, Y., and Vaupel-von Harnack, M. (1969). Zur Feinstruktur und Funktion des Pinealorgans der Taube (*Columba livia*). *Z. Zellforsch. Mikrosk. Anat.* **102**, 1-30.

Oksche, A., Ueck, M., and Rüdeberg, C. (1971). Comparative ultrastructural studies of sensory and secretory elements in different pineal organs. *Mem. Soc. Endocrinol.* **19**, 7-25.

Oksche, A., Kirschstein, H., Kobayashi, H., and Farner, D. S. (1972). Electron microscopic and experimental studies of the pineal organ in the White-crowned Sparrow, *Zonotrichia leucophrys gambelii*. *Z. Zellforsch. Mikrosk. Anat.* **124**, 247-274.

Owman, C., and Rüdeberg, C. (1970). Light, fluorescence, and electron microscopic studies on the pineal organ of the pike, *Esox lucius* L., with special regard to 5-hydroxytryptamine. *Z. Zellforsch. Mikrosk. Anat.* **107**, 522-550.

Owman, C., Rüdeberg, C., and Ueck, M. (1970). Fluoreszenzmikroskopischer Nachweis biogener Monoamine in der Epiphysis cerebri von *Rana esculenta* und *Rana pipiens*. *Z. Zellforsch. Mikrosk. Anat.* **111**, 550-558.

Paul, E., Hartwig, H.-G., and Oksche, A. (1971). Neurone und zentralnervöse Verbindungen des Pinealorgans der Anuren. *Z. Zellforsch. Mikrosk. Anat.* **112**, 466-493.

Quay, W. B. (1965). Histological structure and cytology of the pineal organ in birds and mammals. *In* "Structure and Function of the Epiphysis Cerebri" (J. Ariëns Kappers and J. P. Schadé, eds.), pp. 49–86. Elsevier, Amsterdam.
Quay, W. B., and Renzoni, A. (1963). Comparative and experimental studies of pineal structure and cytology in passeriform birds. *Riv. Biol.* **56**, 393–407.
Ralph, C. L. (1970). Structure and alleged functions of avian pineals. *Amer. Zool.* **10**, 217–235.
Ralph, C. L., and Dawson, D. C. (1968). Failure of the pineal body of two species of birds (*Coturnix coturnix japonica* and *Passer domesticus*) to show electrical responses to illumination. *Experientia* **24**, 147–148.
Ralph, C. L., and Lane, K. B. (1969). Morphology of the pineal body of wild House Sparrows (*Passer domesticus*) in relation to reproduction and age. *Can. J. Zool.* **47**, 1205–1208.
Renzoni, A. (1970). Developmental morphology of the pineal complex in doves and pigeons (Aves; Columbidae). *Z. Zellforsch. Mikrosk. Anat.* **104**, 19–28.
Renzoni, A., Eakin, R. M., and Quay, W. B. (1968). Cilia of modified structure in avian pineal organs, *Electron Microsc. 1968, Proc. Eur. Reg. Conf., 4th, 1968* pp. 563–564.
Rosner, J. M., de Pérez Bedés, G. D., and Cardinali, D. P. (1971). Direct effect of light on duck pineal explants. *Life Sci., Part II* **10**, 1065–1069.
Sayler, A., and Wolfson, A. (1967). Avian pineal gland: Progonadotropic response in the Japanese Quail. *Science* **158**, 1478–1479.
Sayler, A., and Wolfson, A. (1969). Hydroxyindole-0-methyl transferase (HIOMT) activity in the Japanese Quail in relation to sexual maturation and light. *Neuroendocrinology* **5**, 322–332.
Shellabarger, C. J. (1952). Pinealectomy vs. pineal injection in young cockerel. *Endocrinology* **51**, 152–154.
Shellabarger, C. J. (1953). Observation of the pineal in white leghorn capon and cockerel. *Poultry Sci.* **32**, 189–197.
Shellabarger, C. J., and Breneman, W. R. (1950). The effects of pinealectomy on young white leghorn cockerels. *Proc. Indiana Acad. Sci.* **39**, 299–302.
Stalsberg, H. (1965). Effects of extirpation of the epiphysis cerebri in 6-day chick embryos. *Acta Endocrinol. (Copenhagen), Suppl.* 97.
Studnička, F. K. (1905). Parietalorgane. *In* "Lehrbuch der vergleichenden mikroskopischen Anatomie der Wirbeltiere" (A. Oppel, ed.), Part 5, pp. 1–248. Fischer, Jena.
Tilney, F., and Warren, L. F. (1919). "The Morphology and Evolutional Significance of the Pineal Body," Amer. Anat. Mem. No. 9. Wistar Institute, Philadelphia, Pennsylvania.
Ueck, M. (1970). Weitere Untersuchungen zur Feinstruktur und Innervation des Pinealorgans von *Passer domesticus* L. *Z. Zellforsch. Mikrosk. Anat.* **105**, 276–302.
Ueck, M. (1971). Strukturbesonderheiten der Anurenepiphyse nach prolongierter Osmierung und Anwendung der Acetylcholinesterase-Reaktion. *Z. Zellforsch. Mikrosk. Anat.* **112**, 526–541.
Ueck, M. (1973). Fluoreszenzmikroskopische und elektronenmikroskopische Untersuchungen am Pinealorgan verschiedener Vogelarten. *Z. Zellforsch. Mikrosk. Anat.* **137**, 37–62.
Ueck, M., and Kobayashi, H. (1972). Vergleichende Untersuchungen über acetylcholinesterase-haltige Neurone im Pinealorgan der Vögel. *Z. Zellforsch. Mikrosk. Anat.* **129**, 140–160.

Wolfe, D. E. (1965). The epiphyseal cell: An electron-microscopic study of its intercellular relationships and intracellular morphology in the pineal body of the albino rat. *In* "Structure and Function of the Epiphysis Cerebri" (J. Ariëns Kappers and J. P. Schadé, eds.), pp. 332–386. Elsevier, Amsterdam.

Wurtman, R. J., Axelrod, J., and Kelly, D. E., eds. (1968). "The Pineal." Academic Press, New York.

ADDITIONAL REFERENCES

Dodt, E. (1973). The parietal eye (pineal and parietal organs of lower vertebrates). *In* "Handbook of Sensory Physiology" (H. Autrum, R. Jung, W. R. Lowenstein, D. M. McKay, and H. L. Tueber, eds.), Vol. VII/3B, pp. 113–140. Springer-Verlag, Berlin and New York.

Eakin, R. M. (1973). "The Third Eye," pp. 1–157. Univ. of California Press, Berkeley.

Chapter 3

THE AVIAN SKELETOMUSCULAR SYSTEM

Walter J. Bock

I.	Introduction	120
II.	Descriptive Morphology	124
	A. Introduction	124
	B. Current Work and Needed Projects	126
	C. Status of General Descriptions	132
	D. Status of Veterinary Anatomy	135
	E. Nomenclature	136
III.	Functional Morphology	137
	A. Introduction	137
	B. Terminology	139
	C. Skeleton	144
	D. Cartilage	157
	E. Ligaments and Tendons	158
	F. Articulations	162
	G. Musculature	164
IV.	Physiological Adaptation	223
V.	Ecological Morphology and Adaptation	226
VI.	Comparative Morphology and Systematics	228
	A. Introduction	228
	B. Homology	229
	C. Proof and Disproof	234
	D. Breadth of Comparative Studies	235
	E. Phylogenetic Sequence	236
	F. Nonancestral Similarity	237
	G. Comparisons and Paradaptations	238
	H. Taxonomic Value	245
	I. Role of Functional Studies	248

| VII. Epilogue | 250 |
| References | 250 |

I. Introduction

Morphology, Darwin stated in his "On the Origin of Species" (p. 434), "is the most interesting department of natural history and may be said to be its very soul." And to Darwin's assertion, we may add that study of the skeletomuscular system is the very heart of vertebrate morphology. Investigations of the vertebrate skeleton date from antiquity, being among the earliest biological studies. Comparative anatomy began with the skeletomuscular system. Over four hundred years ago, Belon in his *l'Histoire de la Nature des Oyseaux* (1555) illustrated the skeletons of a bird and of a mammal side by side and labeled, with the same names, the corresponding bones. Ever since the rise of the modern science of ornithology, studies of the skeletomuscular system have been an accurate index to the history of avian morphology, heralding periods of triumph and of utter decline.

It is not without reason that study of the skeletomuscular system has occupied the central position in avian morphology. All support for the avian body is provided by the skeletal system, or more correctly stated, by the connective tissue system. And, with few exceptions, all responses by birds to external stimuli are achieved by muscular contraction, be they skeletal, smooth, or cardiac. Thus, the skeletomuscular system is the foundation for locomotion and feeding, which are the two actions underlying all other activities in avian life history, e.g., reproduction and migration. The host of structures required for locomotion and food capture, the ease of preservation and storage, and the size (gross and histological levels) of the skeletomuscular system provide the avian anatomist with an abundance of easily studied features for all types of morphological studies from functional to comparative systematic and evolutionary. Data arising from study of the skeletomuscular system form the major foundation for our system of avian classification, as well as the basis on which theoretical principles of biological comparison, evolution, and systematics have been postulated. And without doubt, the skeletomuscular system will retain its role as the main empirical foundation for investigation of avian evolution and classification well into the future.

No one will dispute the argument that animals are functional wholes and that selection acts upon the individual organism as a unit and not

3. THE AVIAN SKELETOMUSCULAR SYSTEM

upon its individual components. Yet a sensible division into systems and discrete features is necessary for ease of study and communication. Traditionally, the skeletal system and the muscular system have been treated separately, such as the scheme adopted in the "Biology and Comparative Physiology of Birds." This separation of the skeletal and muscular systems follows that used in an overwhelming majority of research papers and represents a most reasonable subdivision of the avian body. I have chosen the alternative scheme of treating the skeletomuscular system as a unit in the belief that the skeletomuscular system comprises a unified adaptive system whose functional and adaptational properties and whose evolutionary and systematic significances can be comprehended far better as a single interwoven complex than is possible if these systems are considered separately. I hasten to add that the skeletomuscular system is not the only possible scheme nor is it without serious limitation. One could choose the neuromuscular system, which has distinct advantages in that the muscles and their motor control are not severed. Recent studies on avian twitch and tonus skeletal muscular fibers and their distinct motor supply emphasize the importance of studying the neuromuscular system as a unit. Frequently, other features must be included in analyses of the skeletomuscular system before the functional and adaptive significances of the muscle–bone complexes can be comprehended. The osseous arch of the radius in owls is associated with a complex of Herbst corpuscles. The preglossale of the tongue in *Passer* supports the thick pad of epidermis and included sensory endings that comprise the "seed-cup." And the evolution of the unique ectethmoid–mandibular articulation in several genera of meliphagids was understandable only after the salivary glands were described and the functional interactions of the jaw apparatus and the tongue apparatus in the feeding methods of these birds were clarified. I do not gainsay the necessity and value of studies of the skeletal system or the muscular system in isolation. Availability of material or time frequently dictates the scope of particular investigations. And occasionally one part of the skeletomuscular system contains most of the useful information. Among passerine birds, the musculature is far more important than the skeleton in studies of the tongue apparatus. I claim only that analyses of skeletomuscular systems will, in general, provide a significantly better comprehension of these features than obtainable from independent studies of the skeleton or the muscles.

Establishment of reasonable limits of skeletomuscular units in particular studies is a far more difficult problem to solve. Some obvious units are easily recognized, such as the hindlimb and girdle, the forelimb and girdle, the neck and the head in terms of regional morphology, or the feeding apparatus, the flight apparatus and the locomotory (hind limb) apparatus in terms of functional systems. Yet, no matter which type of unit is accepted, two major questions exist: (1) What are the exact limits of the unit; should the neck be included in the feeding apparatus? In some birds, such as herons and anhingas, the neck is definitely part of the feeding apparatus, while in many other birds, the neck appears to be of little importance. (2) Do subdivisions of these broad units constitute valid groupings? Can the jaw apparatus and tongue apparatus be separated, as is commonly done, in studies of feeding adaptations? Or, can the muscles and bones of the wing tip be analyzed independently from the remainder of the wing? No simple solution can be offered; rather it depends upon the particulars of each individual case. Quite likely, the jaw apparatus of seed-eating passerines can be investigated separately from the tongue apparatus, although surprises exist as in the tongue of *Passer*. The tongue apparatus of the Old World nectar feeders can be analyzed with considerable success without reference to the jaw mechanism. Yet the adaptive significance of the ectethmoid–mandibular articulation in *Melithreptus* (Meliphagidae) could not be understood without examining the morphology of the entire feeding apparatus. Subdivision of large functional units is a practical necessarity in many studies of the skeletomuscular system, but it must always be remembered that the analysis of a restricted subunit, no matter how detailed the study may be, involves the risk of missing the solution to the functional or the adaptive importance of the features.

Studies of the avian skeletomuscular system, and indeed of all avian morphology, have always occupied a rather curious peripheral position in the field of vertebrate morphology. Major treatises often include detailed anatomy of numerous forms in every class of vertebrates with an inadequate treatment of one or two avian species, if they are mentioned at all. Occasionally, a few peculiar birds were singled out for closer examination; the ratites have always been a favorite of vertebrate anatomists. The development of avian morphology in isolation from the rest of vertebrate morphology resulted in a double disadvantage. Avian morphologists never benefitted fully from an exchange of ideas, knowledge, and concepts with other vertebrate morphologists during the golden age of morphology. Yet, they shared in the disastrous decline suffered by vertebrate morphology beginning in the early years of this century.

After the decline of interest in morphology, studies of the avian skeletomuscular system continued at a low level. Most significant is that no new distinctive schools of morphological thought developed. No major anatomical summary comparable to Fürbringer's (1888) or Gadow's (1891-1893) monographs has appeared since 1900. A large majority of skeletomuscular studies were comparative or taxonomic in scope; and most were undertaken by workers with little, if any, previous experience or morphological guidance. The accepted philosophy was that anyone could observe, describe, and compare morphological features, as if these structures were no more than geometric figures. A disportionately large number of anatomical papers were written by workers with fewer than five contributions to the literature of avian morphology. Many were doctoral dissertations, after which the worker left avian morphology. This pattern of publication simply reflected the high level of frustration and disappointment among students of avian morphology. Greener fields of study lay elsewhere. The downward course of this long decline may have slowed or even stopped in recent years; but it has not been reversed contrary to numerous statements about the recent revival of interest in avian anatomy.

Aside from offering two comments, it is not my intention to rationalize about the decline of avian anatomy. I do believe that the fault for declining interest in any area of scientific endeavor lies internally to that discipline. If biologists are less than enthusiastic about morphology and if students turn to other fields, then morphologists must reconsider the foundations of their science most carefully. Second, avian morphology is one of the most exciting disciplines in ornithology, and I am perplexed at the continuing failure of morphology to achieve its rightful position in ornithology.

For these reasons I decided against structuring this chapter as a detailed morphological review of the avian skeleton and muscular systems, and a summary of past studies. The value of such encyclopedic treatments cannot be gainsaid, but it would not be the most heuristic approach. The reviews by Fisher (1955a) on avian anatomy in general, Bellairs and Jenkins (1960) on the skeleton, and Berger (1960) and George and Berger (1966) on the muscular system provide excellent coverage of anatomical details, prevailing ideas and concepts, and literature citations. Much of the factual information in these reviews is still current and need not be repeated in this essay. To be sure, some systems, such as the feeding apparatus (jaw and tongue skeletomuscular systems) and the neck are still inadequately reviewed. In spite of its desirability, detailed description of a few isolated segments of the skeletomuscular system would be out of place and have been omitted.

The central theme of this chapter will be an inquiry into the foundations, methodology, and goals of evolutionary morphology as exemplified by the avian skeletomuscular system. Evolutionary morphology may be defined as a comparative biological study of the morphological features in accordance with the principles of evolutionary theory. Its foundations are very broad—as well they need be—because the primary goal of this endeavor is the elucidation of the evolutionary history of morphological structures and of animal taxa. To be successful, evolutionary morphology must be a blend of traditional descriptive and comparative morphology, of the newest concepts of functional and ecological morphology, and of theoretical principles of evolution and phylogeny. The debt of evolutionary morphology to classic comparative anatomy cannot be denied. Nor can the need for continued excellent descriptive morphology be gainsaid. Many earlier morphological developments of excellent merits foundered because too little attention had been given to description.

Birds have been all but completely ignored in the past by classic anatomists, largely because of their small size and their morphological uniformity. Yet those characteristics of birds responsible for their rejection by earlier morphologists make them ideally suited for studies of evolutionary morphology. The wealth of species, well known taxonomically and biologically, exhibiting a wide range in their morphological diversity from great uniformity to considerable difference makes them excellent material on which general principles of biological comparison and of evolutionary mechanisms can be formulated. Their small size, diurnal habits, ease of maintenance in captivity and of experimental manipulation, ease of observation both in the field and in captivity, and the abundance of comparative information from all aspects of avian biology permit studies of functional and ecological morphology of birds on a sophisticated level scarcely attainable in other group of vertebrates. Above all else, the primary goal of this chapter is to show that, not only does avian anatomy offer much to the student of functional and comparative morphology, but that studies of the avian skeletomuscular system have set the pace in the development of evolutionary morphology.

II. Descriptive Morphology

A. INTRODUCTION

Descriptive morphology of the avian skeletomuscular system is, by itself, dull for both the writer and the reader, but it is an essential foundation for the more exciting studies of functional, systematic, and evolutionary morphology. A large body of factual information on the skeletomuscular system—the result of two centuries of work—exists and provides ornithologists with a better comparative knowledge than any other aspect of avian biology except species systematics and general ecology and life history. This has led to the conclusion that factual studies of avian morphology are largely completed, a misconception stemming from several sources. First is a lack of knowledge of the degree of detail of the available morphological information. And second is the nature of the questions to be asked about evolutionary morphology of the skeletomuscular system. In reality, a more correct statement is that only a minor part of available descriptions is really suitable for "modern" studies of functional or evolutionary morphology or for systematic analyses. Indeed, only the barest foundations of descriptive morphology of the avian skeletomuscular system are available.

The seeming conflict between a large body of morphological description and the conclusion that much of it is unsuitable for future morphological studies leaves the ornithologist in a quandry as to what sort of descriptive studies should be undertaken and how these should be approached. Pure description for the sake of morphological description has little value, simply because it is not possible to anticipate future morphological questions and thereby to provide the needed descriptive information. All descriptive studies should be undertaken within a framework of definite questions. These may be ones relating to a definite taxonomical problem, but not simply a broad survey in the hopes of discovering some taxonomic value in a particular feature. Or they may pertain to the elucidation of a functional or ecological question or some principle of evolutionary change. Redescription of a morphological system to correct existing errors is justified, but it is always advantageous to improve the detail of description. Previously unknown morphological features or strikingly different configurations should be described, but one should always attempt to go beyond pure description and ask additional questions. New features frequently provide new data or insights with which previously unsolvable problems can be clarified. It is not necessary to state and attempt to solve definite questions in every descriptive paper, but only to conduct these studies within the framework of some set of questions. In systematic work, it is often necessary to establish a foundation of comparative morphological information in a series of papers. They should never be undertaken in the absence of a broad set of questions, because the soundness of future systematic conclusions depends largely upon the comparative data upon which they are based.

The importance of basing descriptive morphology upon concrete questions is emphasized because I believe that a major weakness of avian anatomy stems from a failure to proceed in this fashion. Examination of a large number of morphological studies, including comparative and systematic ones, reveals that a significant percentage are not directed toward any question, not even one posed tacitly. To be sure, one does not begin every morphological investigation with definite questions in mind. Frequently, these arise during the course of dissection and observation, or the original questions are set aside in favor of new and more interesting ones. But in every case, a descriptive morphological study should always be written within the framework of clearly stated questions.

B. Current Work and Needed Projects

It is always possible to think of additional studies in any field. I would like to give some attention to those aspects of descriptive anatomy of the skeletomuscular system necessary for study of function, evolution, and systematics.

1. New or Forgotten Features

Discovery of a truely new morphological feature is always exciting and usually forms the basis for further functional insights and taxonomic conclusions. Numerous new structures in the avian skeletomuscular system remain to be uncovered, judging from the continued rate of new finds. I would like to mention briefly a few recent discoveries.

a. The Basitemporal Articulation of the Mandible. This articulation is formed by the tip of the internal process of the mandible abutting the basitemporal plate (Bock, 1960a); it provides a medial brace of the lower jaw. The basitemporal articulation is found in a number of avian taxa, exhibits considerable diversity in morphological detail, and clearly arose independently at least ten times. It is of considerable interest as an analog to the evolution of the mammalian jaw articulation (Bock, 1959).

b. The Ectethmoid–Mandibular Articulation. This articulation is formed by the dorsal process of the mandible fitting into and articulating with the ventral fossa of the ectethmoid in a peg-and-socket arrangement (Bock and Morioka, 1971). It is found in three genera

(*Melithreptus, Manorina* and *Ptiloprora*) of the Meliphagidae, presumably arising independently in each. The ectethmoid brace, which exists only when the mandible is fully adducted, allows the mandible to be supported against the brain case independently of the quadrate articulation. Analysis of the functional properties of the ectethmoid brace has clarified several important aspects of avian cranial kinesis.

   c. *The Cruciform Splenius Capitis Muscle.* This muscle is a peculiar criss-crossing origin of the M. splenius capitis in swifts, hummingbirds, and a few other groups (Burton, 1971) in which each muscle arises from the contralateral side of the second cervical vertebra. Burton concluded that the cruciform origin of this muscle permits rapid and extensive head turning, but was unable to ascertain its biological role. Most interesting is the presence of this peculiar muscle arrangement in hummingbirds and swifts (birds with quite different feeding habits), and it may argue strongly for the monophyly of the swifts and hummingbirds.

   d. *The Preglossale in Passer.* This feature is a neomorphic skeletal element in the tip of the tongue (Bock and Morony, 1971). It is a medial element articulating with the anterior tips of the paired paraglossalia. The large paired M. hypoglossus anterior takes origin from the posterior fossae of the preglossale and runs to the paired paraglossalia. The preglossale supports the thick epidermal pad comprising the seed-cup of the tongue. This unique neomorph is present only in the passerine finches (*Passer, Montifringilla,* and *Petronia*) and argues strongly for the monophyly of these genera. Moreover, the morphology of the entire preglossal complex suggests that this group is derived from a primitive group within the Ploceidae, if indeed the passerine finches are related to the weaver finches.

   e. *The "Hatching Muscle."* This term is applied to the transient turgid condition of the M. complexus at the time of hatching (Fisher, 1958). Although its exact function during hatching is still unclear (Bock and Hikida, 1969), this muscle plays some definite role in the hatching process; hence, its nickname is an apt one. Among other attributes of the hatching muscle is its possession of tonus skeletal muscle fibers in addition to twitch fibers (Bock and Hikida, 1968; Krüger, 1952). Preliminary studies of the morphology and mechanical properties of avian twitch and tonus muscle fibers needed for comprehension of their function in the M. complexus has resulted in the discovery that tonus fibers are widespread in the avian muscular system and must be considered in any detailed functional anatomical study of the avian skeletomuscular system.

f. The Osseous Arch of Owls. This low bony arch on the radial shaft (Bock and McEvey, 1969b) was originally described by Shufeldt in 1900, mentioned by Pycraft in 1903, and thereafter forgotten. Any projections or indentations on the shaft of long bones are unusual, and an elongated low arch such as found on the radial shaft of owls is unique. Two muscles attach to the arch, but their attachment does not appear to be a prime factor in the evolution of the osseous arch. Histological sections revealed that the space between the arch and the radial shaft is filled with numerous Herbst corpuscles arranged parallel to the shaft of the bone. Herbst corpuscles have been reported along the shafts of long bones of the wing and leg in most birds (Schildmacher, 1931), but never associated with a special bony feature. The complex of Herbst corpuscles and osseous arch appears analogous to many strain gages in which the properties of a metal block determines the range of the gage. No real hints exist as to the exact function of the arch complex or of its biological role in the flight of owls. Nevertheless, it is a unique feature of all owls and argues for monophyly of this taxon.

2. Description of Single Bone–Muscle Systems

Herein are included the basic descriptions of anatomical systems that serve as the foundation for all other morphological studies. The primary descriptions may be done on a single species or on a comparative basis; the latter is preferable, as it reduces the danger of describing a specialized condition as generalized. These studies have become far less frequent in recent decades, testifying to the excellent monographs published earlier; these will be cited in Section II,C. Yet some systems are still not well described even on the gross level, and the finer detail of most well known divisions of the skeletomuscular system must still be described. I would like to mention a few of these poorly known systems as examples.

The glottal apparatus is an excellent example, because it is simple morphologically, has one major function, and yet a really adequate description of the musculature does not exist. Only the vaguest suggestions on the functions of the glottal muscles — how they open and close the glottis — are available. Yet dissection and description of these muscles should be a simple task.

Passerine jaw muscles have been described by several workers, with excellent descriptions provided by Fiedler (1951). Yet during the subsequent two decades, numerous errors and confusing statements have crept into the literature. Numerous questions on subdivisions of individual muscles, relationships between muscles and cranial features, and the relative development of muscles still exist. Similar problems exist in studies of jaw muscles in other avian orders, with a major question being homologizing these muscles and their subdivisions in different orders (Hofer, 1950; Stark, 1959). The relationship between branches of the trigeminal nerve and the finer subdivisions of the jaw musculature is still a matter of controversy (Barnikol, 1951, 1953b).

Closely associated with description of the jaw musculature is comprehension of the structure and mechanism of avian cranial kinesis. One can still find statements in recent publications that some birds possess an akinetic skull; indeed, such statements are combined with descriptions of jaw muscles that function only within the realm of cranial kinesis. All birds possess the basic morphology of cranial kinesis, and no evidence has ever been presented demonstrating an akinetic skull in any avian species (Bock, 1964). Equally problematic are the numerous vague and incomplete descriptions of the form and function of jaw ligaments and musculature that bear upon comprehension of the operation of the avian kinetic skull.

Passerine tongue muscles are still so poorly known on the gross level that no useful functional or taxonomical investigations can be undertaken. The best description of passerine tongue muscles is found in Leiber's (1907) study of the tongue apparatus of woodpeckers, an excellent descriptive anatomy of a skeletomuscular system. Mudge (1903) described clearly the myology of parrot tongues.

Although we possess detailed myologies of appendicular and axial muscles, knowledge of fiber composition, including fiber types, number, length, and arrangement, is very scanty. Fiber arrangement is perhaps best known in the jaw muscles. Cracraft (1971b) presented one of the first detailed accounts of fiber composition in pelvic limb muscles; I know of no similar description of pectoral myology. Cracraft also stressed ligamental and articular structure in his analysis of the pigeon hindlimb, two additional aspects of the avian skeletomuscular system that have received inadequate treatment in the past.

3. *Descriptions of Single Species*

Descriptive anatomies of a skeletomuscular system in a single species are often done as the basis of a functional study or an analysis of the ecological–adaptive aspects of the structural features (e.g., Miller, 1937; Manger Cats-Keunen, 1961; Fisher and Goodman, 1955). Herein, the primary concern in description is for morphological information that may be correlated with functional properties and with ecological factors. Frequently the information desired in these studies differs considerably from that needed in comparative–systematic studies. The analyses by Zusi (1962) on the head and neck of skimmers, by Spring (1965) on climbing and pecking in woodpeckers, by Bühler (1970) on the head of caprimulgids, and by Bock and Morioka (1971) on the ectethmoid–mandibular articulation in meliphagids are examples of single-species studies (or basically noncomparative if several species were included). These studies were done within the framework of a definite set of questions relating mainly to function or adaptation.

4. Comparative Anatomical Descriptions

These studies have been directed primarily to systematic and phylogenetic questions and, with the inclusion of functional and ecological information, to evolutionary questions. Because comparative descriptions constitute a majority of anatomical studies, they are well known to most ornithologists and little need exists to cite representative examples. A few comments should be offered on some special problems and on some less well known approaches.

Success of any comparative study depends primarily upon the choice of the anatomical features to be described and compared. Unfortunately, much less attention has been given to this phase of comparative study than it deserves. Numerical taxonomists, for example, base many of their comparative studies on measurements of structural features with no considerations of whether the measurements are really comparable in the species being studied or whether they have functional and evolutionary importance.

A special problem of this type exists in the recent series of studies by George Hudson and his students (Hudson *et al.*, 1966, 1969; Vanden Berge, 1970). These taxonomic studies depend primarily upon measurements and ratios thereof of the appendicular skeleton and musculature. Unfortunately, the rationale for most of these measurements and ratios is not presented. Careful analysis of these characteristics fails to convince me that most of these measurements and secondary ratios are correlated with functional properties of the skeletomuscular system or with ecological factors and hence are meaningful indicators of phylogenetical or evolutionary relationships. The same conclusion must be applied to the use of muscle weight or mass in many comparative studies of the skeletomuscular system (e.g., Goodman and Fisher, 1962; Klemm, 1969; Raikow, 1970). Muscle weight or mass correlates poorly or not at all with most of the essential functional properties of the skeletomuscular system and is consequently of very restricted value in comparative studies. Unless proper consideration is given first to the features and characteristics to be compared, the tremendous efforts utilized in obtaining the measurements and analyzing the data are often negated in these comparative morphological–systematic investigations.

The detailed comparative study of the skeletomuscular system of two closely related species of murres (*Uria aalge* and *Uria lomvia*) by Spring (1971) deserves special mention as a prototype of a comparative investigation of adaptive differences between rather similar and closely related species. With the use of statistical analysis of a large number of measurements of skeletal elements and muscles, Spring was able to establish correlations between fine morphological differences and ecological differences, including feeding, between these two species of alcids. Future advances in morphology will depend more and more upon comparative studies of the nature and detail set forth by Spring.

The method of single character study (Bock, 1960b) is a special subdivision of comparative investigations. In this approach, attention is focused narrowly upon a single feature rather than an entire functional complex such as the pelvic musculature. The restriction in breadth of morphological comparison must be compensated by a deeper consideration of functional and evolutionary aspects, for otherwise the comparative aspect of these studies is greatly weakened. It is essential to choose the feature very carefully, as it must exhibit meaningful variation within the taxon being studied. Functional analysis of the feature under consideration almost always embraces a larger complex of features as, for example, the single-character study of the lacrimal–ectethmoid complex (Cracraft, 1968), which required a functional analysis of cranial kinesis. Some compromises must be reached during any such study, because it is usually not feasible to undertake a detailed comparative study of the whole functional complex of which the single feature is a component part. Whenever a single character study is undertaken, the inherent limitations must always be kept firmly in mind, since otherwise serious taxonomic errors could be made.

A few general comments should be offered on the interpretative aspect of comparative systematic studies that influence the choice of features and their description. Taxonomic conclusions do not have to be offered at the conclusion of every comparative study. Single character analyses, for example, are taxonomically useful only after a number have been completed. Furthermore, the assumption that a careful, detailed study of every morphological complex will result in taxonomic conclusions is simply not justified. An anatomical feature need not possess the same taxonomic value over the entire taxon being studied. It can easily be shown that a morphological complex, such as the tongue morphology in passerine birds, provides excellent taxonomic data in some families and is useless in others.

Thus, all extensive comparative morphological–systematic studies should be preceded by sufficient preliminary investigations of functional, adaptive, and other aspects of the features and careful consideration of the potential taxonomic usefulness of the comparisons. Otherwise much time and effort will be wasted on relatively meaningless descriptions and comparisons.

C. STATUS OF GENERAL DESCRIPTIONS

A detailed morphological description of the avian skeletomuscular system is not necessary in a general review of this sort; however, reference to monographs treating the descriptive morphology and nomenclature of such a system must be included for the convenience of the avian anatomist and nonanatomist alike. In each case, I have tried to choose the best descriptive monographs, as well as the ones most readily available, and those that present a good synonymy of anatomical terms. Whenever feasible, a recommended nomenclature will be indicated.

1. Skeleton

The skeleton is the best known, by far, of all internal anatomical systems in birds, and yet it is not easy to cite a good general account of avian osteology; the descriptive section in Bellairs and Jenkins (1960) is not adequate. Many extensive comparative analyses of the avian skeleton illustrate the bones thoroughly, but omit labeling and therefore are not suitable as a general reference. The best general account of the avian skeleton is still H. Howard's (1929) introduction to osteological elements in her 'Emeryville Shellmound' paper. Most later works, including paleontological studies, use her paper and terminology as a base. Papers on fossil birds generally treat the long bones of the limbs, especially the details of the articular surfaces and the small processes and foramina close to the ends of the bones, in greater detail than studies on recent birds. A review of the entire skeleton except the vertebral column and a few miscellaneous elements can be found in the osteology of the Turnices (Bock and McEvey, 1969a).

The most nearly complete description and nomenclature of cervical vertebrae is that of Boas (1929), which is followed by most later workers. More recent papers by Zusi (1962) and Zusi and Storer (1969) describe the neck vertebrae in detail and include excellent labeled illustrations. Treatments of the entire vertebral column are rare, and one may have to go back to such papers as Mivart (1874, 1879) or the general works of Gadow and Fürbringer.

The skull has been described in a large number of papers, but a complete description and consistent terminology for all structural features of the cranium are still not available. The difficulty stems simply from the far greater complexity of the skull and the prodigious diversity in cranial morphology of birds as compared with the complexity and variability of the postcranial skeleton.

Details of the passerine skull can be found in Bock (1963a,b, 1964), Bock and Morioka (1971), and Richards and Bock (1973) and of the nonpasserine skull in Bock and McEvey (1969a). These papers contain numerous references to earlier studies. Jollie (1957) presented an excellent review of the structure and development of the skull in *Gallus*; he advocates a rigid terminology based upon the limits of embryological bones. This system is not practical in birds in which the bones of the adult skull are fused extensively and also somewhat differently in diverse taxa. Although knowledge of the limits of embryological bones and of their development is important in comparative and evolutionary studies of the avian skull, I do not advocate Jollie's terminology.

The tongue skeleton has been well treated by Leiber (1907), Engels (1938), George (1962), Bock and Shear (1973), and Richards and Bock (1973).

2. Articulations

Description of articulations, including articular surfaces and articular ligaments, is woefully scant. Stolpe (1932) and especially Cracraft (1971b) have described the articular structures of the avian hind limb. The articulations, including ligaments, of the wing were described thoroughly by Sy (1936). Some of the articulations in the skull were described by Fourie (1955) and Schoonees (1963), but a good general description of all cranial articulations is not available.

3. Ligaments

The ligamental system includes those fibrous bands that are clearly not part of the articulations. We know nothing about the general ligamental system of the appendicular and axial skeletons; a ligamental system in addition to those present at articulations may be absent in the limbs. Ligaments of the jaw apparatus have been studied extensively by a number of authors, namely, Lebedinsky (1921), Starck (1940), Fiedler (1951), Barnikol (1952), Davids (1952), Fisher (1955b), Goodman and Fisher (1962), and Bock (1964); these ligaments were reviewed by Bock (1964). His terminology should be used. Several new ligaments are known in the jaw (Bock and Morony, 1972). At least one, and possibly more, ligaments are present in the tongue apparatus (Bock and Shear, 1973). Boas (1929, pp. 143–148) describes the ligaments in the neck, which are apparently composed of yellow elastic fibers.

4. Muscular System

The gross morphology of the avian muscular system is well known and has been the subject of many excellent monographs during the past several decades. The appendicular musculature has been described by Hudson (1937, 1948), Hudson and Lanzillotti (1955, 1964), Hudson *et al.* (1959), Fisher (1946), Fisher and Goodman (1955), Berger (1952, 1953, 1954, 1955, 1956a,b,c, 1957), Klemm (1969), Raikow (1970), Cracraft (1971b), and others. Cracraft is the only author to pay attention to fiber composition and arrangement.

Berger (1960, 1966; in George and Berger, 1966) presented the most recent and complete reviews of the avian myological system. His coverage is excellent for the appendicular and body muscles, and becomes less authoritative for the jaw and tongue muscles—a reflection of the accuracy of available studies. Because Berger's 1966 monograph is the most complete treatment of avian myology, I would urge that his terminology be accepted. This would help to solve the problem of the conflicting nomenclatures employed by Fisher and by Hudson for pelvic musculature; Berger follows Hudson's system more closely. Another problem that would be avoided is the names of the muscles to the digits of the wing which depend upon whether one concludes that the existing digits are numbers 1, 2, and 3 or 2, 3, and 4. The major difficulty with Berger's review is that he has not relabeled the figures to agree with the muscle terminology that he advocated. If this labeling problem is known, it becomes a minor nuisance rather than a serious difficulty.

Most workers (Zusi, 1962; Zusi and Storer, 1969; Berger, 1966) follow Boas' (1929) treatment and nomenclature of the neck musculature.

Analyses and nomenclature of avian jaw muscles are based upon the monograph of Lakjer (1926). Hofer (1950) discusses jaw muscles in many nonpasserine birds, and Fiedler (1951) covers passerine birds. A few name changes exist for the individual muscles, but most of the disagreements exist in the subdivisions of the individual muscles. The scheme advocated by Beecher (1953) has several serious errors in subdivisions of individual muscles. Berger's (1966) handling of the jaw muscles is considerably less clear than his other sections. A recent review of passerine jaw muscles by Bock (1973b) follows the earlier papers of Hofer and Fiedler. Although it treats only passerine birds, the system of muscle subdivisions and terminology is sufficiently general to serve as a starting point for studies of nonpasserine jaw muscles.

The tongue musculature remains as one of the most inadequately analyzed system in avian myology. The best papers are those of Leiber (1907) on woodpeckers plus a few other species, and Mudge (1903) on parrots. Later studies by Moller (1930, 1931), Engels (1938), and George (1962) are simply too vague in details. An extensive review of passerine tongue muscles (Bock and Shear, 1973) is being prepared. It follows the terminology of Engels most closely and hence differs considerably from Berger's treatment (1966). It is sufficiently general to be used as a starting point for studies of nonpasserine tongue muscles although it is clear that specializations exist in some groups (e.g., the parrots).

The study by Ames (1971) provides descriptions, and terminology that should be followed in studies of passerine syringeal muscles.

The innervation of avian muscles has been neglected in recent years, and workers must refer back to the treatises of Fürbringer and Gadow, or to the recent papers of Fisher (1946) and Fisher and Goodman (1955). Hofer (1950), Barnikol (1951, 1953a,b,c), and Starck and Barnikol (1954) described the innervation of the jaw muscles but disagree on some fine points. This problem is reviewed by Bock (1973b). The details of innervation of avian muscles is in serious need of a careful review.

D. Status of Veterinary Anatomy

An almost complete gap separates zoological and veterinary avian anatomists, as reflected in different interests and approaches in anatomical research and in radically different sets of morphological terms. The major works in veterinary anatomy are those of Bradey (1915), Kaupp (1918), Sisson et al. (1938, and later editions) and Chamberlain (1943). Three recent works on the turkey (Harvey et al., 1968), the Japanese quail (Fitzgerald, 1970), and the Budgerigar (*Melopsittacus undulatus*) (Evans, 1969) also treat the skeletomuscular system in detail. The recent monographs, with the exception of Evans, follows the nomenclature of Chamberlain closely. This has a great shortcoming in being based on a mammalian nomenclature that is unsuited for avian studies without careful evaluation. The detail and accuracy of dissections, descriptions, and figures in these works vary considerably. Unfortunately, they are exceptionally poor in their treatment of head anatomy, which I know best and on which I tend to evaluate an anatomical atlas. Chamberlain has an extensive description of the ligaments associated with articulations which is not covered in equal detail in any zoological monograph on avian anatomy.

My general conclusion is that the monographs on veterinary anatomy are not as useful to ornithologists as are works done on non-domesticated birds. Except for details of the anatomy of the species treated in the veterinary anatomies and sections such as that by Chamberlain on ligaments, I would recommend against using these sources.

Evans' (1969) study is an exception. He approached the description of Budgerigar morphology from the viewpoint of a zoologist giving an accurate account of the anatomy of this parrot and a good bibliography to the anatomical literature. Unfortunately, he was greatly restricted in space and had to present very brief descriptions.

E. NOMENCLATURE

Anatomical nomenclature shares many of the problems of zoological nomenclature, but lacks the excellent mechanisms by which zoological nomenclature achieves stability. Anyone is free to provide new names or choose between existing names, with general usage depending largely upon the quality (and quantity) of the anatomical study and the authority of the worker. The desired goal is to apply the same name to homologous anatomical features throughout birds, or tetrapods, or even vertebrates if that is possible. It is not always possible to agree on homologies, or features are not always comparable as homologs in a convenient way for nomenclature, and even when definite homologies are established, the problem of choosing one of several possible names still exists.

The status of avian anatomical nomenclature is reasonably good, with quite stable sets of names in use for most parts of the skeletomuscular system. Unfortunately, some of the names are of hybrid origin, based upon mammalian names that do not apply to birds. Hence, the names M. stylohyoideus and M. thyreohyoideus are used for tongue muscles in spite of the fact that birds lack a styloid process and a thyroid cartilage; nevertheless, these names have been widely used and should not be changed simply because they are misnomers.

Terminology for the skeletal system is quite stable, with most workers following Howell. Some difficulties exist with smaller processes and foramina. And names for parts of the skull are not as well worked out because of the greater detail and greater diversity in skull morphology. However, the nomenclature for avian cranial features is slowly improving. Two main systems for hind limb muscles are those of Hudson (1937) and of Fisher (1946). A similar problem exists for some of the wing muscles because of disagreement in numbering of the remaining three digits of the avian manus (whether I, II, III or II, III, IV). I would urge that avian anatomists follow the nomenclature for the muscles of the axial skeleton, the leg, and the wing used by Berger (in George and Berger, 1966) and would like to refer to Berger's discussion on nomenclature (pp. 226–227). For jaw muscles, I recommend the terminology advocated by Bock (1973b), which follows Lakjer (1926); and for tongue muscles, the terminology advocated by Bock and Shear (1973), which is based upon Engels (1938). Terminology for syringeal musculature should follow that suggested by Ames (1971).

Another source of difficulties is the differences between the anatomical names used in ornithological morphology and those used in veterinary morphology. A long-term effort to establish a "Nomina Anatomica" has been underway but with little real success. This has been rejuvenated recently as an "International Committee for Avian Anatomical Nomenclature" under the chairmanship of Alfred M. Lucas and A. S. King. Although a central aim of this committee is to provide a system of names suitable for veterinary anatomy, it is hoped that the resulting list will be equally suitable for the needs of ornithological anatomy.

III. Functional Morphology

A. INTRODUCTION

In considering the functional morphology of the avian skeletomuscular system, I shall be concerned with functional analysis and with concepts of adaptation at the level of bony and muscular tissue. Bones and muscles will be discussed as individual organs, but I will not be concerned with the functional and adaptive significance of actual features in particular species. These analyses are largely mechanical, as the skeletomuscular systems can be analyzed in terms of physics and mainly in mechanics and strength of materials. Any college-level textbook in physics can provide the necessary background in physical analysis. Better yet would be an elementary text in mechanics and in strength of materials (e.g., Den Hartog, 1961a,b; Levinson, 1961, 1963) in combination with a physics textbook. Alexander (1968) provides an excellent introduction to biomechanical methods. I must stress that analysis of the skeletomuscular system must include more than a mechanical approach. Several aspects of muscle function and adaptation are better understood in terms of physiological factors rather than mechanical ones. Muscle–nerve interactions are doubtless important, although insufficient comparative information is available to provide a foundation for judgment. Sensory organ–bone interactions also exist, such as the aggregates of Herbst corpuscles along the shafts of long bones (Schildmacher, 1931) and under the span of the osseous arch on the radius of owls. Glands, blood vessels, and other soft organs interact with bones and can modify their shapes via mechanisms of physiological (=somatic) adaptation.

In functional morphology lies one of the major current areas of research on the avian skeletomuscular system, and one in which training is still poor. A thorough knowledge of functional morphology is essential for all types of morphological studies, including those done as a basis for systematic investigations. Some workers have discounted the importance of functional morphological studies especially in systematics, but do so mainly out of ignorance. I know of no case in which taxonomic conclusions were not improved by a functional analysis of the taxonomic features, and can cite a number of examples in which difficult taxonomic problems have been solved or sounder conclusions have been reached after a thorough functional investigation.

Methods for functional analyses of complex muscle–bone systems, such as the avian hindlimb or the jaw apparatus, will not be included in the following discussion because of their complexity and because the development of these methods is still in its infancy. A number of different approaches are covered in Cracraft (1971b), Goslow (1972), Spring (1965, 1971), Bühler (1970), Bock (1964), Bock and Morioka (1971, 1973), Bock and Shear (1973), and in papers cited in these studies. These works should be consulted for possible appropriate methods of functional analyses; however, the reader must be warned that it is quite possible that none of the approaches used and advocated in the above-cited papers may be suitable for a particular problem at hand. Careful judgment must be exercised in the decision of which method of functional analysis to use in any study of a complex muscle–bone system.

B. Terminology

Studies of functional morphology and the adaptive significance of morphological features have been hindered because of vague and/or conflicting definitions of many critical terms. These terms must be clearly defined to avoid confusion in the following discussion and in future studies of avian morphology.

1. Adaptational Terms

I will follow the terms and definitions advocated by Bock and von Wahlert (1965); namely the following.

a. Feature. Any part or attribute of an organism will be referred to as a feature if it stands as a subject in a sentence describing the organism. The feature is the adaptation to a particular aspect of the environment.

b. Form. In any sentence describing a feature of an organism, its form would be the class of predicates of material composition and the arrangement, shape, or appearance of these materials, provided that these predicates do not mention any reference to the normal environment of the organism. It must be emphasized that a feature may possess a range of forms, such as modifications of the lens of the eye, or change in length and diameter of a muscle fiber as it contracts and shortens, or the shape of a flight feather because of changing air pressures as the bird flies.

c. Function. In any sentence describing a feature of an organism, its functions would be that class of predicates that include all physical and chemical properties arising from its form (i.e., its material composition and arrangement thereof) including all properties arising from increasing levels of organization, provided that these predicates do not mention any reference to the environment of the organism. It should be emphasized that features possess many functions that are never utilized by the organism, yet these are valid functions and are worthy of being studied. Moreover, as the form of a feature changes, its function automatically changes. This definition of biological function is free, as it should be, of any form of teleology, Aristotelian or otherwise, or of any form of teleonomy; it does not involve any aspect of purpose, design, or goal. A feature generally possesses a number of functions simultaneously even if it has only one form. Not all of the functions of a particular feature need be investigated by an anatomist, especially if it is clear that the functions are not pertinent to the problems at hand.

d. Faculty. A faculty is defined as the combination of a form and a function of a feature. Or more formally, in any sentence describing a feature of an organism, its faculties would be that class of predicates each of which includes a combination of a form (material composition and arrangement) and a function (physical and chemical properties) of the feature, provided that these predicates do not mention any reference to the normal environment of the organism. The faculty, comprising a form and a function of the feature, is what the feature is capable of doing in the life of the organism and is the unit that bears a relationship to the environment of the organism. The faculty is the unit acted upon by selection and is the aspect of the feature adapted to the environment.

e. Biological Role. In any sentence describing a feature of an organism, the biological roles would be that class of predicates which includes all actions or uses of the faculties (the form–function complexes) of the feature by the organism in the course of its life history, provided that these predicates include reference to the environment of the organism. Essential to the description of a biological role is the observation of an organism living naturally in its environment. A biological role cannot be determined by observations made in the laboratory, a zoo, or under other artificial conditions. Studies of function may be, and usually are, made on captive animals or upon preparations (e.g., a muscle–bone preparation) in the laboratory. This is the basic distinction between study of functions and study of biological roles; in morphology, the former would be functional anatomy while the latter would be biological anatomy.

f. Synerg. A synerg may be defined as a link between the organism and its umwelt formed by one selection force of the umwelt and one biological role of a faculty. (See Bock and von Wahlert, 1965; pp. 279–281, for a discussion of the concepts relating to environmental factors.) Selection acts as individual environmental units on individual biological roles of faculties. It does not act on the form or the function of a feature independently of the other.

g. Niche. The niche of an organism is defined as the sum of all its synergs. Thus, the niche is the total relationship between the whole organism and its complete umwelt. This definition of the niche agrees with that used by some ecologists following Elton. It is important to note that the niche of a species can change without modification in its umwelt (habitat) if the animal uses the factors of its umwelt in different ways.

3. THE AVIAN SKELETOMUSCULAR SYSTEM

h. Adaptation. An adaptation is a feature of the organism that (a) has a role in the life history of that organism (i.e., must perform some definite task, such as obtaining food or escaping from a predator), and that (b) performs this biological role with a certain degree of efficiency. Hence, an adaptation is always part of the organism and it is a feature possessing a faculty (a form–function complex) which has a biological role and interacts with some environmental factor of the umwelt of the organism. The bond between the faculty and the environmental factor is the synerg formed by the biological role interacting with the selection force.

The degree of efficiency of an adaptation must be measured in terms of some system other than survival or differential reproduction. One such measure would be the amount of energy required by the organism in the operation of the adaptation because all organisms require energy to maintain life. It is obvious that the organism must maintain all of its synergs—that is, it must maintain its niche—with its available energy. Because this energy is limited, it is advantageous for the organism to use the minimum amount of energy possible to operate each individual synerg. Hence, the degree of evolutionary adaptation, the state of being, is defined as the minimal amount of energy required by the organism to maintain successfully the synerg if a single biological role of a faculty is considered, or to maintain successfully its niche if the whole organism is considered. Thus the faculty is well adapted to the environment when the operation of the synerg requires less energy and is poorly adapted when the operation of the synerg requires more energy. The less energy used, the more successfully the synerg or the niche will be maintained, and the more energy will be available to the organism for the operation of other synergs or to meet unexpected or strenuous conditions. Last, evolutionary adaptation, the process, is defined as any evolutionary change which tends to decrease the amount of energy required to maintain successfully a synerg, or the niche as the case may be.

2. Physical Terms

Several physical terms should be defined because they are so frequently misused by morphologists. I follow the standard physical definitions. I make no attempt to define all physical terms that must be used in mechanical analysis of the avian sheletomuscular system.

a. Force. Force is a push or a pull, be it a contact force or an action-at-a-distance force, which acts to modify the equilibrium of a mass.

When a force acts on a mass, that body will be accelerated or decelerated. Forces are vector quantities and hence possess a direction as well as magnitude; hence, they can be summed only by vector addition in which both the magnitudes and vector directions are included.

Forces are measured in dynes or newtons. Grams and kilograms are units of mass. A newton is the force that imparts to a mass of 1 kg an acceleration of 1 m/second^2. If grams and kilograms are used as units of weight or force, as is most commonly done, they should be called gram-weight (gm-wt) or kilogram-weight (kg-wt).

When muscles contract, the force developed is correlated with the number of contracting fibers and the relative length of the fibers.

b. Work. Work is the product of the displacement of a mass and the component of force in the direction of the displacement. Work is done only when a force exerted on a body moves that body in the direction of a component of the force. If the body is not displaced, or if the body is displaced, but not in the direction of a component of force, then no work is done. A muscle contracting isometrically does not shorten and hence does no work regardless of the amount of force developed. Because of the operation of muscles in a skeletal lever system, muscles frequently do not shorten and hence do no work. The concept of work is of restricted application in mechanical analyses of the skeletomuscular system.

The metric system units of work are dyne-centimeters (ergs) or newton-meters (joules). Work is a form of energy. Heat is another form of energy, and is important in evaluating relative adaptation of muscles.

c. Power. Power is the rate at which work is done. The unit of power, the watt, is defined as 1 J/second. Power can be measured only if work is being done; if no work is done, as in an isometrically contracting muscle, then no power is developed. Power, like work, is not a useful concept in studies of the skeletomuscular system.

When muscles contract, they develop *force,* but not necessarily *power.* This distinction is not always made by morphologists. Only if a muscle shortens and moves a body in the direction of a component of its force is work done and power developed.

The correlation between power output and weight of muscles as used by some functional anatomists (e.g., Gray, 1968) is of dubious value.

d. Torque. Torque, or moment of a force, is the product of a force and the length of its moment arm. The length of the moment arm of a force is the perpendicular distance from the axis of rotation of a body to the line of action (vector of the force). The consequence of the torque produced by a force is to impart circular acceleration to the body on which the force acts. Torque is a vector quantity and is measured in centimeter-dynes (or meter-newtons).

e. Stress. Stress results when two equal but opposite forces act on a body, and it is measured as force per unit area. If the two forces pull away from each other, then the body is subjected to a tension. If the two forces push toward one another, then the body is subjected to a compression. If the two forces, either a pull or a push, are offset from each other (i.e., their vectors do not lie on the same line of action) then they will place a shearing stress or simply a shear on the body.

When a stress is placed on a body, even one composed of homogeneous material, by a pair of compressive forces, for example, the forces do not pass through the body along the paths of the original vectors. Rather the forces are distributed in a pattern of stresses (a stress field) depending on the shape of the object and the placement of the external forces. Most objects (e.g., avian bones) are subjected to asymmetrical stresses, hence the pattern of internal stresses is very uneven and some parts of the object are subjected to very high stress while other parts experience extremely low stress.

f. Strain. Strain is the relative change in dimension of a body subjected to a stress. Strain is distortion of a stressed body. It is measured as the ratio of change in dimension relative to the original dimension; hence strain is a dimensionless number. Stress and strain are closely related concepts, but they should not be interchanged.

An elastic modulus is the ratio of a stress to the corresponding strain. Young's modulus, for example, is the stretch modulus and is the longitudinal stress relative to the strain; it is the same for compressive and tensile stress on the same material.

g. Breaking Point. Breaking point is the limit to which a material may be strained under a stress without rupturing. The breaking point of an object depends upon its strength at every point in the object relative to the distribution of stress within the object. Objects that are subjected to asymmetrical stresses must withstand differing amounts of stress throughout their volume and must be sufficiently strong to withstand the maximum stress experienced at any point, not the average stress on the object.

h. Elasticity. Elasticity is the ability of a material deformed under a stress to return to its original shape after the deforming stress is removed. A perfectly elastic material will return exactly to its original form after the distorting force is removed; a perfectly inelastic material does not return at all. A steel ball is a highly elastic object while a putty ball is quite inelastic. Avian bones and bands of collagenous fibers (tendons and ligaments) are highly elastic materials.

i. Compliance. Compliance is the ability of a material to be deformed under stress. A highly compliant material deforms readily while an incompliant material deforms little under a stress. Compliancy and elasticity are quite different properties. A rubber band or ball is highly compliant and moderately elastic while a steel ball is incompliant and highly elastic. Collagenous ligaments are incompliant, but highly elastic. Unfortunately, many workers use the term elastic when they mean compliant. Thus, numerous statements can be found in the anatomical literature that collagenous ligaments are inelastic by which the authors mean that these features are noncompliant. Statements referring to "elastic ligaments" are ambiguous, as in most cases it is not clear whether the author means a ligament composed of yellow elastic fibers, a ligament that is compliant, or a truly elastic but possibly noncompliant ligament.

Other terms will be defined as they are used in the following discussions of mechanical analyses. I cannot overstress the importance of employing the precise physical definition in all studies of functional anatomy, especially when the term has a looser general definition, as does elasticity. Unless precise physical definitions are used, the functional statements will often be too ambiguous to permit any real advance in this field.

C. Skeleton

1. Introduction

Functional studies of connective tissue structures, including bones, are centered upon mechanical forces acting upon them and the strength of the structures to withstand these forces without damage (for general discussion, see Koch, 1917; Murray, 1936; Pauwels, 1965; Kummer, 1959a,b, 1962, 1966). It is essential to specify exactly the forces involved at each point in the skeletal framework of the body and to show how the connective tissue structures can withstand these forces. Statements of the sort "that a ligament protects an articulation against damage" or "that a bone is massive enough to withstand forces placed upon it" are much too vague to be of any value either to comprehend the functional significance of the structure in the bird being studied or to use in any comparative study.

Adaptation of skeletal and other connective tissue structures must include, of course, the ability to fulfill successfully its necessary biological roles. This ability must be measured in terms of some degree of efficiency such as energy utilization. Direct measurement of energy consumption is both rather difficult and meaningless in judging adaptation of connective tissue structures because these structures do not utilize metabolic energy directly or in proportion to the demands placed upon them as they carry out their biological roles. A widely used basis for judgment of adaptation of connective tissue features that is in agreement with the definition of degree of adaptation advocated by Bock and von Wahlert (1965) is the maximum–minimum principle. In brief, this means maximum strength of a connective tissue element against the usual forces acting upon it (i.e., the maximum force that may act during the life of the individual) with the minimum amount of material. The maximum–minimum principle must be carefully stated and applied. A structure is not stronger because it has less material, it is actually weaker. A solid bone is stronger than a hollow bone, but it weighs more, and the additional weight either does not add significantly to the strength of the element or adds strength in an unessential way. Energy is saved by the maximum–minimum principle partly through reduction of the metabolic energy needed for the development and the maintenance of the connective tissue structure, but this factor accounts for only a small part of the total energy saved. The major portion of the energy used by an animal for its connective tissue structures is that required for their transportation by the animal. Herein weight becomes important. The heavier these structures are, the more energy is required by the animal to accelerate and decelerate itself as it locomotes. For flying creatures, such as birds, the maximum–minimum principle becomes even more critical than for most land- and water-bound creatures. Not only is more energy required for a heavier bird to transport itself, but it requires an increasingly larger and more precise flying apparatus. Thus, it is no real surprise that a functional–adaptational analysis of the elements in the avian connective tissue system will reveal a precise arrangement of material to provide maximum strength for a minimum amount of material.

Bones possess strength against compressive, tensile, and shearing forces and can be loaded with any of these forces or a combination of them. When a bone is subjected to the action of a single force or a combination of forces, then the bone may be accelerated (moved) and/or stressed. The consequences of these forces require analyses that clarify the pattern of stresses within the bone (internal forces) and the action of the several external forces on the bone.

2. Analysis of External Forces

Several methods of analysis are suitable to investigate external forces; I prefer the method of free-body diagrams (Dempster, 1961; Bock, 1966, 1968a), because it is clear and straight-forward, and because it provides answers to most pertinent questions about the action of external forces on bones. The disadvantages of free-body diagrams are shared by other methods of analyzing forces. Free-body diagrams are advantageous because they can be applied to both static and dynamic (both rotational and linear motions) conditions and they can treat a number of forces acting on a bone including the force at the articulation. These diagrams require few assumptions that can be modified according to the exact conditions of the system being studied. The assumptions necessary for the following discussion of free-body diagrams are: (a) that the skeletal elements are rigid; (b) that no friction or other force-absorbing processes occur at the points where external forces act on the skeletal elements; and (c) that the articulations are ideal joints that are frictionless and do not store energy that may be released later. If bones deform (bend) under the forces usually applied to them, or if friction exists at the articulation or at the point where external forces are applied to the bone, then these conditions must be recognized and the basic assumptions must be modified accordingly.

I present here the method of free-body diagrams using the avian upper jaw (see Bock, 1966) as an example and only for static cases; dynamic cases are better treated below in the discussion of muscles. Indeed, the method of free-body diagrams illustrates clearly that muscles and bones are best considered as a single functional unit.

In the first case, the upper jaw of a crow (Fig. 1) will be considered when the bird is biting on an object and when it is pounding on an object with the tip of its bill. The crow, as all birds, has a kinetic upper jaw with its axis of rotation located at the nasal–frontal hinge. All forces acting on the maxilla exert a torque (or moment) on it depending upon the magnitude of the force and the relationship between the line of action of the force vector and the center of rotation which is the nasal–frontal hinge.

When the bird bites on an object, the force is provided by the M. pterygoideus (mainly) and is transmitted to the upper jaw via the palatines; it is shown in Fig. 1A as F_m. The object being bitten exerts a force against the upper jaw normal to the curvature of the tomium; it is shown in Fig. 1A as F_r. The articular force acts at the nasal–frontal hinge and its vector direction can only be guessed prior to analysis of all forces; it is shown as F_a. The magnitude of the forces, if known, can

3. THE AVIAN SKELETOMUSCULAR SYSTEM

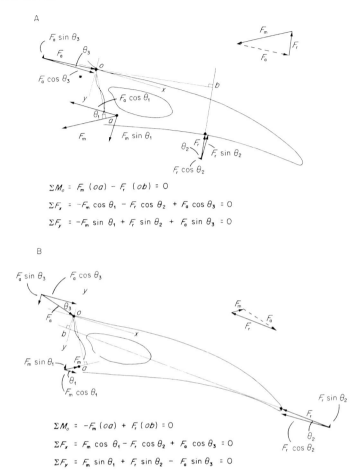

FIG. 1. Analysis of the external forces acting on the upper jaw of a generalized bird, e.g., *Corvus*, using the method of free-body diagrams; static conditions are assumed. (A) Analysis of forces when the bird is biting on an object held at the center of the maxilla. (B) Analysis of the forces when the bird is pounding on an object with the tip of the maxilla. Equations for the sums of moments and of forces are given below each free-body diagram, and the graphic method for vector addition of forces is shown directly above each diagram. (Modified after Bock, 1966, Fig. 2.)

be represented by the relative length of the vector arrows. An arbitrary set of xy axes is placed on the diagram; each force intersects the y-axis at some angle α. Moment arms are shown for each torque producing force; a moment arm is the perpendicular drawn from the center of rotation to the line of the force vector. Clockwise torques are designated arbitrarily as positive, counterclockwise forces as negative. Components of force acting to the right along the x-axis and upward along the y-axis are designated as positive, the opposite forces are negative.

Each force exerts a rotational and linear action on the maxilla; however, the bone is assumed to be in a static state so that the sum of the moments and the sum of the forces must be equal to zero as is shown by the equations below the maxilla in Fig. 1A. Summation of the forces can be done by decomposing each force into its rectilinear components according to the xy axes and adding the forces. Or it may be done graphically as shown in the inset (the unknown force F_a illustrated by a dashed arrow). Graphic vector addition requires that the length of the arrows be drawn proportional to their magnitude.

The arrangement of forces is different when the bird is pecking at an object with the tip of its bill. In this case the object places a compressive force on the bill tip that would tend to depress the maxilla. A protraction force (F_m) provided by the M. protractor quadrati et pterygoidei would be exerted on the upper jaw via the palatines. Again, the sum of the moments and of the forces acting on the maxilla must be equal to zero because it is static; vector summation is shown graphically in the inset.

These free-body diagrams and equations illustrate several points. If the bird wishes to bite harder on an object held in the bill, it must develop more muscle force (increase F_m) or hold the object closer to the base of the bill (decrease ob); either change will increase F_r. Both types of changes would also result in an increase of the force (F_a) at the nasal–frontal hinge and perhaps its vector direction toward the y-axis. The force F_a must be kept below the breaking point of the thin plate of bone constituting the nasal–frontal hinge. When the bird pecks at an object, increased force on the object results in a larger F_r force and consequently a larger muscle force (F_m) is required to balance it. Increase in these forces increases the magnitude of the articular force and changes its vector direction toward the y-axis because of the details of this particular example. The force on the bone of the nasal–frontal hinge would not only increase but also tend toward a shearing stress against which the bone would have the least strength.

Modification in bill structure to one specialized for application of force at the tip can be seen in woodpecker bills (Fig. 2). In these birds, a large force is exerted against the tip of the bill (including the maxilla) when the bird hits the tree with its bill. This force of the tree on the bill (F_r) is opposed by the force of the protractor muscle which is transmitted to the upper jaw via the palatines (F_m). For simplicity, these forces are arranged parallel to the x-axis. Because of the length

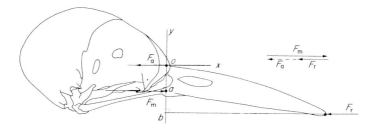

$$\Sigma M_o = -F_m(oa) + F_r(ob) = 0$$

$$\Sigma F_x = F_m - F_r - F_a = 0$$

FIG. 2. Analysis of the external forces acting on the upper jaw of a woodpecker, using the method of free-body diagrams; static conditions are assumed. The forces shown are those existing when the bird is pounding against a tree with the tip of its bill. Equations for the sums of moments and of forces are given below the free-body diagram and the graphic method for vector addition of forces is shown just above it. (Modified after Bock, 1966, Fig. 3.)

of their moment arms, force F_m is greater than force F_r; hence the force at the nasal–frontal hinge is a tensile, not a compressive, force and is probably relatively small. In Fig. 2, force F_a is the force of the brain case on the maxilla. It shows that the brain case is being pulled away from the upper jaw at the instant the woodpecker slams its upper jaw against a tree trunk. The overlap of the frontal bones of woodpeckers over its nasal–frontal hinge appears to be correlated with this mechanism (Bock, 1966, pp. 28–29).

Specialization in the direction of heavy biting forces is seen in finches (Fig. 3). Similar modifications can be seen in parrots, hawks, and owls. In finches, the decurved bill results in the muscle force (F_m) and the resultant force of the seed on the bill (F_r) being almost parallel to each other and of about the same magnitude, which reduces the force at the nasal–frontal hinge. With further decurvature of the bill, the muscle and resultant forces will become more parallel and closer in magnitude, thereby further reducing the force at the nasal–frontal hinge.

Several general observations can be made. The shape of the bone affects the location and direction of the force vectors, but otherwise bone shape has no effect on the analysis of external forces. The magnitude and direction of the force at the articulation is critical. This force must be withstood by the bony structure of the articulation and more importantly by the ligamentous system of the articulation. The small muscles located about the hinge may provide force to counterbalance large forces on the bone that might disrupt the articulation. Because

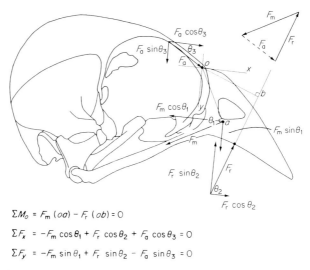

$$\Sigma M_o = F_m (oa) - F_r (ob) = 0$$
$$\Sigma F_x = -F_m \cos \theta_1 + F_r \cos \theta_2 + F_a \cos \theta_3 = 0$$
$$\Sigma F_y = -F_m \sin \theta_1 + F_r \sin \theta_2 - F_a \sin \theta_3 = 0$$

FIG. 3. Analysis of the external forces acting on the upper jaw of a finch, using the method of free-body diagrams; static conditions are assumed. The forces shown are those existing when the bird is shelling a seed held in the center of its maxilla. Equations for the sums of moments and of forces are given below the free-body diagram and the graphic method for vector addition is shown just above it. (Modified after Bock, 1966, Fig. 4.)

these articular muscles have a small moment arm, they produce little torque and hence have little influence on the overall torque placed upon the bone. Additional forces can be included in the analysis by simply adding terms for each additional force. Dynamic cases can be easily analyzed as will be discussed below (Section III,G).

To date, few investigators have included analyses of torque production in their functional investigations of bony elements. I must stress that both the effect of torque produced by each force and the linear effect of each force must be included in any analysis. A discussion of only torque to the exclusion of linear aspects is as erroneous as the usual omission of torque. Almost no consideration has been given to the torque developed by muscles acting on the long bones of the hind leg and the wing.

3. Analysis of Internal Forces

Study of the internal forces in bones involves the analysis of the internal distribution of stress and the arrangement of material of the skeletal element to best withstand these stresses. Almost no studies of this type have been conducted on avian skeletal elements, and there is not even a detailed knowledge of the internal morphology of avian bones. For many years, the classic example, and almost the only example, has been a segment of the metacarpal bone from the wing of an unknown species of vulture showing the internal struts, taken from D'Arcy Thompson. More recently, Manger Cats-Kuenen (1961) and Bowman (1961) illustrated the arrangement of internal struts in the skull and attempted to analyze this aspect of skeletal morphology.

One difficulty is that simple analytic methods, such as free-body diagrams, are not available to study the pattern of stress distribution within solid bodies. Empirical approaches, such as photoelastic techniques, must be used. Kummer (1959a,b, 1962, 1966) discussed the techniques and results of photoelastic analysis of stress distribution in bones (see also the studies of Pauwels, 1965, who did much pioneering work in this area of functional morphology). These methods have been applied to analysis of the avian upper jaw (Bock, 1966; pp. 37–44) and the avian mandible (Bock and Kummer, 1968), the latter including a photoelastic study.

When a pair of forces are applied to a solid body, a pattern of stresses is formed within the body depending upon the position of the applied forces and the shape and arrangement of material of the body. The pattern of stress within the body can be represented by a system of lines of force or trajectories. Each trajectory represents a certain amount of force. In regions of the body that are subjected to great stress, the lines crowd closely together, and in regions of low stress, few are present. Hence, representation of stress with a pattern of trajectories is exactly the same as representing elevations on a map with contour lines. Basically, the trajectories correspond to the two principal stresses (tension and compression) in the object, with the tension trajectories crossing the compression trajectories at right angles. The work of Pauwels, Kummer, and others has shown that the distribution of bony material within a skeletal element corresponds closely to the pattern of stresses as represented by the trajectory system. Where stresses are great, solid compact bone is present. In areas of low stress, the bone is hollow. And the pattern of bony trabeculae corresponds to the distribution of trajectories of force.

A uniform application of compressive or tensile forces on a body would result in a uniform stress pattern with the magnitude of force homogeneous throughout the cross section of the body. The amount of force in any part of the cross section would be equal to the total force divided by the cross-sectional area. Such cases are neither overly interesting nor very common in the avian skeleton. Almost all stressing forces that act on the skeleton are asymmetrical, and usually extremely so. Asymmetrical stresses produce bending moments that may be represented by sets of tensile and compressive trajectories. The magnitude of the stress in different parts of the cross section of the object varies to an extreme degree, with almost no stress in the middle of the object and very high stresses along the periphery. The maximum stress resulting from asymmetrical loading depends on the bending moment produced by the applied forces (their magnitude times the distance between their force vector and the center of bending) and on the placement of the material within the object, namely, the distance between the material and the center of bending. The further the material is placed away from the center of bending, the lower will be the stress for a given asymmetrical loading. The periphery of a pipe would be stressed less than a solid rod containing the same amount of material and loaded asymmetrically in the same way.

The strength of an object to any stress, whether it is a symmetrical or asymmetrical loading, is not dependent on the mean stress across its cross section relative to the strength of its material, but rather on the maximum stress that exists anywhere within the object relative to the strength of the material at that spot. In symmetrical loading, the stress over the cross-sectional area is uniform, so that the strength of the entire object is dependent on the mean stress. But in offset loading, the stress (be it compressive or tensile) varies greatly and is usually highest along the periphery of the object. The strength of the entire object (its breaking point) is dependent upon the maximum stress and the strength of the material to withstand this stress. If the material at any point is strained beyond its breaking point, it will rupture there and the entire object will break if the stressing forces are maintained. Thus, in asymmetrical loading, the problem would be to prevent the build-up of stress at any location to the extent that the material is strained beyond its breaking point at that spot.

The strength of an object subjected to offset loading may be increased by arranging the constituent materials so that the amount present at any spot is proportional to the stress. Thus, the distribution of bony trabeculae follows the pattern of stresses. Maximum stress at the periphery of bones is resisted by compact bone. And the magnitude of this stress is reduced by increasing the diameter of the bone. Another set of mechanisms that reduces the stress of offset loading is that of counterbalances in which additional forces are placed on the object so as to reduce the degree of asymmetrical loading and hence the maximum stress experienced by the object. In a counterbalancing mechanism, the mean stress is increased but, importantly, the maximum local stress is decreased. The distribution of internal forces and the methods for withstanding the maximum stresses or for reducing them can be seen in the following examples.

The maxilla of birds (e.g., a crow as shown in Fig. 4) is attached to the brain case via the nasal–frontal hinge and the palatines. If the bird bites on an object, an offset compressive force is applied to the tomium (see Bock, 1966; pp. 37–43); this is resisted by a tensile force applied to the upper jaw by the palatines as discussed above. The asymmetrical force on the upper jaw establishes a stress distribution approximately as that shown in Figure 4B. This pattern of stress is the same as that found in an asymmetrically loaded cantilever beam (Fig. 4C).

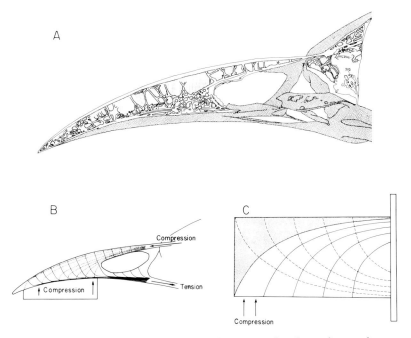

FIG. 4. Internal structure of the upper jaw of *Corvus* and analysis of internal stresses. (A) Midsagittal section of a crow maxilla to show the arrangement of internal trabeculae. The deeper, nontrabecular area of the jaw is shown in fine stippling, and outer bony surfaces are designated by heavier stippling. Compare the arrangement of trabeculae with the pattern of trajectories shown in B, and note the thin sheet of compact bone at the nasal–frontal hinge. (B) Arrangement of trajectories (lines of force representing a stress field) within the maxilla when the bird closes its bill against an object. Solid lines represent tension and dotted lines represent compression. The illustrated pattern of trajectories is schematic, serving to provide an impression of the stress pattern within the upper jaw. (C) The trajectorial diagram of a cantilevered beam drawn upside down. The area outside the outer trajectory is shaded to show the resemblance of the beam to the general shape of the avian upper jaw. (Modified after Bock, 1966, Figs. 7, 8, and 13.)

The trajectories of force within the maxilla correspond approximately to the distribution of bony trabeculae (Fig. 4A). Areas of low stress, as in the center of the basal half of the bill, can be devoid of bone and can house other structures, such as the nasal cavity, and can permit the presence of holes in the surface of the object such as the external nares. Maximum stresses are located in the layer of dense bone along the outer shell of the maxilla.

The avian mandible is hinged at its quadrate articulation and has a series of jaw muscles attached to it. All adductor muscles attach anterior to the quadrate hinge (see Bock and Kummer, 1968). When the bird bites on an object (see Fig. 5), compressive forces are placed on the dorsal edge of the mandible close to both ends, with a tensile muscular force pulling on the mandible somewhere between the compressive forces. This pattern of forces is exactly like that in a beam supported at both ends (Fig. 5C). Photoelastic studies of a mandibular model demonstrate that the maximum stress exists along the dorsal and ventral edges of the ramus (Fig. 5A and Fig. 6). Cross sections of mandibles reveal that thick compact bone forms the dorsal and ventral edges of the ramus, with thin bone and trabaculae composing the central three-fourths of the mandibular height. A large mandibular fossa can be present because the central portion of the ramus experiences little stress (Fig. 6).

Long bones of the wing and leg are subject to extreme offset loading and consequently to large peripheral stresses. These bones have a maximum diameter and are hollow, with thin walls of very dense bone. The large diameter of the bones reduces the stress in the walls which can be withstood by the strength of the dense bone. The thin walls and hollow center of the bones reduce their weight without sacrificing strength.

The outer shape of many bones is correlated with the necessary distribution of material to withstand stress. No advantage is gained by having material outside of the curvature of force trajectories, and hence bills are rounded (see Bock, 1966, pp. 41–43). Curvature of long bones or of the sternum and synsacrum may also correspond to the pattern of stress distribution, but this has not been studied. Any indentations, holes, or projections on bones act as stress concentrators, with the result that stresses are always greater near these features. Quite commonly, more bone is present around the edge of a hole or at the base of a process. I must emphasize that processes serve as stress concentrators and weaken the bone in the area of the projection even though more bony material is present than in the absence of the projection. Thus, the shafts of long bones are smooth and lack both foramina and processes unless such features are absolutely necessary. If any such features are present in the shaft of a long bone, one can con-

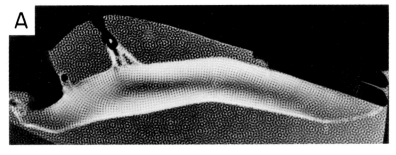
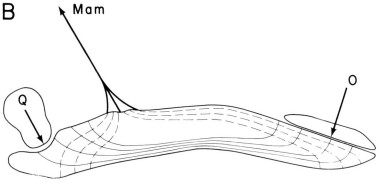
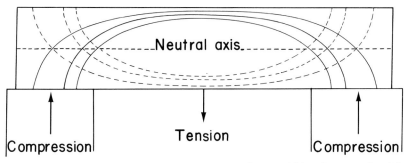

FIG. 5. Analysis of the internal stresses within the mandible of a generalized bird (*Corvus*) based on a photoelastic study of a plexiglass model. The forces shown are those existing when the bird is biting upon an object with the dorsal adductor muscles of the mandible contracting. (A) The original photograph composed of the nine exposures taken during the experiment. It shows the trajectorial pattern of forces within the mandible. (B) Interpretation of the trajectorial pattern with compression shown by solid lines and tension by dashed lines. Note the distribution of forces relative to the dorsal and ventral edges of the lower jaw. (C) The trajectorial diagram of a beam supported at both ends to show its similarity to the avian lower jaw. (Modified after Bock and Kummer, 1968, Fig. 4.)

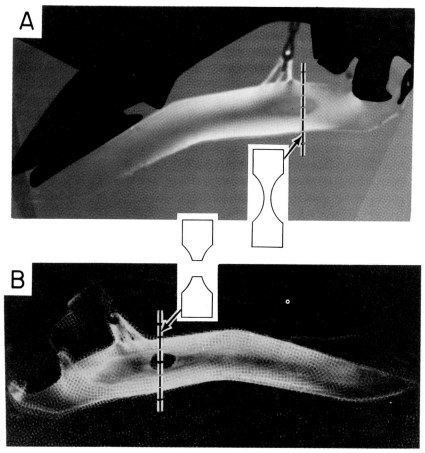

FIG. 6. Analysis of the internal stresses within the mandible of a generalized bird (*Corvus*) based upon a photoelastic study of a plexiglass model. The forces shown are those existing when the bird is biting upon an object with the dorsal adductor muscles of the mandible contracting. (A) The original photograph showing the trajectorial pattern with a partly evacuated mandibular fossa. (B) The original photograph showing the trajectorial pattern with a complete mandibular fossa. This photograph shows the trajectorial pattern especially well. (Modified after Bock and Kummer, 1968, Fig. 9.)

clude with reasonable certainty that they possess important functional roles that deserve study. The osseous arch of the radius in owls is a good example.

Another method of reducing stress in a bony element is to redistribute the force to another part of the bone or another bone that contains sufficient bony material to withstand the force. The kinetic maxilla of birds provides an excellent example of redistribution of stress from a potentially weak area (skull roof) to a strong area (base of the brain case). In birds, the base of the skull must be thick for several reasons. If the large forces resulting from biting forces on the maxilla can be transmitted to the brain case base instead of to the skull roof (as is the case in mammals) it would be possible to thin the bone of the skull roof and lateral walls, and hence reduce the weight of the skull. The kinetic mechanism of the avian skull (Bock, 1966, p. 50) permits this redistribution of stress to the base of the skull, which possesses thick walls.

Large peripheral stresses can be reduced by counterbalancing, as in the use in engineering of counterweights to balance asymmetrical loads. It is easier to carry two heavy pails of water than a single one. Counterweighting devices are not feasible in birds because of the construction of the skeletomuscular system and the ever changing pattern of external forces. But counterbalancing may be achieved by the placement of muscular forces and of ligaments. For example, Pauwels showed (see his collected works, 1965, pp. 226–231; Bock, 1968a, p. 25) that the distribution of muscles in a limb may serve to reduce the maximum stress within the bones while providing them with the same amount of torque. Other muscles, such as the iliotibial tract in mammals, place a direct counterbalancing force on the bone and thus reduce the maximum tensile and compressive stresses within the bone (see Pauwels, 1965). Such systems have not been studied in birds, but there is little doubt that numerous examples will be found once the avian skeletomuscular system is examined with these ideas in mind.

D. CARTILAGE

Aside from its role in development, cartilage is present in the avian skeletomuscular system mainly as articular pads. Cartilage can withstand compressive forces, but unlike bone, cartilage deforms considerably under stress. When it is subjected to a compressive shock, cartilage will yield and hence lengthen the duration of the impact time, which decreases the maximum force on the skeletal element during the impact. Hence the cartilaginous articular pads serve as shock-absorbing mechanisms between bones.

Some articular pads are composed of fibrocartilage. The basitemporal articulation of the mandible in skimmers and presumably other birds (Bock, 1960a) and the ectethmoid articulation of the mandible in *Melithreptus* (Bock and Morioka, 1971) are two examples. Fibrocartilage articular pads are presumably stronger against shearing stress and probably also compressive shocks. However, these specialized articular pads in birds require additional study before their functional significance can be understood.

E. LIGAMENTS AND TENDONS

All fibrous bands possess strength only against pulling forces; they will collapse if loaded with a compressive or a shearing stress and hence are flexible. They are elastic, but the degree of compliance depends upon the material composition. Bands composed of collagenous fibers are highly elastic and quite noncompliant, stretching only 2–3% at the breaking point. Ligaments formed of yellow elastic fibers are elastic and relatively compliant, yielding to about 150% of their resting length at their breaking point. The strengths of collagenous and yellow elastic fibers differ greatly, with collagenous fibers being up to ten times stronger than yellow elastic fibers. I know of no studies of the physical properties of avian tendons and ligaments, so that any general statements on these properties are by analogy from mammalian studies.

Yellow elastic ligaments can be discussed first because so little is known about them in birds. Boas (1929; pp. 143–148), which remains one of the best discussions available, covered yellow elastic ligaments in the avian neck. Other workers describe tendons as "fibroelastic" but are not clear on the exact nature of the fibrous material. The fibrous bands in the leading edge of the propatagium and in the rear border of the metapatagium of the wing may be composed of yellow elastic fibers, but I am not aware of any definite analysis. Much more attention must be given to the distribution of yellow elastic ligaments in the avian skeletomuscular system before valid generalizations can be formulated.

Collagenous fibered bands are flexible inextensible structures that can be loaded only in tension; they may be considered as a rope or chain wherever they are found in the avian skeletomuscular system.

Many muscles are connected to bones by tendons that transmit the muscular force to the site of their action on the bone. The advantages of tendinous insertions are several-fold, apart from the requirement of tendons in pinnate muscles. First is that the surface area of avian bones is insufficient to permit all muscles to insert directly via "fleshy" attachments. Moreover, only a small part of the total bony surface is desirable for muscle attachment in terms of mechanical and/or physiological advantage. Tendons permit a number of muscles to insert on a restricted surface area. Tendons permit muscles to span a large distance between their origin and insertion without requiring excessive length of muscle fibers. They also permit location of muscles in a favorable position in the body away from their site of action, e.g., location of the toe flexors and extensors at the proximal end of the tibotarsus, which reduces the moment of inertia of the hind limb. And they permit change in direction of the muscle pull, e.g., bending of the tendon of the M. supracoracoideus at the triosseal canal. Lastly, tendons may store energy momentarily to be released in conjunction with the force development of other muscles, but little is known about this possible function even in mammals.

The basic properties of collagenous fibers are, not surprisingly, those required for tendons. A tendon must transmit the force developed by the muscle; hence it must be much stronger than muscles if the diameter of tendons are smaller than the cross-sectional area of the muscle fibers. Tendons possess a strength against tensile stress of thirty or more times per unit of cross-sectional area than can be developed by muscle fibers. Flexibility of collagenous fibers is essential, as tendons must curve about bones at articulations and must be able to bend at their bony attachment. The special fibril arrangements of tendons at their bony attachment have been analyzed by Mollier (1937). Lastly, collagenous fibers are highly noncompliant especially when arranged in parallel as in tendons and ligaments. If a tendon stretched as the muscle contracted, the force delivered to the bone would be reduced because part of the muscle force is used to stretch the tendon and because the muscle fibers would shorten and drop into a lower part of the tension–length curve. Tendons do stretch slightly (2–3%), which may be advantageous in smoothening out the onset and termination of force development and thereby eliminating jerky movements; however, little is known of this aspect of tendon function.

Ligaments are collagenous-fibered bands between bones and may be separated into two main groups—articular ligaments and linkage ligaments. In both cases, the basic functions of ligaments depend upon the same functional properties of collagenous fibers discussed above for tendons. Most ligaments are closely associated with articulations, passing from one bone to the other in a complex pattern (see Sy, 1936; Stolpe, 1932; Cracraft, 1971b). The articular ligaments serve two main functions: (1) They bind the bones of the skeleton together and prevent disruption of the articulations when tensile and/or shearing stresses are placed on them. (2) They fix the type and extent of movement between the two bones at the articulation. Few articulations possess bony ridges and processes that regulate the movement of the bones.

Linkage ligaments are uncommon and may be restricted to the head; none have been reported from the postcranial body. These ligaments are characterized by their separation from articulations and (usually) by their spanning two or more joints. The postorbital ligament of the head is an excellent example (Bock, 1964, pp. 9 and 18–25). Linkage ligaments, as all collagenous-fibered bands, are flexible, noncompliant, and can be loaded only in tension. Hence, the two bony attachments of these ligaments can move toward one another and to the side, but cannot move away from one another except for the almost imperceptible stretch of the ligament under tension. Thus, the possible movements of the two bones joined by the ligament are limited or coupled. In the case of the postorbital ligament, the quadrate bone lies between the attachments of the ligament on the brain case and on the mandible. Because this distance must be held constant, the quadrate must swing forward (and hence the maxilla is raised) to reduce the distance between the brain case and mandible before the lower jaw can be abducted (see Bock, 1964, pp. 18–25). This coupling mechanism of the postorbital ligament permits several functions of the kinetic skull that would be lacking in the absence of this ligament or a substitute. Other jaw ligaments, e.g., the occipitomandibular ligament, may possess similar coupling functions. A ligament runs from the dorsal surface of the ceratobranchiale, close to its proximal end, to the dorsomedial surface of the paraglossale, close to its articulation with the basihyale (Bock and Shear, 1973). This ligament spans the basihyale and hence two articulations. Quite possibly, it possesses a coupling function, but this has not yet been analyzed. The postcranial body should be searched for possible coupling ligaments.

One remaining property of ligaments is that of energy storage when stretched and its release (as force) by elastic recoil as the ligament returns (shortens) to its original length. This property may be possessed by one-joint articular ligaments or by linkage ligaments. It must be emphasized that the ligament need only to be elastic to possess this property of elastic recoil; the ability to yield greatly under a tensile stress is not necessary and indeed may not be desirable. The position of a recoil ligament is very essential, as it must lie on the opposite side of the articulation from the force that will stretch the ligament and supply it with energy. Two as yet undescribed ligaments in tyrannid flycatchers and cotingids between the quadrate and the pterygoid and mandible possess a recoil function (Bock and Morony, 1972). These ligaments lie at the anterior side of the quadratomandibular hinge opposite to the M. depressor mandibulae. When the bird opens its mouth widely, the ligament is stretched by the M. depressor mandibulae and the force of this muscle used to stretch the ligament is stored in the taut ligament as energy. Energy, not force, can be stored in ligaments. When the M. depressor mandibulae relaxes, the stretched ligament is released and it shortens by elastic recoil to its normal length. The force applied by the recoiling ligament to the mandible and quadrate closes the jaws. Moreover, the force is applied immediately to the mandible upon relaxation of the depressor mandibulae with the peak force reached almost instantaneously. The force for closing the jaws is provided by the M. depressor mandibulae produced when the jaws are opened. Hence the bird is able to snap its jaws together at great speed immediately after it opens them, which would be advantageous to a flycatcher which must close its bill quickly but with little resistance until its prey is engaged. After the insect is caught, the jaw muscles could provide any additional force required to hold it in the bill. This snap mechanism operates only if the bird opens its bill sufficiently wide to stretch the ligament. When the bird does not open its bill to its maximum extent, nothing happens; hence it can sing without its bill snapping shut during the middle of a phrase.

Elastic recoil mechanisms operate only when the ligament is loaded with energy. A ligament held taut at its normal length would not have a recoil function. Thus, the postorbital ligament would normally not serve in this mechanism.

I must stress that the elastic recoil mechanism and the coupling mechanism of ligaments operate because an external force is placed on the ligament, usually by some muscle. Ligaments act and react passively. They cannot, by themselves, keep two bones of an articulation tightly ajoined, but prevent their separation. The discussion of Manger Cats-Keunen (1961, pp. 28–29) on the function of the jaw ligaments is obscure because she did not clarify whether the ligaments were reacting to external forces placed upon them or acted like stretched springs. (Part of the difficulty in this discussion stems from problems in translation, but it demonstrates the need for exact statement.)

A word should be said about ossified tendons. If a tendon (or ligament) is ossified, the bony segment possesses the general properties of bones. For many ossified tendons, the major change is from a flexible structure to an inflexible one (strength against shearing forces). Thus, the tendon is not able to bend as seen in the ossified platelike tendons of origin of the M. pseudotemporalis superficialis on the rear orbital wall of some finches. The ossified tendons of the leg muscles as well shown in gallinaceous birds (Hudson *et al.*, 1965), present a problem because they could not bend in the ossified segments even if the tendons were not ossified. Hudson showed that these tendons remain unossified in the segments that must curve around articulations. For these tendons, it is possible that the bony portions stretch even less than the collagenous fibers (see Bock and von Wahlert, 1965, p. 274). The length of the ligament suggests that even a small percentage stretch of the tendon would result in a relatively large percentage shortening of the short-fibered leg muscles, and hence the fibers would drop into a low tension portion of the tension–length curve. Ossification of the tendons would reduce the total stretch and help maintain the fibers in a high-tension part of the curve. This suggestion requires further testing.

F. ARTICULATIONS

Whenever two bones come into contact, they form some type of specialized juncture zone. If the bones do not move relative to one another, this juncture is a suture or an ankylosis in which the bones are fused together. Indeed, within a relatively short time in the life of an individual, especially during ontogeny, immobilized diarthroses will modify to a suture or ankylosis (Murray and Drachman, 1969).

Contacts between bones or bony segments in which one element moves or bends relative to the other are articulations or joints. These are usually, but not always, found between embryonic bones; some articulations in the skull are notable exceptions. Articulations are divided into two main groups: (1) diarthroses in which a joint cavity is present; and (2) synarthroses in which a continuous intervening substance is present between the bones. Synarthroses are further divided according to the connecting material and may be (1) a synostosis if the material is bone, (2) a synchondrosis if the material is cartilage, and (3) a syndesmosis in which the material is fibrous tissue. An amphiarthrosis applies to an articulation allowing little movement; this term alludes only to the functional ability of the joint and not to its structure. The nasal–frontal hinge in many birds is an excellent example of a synostosis as this articulation is formed by a thin flexible sheet of bone. The basitemporal articulation of the mandible in plovers and presumably the ectethmoid articulation of the mandible in some species of *Manorina* (Meliphagidae) are syndesmoses.

Articulations permit movement between bones with the type and limit of movement usually restricted by ligaments. Thus, joints serve as the center of rotation of bones and the point from which torques of muscular and other forces are calculated. It is generally difficult to pinpoint the exact center of rotation at a joint, and this rotational axis may shift as the bones move relative to each other. Because bones are able to rotate at articulations and because of the arrangement of muscles, ligaments, and other force producing and transmitting devices on the bones abutting at the articulation, the entire skeletal system is usually stronger than if the articulation were absent and the bones fused solidly together. This consequence stems from the reduction of high local stress concentrations in the bones as shown by Pauwels (1965) and Kummer (1959a,b, 1962, 1966). This effect is especially noticeable where the bones lie at an angle to one another at the articulation, but exists even when the bones lie in the same axis. Moreover, articulations permit the redistribution of a force from one part to another in a complex system, such as the skull, and hence route the force to a portion of the skeleton where the bone is sufficiently thick to resist the strain developed within itself. These functional properties of articulations depend only upon the ability of the bones to move relative to one another and not upon the actual amount of rotation. Pauwels' observations were made on static systems. Similarly, the observations on the properties of the avian kinetic skull and on the mechanism whereby the stress transmitted to the base instead of the roof and sides of the brain case (Bock, 1966) were based upon static conditions. Thus, the presence of an articulation in some position in the avian skeleton does not necessarily imply that much movement can take place at that point. And articulations that permit little movement are not less functional or less of an articulation because little rotation may occur there as compared to other hinges. Many statements can be found in the literature that articulations, e.g., the kinetic hinge between the maxilla and brain case, are nonfunctional because little actual movement exists; such statements are ill founded and these articulations must be analyzed further. Moreover, these general properties of articulations, together with ligaments and muscles, to reduce stress and redistribute forces within the skeleton have not been considered for birds. These properties deserve critical analysis.

Because of their structure, articulations are potentially weak points in the skeletal framework of the body, although it should be noted that breaks of bones appear to be more common than disruptions of articulations. Damage to joints includes actual dislocation and rupture of articular ligaments. Articulations would be decidedly weak points, were it not for the presence of ligaments and muscles that regulate force distribution about and on the articulation. The *de novo* evolution of an articulation within a bone or a complex of fused bones is almost impossible unless a complex of muscles is already present to control forces placed on the newly developing hinge. Otherwise, a great danger of extreme movement and dislocation of bones would exist at the evolving joint. And it can be mentioned that little need exists for a complex of muscles (and ligaments) to be present in a region of the skeleton lacking an articulation. Few, if any, data are available that bear on the question of whether fully developed articulations are actually weak points within the skeleton. Possibly they are not, but considerable study is required before we have even rough answers.

Studies of articular function in the avian skeletomuscular system are rare; those of Sy (1936), Stolpe (1932), and Cracraft (1971b) have been mentioned, and some theoretical considerations have been suggested in Bock (1966, 1968a). This is an area that deserves much additional study including good comparative descriptions on the gross and histological levels, direct observation of articular movement, and experimental studies of possible functions. Most important is to consider articulations as part of a broader skeletomuscular system extending beyond the limit of the bony condyles that are abutting against each other.

G. MUSCULATURE

1. Introduction

Study of the muscular system has always been one of the weakest links in vertebrate morphology, as may be ascertained by perusal of texts in comparative anatomy. The relative lack of attention given to vertebrate musculature may be attributed to several factors. Its study requires a high degree of manual skill and is long, tedious work on messy and frequently ill smelling specimens. Comparative studies of muscular systems require much more time than do similar studies of the skeleton. These features in different species are difficult to compare and recompare directly without the opportunity to return to a dissection and check an overlooked or newly discovered aspect of the musculature. Muscles are much more difficult to describe and to illustrate completely. Studies of individual variation in muscular systems are burdensome even when suitable series of specimens are available, which is rare itself. And comparative myological studies are hampered more than osteological work because of large gaps in species representation in collections of alcoholic specimens. Lastly, many morphologists are also paleontologists or directly interested in comparing the results of their work on living vertebrates with the morphology of fossil forms; hence, their interest would lie in osteology and not with myological studies.

Coupled with the lesser interest of vertebrate morphologists in myology is the greater complexity of muscle physiology that is essential for comprehension of functional morphology of the musculature. Except for studies of human anatomy directly related to medical problems, a considerable gap separated the work of morphologists and of physiologists studying vertebrate skeletal muscles. The unfortunate consequence is that most morphologists are still unaware of the basic aspects of functional morphology of the muscular system and are still uncertain of which morphological parameters are essential to observe and record for functional, and hence adaptational, considerations. We still lack for birds, and indeed all vertebrates, the necessary morphological data with which comparative functional studies of skeletal muscles can be undertaken.

The following discussion is an attempt to clarify the significant physiological properties of muscles and to correlate these properties with morphological parameters as a foundation for future work on the functional morphology and adaptive significance of avian skeletal muscles. The discussion is based upon extrapolation of physiological information from other vertebrates and theoretical analyses of muscle function. Very little has been published on the physiology of the mechanical properties of avian skeletal muscles (see Goslow, 1972). We have completed enough studies of mechanical properties (e.g., tension–length curves, force–velocity relationships) of avian twitch and tonus muscles in my laboratory to convince me that physiological information from other vertebrate groups may be applied to birds. Only striated skeletal muscles will be discussed.

2. Resting State

A muscle receiving no motor nerve stimulation is in a relaxed or resting condition. Such a muscle is soft and yields readily to any stresses placed upon it. The shape of the muscle is maintained by its internal skeleton of collagenous fibers. A resting muscle is not producing any force. If a tensile force is placed on a resting muscle at its origin or insertion, the muscle will stretch depending upon the force placed upon it. Resistance to the tensile stress is provided by the collagenous skeleton of the muscle and the collagenous fibers in the sarcolemmas of the muscle fibers. The stress–strain relationship of a resting muscles follows a definite curve (see Fig. 14A) that appears approximately at the reference length of the muscle, rises slowly at first and then very steeply until the breaking point is reached. Any discussion of passive mechanisms involving muscles, such as the passive perching mechanism (Bock, 1965), must take the resting tension curve of muscles into consideration.

Muscle tone (not to be confused with tonus fibers, see below) is the result of continuous low-level motor nerve stimulation of the muscle. Some muscle fibers are contracting at all times, which provides the low level of force delivered by the muscle. Muscle tone and its consequence of holding parts of the body in place must be separated sharply from passive mechanisms. Maintenance of muscle tone requires a constant, if low, level of energy utilization above the resting metabolism of the muscle, whereas passive mechanisms require no additional energy.

3. Active State

When a muscle is stimulated via its motor nerves, it is activated — it is said to be in the active state — and hence contracts. Each individual twitch muscle fiber is activated by an all-or-none mechanism and will contract if the stimulus is above the threshold value. The extent of the contraction of the twitch fiber, be it the tension developed or the distance shortened, is dependent upon the duration of the volley of stimuli. Thus, the peak tension in a single twitch of a twitch fiber is far less than that reached during a tetanic stimulus. Tonus muscle fibers also exhibit a twitch response, but it is uncertain whether they possess an all-or-none response or contract more with stronger stimuli. Whole muscles, be they of twitch, tonus, or a mixture of fibers, show a contractural response (including a twitch) according to the number of muscle fibers that are stimulated. The steps are actually in terms of motor units, not individual fibers. Under normal nervous stimulation in the body, a maximum of only 35–40% of the fibers of a muscle will be contracting at any one time, even when the muscle is activated maximally. The other fibers are relaxed and resting, to be recruited into the suite of contracting fibers while others then relax and rest; turnover continues during the entire period that the muscle is active. In physiological experiments, maximum stimuli are used to insure that 100% of the muscle fibers contract each time the muscle is stimulated in order to obtain comparative results. Most physiological studies utilize maximum tetanic stimuli to achieve maximum force development. Correlations obtained in these studies between physiological properties and morphological parameters must be applied to *in vivo* muscles with care. The minimum correction factor that should be applied is that muscles *in vivo* can develop at most about one-third the maximum force measured during *in vitro* experiments. This factor must be applied not only to the force a muscle can apply to a bone, but also to its speed of shortening and its distance of shortening.

When muscles are stimulated, they contract and develop force. Thus muscles pull, that is, the two ends of the muscles are pulled toward the belly of the muscle; this force is, by tradition, called a tensile force. The external force placed upon a muscle, which is also a tensile force, is usually termed a load, which will be used here to distinguish it from the force developed by the contracting muscles. Muscles can only develop a tensile force and pull the ends of the fibers together. No internal mechanism is known whereby a muscle can push its ends apart. If a muscle has shortened during a contraction, an external load must be applied to the muscle to lengthen it; only a tiny external tension is required to stretch a relaxed muscle to its normal length.

Because of the particular force-producing mechanism at the level of the sliding filaments, the active muscle not only produces force but also shortens, depending upon the external conditions on the contracting muscle. If the external load on the muscle is equal to the force produced by the active muscle, the contracting muscle will remain at the same length while producing force. Such conditions result in an isometric contraction, which can occur at any relative length of the muscle in which force is produced. If the load on the muscle is greater than the force produced, then the muscle will be stretched until the muscular force equals the load placed on it. If the external load is too great, it will stretch the muscle beyond the breaking point. If the load on the muscle is less than the muscular force then the muscle will shorten until the force being produced equals the external load, at which point the muscle will stop shortening and continue to contract isometrically. Whenever the muscle changes length during contraction, it is termed isotonic contraction; physiologists and morphologists are concerned mainly with those isotonic contractions in which the initial external load is less than the initial force produced by the muscle. In isotonic contractions, muscle length changes and so does the amount of force produced by the muscle; the constant parameter is the external load placed on the muscle.

No internal mechanism exists whereby a muscle can contract isotonically or isometrically. These "types" of contraction depend upon the force of the muscle relative to the particular external load placed upon it. Although frequently expressed, it is wrong to state that a particular muscle contracts and produces a maximum isometric force, and thereby holds a bone in position. A suitable external load must be present. A muscle may, of course, vary its force development to balance a variable external force and hence maintain an isometric contraction.

The importance of the load placed on a muscle cannot be emphasized too strongly, because all mechanical and many other physiological properties of active muscles depend, in part, upon the load on the muscle. These parameters include speed of shortening, distance of shortening, and metabolic efficiency of the muscle.

The relative amount of force developed by a muscle depends upon its relative length. As a muscle shortens, its force development drops until it reaches zero, and as a muscle is stretched, its force production also drops gradually to zero. In comparative studies of muscles, we must be concerned with the absolute lengths of different muscles and with their relative lengths. Force production of a longer-fibered muscle at a 70% relative length may be compared, for example, with force production of a shorter fibered muscle at a 90% relative length. Care must be used in reading the literature on muscle physiology, as length is almost always used in the sense of relative length, but occasionally physiologists confuse relative and absolute lengths. The major problem is establishment of the absolute (100%) length of a muscle from which relative lengths are measured. The absolute length is often called rest length or normal body length and measured as the length at rest or with the bones held in a normal position, but no one specifies the position of bones that is normal, as for example the positions of the opposite members of a pair of antagonistic muscles. At present, the best system is to assign the 100% value to the reference length (termed L_0 in muscle physiology) which is defined as the length of the muscle at which the maximum isometric force is produced. Unfortunately, we do not know whether the reference length of different muscles are really comparable.

Speed of shortening of a contracting muscle depends upon the length of the muscle fibers and also on the relationship between the muscular force and the external load. The decrease in speed of a muscle, as the load increases from zero to the maximum isometric force of the muscle, does not follow a linear relationship as may be expected, but rather a curve very close to a hyperbola.

The distance that a muscle can shorten depends again upon length of its fibers and the load placed upon the muscle. Maximum shortening occurs when no load is placed upon the muscle, but muscles move objects to achieve useful actions and hence have a load on them. The distance that this load can be moved depends upon an interrelationship between the size of the load relative to maximum isometric force of the muscle and the length of the muscle fibers.

It should be obvious that simple statements or broad generalizations about muscle function and about the relationship between physiological properties and morphological parameters of muscles must be avoided. Advances in general understanding of avian skeletal muscles have been severely hampered in the past by oversimple and often erroneous generalizations. For example, the interpretation that contraction of a muscle equals shortening is wrong and must not be used. Unfortunately "contraction" is a poor choice of words, as it means shortening, but we are stuck with it and we should at least understand what muscles do when they pass from the relaxed to the active state following stimulation. When the muscles contract, they develop force and can shorten certain distances at certain velocities depending upon the relationship between the external load placed upon the muscle and the inherent physiological properties of the muscle. Therefore, we must consider the morphological properties of muscles that determine force development, possible excursion (distance the muscle can shorten and be stretched), speed of shortening, and adaptation of muscles.

Space does not permit discussion of the arrangement of the thick (myosin) and thin (actin) sliding filaments and the macromolecular basis for muscle contraction. A general knowledge of muscle structure and function at the macromolecular level of muscle organization is essential to comprehend the following discussion of muscles; good reviews can be found in any recent textbook of physiology.

4. Structure and Function of the Sarcomere

A brief review of the sarcomere is necessary as this muscular unit is basic to many of the properties of muscles at higher levels of organization. The sarcomere (Fig. 7) is a longitudinal segmental unit of the muscle fiber and is bounded by the sarcolemma and a pair of consecutive Z disks (or bands). The muscle fiber is composed of many sarcomeres attached end to end with neighboring sarcomeres sharing a common Z disk. When the sarcomere is at rest length, a pattern of light and dark bands will be seen with light microscopy; these are the striations of the cross-striated skeletal muscles. Each end of the sarcomere contains half of a light (I) band and the center is occupied by a dark (A) band with a somewhat lighter narrow (H) band at the midpoint. We need not be concerned with other bands. The light and dark bands result from the arrangement of the thin and thick filaments and their overlap. Several points should be noted.

The length of the thick filaments (A band) is about 60% of the sarcomeral rest at length. Cross bridges between the thin and thick filaments occur along the overlap segment between these filaments. Bridge sites are uniformly distributed along both filaments with the exception of a short portion at the middle of the thick filament where they are absent. The bridges provide the tension-developing and the

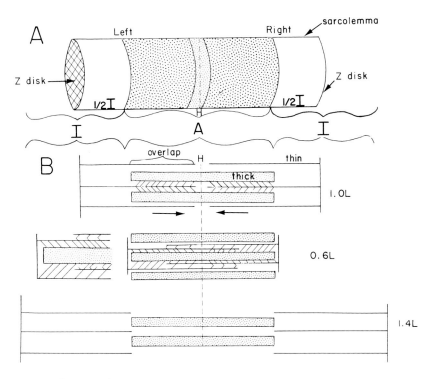

FIG. 7. Schematic diagram showing the banding pattern of a sarcomere and the arrangement of thick and thin filaments. (A) A single sarcomere bounded by the sarcolemma and two successive Z disks showing the pattern of I (light), A (dark), and H (central) bands. The middle of the sarcomere is indicated by the vertical dashed line; note the right–left symmetry of the sarcomere. (B) The arrangement of thin (actin) and thick (myosin) filaments of the sarcomere showing their relative overlap at different relative lengths of the sarcomere and their correlation with banding pattern of the sarcomere. The bridges between thick and thin filaments are shown, with incorrectly oriented bridges shown by an angled line (see enlarged view at 0.6L). The relative overlaps are shown at reference length (1.0L), minimum length (0.6L), and maximum length at which active tension is still recorded (1.4L). Relative tension of the sarcomere depends upon the number of bridges (length of overlap) and percentage of incorrectly oriented birdges. Tension developed in the right half of the sarcomere equals that developed in the left half.

shortening mechanisms of muscle contraction. Each bridge provides the same amount of tension. They are present only when the muscle is in its active state.

Tension is developed and transmitted only through the system of thin and thick filaments and the bridges. No other system of filaments has been demonstrated. Moreover, the collagen fibers found in the sarcolemma do not transmit muscular force in normal fibers. When the muscle is in its relaxed state, the bridges are absent (unformed) and hence tension cannot be transmitted. The thick and thin filaments are reasonably uniformly distributed throughout the cross-sectional area of the sarcomere.

The force of the sarcomere is that which would be recorded, if it were technically possible, at the Z disks located at the two ends of the sarcomere. This tension would be uniform at any point along the length of the sarcomere. Hence the total tension would be distributed among all of the thin filaments in one half of the sarcomere, among all of the bridges between them and the thick filaments, among all of the thick filaments, and so forth until the other end of the sarcomere is reached.

Force development of a sarcomere is dependent upon the number of bridges present in the sarcomere. The force developed in the right half of the sarcomere must be equal to that developed in the left half as each half serves to transmit the force of the other portion to the Z band. And the tension must be uniform at any point along the length of the sarcomere. Hence the maximum force (termed P_0 in muscle physiology) development depends upon the total number of bridge sites present in one half of the sarcomere, which depends on the number of filaments present (only the thick filaments need be counted) and the length of the overlap between the thin and thick filaments. Any modification in the sarcomere that increases the total number of bridges would increase its maximum force. These modifications include increase in density of the filaments, increase in the cross-sectional area of the sarcomere, increase in the length of the sarcomere, and increase in the length of the thick filament relative to the rest length of the sarcomere.

As the sarcomere shortens or is stretched, the thin filaments move (slide) relative to the thick filament and can even move beyond the midline to the opposite half of the sarcomere or beyond the ends of the thick filament. With movement of the thin filaments, the length of the overlap segment changes, and hence the number of bridges is modified with a change of tension developed (Fig. 7). This change in total number of bridges accounts for the decrease of tension (Fig. 8) as the sarcomere is lengthened until the free ends of the thin filament move beyond the ends of the thick filament. When the tips of the thin filaments move into the opposite half of the sarcomere, some bridges are formed between thin filaments from the right half of the sarcomere and the left half of the thick filaments and vice versa. These bridges are incorrectly oriented (Fig. 7B) and presumably develop less force than a correctly oriented bridge between thin and thick filaments on the same side of the sarcomere. Increased double overlap of thin filaments increases the number of incorrectly oriented bridges, and hence there is a steady reduction in force as the sarcomere shortens (Fig. 8). The final and rapid reduction in tension results from the thick filaments abutting against the Z disks and being deformed, hence absorbing force. Thus the relative amount of force developed as the sarcomere shortens or lengthens is dependent upon the maximum force (P_0) and the relative overlap between thin and thick filaments.

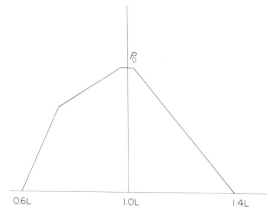

FIG. 8. Generalized tension–length curve for the sarcomere. Tension is measured along the vertical axis and relative length along the horizontal axis. The relative amounts of tension may be compared with differing degrees of overlap between thick and thin filaments and incorrect orientation of bridges (Fig. 7B); the final steep decline of tension during shortening occurs when the thick filaments contact the Z disks. Compare this curve with the tension–length curve of the whole muscle (Fig. 14A).

The maximum distance that a sarcomere can shorten is dependent upon the length of the thick filament relative to the length of the resting sarcomere. The sarcomere shortens until the ends of the thick filaments abut against the Z disks. The relative distance that a sarcomere shortens depends upon the external load placed upon it. Shortening will continue until the sarcomeral force as shown in the tension–length curve equals the external load. The particular shape of the tension–length curve depends upon the factors affecting maximum tension development and the distance the sarcomere can shorten. The latter factors are potentially very complex and no comparative studies are available.

The speed at which a sarcomere shortens depends upon a complex of factors including the maximum tension (P_0) developed, the distance shortened, and the load placed on the sarcomere as illustrated in the graph in Fig. 9. Moreover, empirical observations show that the speed of shortening for a sarcomere (actually demonstrated for a whole muscle and muscle fiber, and extrapolated to sarcomeres) is constant for most of the distance shortened. Acceleration occurs at the onset of shortening and deceleration at the very end of shortening. If the load on a sarcomere is varied, then the sarcomere will shorten faster and over a longer distance for a lighter load than for a heavier load. The speed of shortening is always greater over a fixed absolute distance for a light load than for a heavy load.

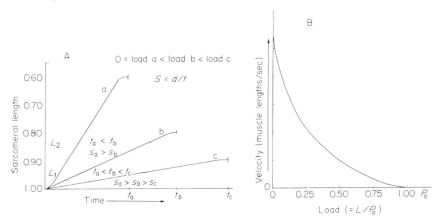

FIG. 9. Speed of shortening of the sarcomere. (A) Hypothetical relationship between length of a sarcomere and the time of shortening of a sarcomere; speed of shortening is shown by the slope of the line. Speed of shortening and maximum distance of shortening decreases as the load on the sarcomere increases. The times indicated along the horizontal axis are those required for the sarcomere to reach maximum distance of shortening for increasing loads. The two lengths indicated (L_1 and L_2) are two arbitrarily chosen lengths of shortening. The relationship shown between relative length and time shown for the sarcomere holds also for the muscle fiber and also for the entire muscles. (B) Hypothetical relationship between load and velocity of shortening of a sarcomere extrapolated from information obtained from studies on whole muscles; hence this relationship is true for muscles. The curve is hyperboloid.

If the morphological properties of the sarcomere are modified to alter the maximum tension developed or the maximum distance shortened, then the speed of shortening will alter. Increase in cross-sectional area to increase maximum tension would increase speed of shortening; the slopes of all the lines shown in Fig. 9A would be steeper. Other modifications, namely, change in sarcomeral length or relative length of the thick filaments would be more complex and we need not be concerned with them here.

The relationship between load and velocity is shown in Fig. 9B;

this is the force–velocity curve of muscle physiologists. The velocity is measured during the linear portion of the shortening curve. The force–velocity curve is hyperboloid.

5. Structure and Function of the Fiber

A muscle fiber is composed of a number of sarcomeres attached end to end, i.e., the sarcomeres are arranged in series. I shall assume that the sarcomeres are identical in their morphological and hence in their physiological properties. This is not true for real muscle fibers, but the general consequences of the variation in sarcomeres along a muscle fiber does not affect the properties of muscle fibers to be discussed. The fiber is bounded by a single, continuous sarcolemma. The sarcolemma transmits the stimulus from the motor end plate to all parts of the muscle fiber as a propagated action potential. Several points should be mentioned.

The tension measured at the ends of a muscle fiber is the same throughout its length. The tension developed by a sarcomere is transmitted to the ends of the muscle fiber via the filament system of other sarcomeres in normal muscle fibers. The sarcolemma does not transmit the tension developed by the sarcomeres. The tension in all sarcomeres of a muscle fiber must thus be the same and equal to the tension for the fiber.

The sarcolemma transmits the stimulus from the motor end plate to all parts of the muscle fiber at a high speed compared with the length of the fiber. The time lag between the stimulus at the motor end plate and the furthest point of the muscle fiber can be measured, but it is so brief that it can be assumed that all sarcomeres along the length of the fiber are stimulated simultaneously. Hence, they start contracting together and finish together. Because the sarcomeres were assumed to be identical, all contract together and the time course of contraction of the muscle fiber is identical to that for the sarcomere. Again, this is not quite so in real muscle fibers, but the difference does not affect the general properties of fibers to be discussed.

The maximum force (P_0) of a muscle fiber is identical to that of an individual sarcomere no matter how many sarcomeres are arranged in series. Thus, the maximum tension of the fiber is independent of its length and hence its mass. Any modifications in the sarcomere that changes its maximum tension, as discussed above, will change the maximum force of the fiber.

The relative tension of a muscle fiber will depend upon its maximum tension and the relative distance the fiber shortened as is true for sarcomeres. If muscle fibers of different absolute lengths are compared, then distances shortened must be considered both in terms of absolute and relative distances. If these fibers shorten the same relative amount, they develop the same tension, but the longer fiber has shortened over a greater absolute distance. If these fibers shorten the same absolute distance, the longer fiber has shortened a smaller relative amount and hence will develop more force at the completion of its excursion than will the shorter fiber. Hence, if a fiber must move a fixed load over a fixed absolute distance, its length must be great enough to allow a force development at the end of its excursion equal to or greater than the load.

The absolute distance that a fiber can shorten is dependent upon its length. If fibers of different absolute lengths shorten to 60% of their reference length, then the longer fiber shortens the greatest absolute distance. And, as just mentioned, longer fibers can move the same load over a greater absolute distance, or a greater load over a given distance. The relationship between fiber length, distance shortened, and relative tension development depends upon the shape of the tension–length curve (Figs. 8 and 14), which shows that an increasing drop-off

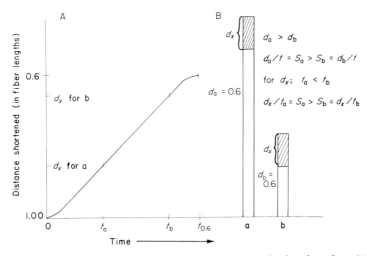

FIG. 10. Speed of shortening of muscle fibers of different absolute lengths. (A) Relationship between relative distance shortened (in fiber lengths) and time. This relationship is the same as that for an individual sarcomere and for fibers of any length, because distance shortened is measured in terms of fiber lengths (a relative measure). (B) Identical muscle fibers of two different absolute lengths with fiber a being longer than fiber b, and d_x an arbitrarily chosen fixed absolute distance. The equations demonstrate that the longer fiber always shortens faster than the shorter fiber with the same load, no matter whether both fibers shorten the same absolute distance (d_x) or the same relative distance (e.g., up to $0.6L$).

in tension occurs as a muscle fiber shortens a greater relative amount; that is, the fiber drops into a lower portion of the tension–length curve.

The speed at which a fiber shortens depends upon its maximum force development, the load on the fiber, and its length. If the first two parameters are kept constant, then longer fibers shorten faster, regardless of whether speed is measured for an absolute or a relative distance of shortening. Muscle fibers possess, as noted above, the same time course for contraction as does an individual sarcomere contained within the fiber. If two fibers, one long and one short, comprised of identical sarcomeres are compared (Fig. 10), then it can be seen that the longer fiber always has the greater speed, regardless of whether the fibers shorten the same absolute or the same relative distance.

If the length of two fibers is held constant, but their maximum tensions are varied, then the stronger fiber will shorten faster for a given load because it can provide more force for acceleration.

Variation of both length and tension development in fibers results in a complex relationship, as shown in Fig. 11 for both speed and distance shortened for varying loads. A crossover point exists somewhere depending upon the exact properties of the two fibers. The basis of the crossover is seen in the tension–length curve of the two fibers, with the length given in absolute values. Comparison of the speed of these fibers for different absolute distances of shortening is very complex and simple generalizations are not possible. These comparisons are mentioned because they are reasonable ones facing morphologists, and yet the solutions are most difficult.

6. Structure and Function of the Muscle

A muscle is composed of a number of muscle fibers lying parallel to each other. A muscle can also be considered as a collection of a number of sarcomeres arranged both in series and in parallel. Some of the properties of the muscle depend upon the number of sarcomeres arranged in series and other properties depend upon the number of sarcomeres arranged in parallel, assuming, of course, the properties of the sarcomeres. In the following discussion, the muscle is composed of fibers having uniform length and containing identical sarcomeres. For the initial treatment, only parallel-fibered muscles will be covered, for simplicity. (See Gans and Bock, 1965, for a general review of muscle structure and function.)

Maximum tension of a muscle depends on the number of sarcomeres arranged in parallel and on the properties of the sarcomere. Hence the total cross-sectional area of all sarcomeres (=area of all fibers) is

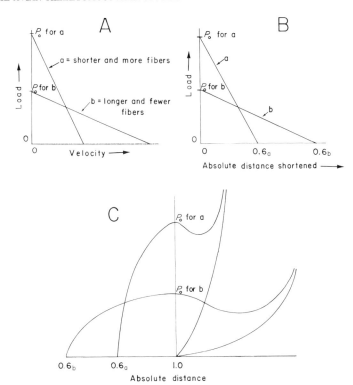

FIG. 11. Comparison of the mechanical properties of two muscle fibers having the same mass, but differing in cross-sectional area and length. Fiber a is shorter and has a greater cross-sectional area than fiber b. The relationships of mechanical properties illustrated for these fibers also holds for two parallel-fibered muscles of equal mass but differing length and cross-sectional area as shown in Fig. 12. (A) Relationship between load on the muscle and velocity of shortening. Note that fiber a shortens faster for heavier loads than fiber b, and that a crossover point exists. (B) Relationship between load and absolute distance of shortening. Again fiber a shortens a greater distance than fiber b for heavier loads and a crossover point exists. (C) Tension–length curves for fibers a and b, with absolute distance shown on the abscissa to show the cross-over point.

proportional to maximum muscular force, but the length of the muscle and the volume or mass of the muscle are independent of the force development (Figs. 12 and 13). At this point, two terms that relate to tension development of a muscle should be defined. The morphological cross section of a muscle is the area formed by passing a plane through the thickest part of the muscle perpendicular to its longitudinal axis. The physiological cross section is the area formed by passing a plane through all fibers of a muscle at right angles to their longitudinal axes. The area of the physiological cross section is equal to that of the morphological cross section in parallel-fibered muscles, and is greater than that of the morphological cross section in pinnate muscles. The physiological cross section may be many times larger than the morphological cross section in multiple pinnate muscles. The maximum force of a muscle is a function of its physiological cross section (multiplied by the cosine of the angle between the muscle fiber and tendon). The gross physiological cross section is only a rough index to maximum force; refinements such as using the fiber physiological cross section should be used if possible. Contrary to assertions in the literature, the maximum force of a muscle is not a function of its morphological cross section.

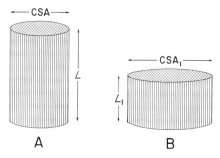

FIG 12. Schematic drawing of two parallel-fibered muscles of equal volume, but unequal in length and cross-sectional area. The shorter muscle, B, can develop more force, but shortens over a smaller absolute distance than the longer muscle, A. See Figs. 11 and 14D for comparison of mechanical properties of these muscles. (Taken from Bock, 1969a, Fig. 13.)

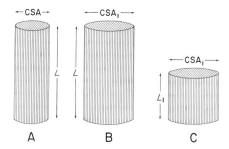

FIG. 13. Schematic drawing of several parallel-fibered muscles. Muscles A and B are of equal length and can shorten the same distance, but B can develop more force. Muscles B and C can develop equal force, but B can shorten more than C. See Fig. 14 for comparisons of tension–length curves of these muscles. (Taken from Bock, 1969a, Fig. 14.)

The relative development of force as a muscle shortens follows the same pattern as for individual fibers; the tension–length curve (Fig.

14A) describes the relationship between relative length and force. The tension–length curves for two muscles of equal length but different cross-sectional areas (Fig. 13A,B) are shown in Fig. 14B; those for two muscles of the same cross-sectional area, but different lengths (Fig. 13B,C) are shown in Fig. 14C, and those for two muscles of equal volume but different lengths and cross-sectional areas (Fig. 12) are shown in Fig. 14D.

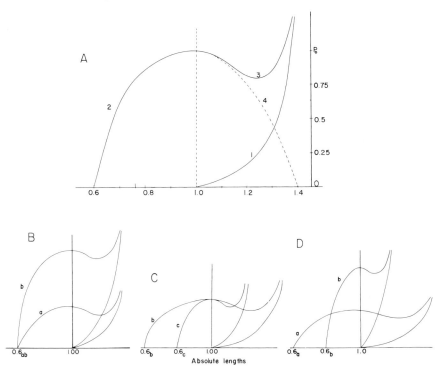

FIG. 14. Tension–length curve of a whole muscle. (A) Generalized tension–length curve with relative length shown on the abscissa and tension on the ordinate. Curve 1 is the resting tension and represents the force required to stretch the relaxed muscle. Curve 2 is the active tension developed by the contracting muscle below reference length. Curve 3 is the tension measured when the muscle is contracting above reference length, being the sum of the resting tension and the active tension (curve 4). The dashed line, curve 4, is the active tension above reference length; it cannot be measured directly and is obtained by subtracting curve 1 from curve 3. Note that active tension reaches 0 at 0.6L and 1.4L. (B) Tension–length curves of two muscles of equal lengths but different cross-sectional areas (as shown in Fig. 13A,B). (C) Tension–length curves of two muscles of equal cross-sectional area but different lengths as shown in Fig. 13B,C. (D) Tension–length curves of two muscles of equal volumes but different lengths and cross-sectional areas as shown in Fig. 12A,B. Absolute length and tensions are used in B, C, and D.

Maximum shortening (under no load) of a muscle is a function of the length of the fibers. The relative shortening is a function of maximum tension, fiber length, and external load. A muscle shortens until its force equals the external load on the muscle; the relative shortening must be converted to absolute shortening. These relationships can be readily understood by reference to Fig. 14. Thus, for two muscles of equal cross-sectional area (Figs. 13B,C and 14C) the longer muscle can move the same load over a greater distance than can the shorter muscle. Or for muscles of equal length, but different cross-sectional areas (Figs. 13A,B and 14B) the thicker muscle can move a heavier load the same distance or the same load over a greater distance. Comparison of muscles of the same volume (Figs. 12 and 14D) is complex. The shorter, thicker muscle can move heavier loads, but the longer, thinner muscle can move a lighter load over a longer distance. The relationship between maximum tension and load and the crossover point is illustrated in Fig. 11.

The maximum speed of either absolute or relative shortening (under no load) of a muscle is a function of the length of its fibers, as is the case for the individual muscle fiber (Fig. 10). A longer-fibered muscle shortens faster. The relative speed of shortening is a function of maximum tension, fiber length, and external load, but the interrelationships of these factors and velocity of shortening is more complex than it is for distance of shortening. The same problem exists as for individual fibers. Generalizations are possible if only one of these factors is varied, but become difficult if two factors are varied independently. Increase in maximum tension and/or fiber length or decrease in external load would increase the speed of shortening. But the consequence of increasing maximum tension while decreasing length, as shown in the two muscles of equal mass in Fig. 12, is difficult to generalize; comparisons may have to be made for each individual pair of muscles.

These comparisons demonstrate that the two most essential gross morphological parameters of muscles are fiber length and number of fibers; the latter may be expressed as the physiological cross-sectional area. The third essential morphological parameter is the angle of pinnation (see below), which exerts its influence mainly as the cosine of this angle; the angle of pinnation of a parallel-fibered muscle is 0° and its cosine is 1.0. Either one or both of these morphological parameters determine the important functional properties of maximum force, relative force during excursion, distance of shortening, and speed of shortening. Increase in any of these functional properties requires an increase in one or both of these morphological parameters. As fiber number and fiber length are inversely related in muscles of constant mass, increase in either must be at the expense of the other. Increase in one of these parameters with the other held constant will increase the mass of the muscle, and other factors limit the overall size of the muscle. Unfortunately almost all reports on avian myology fail to measure fiber length and fiber number, even in relative terms. Thus, the necessary morphological basis for comparative study of function and adaptation of the muscular system of birds, or also for other classes of vertebrates simply does not exist. Amassing this information is crucial in descriptive studies of the avian skeletomuscular system.

The physiological properties of maximum force, relative force, distance of shortening, and velocity of shortening do not correlate at all with the mass or weight of skeletal muscles. This can be demonstrated beyond doubt, and an examination of Figs. 12–14 should suffice. Moreover, we have gathered empirical observations (W. J. Bock, unpublished observations) that provide direct documentation. Mass or weight cannot be used as an index to the functional properties or the adaptive significance of skeletal muscles, as has been common in earlier studies. The data from such studies must be reanalyzed and the conclusions reevaluated. The only possible case in which weight or mass of muscles can be used as an index of muscle function or adaptation is in the comparison, in closely related species, of homologous muscles that do not differ greatly in size or shape (e.g., Richards and Bock, 1973; Spring, 1971). Even in such cases, weight is a poor substitute for fiber number and fiber length and should not be used as a functional index.

A second point to be mentioned is the measurements of muscles used by Hudson and his associates. Few of these or ratios calculated from them can be converted to either fiber length or fiber number (plus the parameters important in pinnate muscles). Hence, they have little meaning in terms of muscle function and adaptation, and are of doubtful taxonomic value.

7. Pinnate Muscles

Fiber length in skeletal muscles is, as a rule, just long enough to permit the required amount of shortening and relative force development. As will be shown below, fiber length in most muscles tends to be short with respect to the length of the muscle, and in numerous cases fibers are very short. It is more advantageous in most muscles to increase the number of fibers rather than their length. One consequence of increased fiber number and decreased fiber length is an awkwardly shaped muscle if the fibers are arranged parallel to each other. Pinnation is a simple mechanism whereby many short muscle fibers can be packed into a muscle of reasonable dimensions (Fig. 15).

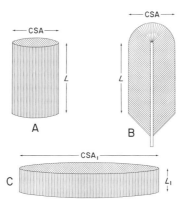

FIG. 15. Schematic drawing to compare parallel-fibered and pinnate muscles. (A) A parallel-fibered muscle. (B) A pinnate muscle of the same volume as muscle A. (C) A parallel-fibered muscle of the same fiber properties (length and number) as the pinnate muscle B. The pinnate muscle has more fibers and can develop more force than the parallel-fibered muscle, A, but it has shorter fibers and can shorten less. The pinnate muscle is equivalent to the short, broad, parallel-fibered muscle, C, but its fibers are arranged in a more "convenient" shape. (Taken from Bock, 1969a, Fig. 15.)

With the exception that the angle between the muscle fiber and the tendon of insertion must be included in all calculations (see Gans and Bock, 1965), pinnate muscles follow the same relationships discussed above for muscles with parallel fibers. It is incorrect to assume, as is frequently done tacitly, that two classes of skeletal muscles—parallel-fibered and pinnate—exist and possess distinct sets of functional properties. Indeed the two types of muscles merge smoothly into one another, with the influence of the angle of pinnateness on the morphological and functional properties of the pinnate muscle increasing gradually. Short-fibered pinnate muscles differ significantly from long-fibered, parallel-fibered muscles, but so do short-fibered, parallel-fibered muscles. Hence, fiber length and fiber number must be obtained for pinnate muscles in addition to the angle of pinnateness. The latter must, of course, be measured when the muscle is at reference length (presumably rest length of the muscle of the body).

Comparison of pinnate muscles with parallel-fibered muscles and even between different pinnate muscles is far more difficult than for parallel-fibered muscles. There is no simple basis for comparison, such as mass, overall shape of the muscle, or angle of pinnation. Certainly modifications in morphological and physiological properties of pinnate muscles do not depend solely upon change in the angle of pinnation. Muscles with the same overall dimensions and same angle of pinnateness can possess quite different functional properties depending on whether they are unipinnate, bipinnate, or multipinnate. In every case, angle of pinnateness, fiber number, and fiber length must be measured. Moreover, a number of measurements of fiber length and angle of pinnateness should be made, since these parameters can vary greatly within a muscle (W. J. Bock and C. R. Shear, unpublished data, for the M. pseudotemporalis superficialis of *Passer*). I will assume that these parameters are uniform for the purposes of this discussion.

The basic physiological properties of pinnate muscles may be ascertained as for parallel-fibered muscles, remembering that summation of angled forces implies vector addition (Gans and Bock, 1965). The force of a single fiber in a pinnate muscle must be separated into two components, a useful one in the direction of the tendon (Fig. 16A),

$$F_m = F_f \cos \alpha$$

and an unused component at right angles to the direction of the tendon

$$F_1 = F_f \sin \alpha$$

As the angle of pinnation increases from 0° to 90°, cos α ranges from 1 to 0, but does not decrease rapidly until the angle becomes greater than 60°. The maximum force of a pinnate muscle (P_0) depends on the number of fibers (physiological cross section) and the cosine of the angle of pinnation, as above; hence,

$$P_0 = k q_p \cos \alpha$$

where k is a constant for the unit of force development (assumed to be 10 kg-wt/cm^2 in human muscles), q_p is the physiological cross-section in square centimeters, and α is the angle of pinnation at the time the force is exerted.

Only part of the total force developed by the fibers of a pinnate muscle can be resolved into an effective component parallel to the pull of the tendon; the component at right angles to the tendon is lost. No other source of force is available in the contraction of a pinnate muscle to compensate for the lost force, contrary to the arguments of Pfuhl (1937), Schumacher (1961), and others (see Gans and Bock, 1965, pp. 135–138).

As a pinnate muscle shortens, the fibers develop less tension, according to the tension–length curve, and the angle of pinnation increases so that its cosine decreases. Both factors act in the same direction, reinforcing each other, so that the force of a shortening pinnate muscle decreases at a much faster rate than does the force of a parallel-fibered muscle.

When a pinnate muscle fiber shortens from f to f_1 (Fig. 16), the distance the tendon travels (h) can be calculated from the initial fiber length (f), the actual fiber contraction coefficient (n, where $n = f_1/f$) and the initial angle of pinnation (α) with the equation (Gans and Bock, 1965, p. 126)

$$h = f[\cos \alpha - (\cos^2 \alpha + n^2 - 1)^{1/2}]$$

In all cases, the distance that the tendon travels is always greater than the distance the muscle fiber shortens ($f - f_1$). This difference is not great, but it is advantageous in pinnate muscles that contain very short fibers.

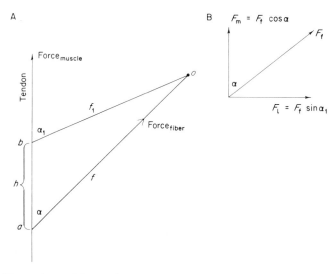

FIG. 16. Shortening and force of a pinnate muscle fiber. (A) A single muscle fiber, oa, of length f attaches to the tendon at an angle α. Upon shortening, the fiber, ob, now has length f_1 and inserts at a larger angle α_1. Force along the tendon (muscle force) depends upon the cosine of the angle of pinnation. Distance h, the distance the tendon moves from a to b is always greater than the distance the muscle fiber shortens ($f - f_1$). Thus, at a, $F_m = F_f \cos \alpha$, at b, $F_m = F_{f_1} \cos \alpha_1$; $F_f > F_{f_1}$; $\cos \alpha > \cos \alpha_1$; (B) Decomposition of the force of a pinnate muscle fiber (F_f) into the useful muscle force (F_m) and the "lost" force (F_1).

Speed of shortening of a pinnate muscle is equal to h/t and is always slightly greater than the speed at which the muscle fiber shortens $[(f - f_1)/t]$, but again this difference is not great. The shortening velocity of a pinnate muscle depends on the muscular force (measured at the tendon), the load on the muscle, and the relative shortening of the fibers. These muscles generally provide great force for acceleration of the load, but the relative shortening of the fibers must be great; hence it is difficult to estimate speeds of shortening. Such estimates matter little, at least for short-fibered pinnate muscles, which have a minimum excursion.

As a pinnate muscle contracts and shortens, its physiological properties change with change in the angle of pinnation; the modification is toward decrease in magnitude of each property, especially force development. Moreover, the magnitude of the change depends not on the increase in the angle of pinnation, but in the decrease in the cosine of the angle, which changes at a faster rate as the angle of pinnation becomes larger. Hence a 15° change in the angle of pinnation from 30° to 45° results in a greater decrease in the physiological properties of the muscle than a change from 15° to 30°. However, the problem of reduction in functional properties with shortening is not serious. Pinnate muscles, in general, and especially short-fibered pinnate muscles are found where the basic requirement is for great force development coupled with little or no excursion.

Comparison of pinnate muscles with each other and with parallel-fibered muscles can be made with the help of Figs. 15 and 17. In this comparison, the muscles are of equal mass and overall dimensions (approximately); the pinnate muscle is a bipinnate one. Change in morphological and physiological properties are examined with respect to increase in the angle of pinnation; the graphs represent guesswork and serve only to illustrate trends. A parallel-fibered muscle has an angle of pinnation of 0°.

   a. Fiber number or physiological cross section begins with that for the parallel-fibered muscle and increases steadily as the angle of pinnation goes to 90°, at which point the maximum number of fibers is reached.

   b. Fiber length is maximum in the parallel-fibered muscle, and drops sharply as the muscle becomes pinnate, followed by a steady decrease in length until the minimum length is reached with an angle of pinnation of 90°.

   c. Total force of the muscle fibers follows the curve of physiological cross section and is lowest for the parallel-fibered muscle and reaches its peak value when the angle of pinnation is 90°. The maximum force of the muscle (P_0), however, follows a curve beginning with the force of the parallel-fibered muscle, increasing to a peak value at an angle of pinnation, estimated to lie in the range between 45° and 60°, and then drops off sharply to reach a value of zero at an angle of pinnation of 90°, where all of the force is directed at right angles to the tendon of insertion.

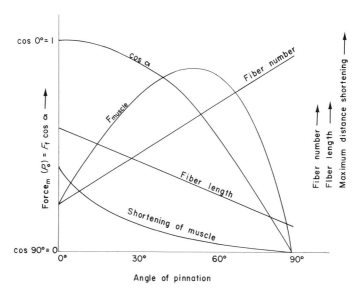

FIG. 17. Graph illustrating the change of several properties of a pinnate muscle with a change in the angle of pinnation from 0° (a parallel-fibered muscle) to 90°. The slopes and exact shape of each curve are educated guesses and should be taken only as an indication of the existing relationship. The exact details of each curve will change as the outside dimensions of the muscle change and as the pinnate muscle is modified from an unipinnate to a bipinnate to a multipinnate muscle. Note that a peak exists for the muscle force, presumably about 60°, although the fiber force continues to increase (as the fiber number increases) as the angle of pinnation increases to 90°.

d. The maximum shortening with no load on the muscle begins with the value for the parallel-fibered muscle. Possibly, pinnate muscles with a very low angle of pinnation may shorten more than parallel-fibered muscles because of the relationship between fiber excursion and tendon excursion discussed above. In any case, the distance the muscle can shorten decreases with an increasing angle of pinnation and reaches zero at an angle of 90°.

These relationships shift if the muscle is unipinnate with fewer and longer fibers than in the comparable bipinnate muscle (same angle of pinnation). The force would decrease, but the muscle could shorten over a greater absolute distance. If the muscle is multipinnate, such as the M. pseudotemporalis superficialis in most passerine birds, the fibers would be even shorter and more numerous than in the comparable bipinnate muscle. The multipinnate muscle is stronger and able to shorten less than could the bipinnate muscle.

The angle of pinnation in several very short-fibered, multipinnate muscles examined is relatively large, between 30° and 45°. These muscles appear to be so arranged in the bone-lever system that they must undergo an absolute minimum or even no excursion. They appear to be specialized for maximum force development.

Mention must be made of the dual assumption of some authors that pinnate muscles are inherently better and more adaptive than parallel-fibered muscles, and that pinnate muscles are derived as compared with primitive parallel-fibered muscles with increase in the angle of pinnation serving as a useful index to the degree of advancement of a particular group (e.g., Beecher, 1953). All parts of this assumption must be rejected. Parallel-fibered and pinnate muscles each possess functional advantages lacked by the other type and would be selectively advantageous under different sets of conditions. No clear evidence exists for the phylogenetic antiquity of any muscle arrangement in vertebrates, especially in the tetrapods. And considerable variation exists in the angle of pinnateness within a single muscle, so that the establishment of a simple numerical index appears dubious. Lastly, it should be mentioned that most of the jaw muscles described as parallel-fibered by Beecher are actually pinnate. All fibers of parallel-fibered muscles, by definition, run from the origin to the insertion of the muscles and are parallel to the longitudinal axis and force vector of the muscle. Fibers of pinnate muscles insert onto a tendon at an angle to the longitudinal axis of the muscle and force vector of the muscle, as in most passerine jaw muscles (Bock, 1973). The fact that parts of a pinnate muscle may appear to be parallel-fibered, as in some of Beecher's figures, is quite irrelevant.

8. *Muscle Fiber Types*

Up to this point, I have assumed that striated skeletal muscle fibers are uniform in their cytological and biochemical characteristics. Actually there are considerable variations in these properties, but much confusion still exists on the details of the diversity of fiber types. The existence of red and white muscles has been known since the nineteenth century, along with their general correlation with rate of fatigue in muscles. Small differences in cytological structure in muscle fibers plagued biologists until they were analyzed in painstaking detail by

Krüger (1952), who laid the foundations of our knowledge of *Fibrillenstruktur,* or twitch muscle fibers, and *Felderstruktur,* or tonus muscle fibers. Considerable work has been done on histochemical analysis and electron microscopic examination of muscle fibers. Much of the available work on histochemical properties of avian muscle fibers has been done by J. C. George, who presented an extensive and detailed review of his work in George and Berger (1966).

In spite of the work done in this area (see George and Berger, 1966, for literature), no generalizations can be presented at this time. Sufficient comparative studies on fiber types present in a number of individual muscles in the same species, and in a sampling of muscles in a number of species, are not available. Correlation of the many individual morphological and histochemical properties have had limited success to date; we do not know which of these parameters can vary independently and to what degree. And very few empirical studies on the correlation between the morphological and histochemical properties and the physiological parameters are available. Most of the statements found in the literature on relationships between morphological–histological properties, and function of fiber types are based on supposition, not direct observations. Thus, it is not clear which of the generalizations advocated by George (in George and Berger, 1966; see the review by Bock, 1967b) on the correlation between various morphological and histochemical characteristics of avian skeletal muscle fibers and the correlation between these parameters and functional properties are valid. Moreover, extrapolation of properties and correlations known for mammalian muscle fibers to avian muscles should not be attempted without a very careful evaluation of the evidence. Much of the earlier correlation of properties in avian muscles by extrapolation from mammalian studies has been made without any precautions.

Establishment of muscle fiber types is a necessary step in the development of our knowledge and understanding of these subtle properties of avian skeletal muscle fibers. But it must not be forgotten that the procedure of establishing fiber types is typological, based on the principles of classic ideal morphology. Types that are established for convenience of study and communication must be abandoned as soon as available evidence demonstrates a continuum of characteristics. The establishment of an intermediate fiber type (I) between red (R) and white (W) muscle fiber types (George and Berger, 1966; pp. 75–76) is, in itself, strong evidence that the distribution of these characteristics varies in a continuum rather than in discrete types, and that the original description of red and white fiber types characterized only the two extremes of this continuum.

The two major sets of fiber types described for birds are: (a) red and white muscle fiber types differentiated by their myoglobin content; and (b) twitch and tonus muscle fibers differentiated by their pattern of innervation, sarcotubular system (reflected as *Fibrillenstruktur* and *Felderstruktur*) and other cytological structural features. These two sets of fiber types should be kept separate until additional information indicates a correlation between them. Any attempt to force all dichotomies of muscle fiber characteristics into a single set of fiber types is doubtless doomed to failure. George (in George and Berger, 1966; pp. 75, 138, and 183) equates red with tonus and white with twitch without supporting evidence. Indeed, recent evidence demonstrates that the tonus-fibered M. latissimus dorsi anterior and posterior slip of the M. serratus metapatagialis of the pigeon (W. J. Bock, unpublished results) are white, and the anterior two slips of the M. serratus metapatagialis are red and twitch fibered. Recent assertions of Gringer and George (1969) that red fibers in the M. pectoralis of the pigeon are "slow twitch" and white fibers are "fast twitch" are without direct physiological support and are based on extrapolation from mammalian muscles that may not be valid.

Red and white fibers are distinguished from one another by the amount of myoglobin they contain. Histochemical staining techniques for myoglobin are difficult, give inconsistent results, and do not provide a good resolution between fine degrees of differences. Most important is that these staining techniques do not distinguish consistently between avian white and red muscle fibers. These fiber types are identified (by George and his coworkers) in mixed muscles by other characteristics, such as size, which are believed to be correlated with amount of myoglobin. George (in George and Berger, 1966) is vague about the staining techniques used to demonstrate myoglobin content in red and white fibers. Because of the difficulty of independently ascertaining myoglobin content, correlations between this property and others are elusive.

Functional correlations of red and white fibers can be stated only in the broadest terms. They should not be called slow and fast (which is also used for twitch and tonus fibers), nor should they be regarded as slow twitch and fast twitch unless proven so by direct physiological observations. Long-established evidence suggests that the important functional basis for the degree of myoglobin content is rate of fatigue, especially long-term fatigue; much of this evidence comes from general observations, such as noting the color of the M. pectoralis and relating it with the habits of the bird. This correlation will probably have to be refined further, as it may relate only to fatigue rates in phasic muscles that move bony segments actively and not to tonic or holding muscles (be they tonus muscles or twitch muscles). Hence, correlations would be expected between red fibers, which fatigue slowly, and features essential for aerobic metabolism. White fibers, which fatigue rapidly, should possess properties for anaerobic metabolism. Such correlations may, however, not be absolute, since some available evidence suggests that tonus muscle fibers, which are holding muscles and do not fatigue (as compared with twitch fibers, unpublished physiological observations), may possess many of the same characteristics for aerobic metabolism as do red fibers, even though the tonus fibers possess little myoglobin and are white.

Twitch and tonus fibers were first investigated by Krüger (1952) who showed the correlation between *Fibrillenstruktur* and twitch fibers and *Felderstruktur* and tonus fibers (see Bock and Hikida, 1968, for a review; also Hikida and Bock, 1970, 1971, 1972; Lee, 1971; Canfield, 1971). Unfortunately the term fast has been applied to fast twitch fibers and all twitch fibers, and slow to slow twitch fibers and tonus fibers. This overlap of terms has been largely responsible for the confusion of the properties of vertebrate striated muscle fibers that abounds in the literature. It would be best to always use twitch and tonus or fast twitch and slow twitch, or at least to identify clearly the system of fiber types being studied. Mammals possess fast twitch and slow twitch fibers (tonus fibers have been demonstrated only in a few muscles, e.g., extraocular muscles); hence, slow and fast fibers in mammals refer to these types. Birds possess twitch and tonus fibers, but no one has, to date, demonstrated the existence of fast twitch and slow twitch fibers, although such a distribution of twitch contraction times may well exist in avian muscles. Thus, fast and slow fibers in birds refer to twitch and tonus fibers, respectively. Correlation of properties of fast and slow fibers in mammals and birds is hence not possible, and earlier studies must be scrutinized to insure that the author has not attempted such comparisons.

Twitch fibers (see review in Bock and Hikida, 1968; Canfield, 1971) are characterized by a well developed sarcotubular system with triads present at every sarcomere, sharply delimited fibrils (myofibrils), an M band, a thinner, straight Z band with well ordered thin filaments attached to it, regular banding pattern, and a large number of mitochondria. They have one (usually, but several are present in the pigeon M. serratus metapatagialis) large *en plaque* end plate of large-fibered nerves with junctional folds of the muscular plasma membrane. Tonus fibers are characterized by a greatly reduced sarcotubular system with the triad system greatly reduced or absent, no fibrils (the "fibrils" seen by light microscopy are clumps of filaments that should be termed "pseudofibrils"), no M band with its thin connections between thick filaments (this feature may be present in some tonus fibers), a thicker amorphic, irregular Z band with the thin filaments attached to it irregularly, very irregular banding pattern, and few mitochondria. They have numerous small *en grappe* end plates of small-fibered nerves with an almost complete absence of junctional folds of the muscular plasma membrane; the end plates are distributed along the entire length of the sarcolemma.

No consistent pattern in the distribution of glycogen, lipids and succinic dehydrogenase could be found in twitch and tonus fibers of avian muscles (Lee, 1971). Twitch fibers contain a variable amount of heart lactic dehydrogenase (H-LDH), varying from 1 to 25% certainly and perhaps up to 50%. Tonus fibers (of the M. latissimus dorsi anterior) contain 99% H-LDH (Kaplan and Cahn, 1962). The tonus–fibered M. latissimus dorsi anterior and the posterior slip of the M. serratus metapatagialis of the pigeon are myoglobin poor.

The most important known physiological differences between twitch and tonus fibers are the speed of contraction and relaxation, rate of fatigue, and the propagation (depolarization) ability of the membrane. Twitch fibers exhibit the well known twitch—a rapid rise of tension (about 25 milliseconds)—following stimulation and a rapid (almost instantaneous) drop of tension following cessation of stimulation. Fusion of twitches into a tetany requires a high rate of stimulation, generally twenty or more pulses per second. Fatigue in a tetanic contraction occurs within a few seconds, with tension dropping to 40% or less of peak tension. The sarcolemma of twitch fibers propagates action potentials along the length of the fiber. The twitch fiber sarcolemma can be depolarized with an alternating current oriented either along the length of the fiber or at right angles to the fiber axis, but twitch fibers do not maintain a contracture when exposed to depolarizing agents such as acetycholine and potassium chloride or to direct-current stimulation. Tonus fibers also respond to a single electrical stimulus with the characteristic muscle twitch with a slow rise and dropoff of tension. Following cessation of stimulation, tonus fibers lose tension slowly, up to seconds or many minutes. Fusion into a tetany occurs with a low rate of stimulation, about one pulse per second or even slower. Tonus fibers continue to maintain tension with a continuous stimulation for minutes without any signs of fatigue, and tension does not drop below peak tension. A long stimulation of tonus fibers via electrical stimulation of its nerves does result in a steady drop of tension to zero after several minutes, which demonstrates depletion of acetylchloline at the motor end plates.

Immediate stimulation of the sarcolemma of tonus fibers after lengthy nerve stimulation results in an immediate increase to maximum tension (unpublished results). The sarcolemma of tonus fibers cannot propagate action potentials (Ginsborg, 1960) and can be depolarized with an electrical stimulation only when it is oriented parallel to the axis of the muscle fibers. Tonus fibers respond with a prolonged contracture to depolarizing agents such as acetylocholine and potassium chloride or to direct current.

The characteristics of the mechanical properties of tonus fibers in birds are still unknown in detail. They possess tension–length curves and load–velocity curves similar in shape to those of twitch fibers (Canfield, 1971; W. J. Bock, unpublished observations). Tonus fibers were reported to hypertrophy following denervation on the basis of studies on the anterior latissimus dorsi muscle of the chicken. Other studies on the same muscle in pigeons (Hikida and Bock, 1972, and unpublished results) show that tonus fibers atrophy in histological and physiological (maximum tension) properties, but at a much slower rate than tonus fibers. Tonus fibers also atrophy following tenotomy, but at a slower rate than twitch fibers (Hikida and Bock, 1970).

Analysis of the properties of avian fiber types requires more precise determination of such individual characteristics, after which correlations may be made to ascertain the features that vary together. Fiber size must be measured more carefully on material fixed and stained to avoid shrinking. Apparent differences in concentrations of lipids, glycogen, myoglobin, and various enzymes must be examined carefully to ascertain whether these differences may not be a reflection of fiber size. The correlation between morphological and biochemical properties and physiological ones is difficult, as muscles of pure fiber composition are required for study. Only two pure tonus fibered muscles are known in birds. Mixed muscles, even of fiber types as different as twitch and tonus, present great problems for physiological studies. Some evidence suggests that even twitch and tonus fibers may be only the extremes in a continuous range of muscular properties, but this cannot be demonstrated easily with mixed-fibered muscles. Studies are needed on the characteristics of fibers in a variety of muscles present in one species and comparative studies of homologous muscles in different species. Seasonal and short-term changes in these properties during the life of a bird have not been investigated, although many are quite probable. Although many difficulties and pitfalls exist in this area of avian myology, it is a most important aspect of study and deserves much more attention than it has received in the past.

9. Muscle–Bone Systems

Muscles develop tension and pull in a linear direction, but most bones rotate about articulations. Functions and adaptations of skeletal muscles, therefore, are not to be measured in terms of forces, masses, and linear movements, but within a world of torques, moments of inertia, circular motion, and centripetal inertias. Muscles cannot be regarded as isolated structures, but as units within a skeletomuscular system in which their position in the bone-lever system becomes most important. Concepts of one-joint and two-joint muscles become critical. Movement and velocity of shortening of the contracting muscle must be considered also in terms of rotation of the bony segments. The position of muscle within the bone-lever system must be appraised in terms of mechanics (and hence mechanical advantage) and in terms of muscle physiology (and hence a notion of physiological advantage). A definite relationship exists between the architecture of a muscle, including fiber number and fiber length, and the arrangement of the muscle within the bone-lever system. No attempt has ever been made to analyze this relationship.

Several methods are available for analysis of muscles within muscle–bone systems; all will give the same answers. I chose the method of free-body diagrams (Bock, 1968a) because it is well suited to the

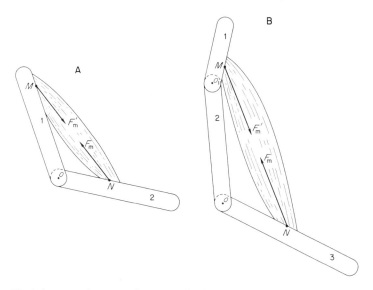

FIG. 18. Schematic drawing of two simple flexor muscles to show the relationship between the muscle, the force vectors of the muscle (F_m and F'_m), and the bones. (A) A one-joint muscle. (B) A two-joint muscle. (Taken from Bock, 1968a, Fig. 1.)

particular demands of muscle bone systems and because it provides a more general solution for both theoretical and practical problems than does any other system widely used by morphologists. Some of the basic assumptions and properties of free-body diagrams have been discussed above, and the reader is referred to Bock (1968a) and the papers cited therein, especially Dempster (1961), for further details. I should note only that a free-body diagram must be drawn for each bone and other object to which forces are applied, and that all of the forces acting on each free body must be included in the analysis. This method does not provide any information about the stresses within the free body.

a. One-Joint Muscles. The simple case of one-joint muscles (Fig. 18), in which the muscle spans one articulation, will be considered first; the free-body diagrams for the bones is shown in Fig. 19. I shall consider the simplest case of a pair of articulated bones with a single muscle acting on them as illustrated in Fig. 20. The one-joint flexor muscle inserts on bone 2 at point N; in this and all other examples,

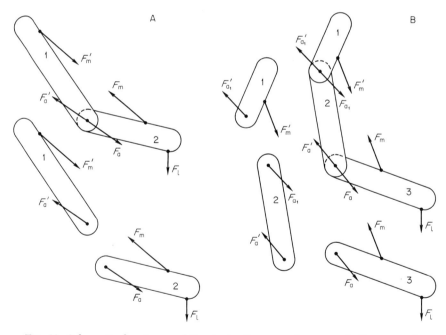

FIG. 19. Schematic drawings and free-body diagrams of two simple flexor muscles, following Fig. 18, but including the force at the articulation (F_a) and a load force (F_l) on the distal bone. (A) A one-joint muscle. (B) A two-joint muscle. Note that a free-body diagram is drawn for each bone and that all external forces on each bones are shown. (Taken from Bock, 1968a, Fig. 2.)

3. THE AVIAN SKELETOMUSCULAR SYSTEM

the mass of the muscle has been ignored. When the muscle contracts, an equal and opposite tensile force F_m and F'_m is exerted on the two bones; force F_m is the only torque-producing force acting on bone 2 and hence imposes a rotational effect on the bone. The consequences of this force on bone 2 can be described completely by the following three equations (considering only the instantaneous conditions shown in Fig. 20):

$$\Sigma M_o = I_o \alpha \qquad (1)$$

$$\Sigma F_x = ma_x = m\omega^2 R_G \qquad (2)$$

$$\Sigma F_y = ma_y = m\alpha R_G. \qquad (3)$$

Equation (1) is the sum of the moments or torques acting on the bone, in which I_o is the moment of inertia of the bone about o, and α is the angular acceleration. Equation (2) is the summation of forces in the x direction in which a_x is the radial acceleration; hence, $m\omega^2 R_G$ is the centripetal inertia. Equation (3) is the summation of forces in the y direction in which a_y is the tangential acceleration, hence $m\alpha R_G$ is the tangential inertia. The term ω is the angular velocity of the bone, m is

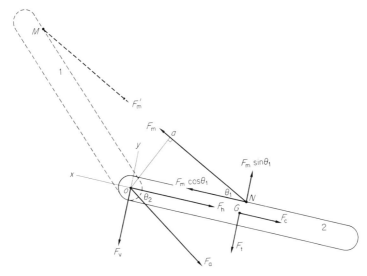

FIG. 20. Schematic drawing of a one-joint flexor muscle and free-body diagram of bone 2 onto which the muscle inserts. All structures and forces shown by dashed lines are not part of the free-body diagram. The system is under gravity-free conditions, and bone 2 rotates under the action of the muscular force. (Taken from Bock, 1968a, Fig. 3.)

the mass of the bone, and R_G is the radius of the center of gravity of the bone measured from the center of rotation o. I shall write these equations in the form of d'Alembert's principle, hence the terms $I_o\alpha$, $m\omega^2 R_G$ and $m\alpha R_G$ will appear as fictitious "inertia torques" or "inertia forces," and all equations will be set equal to zero. The fictitious "inertia forces," $m\omega^2 R_G$ and $m\alpha R_G$, act on the body as if the entire mass of the body were concentrated at its center of gravity. The convention for plus and minus signs is the same as used in the analysis of external forces on bones (p. 147).

In the free-body diagram shown in Fig. 20, the vector of the muscle force does not pass through the center of rotation of bone 2, and hence it exerts a torque on the bone. The torque is the product of the force and the moment arm of the force. The moment arm, oa, of the force vector F_m is obtained by measuring the perpendicular line dropped from the center of rotation to the line of the force vector. The magnitude of the torque becomes greater with an increase in either the force or its moment arm, or both. The equation of the forces acting on bone 2 are (in d'Alembert's notation):

$$\Sigma M_o = -F_m (oa) + I_o\alpha = 0 \qquad (4)$$

$$\Sigma F_x = -F_m \cos\theta_1 + F_h + F_c = 0 \qquad (5)$$

$$\Sigma F_y = F_m \sin\theta_1 - F_v - F_t = 0 \qquad (6)$$

Force F_c is the centripetal inertia and is equal to $m\omega^2 R_G$ and increases as the angular velocity increases; it is present as long as bone 2 is moving with angular velocity. Force F_t is the tangential inertia and is equal to $m\alpha R_G$; it is present when bone 2 is undergoing angular accleration. The forces F_h and F_v are components of the force exerted by bone 1 on bone 2 at the articulation; they are equal to $F_a \cos\theta_2$ and $F_a \sin\theta_2$, respectively. The force at the articulation is equal to the vector summation of the muscular force, the centripetal inertia, and the tangenital inertia. When a muscular force causes a bone to undergo angular acceleration, the force at the articulation is always changing because of changes in the magnitude of the centripetal inertia and possibly the tangential inertia, and in the vector directions of the muscle force and the two inertias.

In the second simple case of a one-joint muscle, the muscle force is balanced by a second force so that static equilibrium, both rotational and linear, exist as shown in Fig. 21. For this static condition, the forces and torques on bone 2 are described by the following equations.

3. THE AVIAN SKELETOMUSCULAR SYSTEM

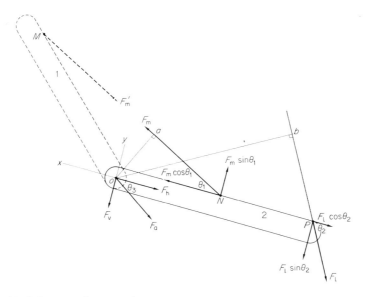

FIG. 21. Schematic drawing of a one-joint flexor muscle and free-body diagram of bone 2. The system is the same as in Fig. 3 except for the addition of a load force which maintains bone 2 in a static condition. (Taken from Bock, 1968a, Fig. 4.)

$$\Sigma M_o = -F_m (oa) + F_1 (ob) = 0 \tag{7}$$

$$\Sigma F_x = -F_m \cos \theta_1 + F_1 \cos \theta_2 + F_h = 0 \tag{8}$$

$$\Sigma F_y = F_m \sin \theta_1 - F_1 \sin \theta_2 - F_v = 0 \tag{9}$$

The articular force is represented by its two components, F_h (= $F_a \cos \theta_3$) and F_v (= $F_a \sin \theta_3$). Note that the terms $I_o \alpha$, F_c and F_t are absent because bone 2 is not undergoing angular movement. The force at the articulation is obtained by vector addition of the forces F_m and F_1. A number of differently placed forces on bone 2 could balance the torque of the muscular force, thereby satisfying the conditions of static rotational equilibrium. For each of these load forces, the articular force would differ both in direction and magnitude. If the load force lay exactly on the line of action of the muscular force, but opposite in direction, then no force would exist at the articulation; but this is a specialized and quite trivial case.

Attention must be given to the relationship between the muscular force and the articular force because numerous workers (e.g., Gray, 1956, p. 203, 1968, p. 6) state that these forces constitute a force couple, i.e., a pair of parallel forces, equal in magnitude but acting in opposite directions. This notion becomes important in comprehending the mechanics of two-joint muscles. Examination of the two cases just discussed should be sufficient to prove that this notion is false (Bock, 1968a, pp. 33–36). These examples represent the simplest possible conditions when a single muscular force acts on a bone; namely, (a) the bone rotates under the acceleration of the torque placed upon it, or (b) the bone is held in a static position by a second force whose torque balances that of the muscular force. In the first possibility, the centrifugal and tangential inertias must be added vectorally to the muscular force to ascertain the size and direction of the articular force. In the second case, the two torque-producing forces must be added vectorally to obtain the size and direction of the articular force. Because another force or inertia must always be added to the muscular force to calculate the articular force, this latter force can never be equal, parallel to, and opposite in action to the muscular force – the necessary requirement for these forces to constitute a force couple. In

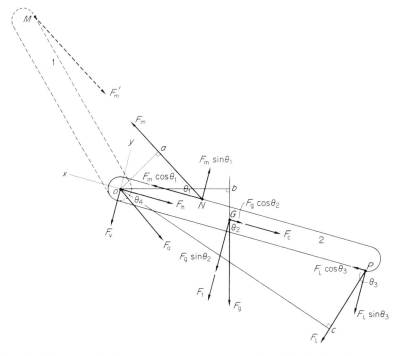

FIG. 22. Schematic drawing of a one-joint flexor muscle and free-body of bone 2. The system is the same as in Figs. 20 and 21, except that it is under gravity conditions. The bone is either under static or dynamic conditions depending upon the size of the load force (F_l). (Taken from Bock, 1968a, Fig. 5.)

3. THE AVIAN SKELETOMUSCULAR SYSTEM

more complicated cases, more forces act on the bone; all must be added together to ascertain the articular force. Perhaps by rare chance, the muscular and articular forces may constitute a force couple because of the particular arrangement of all forces acting on the bone; but this is a special and trivial case that does not violate the general conclusion that the muscular and articular forces resulting from the contraction of the muscle do not constitute a force couple.

Any mechanical analysis of avian bone–muscle systems based on the earlier concept of force couples is inherently erroneous. These analyses must be checked carefully, because the author may state that the muscular and articular forces constitute a force couple, and then never use this assumption, but rather base his mechanical analysis on the opposite assumption. Thus, the results may be correct and the actual analysis may be valid, although the statements of basic assumptions are wrong.

A more realistic example of a one-joint muscle acting on a bone with a resisting force and the pull of gravity also acting on the bone is shown in Fig. 22. In such cases, one may know by direct observations whether the bone is static or is undergoing angular acceleration (plus possible linear acceleration which will not be discussed here; see Bock, 1968a), or one may know the magnitudes and vectors of the forces from which the static or rotational movement of the bone can be determined. If the bone is under static conditions, then the following equations describe the forces and torques.

$$\Sigma M_o = -F_m (oa) + F_g (ob) + F_1 (oc) = 0 \qquad (10)$$

$$\Sigma F_x = -F_m \cos \theta_1 + F_g \cos \theta_2 - F_1 \cos \theta_3 + F_h = 0 \qquad (11)$$

$$\Sigma F_y = F_m \sin \theta_1 - F_g \sin \theta_2 - F_1 \sin \theta_3 - F_v = 0 \qquad (12)$$

where $F_h (= F_a \cos \theta_4)$ and $F_v (= F_a \sin \theta_4)$ are the components of the articular force. If bone 2 is undergoing angular acceleration under the action of the muscular force, then the forces and torques are described by the following equations.

$$\Sigma M_o = -F_m (oa) + F_g (ob) + F_1 (oc) + I_o \alpha = 0 \qquad (13)$$

$$\Sigma F_x = -F_m \cos \theta_1 + F_g \cos \theta_2 - F_1 \cos \theta_3 + F_c + F_h = 0 \qquad (14)$$

$$\Sigma F_y = F_m \sin \theta_1 - F_g \sin \theta_2 - F_1 \sin \theta_3 - F_t - F_v = 0 \qquad (15)$$

F_c $(= m\omega^2 R_G)$ is the centripetal (radial) inertia and F_t $(= m\alpha R_G)$ is the tangential inertia. The moment of inertia and the size of the centripetal and tangential inertias depend on the distribution of the mass of the bone and any other mass that may be attached to the bone. If the force F_1 is associated with a mass (such as if this force were the pull of gravity on an object held at point P), then the mass of this object and its distribution relative to the center of rotation must be included in calculations of the moment of inertia and centripetal and tangential inertias.

The relationship between the magnitude of the muscle force and its position is inherent in the concept of mechanical advantage of the muscle. If the magnitude of the muscular force is held constant, then its torque depends upon the length of its moment arm. This parameter depends upon the insertion of the muscle and the angle between the muscle vector and longitudinal axis of the bone. Increase in the torque of the muscular force would increase the magnitude of the force the bone could exert at point P along the depicted vector direction (an equal and opposite force is exerted on the bone as shown in Fig. 22). This load force can be increased by increase in the torque of the muscular force or decrease in the length of the moment arm of the load force.

Exactly the same holds when the bone is undergoing angular acceleration. Speed of rotation depends upon the angular acceleration. A high acceleration is required for high velocity of rotation, because bones swing in a limited arc before they must stop and reverse their direction of motion. Examination of the above equations shows that torque F_m (oa) of the muscular force equals $I_o \alpha$. If the moment of inertia (I_o) of a limb segment is constant, then speed of rotation depends on the magnitude of the angular acceleration (α), which is dependent on the torque applied to the bone. Any factor that increases the torque of the muscular force will increase the angular acceleration and thus the speed of rotation. Hence, increase in the moment arm of the muscular force would increase the torque produced by this muscle and the speed of rotation of the bone. A decrease in the moment of inertia of the bone would also increase its angular rotation; hence, concentration of the mass of a limb segment close to its center of rotation would decrease the moment of inertia of the limb and its angular rotation.

Thus, increase in the mechanical advantage of a muscle, be it increase in its torque by increase of its force or by increase in the length of its moment arm, or be it decrease in the moment arm of a resisting load, or be it decrease in the moment of inertia of the bone, would result in a larger force applied by the bone to a resistance or in a higher speed of rotation of the bone. Insertion of the muscle further away from the articulation is advantageous for muscles applying a static force against a resistance and for muscles moving bones with maximum rotational velocity. On grounds of mechanical advantage, it is not true that insertion close to the articulation is advantageous for "speed" muscles that move bone segments at high rotational velocity. This aspect of mechanical advantage as applied to muscles moving limb segments has been repeated commonly in literature since the beginning of biomechanical studies of bone–muscle systems, but never with proper analysis. I must repeat that on mechanical grounds, it is advantageous for muscles rotating bones at high velocities to insert as far as possible from the center of rotation, not close to it. Yet, as most ornithologists know, such muscles in birds insert close to the articulation serving as the center of rotation, and hence these muscles have a lower mechanical advantage. They produce less torque and hence impart a lower angular acceleration onto the bone for a fixed moment of inertia. It should be obvious that factors other than mechanical advantage have an important role in the function and adaptation of muscles in bone–lever systems; these will be examined below.

b. Two-Joint Muscles. The mechanics of two-joint muscles must be considered at this point, because these muscles include most of the important strong muscles in the avian body and because they exhibit most clearly the dilemma between mechanical advantage and other factors that bear upon adaptiveness of skeletal muscles. Two-joint muscles are those that span two articulations between their origin and insertion (Fig. 18), and hence can exert an influence on two bones upon contraction. Excellent examples of two-joint muscles are the M. depressor mandibulae of the M. pseudotemporalis superficialis of the jaw musculature; the M. pseudotemporalis profundus is a one-joint muscle (Fig. 27). In the two-joint muscle illustrated, the muscle exerts equal and opposite tensile forces, F_m and F'_m, on bones 1 and 3 when it contracts; the consequence of this force depends on all of the other forces acting on the bones and the arrangement of the bones. Only bones 2 and 3 are free to rotate; bone 1 is held stationary. Note that the orientations of the xy axes differ for bones 2 and 3; one of these axes lies along the longitudinal axis of each bone.

I shall consider only one case, which is sufficient to understand the basic properties of two-joint muscles. Bone 3 is assumed to be initially under static conditions relative to bone 2 at the onset of the muscular contraction. The moment-producing forces, F_m, F_g, and F_1, act on bone 3 as shown in Fig. 23A; the free-body diagram for bone 3

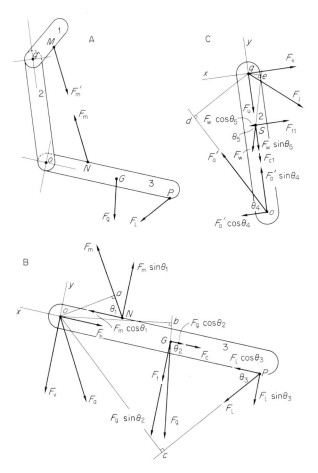

FIG. 23. Schematic drawing and free-body diagram of a two-joint flexor muscle. (A) A two-joint flexor muscle. (B) Free-body diagram of bone 3. (C) Free-body diagram of bone 2. (Taken from Bock, 1968a, Fig. 9.)

is shown in Fig. 23B. The equations describing the forces and torques at the initial period of muscle contraction are

$$\Sigma M_o = -F_m\,(oa) + F_g\,(ob) + F_1\,(oc) = 0 \tag{16}$$

$$\Sigma F_x = -F_m \cos\theta_1 + F_g \cos\theta_2 - F_1 \cos\theta_3 + F_h = 0 \tag{17}$$

$$\Sigma F_y = F_m \sin\theta_1 - F_g \sin\theta_2 - F_1 \sin\theta_3 - F_v = 0 \tag{18}$$

The magnitude and direction of the articular force, F_a, depends on the vector summation of the other forces acting on bone 3. The details of this force are not important, only that some force exists at the center of rotation of bone 3 under the action of the two-joint muscle except in the special and trivial case in which the torque-producing forces add vectorally to zero. Thus, one consequence of the contraction of a two-joint muscle is a force exerted by bone 3 on bone 2 at their articulation. The action of this force, F'_a, must now be analyzed.

A free-body diagram of bone 2 is shown in Fig. 23C; this bone rotates about bone 1 at q. It has two torque-producing forces acting upon it, F'_a and F_w, the latter being the pull of gravity. The equations describing the forces and torques on bone 2 at the initial period of muscle contraction are

$$\Sigma M_q = F'_a (qd) + F_w (qe) - I_q \alpha_1 = 0 \tag{19}$$

$$\Sigma F_x = -F'_a \cos \theta_4 - F_w \cos \theta_5 + F_{tl} + F_k = 0 \tag{20}$$

$$\Sigma F_y = F'_a \sin \theta_4 - F_w \sin \theta_5 - F_{cl} - F_u = 0 \tag{21}$$

New terms in these equations are the force of gravity on bone 2 (F_w), the components of the articular force at q on bone 2 (F_k and F_u), and the tangential inertia (F_{tl}) and the centripetal inertia (F_{cl}) of bone 2 as it undergoes angular acceleration under the action of force F'_a. With the exception of two rare cases, bone 2 will rotate because no torque-producing force exists to balance the torque of force F'_a. These two cases are: (1) when the vector line of F'_a passes through q and thus produces no torque; and (2) when the torque of F'_a is equal and opposite to the torque of F_w, the pull of gravity on bone 2. A one-joint muscle between bones 1 and 2 could balance the torque of F'_a, transforming bones 1 and 2 into a single "rigid" body, and the two-joint muscle into a one-joint muscle.

The result of all torque-producing forces acting on bone 3, which is initially stationary relative to bone 2, is a force, F'_a, acting on bone 2 at point o and causing bone 2 to rotate relative to bone 1 (except in three rare but possible conditions). When bone 2 begins to rotate about point q, the relationships of forces F_m, F_g, and F_l to bone 3 are altered so that the sum of their torques about point o will no longer be equal to zero, except by rare chance. Hence bone 3 is no longer static and will start to rotate about its center of rotation, requiring a revision of the equations describing the sums of forces and their torques. Except for a few rare arrangements of forces, the contraction of a single two-joint muscle cannot by itself maintain the bone-lever system in a condition of static equilibrium—a two-joint muscle—bone system is an inherently nonstatic system. Even the few rare exceptions are cases of unstable equilibrium.

Solution of the forces and movements of bones in an actual case for a two-joint muscle, and dynamic conditions of a one-joint muscle, is difficult as many of the factors change in nonlinear and irregular fashions. The equations cannot be integrated easily, if at all. It is necessary to use simplifying assumptions that may cause errors far smaller than the errors in measuring forces and their moment arms. Or, it may be best to abandon theoretical treatment and approach the problem empirically (e.g., high-speed motion picture analysis).

An important consideration in two-joint muscles is the relationship between (a) the line of the force vector F_m of the muscle and the longitudinal axis of the central bone (bone 2 in Fig. 24), i.e., where the two lines intersect, and (b) the direction of the resultant force exerted by the distal bone (bone 3) upon the central bone — the vector direction of the F'_a force — and hence the rotation of the central bone (Fig. 24). It

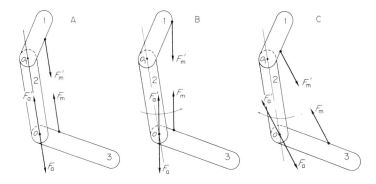

FIG. 24. Schematic drawings of a two-joint muscle to show the (presumed) relationship between the vector of the muscle force, the longitudinal axis of bone 2, and the rotation of bone 2. (A) The vector of the muscle force and axis $o-o_1$ are parallel; bone 2 does not rotate. (B) These lines intersect below point o. (C) These lines intersect above point o_1. The thin curved arrow shows the rotation of bone 2. The analysis illustrated in this figure is incomplete and incorrect, and should not be followed. (Taken from Bock, 1968a, Fig. 10.)

is very tempting, but wrong, to conclude that, if F_m is parallel to axis $o-o_1$ (Fig. 24A), the central bone suffers pure compression and does not rotate; if F_m intersects axis $o-o_1$ below point o (Fig. 24B), the central bone rotates counterclockwise; and, if F_m intersects axis $o-o_1$ above point o (Fig. 24C), bone 2 rotates clockwise. This conclusion follows the notion that contraction of a muscle places a force couple on the bone and has been almost universally accepted by morphologists (see Bock, 1968a; pp. 29–30). However, force couples never occur except by rare chance as discussed above for the one-joint muscle. No direct relationship exists between the direction of the vector of F_m and the movement of the central bone. The free-body equations for bone 3 must be solved to ascertain the magnitude and direction of the articular force F_a, after which the free-body equations for bone 2 must be solved before the rotation of this bone is known. Examination of the morphological diagram alone is insufficient to reveal reliably the effect of a two-joint muscle on the central bone (and hence also on the distal bone) of a three-bone system. This limitation is not a weakness of the method of free-body diagrams, but a consequence of the fact that certain calculations must be made before the diagrams can be completed and all conditions known. The same limitations exist for any method of analysis.

Many authors have discussed the functions and advantages of two-joint muscles as compared with one-joint muscles, the details of which need not be repeated here (see Bock, 1968a; pp. 36–39, for a review). A two-joint muscle can move the distal and central bones of the bone linkage system, while a one-joint muscle could move only one bone. Some advantages in torque and perhaps speed may exist in two-joint muscles, but these advantages appear to be slight. Most of the large muscles in the avian body that move limb segments or the jaws, either for speed or strong force application, appear to be two-joint muscles that are pinnate and possess a large fiber number at the expense of fiber length. Definite summaries cannot be offered, because most descriptions of avian muscle systems neither state whether the muscles are one- or two-joint (although this information can be obtained indirectly from statements of muscular origins and insertions) nor provide details on the fiber composition and arrangement of the muscles. These pinnate, two-joint muscles insert close to articulations and appear to lack the capacity for extensive excursion; they appear to be specialized for large force production, be it for speed of rotation of bony segments or large force application.

A major disadvantage of two-joint muscles is that they lack independent control over the distribution of forces about each articulation and over the movement of each bone. Whenever the muscle contracts and develops a certain tension, the movements of the bones and the forces will depend on the whole system and on all forces acting on it. The effects of the muscle on the individual bones cannot be varied independently. Moreover, the two-joint muscle–bone system is inherently nonstatic, so that a two-joint muscle by itself could not serve as a holding muscle for static conditions. Series of one-joint muscles are required about each articulation to provide the control and necessary static conditions in the bone-lever system.

Very little is known about two-joint muscles in the avian muscular system, because workers simply did not concern themselves with this aspect of muscle–bone systems. Two-joint muscles are discussed in the passerine jaw musculature (Bock, 1973b) and their basic mechanics described (Bock, 1968a), but little else. Yet, comprehension of two-joint muscles is critical for analyzing adapation of skeletal muscles (see below). Simply, a muscle of the same fiber composition and arrangement will have radically different functional properties and adaptive significances depending on whether it is a one-joint muscle or a two-joint muscle. The only morphological difference between these muscles would be the point of attachment relative to the joint (see Fig. 18) – a simple morphological difference causing profound functional and adaptational results.

10. Adaptation of Muscles

Development of a general system of evaluation of the degree of adaptation of skeletal muscles is one of the most fascinating studies in the avian skeletomuscular system, because every factor of muscle structure and physiology on all levels of organization from the molecular to the bone–muscle system must be included. Many of these factors have mutually contradictory effects, the optimum combination for maximum adaptiveness being a complex compromise and balance. An adequate analysis would not have been possible without the contributions of A. V. Hill. His discoveries on the energy budgets of muscles during isometric and isotonic contractions are fundamental to concepts of the degree of muscle adaptation. The basis for his work is precise measurement of heat released from muscles during contractions, including twitches, and a combination of highly skilled experimentation and profound theoretical analyses and conceptualizing. A summary of Hill's work, spanning six decades, may be obtained from his book "Trails and Trials in Physiology" (1964).

Adaptation of skeletal muscles depends, at the minimum, on their functional properties, on the organizational level of whole muscles, and on their functional properties as units within muscle–bone systems. In the following analysis of muscle adaptation, I shall assume that the muscles are composed of fibers containing identical sarcomeres; otherwise the analysis would become hopelessly complex and would lose its heuristic value in providing a basis on which morphologists can estimate adaptiveness of avian skeletal muscles using observations obtainable from gross dissection of museum specimens (any features that can be seen and measured with the assistance, at most, of a dissecting microscope). Other factors, such as the physiological properties of the various fiber types, are certainly critical to the evaluation of the adaptiveness of skeletal muscles. The relative ability to pick up and transport oxygen in red and white muscle fibers, the ability to contract for longer periods without fatigue, and the ability to metabolize fats are important in judging adaptation of red muscles and white muscles. Equally important is the ability of twitch muscles for rapid phasic contractions and the ability of tonus muscles to maintain tension for long periods without fatigue and apparently with relatively low energy expenditure. Yet comparative information on these biochemical and cytological features of avian skeletal muscles is so poorly known at present, that analysis of their adaptive significance is better postponed.

As a component in the bone–muscle system, the muscle must move (rotate) the bony elements, which possess mass and moments of inertia, over the needed distance at the required speed, or must enable the bone to apply a certain force on another object. Hence, the muscle must develop sufficient force, but of far greater importance is the torque developed by the muscle and whether the muscle can continue to develop this torque over the needed arc of rotation to maintain the necessary speed. If certain parameters are established for the bone–muscle system, such as the $I_0\alpha$ term and the required arc of rotation (hence distance of shortening of the muscle) or the force to be applied by the bone at a definite point—the $F_r(om)$ term— then a number of possible muscles of different fiber number, fiber length, and hence muscle mass, would possess the necessary functional properties to satisfy the functional parameters of the bone–muscle system. Two one-joint muscles inserting at differing distances from the articulation (Fig. 25A) could fulfill the same functional parameters of the bone–muscle system; their torque development can be the same [i.e., $F_A(om) = F_B(on)$], according to the magnitude of their moment arms and their force development. A one-joint muscle and a two-joint muscle of equal strength (Fig. 25B) originating from opposite sides of the articulation between bones 1 and 2, and inserting at the same point on bone 3, would provide similar torques on bone 3 relative to its center of rotation at point o because their moment arms are almost identical. The consequences of the actions of these muscles can be quite different, as seen in Figs. 25C and 25D. The one-joint and the two-joint muscles illustrated in Figs. 25C and 25D develop equal maximum isometric force (P_o) and have moment arms of equal lengths at the onset of contraction. The one-joint muscle would have to shorten (Fig. 25C) by a considerable amount to rotate bone 3 and hence must possess fibers of sufficient length to permit this

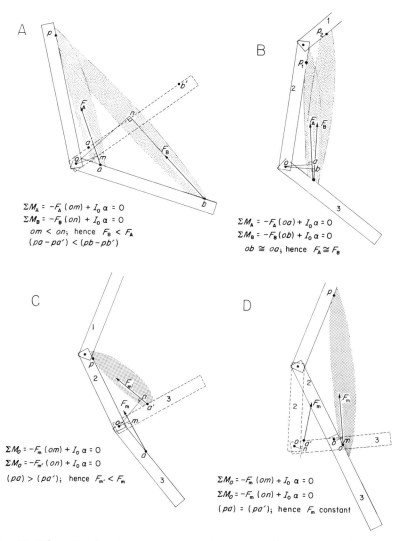

FIG. 25. Schematic drawings to compare the actions of one-joint and two-joint muscles. (A) Comparison of two one-joint muscles developing the same torque. Muscle B is longer and must shorten more, but with less tension than muscle A. (B) Comparison of a one-joint muscle, A, and a two-joint muscle, B, that develop the same approximate tension and torque. (C) Contraction of a one-joint muscle, rotating bone 3 about bone 2 with shortening of the muscle from pa to pa'. (D) Contraction of a two-joint muscle, rotating bone 3 about bone 2 and bone 2 about bone 1 without shortening of the muscle; hence pa equals pa'. Additional muscles may be required to control rotation of bone 2.

shortening. Moreover, the force developed by the muscle decreases as the muscle shortens, but the moment arm increases in length so that the torque produced by the muscle may actually increase. This is advantageous, although this system also contains some serious disadvantages because of the shortening of the muscle. The two-joint muscle could maintain its initial length while rotating bone 3 about bone 2 if bone 2 rotates simultaneously in the opposite direction (Fig. 25D); hence, its force development remains constant, as would its torque development, because its moment arm remains at the same (approximate) length. Moreover, since the two-joint muscle does not have to shorten or shortens much less than the comparable one-joint muscle, its fibers may be very short. This comparison cannot be pushed to its logical extreme of a muscle that need not shorten and hence may possess fibers of no length, because a minimum fiber length is required for internal shortening and adjustment as the muscle contracts. The two-joint muscle shown in Fig. 25D possesses roughly the same number of fibers as the comparable one-joint muscle shown in Fig. 25C, because initial force development is equal; any difference in fiber number would result from pinnate arrangement of the fibers and from the influence of the cosine of the angle of pinnation. Because the two-joint muscle possesses shorter fibers, it would have a smaller mass than the one-joint muscle. Both the smaller mass of the two-joint muscle and its ability to rotate bone 3 with a minimum amount of shortening are definitely advantageous.

An interesting but more complex comparison is between a long-fibered, one-joint, parallel-fibered muscle and a short-fibered, pinnate, two-joint muscle (Fig. 26), both of which develop the same torque and impose the same angular acceleration onto bone 3 which rotates about point o, its articulation with bone 2. The moment arms of the force vectors of these muscles differ considerably, with that of the two-joint muscle being shorter than that of the one-joint muscle. Thus, the required force of the one-joint muscle, A, is lower than that of muscle B; the physiological cross section of muscle A is smaller than that of muscle B, but fiber length of A is greater than that of B. The external load placed upon A is less than that placed on B because of mechanical relationships between the bone and each muscle. The speed of shortening of muscle A is higher than that of muscle B because of its lower external load and its longer fibers. Muscle A must shorten faster because of the high speed of its point of insertion close to the distal end of bone 3. Because of the ability of bone 2 to rotate in a direction opposite to that of bone 3, muscle B shortens very little, if at all, as it

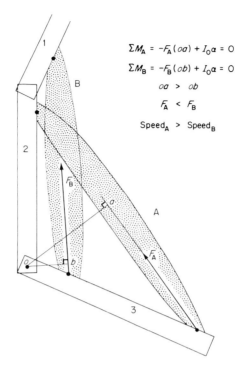

FIG. 26. Schematic drawing to compare the action of a long-fibered, parallel-fibered, one-joint muscle (A) and a short-fibered, pinnate, two-joint muscle (B), both of which develop the same torque and impart the same angular acceleration onto bone 3. These muscles are illustrated by the M. pseudotemporalis profundus and the M. pseudotemporalis superficialis of the jaw musculature (see Fig. 27).

contracts and provides the force to rotate bone 3; hence its fibers may be very short. The masses of muscles A and B would differ in most cases, but it is difficult to predict even their relative masses.

A nice example of a comparable pair of one-joint and two-joint muscles is the M. pseudotemporalis profundus and the M. pseudotemporalis superficialis of the jaw musculature in passerine birds (Fig. 27). The M. ps. profundus is a long-fibered, essentially parallel-fibered, one-joint muscle that inserts on the mandible far from the articulation. The M. ps. superficialis is a very short-fibered, multipinnate, two-joint muscle that inserts on the mandible very close to its quadrate articulation. Both muscles serve to adduct the mandible. The M. ps. profundus has a much smaller physiological cross section than the M. ps. superficialis and hence develops much less force. However, its moment arm is several times longer than that of the M. ps. superficialis and it is quite possible that each muscle develops the

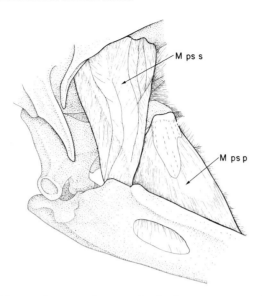

FIG. 27. The M. pseudotemporalis superficialis (a two-joint muscle) and the M. pseudotemporalis profundus (a one-joint muscle) of *Passer*.

same torque (approximately). These muscles are not greatly dissimilar in mass; in *Passer*, which is illustrated in Fig. 27, the muscles are about equal in mass, with the M. ps. profundus being slightly heavier than the M. ps. superficialis.

One important difference between these muscles is the distance each must shorten if the mandible is adducted through the same arc (Fig. 28). Because of the mechanics of the avian kinetic skull, the quadrate usually rotates posteriorly (backwards) as the mandible is adducted, i.e., the two bones rotate in opposite directions. In Fig. 28, the axes of the mandible and of the quadrate are shown with the mandible partly opened and fully closed. The M. ps. superficialis and the M. ps. profundus have been drawn on the two skull diagrams, and their origins and insertions shown with heavy dots. Examination of the figures demonstrates that the M. ps. profundus shortens considerably from a stretched length to an approximate resting length as the mandible is closed, but that the M. ps. superficialis remains at the same length throughout adduction of the lower jaw. This difference in the distance that muscles shorten while moving (rotating) bones is crucial for the evaluation of the relative degree of adaptation of skeletal muscles.

These comparisons emphasize the very important aspect of increasing levels of organization, namely that the morphological–functional

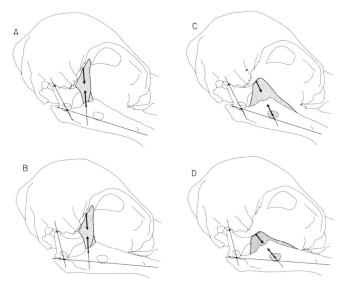

FIG. 28. Comparison of the M. pseudotemporalis superficialis (A and B) and the M. pseudotemporalis profundus (C and D) in a generalized bird (*Corvus*) to show that the length of the two-joint muscle can remain constant as it contracts and adducts the mandible, while the one-joint muscle must shorten. The bones and axes in C and D are identical to those shown in A and B.

properties of a system cannot be guessed or understood completely after extensive study of the properties of individual components at lower levels of organization. Two different approaches were used in the above comparisons. One is to compare two muscles of similar morphological–functional properties that are arranged differently in a bone–muscle system so that the properties of each complex system differ considerably. The second is to arrange muscles of greatly differing properties in a bone–muscle system so that the resulting properties of each bone–muscle system are the same.

The basic problem is how can the relative adaptiveness of these muscle components be measured and compared. If a number of muscles of different morphological parameters impart the same functional characteristics to the muscle–bone system (e.g., provide the same angular acceleration), then it seems reasonable to assume that they vary in their degree of adaptation and that, given a set of requirements for a muscle–bone system, then some particular muscle will be the best adapted one. Using energy requirements as the index to the degree of adaptation (Bock and von Wahlert, 1965), the muscle requiring the least metabolic energy to provide the muscle–bone system with the required characteristics would be the best adapted muscle. A minimum–maximum system works for muscles, as for bones, in that the combination of physiological properties requiring minimum energy would be best. It is not possible, as shown above, to develop a single morphological index for physiological properties, since maximum tension development and shortening ability are in direct conflict. Minimum mass or weight of the muscle would be a factor in estimation of its adaptation, but it is not the only factor or even the chief one. Indeed minimum weight, by itself, is not a valid index to the degree of adaptation of muscles, even those that provide the same functional properties of a muscle–bone system. Support for this statement will be provided below. The degree of adaptation of muscles is measured, instead, directly in terms of metabolic energy required by the muscle in producing the desired functional properties. Note that metabolic energy rather than force, work, or power is used as a measure of adaptation.

Energy utilized by a muscle during contraction may be measured by the amount of heat released from the muscle (plus the amount of mechanical work done by the muscle if it contracts isotonically). Reference to most of the basic literature in this field may be found in Hill (1964).

Resting muscles require a low, constant level of metabolic energy proportional to the weight of the muscle. Hence, the smallest muscles will be most adaptive, and selection should reduce muscle size as much as possible. If a resting muscle is loaded and hence stretched, its metabolic rate increases as measured by heat production (Feng, 1932) and also oxygen consumption. The basis for this effect is not known, but it suggests that the resting (100%) length of a muscle may be the maximum length before the onset of increased metabolism.

When a muscle contracts isometrically it provides force at a fixed length, whether at the reference length or at some shorter or longer relative length. Recordings of heat released by an isometrically contracting muscle (Fig. 29A) show an initial rapid rise of heat followed by a slower rise as the muscle is being stimulated. The heat released during the period of active contraction is heat of activation (H_A); the slower rise of heat had been termed heat of maintenance, but this has been shown to result from successive heats of activation. After stimulation has ceased, the muscle continues to release heat at a slow rate for a long time – up to one thousand times the length of active contraction. The heat released during this period is heat of recovery (H_R),

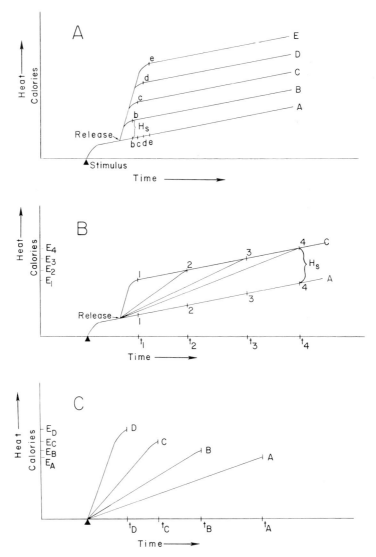

FIG. 29. Heat development in isometric and isotonic contractions taken from the work of A. V. Hill (see 1964 for references). (A) Curve A shows heat development following stimulation in an isometric contraction. Curves B, C, D, and E show heat development in isotonic contractions of increasing distances. The heat of shortening for each isotonic contraction is indicated. (B) Curve A shows heat development during an isometric contraction. Curves 1, 2, 3, and 4 show heat development for isotonic contractions of a distance similar to that of curve C in Fig. A, but with increasing loads; load 4 is the heaviest. Note that the heat of shortening is the same for each curve although the load varies. (C) Curves showing heat development of different muscles inserting at varied distances from the articulation and developing different forces, but the same torque. Muscle A inserts closest to the articulation, develops the most force, shortens the least distance, and slowest, but requires the least amount of energy. Note that all curves drawn are estimates and serve only to show general relationships and trends.

with the total heat of recovery approximately equal to the total heat of activation. The energy budget of an isometric contraction is

$$E_{isom} = H_A + H_R$$

and is given in calories, or commonly in gram-centimeters. In comparative studies, the energy should be expressed in terms of calories per unit force of the muscle per time in seconds, or in terms of calories per unit torque of the muscle per time in seconds. Thus, if two muscles are producing the same force or the same torque, their energy utilization can be compared directly.

Unfortunately, few comparative studies of heat production in muscles are available. Most of the studies are on the sartorius muscle of frogs and toads. Yet theoretical comparisons are possible. Tension developed by a muscle is proportional to its physiological cross section, i.e., the number of muscle fibers. Greater tension requires more fibers actively contracting. The amount of heat will depend on the number of contracting fibers and their lengths. Thus, two muscles of equal fiber number, but of uneven fiber lengths, will produce the same force, but the longer-fibered muscle will require more energy and would be less adapted. Reduction in length of fibers and hence in the weight of the muscle is clearly advantageous in muscles that contract isometrically and develop tension only, without shortening. The major part of the distance between the origin and insertion of the muscle is spanned by tendons (as in pinnate muscles) with the muscle fiber spanning only a small part of the total. Another consequence is that a muscle of a certain mass contracting isometrically will require a certain amount of metabolic energy that is recorded as heat; as the number of fibers in the muscle increases with a decrease in fiber length, the force of the muscle increases for the same energy expenditure. If the primary requirement is for maximum force for a given size of muscle and hence energy expenditure, then muscles with more, shorter fibers will be the better adapted. Those muscles serving primarily for force application of the bone lever system, such as the large adductor jaw muscles in finches and parrots, are very short-fibered, pinnate muscles. These muscles can shorten very little and hence must insert close to the articulation, where their displacement would be small even if the bone rotates through a large arc. Under certain configurations of a bone-lever system, "force" muscles may possess short fibers and also insert far from the center of rotation of the bone if the muscle undergoes little excursion, either stretch or shortening, as the bone rotates. Such muscles possess the double advantage of large force development and large moment arm, hence they can develop a very large torque without suffering the disadvantages of shorting during contraction. But occurrance of muscle–bone systems which permit this pattern of advantages may be rare in the avian body; a two-joint muscle is an absolute requirement of such systems. Generally those muscles that insert far from articulations (center of rotation) must undergo large excursions and hence must have sufficiently long fibers to permit stretch of the muscle as the bone rotates, even if these muscles are not providing the force to rotate the bone.

When a muscle contracts isotonically, the muscle must provide the needed force throughout its excursion and must have fibers of sufficient lengths to permit shortening of the muscle. Recording of heat production during isotonic contractions provided several surprises (Fig. 28A,B). If a muscle contracting isometrically at reference length is released with a constant load, the muscle will shorten isotonically, and a sharp increase of heat will appear and continue to rise as long as the muscle is shortening. The amount of extra heat is proportional to the absolute distance of shortening. If the muscle continues to contract isometrically at its shortened length, the increase of released heat will parallel that of an isometric contraction at reference length. If a muscle is allowed to shorten over a fixed distance with a varying load (Fig. 28B), the rise of the extra heat will be slower as the load increases and the muscle shortens with lower velocity, but the amount of extra heat produced while the muscle shortens is constant and is independent of the load. This extra heat that appears as the muscle shortens is termed the heat of shortening (H_S). Presumably, it appears because of the continual breaking and reforming of bridges between the thin and thick filaments as the muscle is shortening. When the stimulus ceases, heat is released at a low rate, which is the recovery heat and which is roughly equal to the initial heat ($H_A + H_S$). In addition to the utilized energy that appears as heat, isotonically contracting muscles produce mechanical work (W). The total energy budget of an isotonic contraction is

$$E_{\text{isot}} = W + H_A + H_S + H_R \qquad (22)$$

in which recovery heat equals roughly the initial heat. In comparative studies, the energy should be expressed in terms of calories or gram-centimeters per unit torque of the muscle per time in seconds. Generally, little is gained by comparing energy expended with the force of isotonically contracting muscles. Thus if two muscles are producing the same torque, their energy utilization can be compared directly.

Energy comparisons of isotonically contracting muscles are more complex, because most of these muscles are limb muscles and their action involves a cyclic pattern of contraction over a long period of time as the bird is flying or running. Total energy depends upon energy for each contraction multiplied by the number of cycles. The energy per cycle is that utilized up to the point where the muscle stops shortening isotonically. The energy budget of isotonic muscles as shown in Fig. 28B and in Eq. (22) suggest that the most efficient arrangement for isotonic muscles is to shorten slowly with a heavier load to reduce the number of cycles per minute (see papers of A. V. Hill cited in 1964). This can be achieved by increasing the length of the limb so that the load will be larger and the stride will be longer in the case of the avian hindlimb or so that the force exerted on the air will be larger in the case of the wing. The animal can then cover the same distance in the same time—hence its speed will be the same—but the number of cycles of muscular contraction will decrease, and the muscle will contract more slowly against the larger load over the same distance.

A series of muscles that insert at varying distances from the articulation can be compared, with the $I_0 \alpha$ term constant and the number of cycles per unit time constant. Thus as the insertion of the muscle is further from the articulation, the required force decreases, but the distance of shortening increases. The muscles gain mechanical advantage as they insert further from the articulation. Their fiber composition and arrangement changes from a pinnate muscle with numerous short fibers close to the articulation to a parallel-fibered muscle with fewer, longer fibers inserting at the point furthest from the articulation. It is possible that the mass of the pinnate muscle closest to the articulation may be larger than the parallel-fibered muscle attaching to the bone furthest away from the center of rotation. If the energy budgets of these muscles are compared (Fig. 28C), the muscle that shortens least requires the least energy because it has the lowest heat of shortening. In the energy budget [Eq. (22)], the work done and the heat of activation are the same (the latter may vary if the weight of the muscle or the duration of the cycle varies), but the heat of shortening increases as the muscle shortens over a greater absolute distance and the heat of recovery increases as the initial heat, including the heat of shortening, increases. With less energy per cycle and fewer cycles per unit time, the muscle inserting closer to the articulation would be better adapted for isotonic contraction, e.g., rotating limb segments cyclically during locomotion.

Heat of shortening is important because it is large compared with the heat of activation and because it increases proportionally as the distance of shortening increases. And each part of initial heat appears again in the heat of recovery. Thus, it can be concluded that it is advantageous for a muscle to shorten as little as possible to achieve the desired movement of the skeletal system. This advantage may be termed physiological advantage, and it increases as the muscle inserts closer to the articulation and hence must shorten less. The physiological advantage of skeletal muscles contrasts directly with the mechanical advantage. Apparently greater physiological advantage is more important in muscle adaptation because skeletal muscles generally insert close to the articulation, especially if they function to move the limb segments.

Insertion close to the articulation has a second advantage in the outer shape of the limb as it permits a narrow compact shape without large flaps of flesh spanning the angle within an articulation. Moreover, it is not possible for extensor muscles to insert far from the articulation except in the presence of an extensor process (which is usually short compared to the total length of the bone). And the same mechanisms affecting muscle adaptation must exist for flexor and extensor muscles which favor any changes that improve the mechanisms underlying the physiological advantage.

Thus increased adaptation in muscles is correlated with increased physiological advantage, with a larger number of fibers compensating for the loss of mechanical advantage. Even those muscles that possess, as their primary function, static application of force via the skeletal lever system must also lengthen and shorten as the bones rotate and would be subjected to the same requirements of physiological advantage. The morphological reflection of the importance of physiological advantage in muscle adaptation is fiber composition in muscles, i.e., numerous short fibers, inserting close to the articulation. Muscles possess fibers only long enough to shorten the required distance without undue loss of force, generally not shortening below 80% of rest length. If muscles possess very long fibers, it is safe to assume that the muscle possesses some function requiring great excursion of the muscle, or in some cases very rapid shortening. The latter is a special and complex case that deserves analysis. An important consequence of this analysis is that it suggests that the importance placed by many morphologists on mechanical advantage as a positive factor in judging relative adaptation of skeletal muscles is probably invalid; at the minimum, the basic assumptions and the logical arguments supporting the concept that increased mechanical advantage improves muscle adaptation must be reevaluated.

If physiological advantage is important in judgment of muscle adaptation, then it is obvious that the best adapted muscles are those that do not shorten. An isometrically contracting muscle lacks the work term and the heat of shortening term (plus its contribution to the heat of recovery) in its heat budget. A system in which muscles rotating bony segments accomplish this while contracting isometrically would be the best adapted as the muscles would utilize the least possible amount of energy. Decrease of energy utilization below the isometric level is not possible. The two-joint muscle–bone system permits such action (see Figs. 25–28). If the distal and central bones in the two-joint muscle–bone system rotate in opposite directions as the muscle contracts and applies force to the distal bone, the muscle can contract isometrically as the bone rotates. The muscle does no mechanical work and produces no heat of shortening. This system apparently operates in the jaw muscles (Figs. 27 and 28) with the movement of bones involved in cranial kinesis. Almost certainly many more examples will be discovered among the two-joint muscles of the limbs once they are analyzed with these ideas in mind. Thus, two-joint muscles appear to be the best adapted skeletal muscles that move the bony segments as they can contract isometrically when rotating bones and hence require the least amount of energy. The major advantage of two-joint muscles as compared with one-joint muscles appears to lie in the realm of physiological advantage, and not for the reasons usually given (see Bock, 1968a, pp. 36–39).

If the muscles that provide the major share of the force needed to rotate bony segments are two-joint muscles that contract isometrically (or close to isometrically), then these muscles do no mechanical work, and hence produce no power. Analysis involving work and power of muscles operating within a bone–muscle system appears to be invalid as a general approach. Work and power of the entire animal during locomotion, for example, may be estimated and compared with those of other species, but such computations cannot be extrapolated simply to muscles as components in muscle–bone systems. This illustrates the difference existing between the functional properties of muscles when regarded as isolated whole muscles (as when they are studied in testing devices) and when they are regarded as components within bone–muscle systems.

Lastly, it should be mentioned that even if a muscle might be larger in mass to contain sufficient fibers to produce enough force to compensate for the loss of mechanical advantage, it may have gained a large enough physiological advantage that the muscle uses less metabolic energy. Thus, minimum size is not necessarily a good index to adaptiveness of muscles, although muscles will still tend to be as small as possible under the conflicting demands placed upon them by mechanical and physiological advantages.

Before leaving the discussion of muscle adaptation, I should emphasize that the lesser importance of mechanical advantage in judging adaptation of muscles does not imply a lessening of the importance of analyzing the mechanics of muscles within bone–muscle systems. This is a crucial part in the study of muscle function and cannot be omitted. The importance of the concept of physiological adaptation to the overall comprehension of muscle adaptation (or of any other aspect of the skeletomuscular system) is to show that adaptation depends usually on a series of interacting and frequently conflicting factors. Special attention must be given to study of the complete biology of a system to ascertain that some crucial factor has not been overlooked.

11. Special Muscle Systems

Muscles, in combination with other structures, can participate in several types of special systems. These unusual properties of muscular systems are not well known simply because many ornithologists were not aware of the manifold possibilities of these complex systems. To date, there are few actual examples of these special muscle systems known in birds, and I would like to mention them briefly in the hope that additional cases will be discovered.

a. Linkage Systems. The role of ligaments in linkage systems and as a recoil element in snap mechanisms has been discussed above. Muscles can serve in the same capacity, but with slightly differing parameters.

Muscular ligaments are muscles that serve the same functions as ligaments, but are found in those positions within the skeleton where some separation between the bones is required or where some control over tensile force is necessary instead of a simple resistance to the load placed on the ligament. Muscular ligaments are usually short-fibered multiple pinnate muscles with emphasis on force development. Muscles can serve both in the capacity of ligaments and of normal skeletal muscles, as a sharp gap does not distinguish between the properties of ligaments and skeletal muscles. The M. pseudotemporalis superficialis of many passerine birds (exceptionally well shown in finches) may serve as a muscular ligament. The properties of muscular ligaments and collagenous fibered ligaments may be compared as follows:

1. Collagenous ligaments resist tension immediately with great strength and without expenditure of energy, while muscular ligaments require time to develop the force to resist a load placed on

them, are weaker, and must expend energy. The collagenous fibered ligament is advantageous in these properties.

2. Muscular ligaments can stretch, can shorten and resist a load with varying force, hence exerting a degree of control. Collagenous ligaments cannot stretch, resist a load at a shorter length, and can only resist the load placed upon them. Muscular ligaments are clearly advantageous in these properties. The ability to stretch is a major distinction between the two types of ligaments and this factor may be the major basis determining the presence of a muscular or collagenous fibered ligament in a particular spot in the skeletal system.

A combined muscular–collagenous-fibered ligament is possible and exists, for example, in the arrangement of the occipitomandibular ligament and depressor mandibulae muscle in passerine birds. This ligament lies within the medial portion of the muscle and is surrounded by muscle fibers. The properties of this complex ligament are a combination of those of muscular ligaments and collagenous ligaments, but not a simple addition of these properties. The muscular–collagenous ligament cannot stretch, acts immediately and with no energy expenditure at the rest length, but can shorten and apply tension at any length from rest to the maximum shortening of the muscle. Thus, it serves as a collagenous ligament at rest length and as a muscle or muscular ligament at rest and shorter lengths.

As ligaments, muscles can serve as coupling mechanisms, but may do so with controlled tension. They can serve as snap mechanisms, but with slightly modified properties from collagenous fibered ligaments. A muscle that acts as a snap mechanism in conjunction with its antagonistic muscle starts to contract and develop tension while the antagonistic muscle is still contracting. The bone is thus balanced between the forces developed by the two muscles. Upon relaxation of the antagonist, the snap-mechanism muscle can apply its force immediately to the bone and can accelerate it much faster than if the muscle had to develop tension from a relaxed state. Such muscular snap mechanisms may exist in any pair of antagonistic muscles. It has a great advantage over collagenous ligament snap mechanisms in that the muscles can exert control, but a disadvantage in that more energy is required.

b. Rigid Struts. The combination of a muscle and a bone permits the existence of struts that are rigid only when the muscle is contracting. Bony struts have a disadvantage as they are permanently rigid, which precludes their presence in parts of the body.

When a muscle contracts, it develops tension, and also becomes rigid, i.e., possesses great strength against shearing forces perpendicular to the longitudinal axis of the muscle fibers. A muscle that spans the gap between two bones — e.g., the rami of the lower jaw — can serve as a rigid strut when it is contracting, but is flexible and permits expansion when it is relaxed. Definite examples of such muscles are not known in birds, although the M. mylohyoideus and M. serpihyoideus of the hyoid musculature may possess this property.

A second mechanism results from the combination of a thin flexible bony rod which has a longitudinal muscle wrapped about it. When the muscle is relaxed, the rod is flexible according to the properties of the bony rod. Upon contraction of the muscle, the bone resists the tension of the muscle and the structure becomes rigid, possessing strength against compression and shear. The combination of the rod-like hyoid horns and the M. branchiomandibularis has this capacity. The bone is so thin that it would bend if a compression were placed upon it. When the M. branchiomandibularis contracts it transforms the bone–muscle rod into a rigid strut permitting the tongue to be pushed forward by the force of this protractor muscle, which is applied at the distal tip of the hyoid horn.

c. Shock Absorbing Mechanisms. Muscles may serve to reduce stress on the skeleton in several ways, all of which can be grouped together under a broad heading of shock-absorbing devices. These may be divided into three major types.

A muscle can act directly against the force placed on a bone by providing a torque to resist the torque provided by the external force. If a bone is subjected to an impact with a large potential impulse, the important factor is, generally, the maximum force that will act on the bone during the period of the impact. If this time period is shortened, the maximum force will increase and be closer to the breaking point of the bone. Lengthening of the impact time reduces the maximum force on the bone. If the muscle resisting the external force on the bone stretched gradually during the impact, it would serve to lengthen the impact time and reduce the maximum force. It must be stressed that this shock-absorbing mechanism always reduces the maximum force and is not advantageous when the bird wishes to apply a large force, as when a woodpecker pounds on a tree (see the discussion by Spring, 1965; Bock, 1966, p. 50).

A second method by which muscles can reduce forces on bones is by redistribution of the force from one part of the skeletal system to another part that is better suited to withstand the force. Redistribution of force is best shown by the kinetic mechanism of the upper jaw (Bock, 1966; and above) in which an external force on the maxilla is resisted by a protractor or retractor muscle. The force is transferred from the thin roof of the skull to the thick-boned base of the brain case, which is better able to resist this force. It should be noted that the force acting on the maxilla and the brain case is not reduced by this mechanism (it may be increased because of the action of the muscle), but it is transferred to a part of the skull that can better resist the external force.

A third mechanism by which muscles can reduce the stress within bones has been demonstrated by Pauwels (1965). The exact distribution of the muscular forces on a bone will affect greatly the pattern of stress within the bone. Different arrangements of muscular forces which provide the same torques for the bone will result in drastically different patterns of stresses within the bone from one extreme that may be well above the breaking point of the bone to the other extreme that may induce a negligible stress in the bone. Thus, it is possible that the addition of another muscular force on a bone, if strategically placed, would reduce the stress within the bone. Almost nothing is known about this mechanism of force reduction in avian bones aside from the possibility that it exists and may be of major importance in reducing stress. It may be the most important force-reducing mechanism and may operate on all of the long bones of the limbs where the magnitude of stress on the skeleton is highly critical.

None of these special muscular systems and properties are well understood in the avian skeletomuscular system. All deserve extensive study, as they may represent a major factor in the adaptiveness of the entire skeletomuscular system.

IV. Physiological Adaptation

The ability of tissues of the skeletomuscular system to modify their morphological and physiological properties under the influence of changing environmental factors is of profound importance in understanding the adaptation, evolution, and taxonomic significance of the avian skeletal and muscular systems, but little attention has been given to the property of physiological adaptation in most taxonomic–evolutional studies of the avian skeletomuscular system. Physiological adaptation (= somatic adaptation) is the ability of a feature to modify its phenotypical properties following a change in the environmental factors acting on it; environmental factors include interactions between features in the body in addition to aspects of the external environment. Physiological adaptation is one manifestation of the general principle that phenotypical characters are the consequence of the interaction between the genotype and the environment of the organism. A sharp line does not separate physiological adaptation from development lability, with physiological adaptation being applied to modification of features after the individual (or feature) has reached its adult condition. Nor is there a sharp demarcation between physiological adaptation and a simple physiological response. The best single criterion for distinguishing these two types of change is the duration of the stimulus on the organism. If the stimulus is, and indeed must be, steady for a long time in terms of the lifespan of the organism, then the response, if any, would be a physiological adaptation. If the stimulus is brief, changeable (especially cyclic) and quickly reversible, then the reaction would be a simple physiological response (see Bock and von Wahlert, 1965, pp. 284-285).

Modification in features that are the consequence, either entirely or in part, of physiological adaptation are valid evolutionary changes just as much as those modifications were based completely upon genetically controlled variations. Moreover, it is impossible to determine on the basis of comparative observations of adult morphology whether differences in the anatomical features found in different groups of birds are the consequence of evolutionary change involving only physiological adaptation, only genetically based modification, or a combination of both.

The shape of bones and the amount of bony material present in a skeletal bone depends to a great extent upon the nature and magnitude of the forces acting upon the skeleton (Murray, 1936). The arrangement of internal trabeculae and the thickness of the walls of bones modify readily with change in stress patterns. Weak but steady compression will result in erosion of the bone, as seen in the extent of the supraorbital rims and size of the nasal glands in many marine birds (see Bock, 1958, pp. 37-45). A steady tension may result in deposition of bone, such as in the formation of bony porcesses at muscle attachments. Or the tension of a muscle may cause erosion of the bone resulting in a marked fossa.

Collagenous tissues, such as tendons or broad sheets spanning spaces in bones, may ossify under particular conditions of force patterns. A sheet of collagenous fibers can withstand tensile forces, but not compression or shear. Should these latter stresses increase within the structure, such as in the basal part of the maxilla, then ossification will proceed within the existing collagenous fibered structures. The nasal septum, external nares, and space between the bilateral premaxillae and palatines may ossify under such conditions.

The appearance of ossified tendons and sesamoid bones of all types can result from physiological adaptation, although the relationship between force and appearance of these bony structures is not entirely clear. In most cases, ossification results because of increased tension, but it could also occur to reduce shear or stretch in the collagenous fibered structure.

New articulations can develop via physiological adaptation whenever two bones contact one another and continue to rub together. The basitemporal articulation and the ectethmoid articulation of the mandible arose in this fashion. Other articulations such as the pterygoid–palatine joint (which actually lies within the pterygoid bone) of many birds, the hinge between the anterior end of the jugal bar and maxilla in some finches and parrots, and the nasal–frontal hinge in many avian groups could have appeared through another process of physiological adaptation in which a specialized articulation will develop within the limits of a bone at a point subjected to constant bending.

Muscles will atrophy and hypertrophy according to the amount and type of exercise placed upon them. If a greater demand for maximum force development is placed on a muscle, such changes as increase in fiber diameter and increase in filament density will occur. If the demand is for greater distance of shortening, then the muscle fibers will increase in length. Doubtlessly changes in biochemical characteristics will occur with modifications of the demands on muscles, but nothing is known about this possibility in avian muscles.

Little is known about potential modification in the properties of avian skeletal muscles during the life of an individual, and possibly such changes are rare and of minor magnitude. Nevertheless, evolutionary modifications that result in major differences in the myology of different taxa may have a major component of physiological adaptational change in the total differences observed. Should this be the case, our analysis and interpretation of changes in the avian muscular system must be reconsidered completely.

The importance of physiological adaptation is that the various mechanisms for change in the skeleton and in skeletal muscles are universally present in all birds, as far as is known. Thus, similar changes will occur in any species subjected to the same environmental demands. Changes will occur easily and quickly, and similar changes may take place in unrelated taxa of birds subjected to the same environmental stresses. Similarity of these independent modifications in unrelated species will increase as taxa of lower categorical rank are studied. Thus, appreciation of the operation of physiological adaptation in evolutionary changes and its consequence in distribution of morphological similarity in taxonomic features is of critical importance in morphological–systematic studies of birds. Almost no attention has been given to this mechanism of evolutionary change in earlier avian morphological–systematic investigations. Unless due consideration is paid to this evolutionary mechanism, future taxonomic studies, especially those on familial-level taxa, will continue to contain dubious conclusions.

V. Ecological Morphology and Adaptation

After the morphological features of the avian skeletomuscular system are described and their functions are ascertained, several steps must be completed before the adaptive significance of these features can be determined and their evolutionary history and taxonomic significance suggested. The next step is investigation of the biological roles of these features and of the ecological factors that provide selection forces interacting with features. Only after the biological roles have been ascertained is it possible to speculate on the adaptiveness and on the evolutionary history of any anatomical features; a knowledge of their structure and functions is not sufficient. These studies are usually grouped together under the heading of ecological or biological morphology (see Böker, 1935, 1937, for an early review). Considerable confusion still exists between the working techniques and theoretical concepts of ecological morphology and those of functional morphology (Bock and von Wahlert, 1965). A major difficulty lies in the ambiguity between the meanings of function and of biological role (see p. 139). The importance of ecological morphology to our understanding of the skeletomuscular system is matched only by the inherent difficulties of these studies and by our gross ignorance of this facet of avian anatomy. The numerous studies relating features of the skeletomuscular system of birds to their life history and ecology have merely touched the superficial aspects of this topic. At present, it is possible to sketch the barest outlines of ecological morphology and to urge more sophisticated future studies.

Observations of biological roles of morphological features must be made on free-living birds in the environment in which these features evolved. Some observations, mainly those to check details of observations on free-living birds, may be made on captive individuals, but these results are of very limited value in terms of ascertaining biological roles. Field studies of birds in their normal environment are critical, as emphasized by Lack (1965). A feedback exists between description of structural features, determination of functional properties by observations and analysis, and observation of the biological roles in free-living individuals. It is frequently necessary to alternate between field observations and laboratory work several times to check new findings made in each phase of the study. Although this approach is most desirable, it may not be possible if the birds dwell in far off lands or in difficult-to-reach areas. Inability to repeat field work or laboratory observations can result in incomplete conclusions, such as the suggestion for possible biological roles of the crossed rhamphothecal tips of *Loxops coccinea* (Drepanididae) by Richards and Bock (1973).

Investigation of ecological morphology and adaptation provides an excellent opportunity for cooperative studies between morphologists and students of avian behavior and ecology. Such joint undertakings not only combine the specialized abilities of workers from different ornithological disciplines, but permit a maximum feedback between observations made in the laboratory and in the field. An example of such a cooperative study is the recent description of the sublingual pouch and its adaptive significance in the Clark's Nutcracker (*Nucifraga columbiana*) by Bock et al. (1973). Another is Kear and Burton's (1971) paper on the New Zealand Blue Duck in which they showed that decrease in population density of this duck probably resulted from competition for food with the introduced brown trout.

Many observations on biological roles of structural features in free-living birds are exceedingly difficult because of rapid movements, small size, and habitats that hinder observations. Many movements are so complex and rapid that they must be recorded by high-speed motion picture cameras or other devices that are difficult or impossible to use under many field conditions. Consequently, most statements on biological roles are suggestions or deductions, and many widely accepted conclusions are still quite elementary. The adaptive significance of most features in the skeletomuscular system cannot be evaluated with the same precision as is possible in morphological description or functional analysis. Correlations between ecological factors and biological roles are simply not known in any detail, and it is doubtful that the situation will change radically in the near future. It is important to realize the limitation of field observations and the frailty of conclusions about the adaptive significance of anatomical features reported in most papers.

At this time, no coherent theory or set of practical rules for the study of ecological morphology can be offered, only the strong recommendation that continued attention be given to this essential attribute of avian anatomy. Hopefully, advances in this area will be sufficiently rapid in the next decade that a comprehensive review of the ecological aspects and adaptation of the avian skeletomuscular system will be feasible in the next edition of "Avian Biology."

VI. Comparative Morphology and Systematics

A. Introduction

Among the most important contributions of morphological studies of the avian skeletomuscular system are the data with which general principles of comparison can be established and the evidence providing the greatest support for our schemes of avian classification. Although most ornithologists possess a broad general familiarity with the data of comparative avian anatomy, most lack a profound comprehension of the morphological facts and of the basic principles with which systematic and phylogentic conclusions are obtained. Sadly this is true even for most systematists who have dealt with classification of the higher taxa. Numerous earlier anatomic–systematic schemes have not endured the test of time, some did not possess the needed firm descriptive basis, and others did not employ valid methods for comparisons and subsequent systematic interpretations. The result is widespread confusion about the methodologies and results of anatomic–systematic studies having conflicting taxonomic conclusions. Among the invalid notions that are frequently expressed is the idea that a feature possesses equivalent taxonomic value throughout a large taxon, such as the Passeriformes, or the concept that a decision can be reached about the affinities of two taxa by counting the number of characters in which the groups are similar or dissimilar—a sort of vote.

Before examining the individual comparative principles, several general points must be made. The first is that the foundation for all systematic work is accurate descriptive morphology, with the choice of the optimal morphological system and the exact details of each system to examine being part of the art that distinguishes outstanding systematists from others. At times, the need exists to describe the morphology in greater detail. The occasionally heard objection against the "electron microscopic approach" to description of taxonomic features is usually expressed by workers who do not understand morphology and is almost always invalid. The second principle is that comparative examination of taxonomic features throughout the taxa under consideration is essential. Third, it should be possible to decribe the feature in each species independently of other species and to repeat comparisons readily as new information from additional species becomes available. Any feature of the skeletomuscular system that cannot be collected and stored readily for future comparative study is a poor taxonomic character no matter how potentially valuable it may appear to be. Most aspects of skeletal muscle fiber types fall into this category of poor taxonomic features because of the difficulties of preservation, storage, and making comparative observations. Fourth, a realistic attitude must be exercised in judging the usefulness of skeletomuscular features as taxonomic characters and in evaluating the amount of actual support that is available from the skeletomuscular system for currently accepted schemes of avian classification. A common statement is that the accepted schemes of classification are supported by an extensive and detailed comparative knowledge of morphological features. Morphologists have offered this idea to strengthen their position. And it has been used by systematists to argue that an extensive knowledge of morphology has failed to provide a valid classification of birds, a mechanism to discredit morphological characters and to strengthen other taxonomic features. A reasonable position, in my opinion, is that all classification of the higher categories of birds is based essentially on information from comparative studies of the skeletomuscular system. Many of the comparisons and derived taxonomic conclusions are indisputable, but a goodly number of other conclusions will not stand the test of time. Obviously, many earlier studies must be verified, but of greater immediacy is a reanalysis of the basic principles of comparison, some of which remained unchanged for decades despite considerable advance in our knowledge of evolutionary mechanisms.

I would like to examine some of the basic principles of comparison and systematics using features of the skeletomuscular system as examples. A discussion of the theoretical bases for many of these ideas can be found in Bock (1973a). However, I shall not attempt to evaluate the taxonomic value of particular morphological features or to discuss the merits of taxonomic conclusions based upon morphological or other taxonomic features.

B. HOMOLOGY

Although homology is the central concept in comparative studies, it is probably the most poorly understood, simply because its comprehension is universally assumed. Few assumptions are further from reality.

Any comparative study or any classification must be founded on a previously established philosophical basis; hence, the comparative conclusions and classifications are a reflection of underlying philosophical concepts. A fundamental point is what is meant by equivalent attributes of organisms (either features or aspects of features) and how these equivalent attributes are recognized. Biologists accept, almost universally, organic evolution as the philosophical basis for relationship between groups of organisms and for the scheme of classification that reflects the pattern of relationships. Equivalent attributes in different taxa are those that have remained unchanged from those present in the common ancestor during the evolution of each phyletic lineage leading to the taxa being examined. Such equivalent attributes are termed homologs. The formal definition of homology that I advocate is (Bock, 1969a,b): "Homologous features (or conditions of features) in two or more organisms are those that can be traced back phylogentically to the same feature (or condition) in the immediate common ancestor of these organisms." This definition does not provide any clues or methods as to how particular features in two or more organisms are ascertained to be homologous. Moreover, this definition of homology is not circular in any aspect. Homology is defined in terms of phylogeny, and a completely independent definition may be given for phylogeny, namely that: "Phylogeny is the lineages of animals and and plants resulting from their descent through time."

Nonhomology, the opposite of homology, applies to attributes in two or more organisms that cannot be traced phylogentically to the same attribute in the immediate common ancestor of these organisms. I advocate the term "nonhomology" instead of "analogy" which was recommended in an earlier paper (Bock, 1963c; p. 269) because of the continued ambiguity and multiple use of analogy. Analogs are usually defined as structures that are similar because of similar function, but the term is so vague that I recommend against its continued use.

Homology is not an intrinsic property of a feature, such as its mass or color, but a relationship depending upon the existence of corresponding features in other organisms. Because homology indicates a certain relationship, statements about homologous features must always be put in the form that expresses the conditions of the relationship. Thus, one should say "the basipterygoid articulation of galliformes is homologous to the basipterygoid articulation of anseriformes as an articulation between the anterior end of the pterygoid and a broad, low facet on the basisphenoidal rostrum;" or "the splenius capitis muscle of hummingbirds is homologous to the splenius capitis muscle of swifts as a cruciform splenius capitis muscle, i.e., one with a crossover origin from the second cervical vertebra (Burton, 1971)." A phrase stating the conditions of homology must always be included. Statements such as "the syringeal musculature of *Corvus* is homologous to the syringeal musculature of *Pica*" or "the osseous arch of the radius of *Tyto* is homologous to the osseous arch of the radius of *Bubo*" are meaningless. Simply to say that two features are homologous or nonhomologous conveys no information.

The conditional phrase describes what is believed to be the condition of the feature in the common ancestor from which the homologous features had descended. A hierarchical series of homologs can be constructed by repetitive determinations of the homologies of features in a set of organisms using ever increasing narrower and more precise conditional phrases.

No mention of resemblance in adult structure or of similarity in ontogenetic development appears in the definition of homology. Contrary to common opinion, the concepts of homology and nonhomology have nothing to do with similarity of features, but are associated only with common origin versus noncommon origin of taxa. Degree of resemblance and common origin of features (= homology) are quite distinct aspects of phylogentic study and must be kept separated in theoretical discussion and in practical work.

The methods by which homologous features are recognized must be consistent with the general principles of evolution and phylogeny, but these methods must not, of course, be based upon earlier conclusions on the phylogeny or relationships of the particular taxa under study. Determination of homologs starts with a series of organisms and their features, which must be studied and compared in such a way that conclusions about the homologies of features may be drawn. General resemblance is one of the methods used to establish homology between features. Other methods include general position of structues in the body, interrelationships between features, such as innervation of muscles or attachment of muscles on bones, similar pattern of ontogenetic development, and so forth. All of these methods are similarities between features of different organisms which is the only recognizing criterion for homology (Bock, 1973a). Following the comparative study, decision is made on the homologies of features in different organisms, taking care always to give the conditional phase. A conclusion that features in two or more organisms are homologous assumes the hypothesis that these features can be traced phylogentically to the same feature in the immediate common ancestor. As a scientific hypothesis, any conclusion on homology can be disproven, but never proven (see below). Depending upon the nature of the homology and the amount of evidence available from the comparative studies, one would have a greater or lesser amount of confidence in the conclusion.

The sequence of phylogenetic study is simple, but deserves to be restated briefly. After features in a series of organisms are compared, decisions about their homologies are reached using appropriate comparative methods that exclude previous conclusions about the possible phylogeny of the organisms. Using these established homologies, conclusions are reached on the relationships and classifications of the organisms. Taxa of related organisms are established and their possible phylogeny deduced. Thus, the sequence of study is first to establish homologies of features based upon comparative study and then to establish phylogenies and classification based on the homologous features. This sequence of determining homologs first and phylogenies second has nothing to do with the sequence in theoretical definitions of homology and phylogeny. Although particular phylogenies are based on definite homologs, the definition of phylogeny is not expressed in terms of homology. Rather, homology is defined in terms of phylogeny.

A hierarchical series of homologs can be constructed by asking more and more precise statements about the features; that is, by making the conditional phrase describing the homology more and more restricted, and testing whether the features can still be regarded as homologous. Once this is completed, a hierarchy of relationships between the organisms (based upon the hierarchy of the homologs) can be established.

A serious theoretical difficulty and practical methodological problem stems from the phrase "the same feature in the immediate common ancestor" in the definition of homology. The immediate common ancestor is identifiable (whether it is an ancestral species or a taxon of higher categorical rank) only vaguely, and the degree of similarity in the feature to be described in both the ancestoral and descendent taxa may likewise be inexact. This topic was discussed by Bock (1963c) under the notion of pseudohomology, and later (1969a) under the notion of the resolving power of available methods for recognizing homologous features. The resolving power of methods for recognizing homologs is closely associated with the degree of uniformity of the organisms being compared. The greater the uniformity, the less the resolving power of these methods becomes. Groups that are heterogeneous at the present time were much more uniform at the time of their origin or shortly thereafter (before major adaptive radiation of the group had occurred). Numerous examples of pseudohomologous features can be found in avian groups (Bock, 1963c, 1969a).

Homology of a feature can be concluded only after study of that feature and not in reference to patterns of homology of other features. If a feature in a uniform group is shown to be nonhomologous after information about the phylogeny of the group (based upon other evidence, namely the homology of other features) has been introduced, the new evidence really does not invalidate the original conclusion of homology, as shown by Bock (1969a). Instead, one is forced to conclude that the working methods available for ascertaining homologies of any individual feature have relatively low resolving power; hence, it is not possible to establish precise homologies (i.e., to distinguish between true homologies, pseudohomologies, and often nonhomologies) by the use of these methods. Most important is that a broad gray zone of pseudohomology exists between true homology and nonhomology and that most working methods do not permit clear discrimination between parts of this gray continuum. Features falling in this intermediate area between homology and nonhomology must be regarded as homologous if they meet the established criteria because the resolving power of our working methods does not allow a more precise decision. Hence, the demonstration that one or a few features in several organisms are homologous is usually not a sufficient basis on which to deduce phylogenetic relationships; these features may be pseudohomologous and may not provide valid evidence for a monophyletic relationships of these organisms. The homologies of many features must be pieced together before a solid basis for phylogenetic speculation is available. A large number of features will compensate for the low resolving power of the methods for recognizing homologies. If the level of resolving power of these methods can be raised and if one can be confident of the homologies, then few homologous features are required for phylogenetic speculation.

Classification must be based upon established homologous features and herein lies the strength and weakness (perhaps even the pitfalls) of comparative morphology of the avian skeletomuscular system because of the great wealth of information published over the past 100 years. Older works are generally excellent descriptively, but one must know the foundation on which the comparisons and conclusions rest. Many of the ideas in these works are no longer valid, and a worker today should not cite conclusions from older morphological papers without knowing their bases. Although regrettable, utilization of the available morphological information for systematics is neither simple nor obvious.

In every case, the reasons given by the earlier worker for concluding that features in the several organisms are homologous must be re-investigated. Often this is an easy task, but occasionally the homologies must be reestablished. And one must always verify that the same anatomical name really connotes homologous attributes in the different organisms. Often it does not, and this can easily lead the systematist astray. The os prominens in hawks and owls is a nice case in point (Bock and McEvey, 1969b). The same name had been given to an enlarged sesamoid bone in the wrist of hawks and owls with the clear implication that it is homologous and could be used as support for relationship of these groups. Yet, closer analysis suggests that this feature is nonhomologous in hawks and owls as an enlarged sesamoid bone in the tendon of the M. tensor patagii longus and that separate names should be given to these two conditions.

Numerous additional examples could be provided. I should like to stress again that the exact conditions of homology should always be specified in conclusions on homologous attributes, that the sequence of comparative study of features followed by conclusions of homologies followed by conclusions of relationships of taxa should always be observed and that caution must be exercised whenever using the older morphological literature of the skeletomuscular system.

C. Proof and Disproof

Confusion continues to exist on the philosophical and practical meaning of validation or confirmation of systematic conclusions based upon comparative study of morphological features. Many workers continue to argue that taxonomic conclusions can be proven. Substitution of terms such as "suggest" or "indicate" or probability statements for proof or proven do not alter the basic philosophical notions underlying the idea of validation of taxonomic conclusions. This philosophical approach is not valid according to the ideas of Karl Popper (1959), who states that scientific conclusions are never proven, but are hypotheses that are subject to disproof. The strength (degree of corroboration) of a hypothesis increases with the number of failed attempts to disprove it. Any conclusions offered on the homology of particular attributions in a set of organisms, on the establishment of taxa, and on the relationships and phylogeny of taxa—any taxonomic conclusions—are scientific hypotheses offered for disproof. If they are disproven, the conclusions will be discarded. If the attempts to disprove these conclusions fail, then they are retained until a future attempt to disprove them is successful. Several methods are available to disprove taxonomic hypotheses such as the "consistency text" (Wilson, 1965; Bock, 1969a,c, 1973a). Distribution of different paradaptations can also provide strong evidence for disproof of taxonomic statements.

At first glance, no distinction appears to exist between proving a scientific hypothesis and failing to disprove it, yet these two approaches are absolutely different. The philosophical notion that increased confidence is attained by repeated failure to disprove a hypothesis is especially important in systematics in which we deal with taxa that are the result of a long series of historical events none of which can ever be observed directly. With the realization that taxonomic conclusions are hypotheses offered for disproof, we can formulate methods of comparing morphological features that can provide the necessary disproof. Unfortunately, most workers had not approached morphological–systematic studies with this point of view, so that we have few good methods of invalidation, and indeed most taxonomic statements are not even formulated in the correct way for easy disproof.

Under this philosophical approach, it is completely valid to offer a taxonomic conclusion without a shred of supporting evidence. Such a conclusion is as valid as one with much evidence, but most workers will have little confidence in it, and quite rightly so. The supporting evidence provided with a taxonomic conclusion is actually the result of a series of tests conducted by comparative observations, that have failed to invalidate the hypothesis. As such, presentation of the evidence should be in the form of attempts to disprove the hypothesis. And contrary to usual practice, the taxonomic conclusions may be better presented in the introduction of the study rather than at the conclusion. Thus, the reader will know the questions to which the comparative anatomical observations pertain.

D. Breadth of Comparative Studies

Many morphological–systematic studies of the higher taxa, especially those attempting to ascertain the proper position of a problem group, or to determine ancestral–descendent relationship, or to resolve "sister-group" relationship, suffer from the difficulty that a complete study should properly include comparison of all members of the next higher taxon. And yet the extent of the morphological description and the functional study may be so great that a comparative study of all taxa is not feasible, or in many cases not even possible because of the absence of materials. Clearly, a compromise position must be reached which depends upon the nature of the study, the extent of the morphological comparison, and the exact questions being asked.

A comparative study limited to a few taxa is completely justified if the taxa are chosen carefully and if the morphological comparison is quite extensive. Thus, a conclusion about a problem genus may be offered after study of a few (or even only one) families. It is easier to disprove the membership of a genus in a family to which it was assigned and to leave its familial relationships as hypothetical for future studies. Removal of *Promerops* from the Meliphagidae without suggesting its possible affinities is a good example (W. J. Bock, unpublished results). Or it may be possible to move a genus from one family to another after study of only those two families, as removal of *Toxorhamphus* from the Meliphagidae to the Nectariniidae (W. J. Bock *et al.*, unpublished results). It is also possible to suggest membership of a problem genus in some family even after a comparison with only members of that family, such as retaining *Moho* in the Meliphagidae (W. J. Bock and H. Morioka, unpublished), with the evidence used in attempts to disprove the hypothesis. It is not necessary to disprove all other possible hypotheses before offering the hypothesis of choice. However, it must be remembered that the other possible hypotheses were not disproven, so that repeated failure to disprove a favored conclusion may improve our confidence in it, but it does not make it the only, or the best possible, conclusion about the relationships of some problem taxa.

E. Phylogenetic Sequence

Systematic conclusions include (1) the amount of evolutionary change (= differences) and (2) the branching sequences or the actual pattern of phylogenetic pathways. Both factors must be included for a complete comprehension of evolutionary relationships of any group. I am not concerned here with arguments of whether classification should be based purely on phenetics, or purely on cladistics, or on a combination of the two, but I would advocate the last, which is the basis of classic evolutionary systematics (Bock, 1973a). Thus, I accept the concept that information about phylogenetic relationship or sequences of characteristics is possible and that this information can be used in the construction of classifications (see Hennig, 1966; Bock, 1968b, 1969c). Such information includes arrangement of morphological features into possible sequences, distinguishing between primitive and advanced conditions, and ascertaining the directions of evolutionary changes.

Methods for ascertaining information about phylogenetic sequences of morphological features are limited largely to distributional patterns of these features in the taxa under consideration. Hennig (1966) presents an excellent discussion of this methodology and the interpretations permitted. Although these methods have proven to be quite useful, some vagueness still exists on their theoretical basis and on whether circular argument is avoided in the methodology and possible interpretation. Little consideration has been given to morphological methods that will provide conclusions on phylogenetic sequences.

One method for ascertaining phylogenetic sequences of morphological features is the establishment of a hierarchy of homologies as discussed above. Homologies with broad conditional phrases are often the primitive state and those with very restricted conditional phrases the advanced conditions. However, it must be stressed that the correlation between broad conditional phrases and primitiveness and restricted phrases and advancedness is rough at best. This method should generally be used in association with other methods.

Analysis of functional and adaptive significances of morphological features will often provide information on possible sequences of evolutionary changes and of primitive and advanced condition. This was attempted by Bock and Miller (1959, pp. 28–29) for the evolution of perching and climbing foot types, by Bock (1964, p. 33) for avian cranial kinesis, and by Cracraft (1971a) for morphological features in the three families of rollers. In general, confidence in conclusions on phylogenetic sequences of morphological features increases markedly as the depth of functional analysis increases. Too few detailed investigations of phylogenetic sequences based on comparative functional analyses are available to formulate any general principles, or even to guess whether any general principles exist. Herein exists an important facet of morphological–systematic study that is scarcely touched and for which the skeletomuscular system is especially well suited. Birds may provide excellent examples for these studies because they are readily observed diurnal creatures, permitting the necessary observations on the functions and biological roles of morphological features.

F. Nonancestral Similarity

Herein I mean similarity in morphological features that is not attributable to descent from a common ancestor, i.e., nonhomologous similarity, but to evolutionary processes of convergence, parallelism, and reversal. The ease of discovering nonancestral similarity depends greatly upon the degree of homogeneity of the group of organisms (Bock, 1963c). Birds, unfortunately, are quite uniform, so that convergence, parallelism, and reversal constitute a serious problem that underlies the difficulty of distinguishing between homologous and pseudohomologous features. Increase in the detail of the morphological comparison increases the ability to resolve nonancestral similarity from homology. But it must be recognized that comparative anatomical study of birds will be plagued by this problem. Although this problem is widely recognized, attempts to provide methodology have been unsuccessful. A major barrier has been the lack of detail in comparative studies, especially of the muscular system. At present, the greatest need is for a number of comparative morphological studies of the skeletomuscular system, especially of taxa such as the Passeriformes, with emphasis on anatomical detail. Once this information is available in sufficient quantity, it will be possible to reexamine the problem of nonancestral similarity and formulate general principles and methodology.

G. Comparisons and Paradaptations

Comparative anatomists and systematists, in general, have assumed that comparisons between any pair of taxa and that all interpretations which may be deduced from these observations and comparisons are equivalent no matter which taxa are being compared. Thus the interpretations deduced for a particular difference between one pair of taxa would be applied almost automatically to a similar difference between a second pair of taxa. This assumption is, of course, not only invalid, but has been responsible for considerable misapplication of comparative observations to taxonomic problems. Moreover, it has been chiefly instrumental for the origin of some persistent theoretical concepts, such as, for example, the idea that adaptive features are useless as taxonomic characters (Bock, 1967a, 1969a).

A clear distinction must be made between two main classes of comparisons between taxa, namely horizontal and vertical comparisons (Fig. 30), because different sets of conclusions must be drawn about the observations made in each type of comparison. Vertical comparisons are between members of the same phyletic lineage, i.e., between ancestral and descendent taxa. Horizontal comparisons are ones between members of different phyletic lines, no matter whether or not the taxa being compared exist at the same time level. In horizontal comparisons, one attempts to compare not only those features concluded to be homologous, as is also done in vertical comparisons, but also different adaptive answers to the same selection force (Fig. 31). The different adaptive answers may involve homologous features or nonhomologous features.

Vertical and horizontal comparisons differ from each other in whether the differences observed can be concluded to be adaptive in terms of the selection forces controlling the evolution of the features being compared. If differences observed in either vertical or horizontal comparisons cannot be characterized as adaptive in terms of the selection forces controlling their evolution, then we must ask how should these differences be described and what are the evolu-

3. THE AVIAN SKELETOMUSCULAR SYSTEM 239

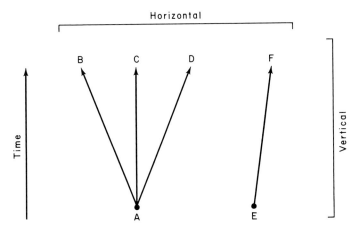

FIG. 30. Schematic diagram to show the difference between horizontal and vertical comparisons. Vertical comparisons are those between members of the same phyletic lineage, e.g., between A and B, or A and C, or A and D, or E and F, along the time axis. Horizontal comparisons are those between members of different phyletic lineages, e.g., between B, C, D, and F, or A and E, or B, C, D, and E, or A and F, no matter if the forms being compared are or are not at the same time level. (Taken from Bock, 1967a, Fig. 1.)

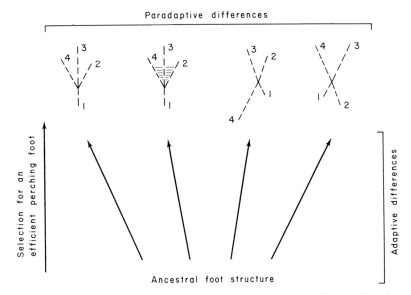

FIG. 31. Schematic diagram to show the pattern of multiple pathways of evolution of perching feet (from left, anisodactyl, syndactyl, zygodactyl, and heterodactyl). The evolution of the four different arrangements of the toes from the ancestral arrangement is under the control of the same selection force for a more efficient perching foot. Differences observed in vertical comparisons are adaptive, while those in horizontal comparisons are paradaptive with respect to this selection force. (Taken from Bock, 1967a, Fig. 2.)

tionary mechanisms underlying these differences. Lastly, it is necessary to ascertain the contribution of each type of comparison and the evolutionary interpretations of the observed differences and similarities to the taxonomic value of morphological features.

The last is of critical importance because of the widespread, but erroneous, assumption that adaptiveness and taxonomic usefulness are conflicting properties of morphological features (see Bock, 1967a, 1969a). This assumption is rather meaningless, because, on theoretical grounds, all existing morphological features that can be studied by morphologists are adaptive to a reasonable degree (see Bock and von Wahlert, 1965, for a definition of adaptation and a discussion of the degree of adaptiveness). At least, it may be argued that all morphological features (not differences between features) used in the classification of birds are adaptive. In an informal survey of the numerous morphological features used in avian classification, I found none that were not clearly adaptive although the details of the adaptive significance of many of these features had not yet been determined. Obviously, at least some useful taxonomic characters must exist, otherwise no system of classification could be established. A dilemma therefore exists between the belief that adaptive features are useless as taxonomic characters and the demonstration that all morphological features used in classification are adaptive, a dilemma that allows only the conclusion that no useful taxonomic characters exist, and consequently, that further attempts to establish classifications are futile. This conclusion can and must be rejected at once as absurd. It serves only to indicate the existence of flaws in the argument upon which it is based.

This dilemma has resulted from an incomplete analysis of the evolutionary mechanisms by which morphological features originate and change, and from an artificial and overwhelming emphasis upon natural selection. I advocate the following dual hypothesis: that all morphological features used in avian classification are adaptive, and that adaptive features may be exceedingly useful taxonomic characters. Adaptiveness in itself does not negate the taxonomic value of a morphological feature; the value depends upon considerations quite apart from the simple one of whether or not a feature is adaptive.

Support for this hypothesis comes from the consideration of all evolutionary mechanisms by which features originate and change (see Mayr, 1962; Bock, 1959, 1967a, 1969a). A clear distinction must be made between (a) those evolutionary mechanisms associated with the origin of new features and with the appearance of all modifications in their later evolution (e.g., with the formation of all types of genetical variation which give rise to the phenotypical variation involved in the origin of new features and in their later modifications) and (b) natural selection, which is that phase of the interaction between the environment and the organism resulting in differential survival of individual organisms and in nonrandon differential reproduction of genetical material. The consequences of both sets of evolutionary mechanisms upon the observable aspects of taxonomic features must be distinguished clearly and analyzed separately. Only then can their taxonomic usefulness be assessed.

Adaptation is associated only with the evolutionary mechanism of natural selection. By definition, features that are favored and maintained by selection are adapted; the degree of adaptation may vary, however (see above, and Bock and von Wahlert, 1965). Features that are rejected by selection are nonadaptive and are conspicuously absent. Hence, those aspects or properties of features responsible for the features being favored by selection would be the adaptive aspects of the features.

The differences in any features observed in vertical comparisons, i.e., between ancestral and descendent groups, are largely dependent upon the action of natural selection and hence would be adaptive in terms of the selection forces controlling the evolution of the features. Adaptive aspects of features would be those aspects that have changed between the ancestral and descendent taxa under the action of natural selection.

Selection can only accept or reject features of organisms exposed to it, establishing limits on the range of features that are acceptable. All features within these limits will be favored, and those outside of these limits will be rejected. Usually, selection can accept a number of possible features ("different adaptive answers") offered to it, in which case selection is said to have broad limits. If it accepts only one or a few restricted features, selection is considered to have narrow limits. Selection can operate only according to rather fixed interactions between the organism and its environment. Selection and, hence, adaptation are the design aspect of evolution in the sense of Mayr (1962).

Although selection can establish limits on the range of acceptable features, selection cannot predetermine the exact properties of the features exposed to it. Different features (adaptive answers) lying within the limits of a particular selection force (see, for example, Bock, 1959; Bock and Miller, 1959) will differ morphologically, some quite radically. Selection can accept each of these morphologically differing features, but selection did not have any role in the appearance of these differences. They must be explained on the basis of other evolutionary mechanisms.

It is essential to recognize that aspects of a feature are not solely adaptive—that is, adaptiveness is not the only evolutionary property of features—and that natural selection is not the only mechanism controlling the evolution of features, underlying the observable differences, and determining their taxonomic usefulness.

The appearance of new features, be they adaptive or nonadaptive, and the origin of all later modifications in these features lies outside of the control of natural selection. Selection can accept or reject features, but it cannot determine the exact properties of form and function possessed by these features. The origin of these features, e.g., the basis for the phenotypical variation to be acted upon by selection, lies under the control of a different set of evolutionary mechanisms and phenomena, namely: (a) mutations, recombinations of all sorts, gene flow, and other chance-based genetic processes that generate the genotypical variations causing the phenotypical variation acted upon by selection; (b) the nature of the preexisting features of the ancestral taxon; (c) the geographical and ecological location of the ancestral group; and (d) the timing of events, such as which group is first to acquire a new adaptation and thus to be able to exploit first a new adaptive zone. Only the first of these could be regarded as a true mechanism; the others can be grouped together under the heading of the "evolutionary situation" (von Wahlert, 1965). The common and significant property of these factors, be they a mechanism or part of the evolutionary situation, is that they are all chance-based (stochastic) with respect to (a) the demands of selection forces that will accept or reject the feature, (b) the future selection forces that will interact with the feature, and (c) the future evolution of the feature and organism possessing it. These factors constitute the accidental aspect of evolutionary change in the sense of Mayr (1962).

Those aspects of a feature that are dependent on, that result from, or that are under the control of, chance-based evolutionary mechanisms may be termed paradaptive, meaning "besides adaptive" in the sense that these aspects are not dependent on selection and hence cannot be judged in the range of adaptive to nonadaptive (Bock, 1967a). Paradaptive aspects of a feature are dependent only upon the accidental (chance-based) evolutionary mechanisms. They may be either adaptive or nonadaptive according to whether they are accepted or rejected by selection.

The most essential property of paradaptive aspects of features is that their occurrence is chance-based, and thus whether a particular paradaptation will appear (and be exposed to selection) depends upon the probability of its occurrence. This probability is based upon a series of complex factors and cannot be ascertained easily, if at all. In most cases, only an approximate guess of this probability can be reached.

The differences in any feature observed in horizontal comparisons, i.e., between members of different phyletic lineages, are generally dependent upon the action of chance-based evolutionary mechanisms and hence would be paradaptive in terms of the selection forces controlling the evolution of these features. Such differences lie outside the realm of the adaptiveness of these features with respect to selection that had favored their evolution. Paradaptive aspects of features are associated with mechanisms controlling their appearance, and not whether they are adapted or not. To be sure, each of the features being compared horizontally is adapted because it has been accepted by selection. If these features were rejected by selection, they would disappear and would not be available for comparison; but the important factor is that the observed horizontal differences are paradaptations. It must be emphasized that different paradaptations may be differentially adaptive with respect to other sets of selection forces, but this lies outside the scope of this discussion.

Adaptive and paradaptive aspects of a feature are not necessarily different ones. An aspect of a feature may be adaptive and paradaptive simultaneously; these properties do not exclude one another. Indeed, each feature or each aspect of a feature would be paradaptive and adaptive. This must be so, at least theoretically, because of the dual and simultaneous action of the chance-based mechanisms of origin and of the design mechanism of natural selection during the evolution of every feature. Because each existing feature has originated (and been modified during its subsequent evolution) and has been accepted by selection, it is both paradaptive and adaptive. This conclusion is not new, but has been emphasized by Mayr (1962) under the heading of accident versus design in evolution. What is new is the identification of the types of differences observed in comparisons of morphological features and correlation of these types of differences with the evolutionary mechanisms underlying their formation.

The relationship between adaptive and paradaptive differences is nicely shown in the comparison of different types of perching feet in birds (Fig. 31) (Bock and Miller, 1959). The horizontal differences between these morphologically different toe arrangements are paradaptive, all types are adaptive to the selection forces for a perching foot. If attention is focused on the heterodactyl foot of trogons, the distinction between adaptive and paradaptive aspects is clear. In this taxon, the rear position of the second toe is paradaptive because this toe happened to reverse to the rear of the foot. The reversed position of the second toe is adaptive because it opposes the remaining anterior toes and has been accepted by the selection forces as a perching foot.

Although adaptive differences and aspects of features are associated more with vertical comparisons, and paradaptive differences and aspects of features are associated more with horizontal comparisons, it must be remembered that the evolution of every feature and hence its observable properties are both adaptive and paradaptive.

Classification is the recognition and distinction among monophyletic (vertically based) groups of organisms, and as such, it is the distinction between groups of organisms possessing different paradaptive properties of taxonomic characters. Paradaptive aspects of features, being associated with the origin of features, will also be associated with the origin of taxonomic groups. Thus, study of paradaptations would provide a good guide to the existence and limits of groups of organisms. Many paradaptations also possess value as diagnostic or key characters. It must be stressed that these paradaptations are also adaptations. Thus, whether a taxonomic feature is adaptive is irrelevant to its value as an indicator of different taxa. Taxonomic usefulness of morphological features depends largely on the nature of their paradaptive attributes.

Paradaptive differences are defined as those observed in horizontal comparisons, and adaptive differences as those observed in vertical comparisons. But when features in a series of organisms are being compared, it is not known which comparisons are horizontal and which are vertical. Instead, it is necessary to conclude on the basis of the study which differences appear to be adaptive and which appear to be paradaptive in terms of the selection forces controlling the evolution of these features. From these conclusions, it is possible to decide which comparisons are horizontal and which are vertical and to sort out monophyletic taxa. The different paradaptations are expressed as different multiple pathways of adaptation (Bock, 1959), hence one fundamental objective of comparative study is to sort out the several multiple pathways of adaptation to the same selection force.

Adaptive and paradaptive aspects of features, and hence adaptive and paradaptive differences in morphology, can be distinguished only after study of the functions and biological roles of the morphological features. Adaptive aspects of features cannot be deduced in the absence of a knowledge of these properties of morphological structures. The methods of evolutionary morphology provide a sound approach to the recognition, separation, and elucidation of the adaptive and paradaptive aspects of morphological features.

H. Taxonomic Value

It is obvious that morphological features differ in their usefulness in constructing classifications, or in their taxonomic value. What is not obvious are the factors that influence the taxonomic value of morphological feature, how this value is measured, and whether it is consistent throughout a large taxon. Moreover, the entire concept of taxonomic value depends upon the philosophical basis for the methodology used in systematics. If one accepts the ideas of Karl Popper and holds that taxonomic conclusions are scientific hypotheses offered for disproof, then the notion of taxonomic value has little meaning. The essential attribute of taxonomic characters and their comparative study is whether the conclusions are useful in disproving particular taxonomic statements, which has little, if anything, to do with any concept of taxonomic value.

Some general statements may be offered on the use and misuse of the concept of taxonomic value.

a. It is erroneous to conclude that a particular morphological feature (e.g., the palatine process of the premaxilla) or of a complex of morphological features (e.g., the tongue musculature) possess the same taxonomic value throughout a large taxon such as the Passeriformes. A feature may be useless for demonstrating relationships in some passerine families and exceedingly useful in others. The tongue musculature is of great use in placing problem genera in the Meliphagidae, Nectariniidae, and perhaps the Dicaeidae, but of little use in many others.

b. The lack of uniformity of the taxonomic value of morphological features throughout a large taxon reduces the usefulness of extensive surveys of a morphological feature or complex as a basis for taxonomic conclusions, which is the common approach of most workers. Such studies do not pose specific questions that can be disproven, but depend greatly upon the assumption of uniform taxonomic value of the feature being studied. If a varying taxonomic value is permitted, then one must demonstrate the exact value of the morphological feature for each comparison between every pair of taxa. Few workers have even tried to show a difference between taxonomic value in any precise way.

c. Taxonomic value of morphological features does not depend on the adaptiveness of the feature, as shown above.

d. Taxonomic value of morphological features does not depend on the uniformity (interpreted frequently as conservativeness of the feature) of the feature throughout the taxon. One problem is that the taxonomic value of the morphological features must be established first because this information is used to establish the taxa and their limits, not the opposite. The use of morphological uniformity of a feature in a taxon as an estimate for taxonomic value leads to circular reasoning.

The taxonomic usefulness of a feature depends on its ability to indicate monophyletic taxa, hence features having a unique origin should possess the greatest taxonomic value. Thus, it is the paradaptive aspects of adaptive morphological features that are important to classification, and it is basically the nature of paradaptations that determines the usefulness of a morphological feature. Unique paradaptations possess great taxonomic usefulness as indicators of monophyletic taxa and the verification of their unique origin is the real feat in systematic studies.

If the limits of the selection force are narrow—i.e., only features with a restricted morphology are accepted by selection—then the paradaptation will usually have little taxonomic usefulness. Under these conditions, it would not be possible to distinguish between independent origins of morphological features because selection favors only a certain phenotype, all others outside of the limits of selection are rejected. Thus independent origins of features cannot be distinguished from a single origin. Among birds, only one muscle serves to protract the tongue. If the selection favors greater protraction of the tongue, this protractor muscle, the M. branchiomandibularis, must lengthen and it can lengthen in only one way. Hence, it is impossible to distinguish between paradaptations of independent origin, and thus this feature has very little taxonomic usefulness.

When the limits of selection forces acting upon the paradaptations are broad (as appears to be the usual condition) then whenever the probability for unique occurrence of a paradaptation is high, the feature possessing that paradaptive aspect would be useful taxonomically. And when the probability for unique occurrence is low, the paradaptation would have little taxonomic value (see Bock, 1963c, for a discussion of features possessing paradaptations with low taxonomic value). I must emphasize that this probability does not depend on how many times a particular paradaptation has appeared, but only on the probability of unique origin. It is not possible to prove that a feature had a unique origin, although such a deductive hypothesis may be suggested after a comparative study of the features using the methods of evolutionary morphology. For this reason, the arrangement of the toes possess little taxonomic value (Bock, 1967a; the reasons given in Bock and Miller, 1959, p. 42, are not really valid). Each of these types possesses little taxonomic value, no matter if it is the zygodactyl foot, which apparently arose independently in nine groups of birds, or the heterodactyl foot that seems to have originated only once in the ancestor of the present day trogons. Although the arrangement of the toes has little taxonomic value, other anatomical features associated with toe arrangements may have considerable usefulness. Whenever a toe is reversed, certain modifications must occur in the distal condyles to change the direction of the tendons running to this toe (see Steinbacher, 1935, for details). The detailed paradaptive aspects of morphological modifications in these condyles may be such that the probability for their unique occurrence is very high, and hence these features might possess considerable taxonomic value. Therefore, the zygodactyl foot of the Pici, Psittaci, and Cuculidae would have little taxonomic value, but the detailed morphology of the distal condyles of the tarsometatarsus may possess great value. Similarly, the heterodactyl foot of the trogons would have little value, but the details of the tarsal condyles may be of considerable usefulness.

As a general rule, more complex structures would have paradaptive aspects with a higher probability for unique origin and hence have greater taxonomic usefulness than simple structures. This notion has been used for many decades by taxonomists working with the higher categories of animals. Examples of such paradaptations would be the osseous arch of the radius of owls, the cruciform M. splenius capitis of swifts and hummingbirds, and the preglossal complex of passerine finches.

Paradaptations, even those possessing a low probability of unique origin, are exceedingly useful taxonomically, as they often provide strong evidence for disproof of a taxonomical hypothesis. If the conclusion is offered that two organisms are members of the same monophyletic taxon, then one expects that they will possess the same paradaptation that was favored by the selection forces in operation during the evolution of the taxon. If upon comparative study of the morphology of the two organisms, they are discovered to possess different paradaptations, one may conclude that these organisms are not members of the same taxon. Thus, different adaptive answers to the same selection forces (multiple pathways of evolution) provide strong evidence for disproof of taxonomic conclusions.

The retractor tongue muscles provide a good example. If selection favors greater excursion of the tongue — greater protraction and retraction — the retractor muscles must lengthen. The limits of this selection force are very broad, with at least five different adaptive pathways being known (W. J. Bock, personal observation). In specialized meliphagids, the usual retractor muscle, the M. stylohyoideus, is greatly reduced and the M. thyreohyoideus serves as the main retractor muscle. The African genus *Promerops* has usually been included in the Meliphagidae, but this bird possesses a greatly lengthened M. stylohyoideus like that found in the Nectariniidae. Possession of this paradaptation by *Promerops* suggests strongly that it is not a member of the Meliphagidae, but it does not provide good evidence that it is a member of the Nectariniidae, because this paradaptation has a simple morphology and probably has a low probability of unique origin. The Hawaiian genus *Moho* has also been placed in the Meliphagidae, and again has a different retractor muscle with a large M. stylohyoideus plus a distinct segment originating from the insertion of the M. serpihoideus. This peculiar paradaptation would have argued that *Moho* did not belong to the meliphagids, had not indications of a similar configuration been found in the Papuan genus *Oedistoma*. In this case, the paradapation possessed by *Moho* and *Oedistoma* suggests that they constitute a monophyletic group within the Meliphagidae because of its complex morphology.

If taxonomists rejected the notion of proving taxonomic conclusions even on a probability basis and adopted the philosophical concept that scientific conclusions are hypotheses offered for disproof, then the entire controversy of taxonomic value of morphological features would vanish. Paradaptations provide excellent evidence for disproof of statements about monophyletic groups. And analyses of the adaptive aspects of features, which involve vertical comparisons, would provide considerable evidence for disproof of conclusions about the primitive–advanced relationships of features and hence groups.

I. Role of Functional Studies

Almost all earlier and most present-day morphological–systematic studies of birds are based purely upon comparison of morphological form. This methodology has permitted many taxonomical conclusions in the past and will allow the formulation of many more in the future. It is a useful approach as it allows rapid comparative surveys of the properties of form of anatomical features by which patterns of distribution of these formal properties can be recognized. These patterns of distribution can serve as the basis for initial and frequently very rough taxonomic conclusions such as the first ordering of species into taxa of higher categorical rank. But this method is a very poor one for testing taxonomic conclusions, as it cannot provide convincing evidence for disproof of such hypotheses. Even in the failure of disproof, confidence in the taxonomic conclusion does not increase because of the weakness of the evidence.

I would argue that the necessary confidence in failure to disprove taxonomic conclusions can be gained only with studies of evolutionary morphology in which functional anatomy plays a major role. And I would argue that the significant step in the establishment of classification is not the grouping of species into taxa of higher rank, but in the attempt to disprove the existence of these taxa. Thus I believe that the best practical approach to systematic studies of the higher categories is to include investigations of functional morphology at the onset. This approach would permit significant attempts to disprove hypotheses earlier in the course of study. Functional studies should not be delayed until the late stages of taxonomic studies as an adjunct method to be applied in cases of problem taxa. Often we do not even recognize problem groups until after completion of a detailed functional analysis in the course of an evolutionary morphological study.

For classification of birds, the importance of functional study is clear. Comparative investigation of morphological form, at least on the rough gross level, of most groups of birds has been completed and numerous taxonomic conclusions — almost more than can be imagined for a group of less than 9000 species — have been offered. At this stage, the important task is to attempt to disprove these conclusions and to discard those for which strong evidence for disproof is available. I know of no case for which better evidence for disproof has been obtained in a comparative morphological study omitting a functional analysis than could be obtained with a functional investigation. And indeed, most studies of comparison of pure morphological form have not provided convincing disproof of taxonomic hypotheses. To date, few attempts to disprove systematic conclusions in ornithology using functional anatomy are available. One example is the failure to disprove monophyly of the ratites using cranial osteology (Bock, 1963b), which provided considerable credence for a hypothesis previously rejected by most ornithologists.

The importance of using different paradaptations in attempts to disprove taxonomic conclusions has been stressed above and need not be repeated. It is impossible to distinguish between possible paradaptations without a functional analysis. Thus, I conclude that functional anatomical studies are a critical component of morphological–

systematic studies, without which further significant improvement in avian classification is not possible.

VIII. Epilogue

Study of the avian skeletomuscular system has come a long way from the time in which all that was needed was the ability to dissect, describe, and compare. To be sure this ability is still essential and is even more important today than in former times, but it is not sufficient. The training required of avian anatomists today must be as rigorous as in the most difficult fields of ornithology. But with the increased training and discipline demanded of avian anatomists has come greatly enhanced rewards. Numerous new and fascinating features of the skeletomuscular system remain to be discovered. Frequently, these new finds permit a sounder comprehension of long-known features. Most of the functional, ecological, and evolutionary attributes of the entire avian skeletomuscular system must still be analyzed; herein lies a vast, fruitful area of inquiry for both skilled experimenters and theoreticians. And, in spite of the large literature on classificatory schemes, our understanding of the phylogeny and relationships of avian orders and families is still in its infancy. Most of the evidence supporting schemes of classification will continue to be sought in the skeletomuscular system. Although avian morphology may not be in the midst of a revival, it is certainly at the threshold of a potentially great advance. It is not possible at this time to guess the multiple directions of future advances, because we are still ignorant of many of the proper questions to ask. Yet it may be predicted safely that the future successes expected in the study of avian morphology and especially of the skeletomuscular system will be spectacular and will eclipse those achieved during the golden age of vertebrate morphology.

ACKNOWLEDGMENTS

I would like to express my thanks and appreciation to Miss Dorthea Goldys for preparing and arranging the illustrations and to Miss Dana Howell for typing the manuscript. Much of the research that formed the foundation of this review was supported by several grants from the National Science Foundation and from the National Institutes of Health (5RO 1 AN 10856). Final writing and preparation of the manuscript and illustrations was supported by grant GB-6909X from the National Science Foundation.

REFERENCES

Most of the studies published prior to 1900 that are cited in this chapter have been omitted from the following list of references to save space; reference to these papers may be found in Fürbringer (1888) and Gadow (1891–1893). Important review articles, either of avian anatomy or a particular morphological system, which provide accurate descriptions and terminology and extensive bibliographies, are marked with an asterisk (°) in front of the author's name. Whenever possible reference to the collected papers or to a list of published works of morphologists who have written extensively on the avian skeletomuscular system has been included; such references are also marked with an asterisk.

Alexander, R. McN. (1968). "Animal Mechanics." Univ. of Washington Press, Seattle.
°Ames, P. (1971). The morphology of the syrinx in passerine birds. *Bull. Peabody Mus.* **37** (i–vi), 1–194.
Barnikol, A. (1951). Über einige Gesetzmässigkeiten in Aufbau der motorischen Trigeminusäte bei Vögeln. *Anat. Anz.* **98**, 217–223.
Barnikol, A. (1953a). Korrelationen in der Ausgestaltung der Schädelform bei Vögeln. *Morphol. Jahrb.* **92**, 373–414.
Barnikol, A. (1953b). Zur Morphologie des Nervus trigeminus der Vögel unter besonderer Berücksichtigung der Accipitres, Cathartidae, Striges und Anseriformes. *Z. Wiss. Zool.* **157**, 285–332.
Barnikol, A. (1953c). Vergleichend anatomische und taxonomisch phylogenetische Studien am Kopf der Opisthocomiformes, Musophagidae, Galli, Columbae und Cuculi. Ein Beitrag zum *Opisthocomus* problem. *Zool. Jahrb., Syst.* **81**, 437–624.
Beecher, W. J. (1953). A phylogeny of the oscines. *Auk* **70**, 270–333.
°Bellairs, A. D'A., and Jenkin, C. R. (1960). The skeleton of birds. *In* "Biology and Comparative Physiology of Birds" (A. J. Marshall, ed.), Vol. 1, pp. 241–300. Academic Press, New York.
Berger, A. J. (1952). The comparative functional morphology of the pelvic appendage in three genera of Cuculidae. *Amer. Midl. Natur.* **47**, 513–605.
Berger, A. J. (1953). On the locomotory anatomy of the Blue Coua, *Coua caerulea. Auk* **70**, 49–83.
Berger, A. J. (1954). The myology of the pectoral appendage of three genera of American cuckoos. *Misc. Publ. Mus. Zool., Univ. Mich.* **85**, 1–35.
Berger, A. J. (1955). On the anatomy and relationships of Glossy Cuckoos of the genera *Chrysococcyx, Lampromorpha,* and *Chalcites. Proc. U. S. Nat. Mus.* **103**, 585–597.
Berger, A. J. (1956a). The appendicular myology of the Pygmy Falcon (*Polihierax semitorquatus*). *Amer. Midl. Natur.* **55**, 326–333.
Berger, A. J. (1956b). On the anatomy of the Red Bird of Paradise, with comparative remarks on the Corvidae. *Auk* **73**, 427–446.
Berger, A. J. (1956c). The appendicular myology of the Sandhill Crane, with comparative remarks on the Whooping Crane. *Wilson Bull.* **68**, 282–304.
Berger, A. J. (1957). On the anatomy and relationships of *Fregilupus varius,* an extinct starling from the Mascarane Islands. *Bull. Amer. Mus. Natur. Hist.* **113**, 225–272.
°Berger, A. J. (1960). The musculature. *In* "Biology and Comparative Physiology of Birds" (A. J. Marshall, ed.), Vol. 1, pp. 301–344. Academic Press, New York.
°Berger, A. J. (1966). See George and Berger (1966).
°Boas, J. E. V. (1929). Biologisch-anatomische Studien über den Hals der Vögel. *Kgl. Dan. Vidensk. Selsk., Skr., Natur. Math. Afd.* [9] **1**, 101–222.
Bock, W. J. (1958). A generic revision of the plovers (Charadriinae, Aves). *Bull. Mus. Comp. Zool., Harvard Univ.* **118**, 27–97.
Bock, W. J. (1959). Preadaptation and multiple evolutionary pathways. *Evolution* **13**, 194–211.
Bock, W. J. (1960a). Secondary articulation of the avian mandible. *Auk.* **77**, 19–55.
Bock, W. J. (1960b). The palatine process of the premaxilla in the Passeres. A study of the variation, function, evolution and taxonomic value of a single character through-

out an avian order. *Bull. Mus. Comp. Zool., Harvard Univ.* **122,** 361-488.

Bock, W. J. (1963a). Relationships between the birds of paradise and the bower birds. *Condor* **65,** 91-125.

Bock, W. J. (1963b). The cranial evidence for ratite affinities. *Proc. Int. Ornithol. Congr., 13th, 1962* Vol. 1, pp. 39-54.

Bock, W. J. (1963c). Evolution and phylogeny in morphologically uniform groups. *Amer. Natur.* **97,** 265-285.

°Bock, W. J. (1964). Kinetics of the avian skull. *J. Morphol.* **114,** 1-42.

Bock, W. J. (1965). Experimental analysis of the avian passive perching mechanism. *Amer. Zool.* **5,** 681.

°Bock, W. J. (1966). An approach to the functional analysis of bill shape. *Auk* **83,** 10-51.

Bock, W. J. (1967a). The use of adaptive characters in avian classification. *Proc. Int. Ornithol. Congr., 14th, 1966* pp. 61-74.

Bock, W. J. (1967b). Review of George and Berger "Avian Myology." *Auk* **84,** 138-140.

°Bock, W. J. (1968a). Mechanics of one- and two-joint muscles. *Amer. Mus. Nov.* **2319,** 1-45.

Bock, W. J. (1968b). Review of Hennig "Phylogenetic systematics." *Evolution* **22,** 646-648.

°Bock, W. J. (1969a). Comparative morphology in systematics. *Systematic Biology Nat. Acad. Sci.—Nat. Res. Counc., Publ.* **1692,** 411-448.

Bock, W. J. (1969b). The concept of homology. *Ann. N.Y. Acad. Sci.* **167,** 71-73.

Bock, W. J. (1969c). Nonvalidity of the "phylogenetic fallacy." *Syst. Zool.* **18,** 111-115.

Bock, W. J. (1973a). Philosophical foundations of classical evolutionary classification. *Syst. Zool.* **22,** 375-392.

°Bock, W. J. (1973b). A redescription of passerine jaw muscles. In preparation.

Bock, W. J., and Hikida, R. S. (1968). An analysis of twitch and tonus fibers in the hatching muscle. *Condor* **70,** 211-222.

Bock, W. J., and Hikida, R. S. (1969). Turgidity and function of the hatching muscle. *Amer. Midl. Natur.* **81,** 99-106.

Bock, W. J., and Kummer, B. (1968). The avian mandible as a structure girder. *J. Biomech.* **1,** 89-96.

Bock, W. J., and McEvey, A. (1969a). Osteology of *Pedionomus torquatus* (Aves: Pedionomidae) and its allies. *Proc. Roy. Soc. Victoria* **82,** 187-232.

Bock, W. J., and McEvey, A. (1969b). The radius and relationship of owls. *Wilson Bull.* **81,** 55-68.

Bock, W. J., and Miller, W. DeW. (1959). The scansorial foot of the woodpeckers, with comments on the evolution of perching and climbing feet in birds. *Amer. Mus. Nov.* **1931,** 1-45.

Bock, W. J., and Morioka, H. (1971). Morphology and evolution of the ectethmoid-mandibular articulation in the Meliphagidae (Aves). *J. Morphol.* **135,** 13-50.

°Bock, W. J., and Morioka, H. (1973). Functional analysis of the jaw muscles in passerine birds. In preparation.

Bock, W. J., and Morony, J. (1971). The preglossale of *Passer* (Aves)—a skeletal neomorph. *Amer. Zool.* **11,** 705.

Bock, W. J., and Morony, J. J. (1972). Snap-closing jaw ligaments in flycatchers. *Amer. Zool.* **12,** 729-730.

°Bock, W. J., and Shear, C. R. (1973). A redescription of passerine tongue muscles. In preparation.

°Bock, W. J., and von Wahlert, G. (1965). Adaptation and the form-function complex. *Evolution* **19,** 269-299.

Bock, W. J., Balda, R. P., and Vander Wall. S. B. (1973). Morphology of the sublingual pouch and tongue musculature in Clark's Nutcracker. *Auk* **90**, 491–519.

°Böker, H. (1935). "Einführung in die vergleichende biologische Anatomie der Wirbeltiere," Vol. 1, Fischer, Jena.

Böker, H. (1937). "Einführung in die vergleichende biologische Anatomie der Wirbeltiere," Vol. 2, Fischer, Jena.

Bowman, R. I. (1961). Morphological differentiation and adaptation in the Galapagos Finches. *Univ. Calif. Berkeley, Publ. Zool.* **58**, (i–vii), 1–302.

Bradey, O. C. (1915). "The Structure of the Fowl" (revised by T. Grahame), 4th ed. Oliver & Boyd, Edinburgh.

Bühler, P. (1970). Schädelmorphologie und Kiefermechanik der Caprimulgidae (Aves). *Z. Morphol. Tiere* **66**, 337–399.

Burton, P. J. K. (1971). Some observations on the splenius capitis muscle of birds. *Ibis* **113**, 19–38.

Canfield, S. P. (1971). The mechanical properties and heat production of chicken latissimus dorsi muscles during tetanic contractions. *J. Physiol.* **219**, 281–302.

Chamberlain, F. W. (1943). Atlas of avian anatomy. Osteology-arthrology-myology. *Mich. Agr. Exp. Sta., Mem. Bull.* **5**, (i–xi), 1–213.

Cracraft, J. (1968). The lacrimal-ectethmoid bone complex in birds: A single character analysis. *Amer. Midl. Natur.* **80**, 316–359.

Cracraft, J. (1971a). The relationships and evolution of the rollers: Families Coraciidae, Brachypteraciidae, and Leptosomatidae. *Auk* **88**, 723–752.

°Cracraft, J. (1971b). The functional morphology of the hind limb of the domestic pigeon, *Columba livia*. *Bull. Amer. Mus. Natur. Hist.* **144**, 171–268.

Davids, J. A. G. (1952). Etude sur les attaches au crâne des muscles de la tête et du cou chez *Anas platyrhyncha platyrhyncha* (L.) I, II, III. *Proc., Kon. Ned. Akad. Wetensch.*, **55**, 81–94, 525–533, and 534–540.

Dempster, W. T. (1961). Free-body diagrams as an approach to the mechanics of human posture and motion. *In* "Biomechanical Studies of the Musculoskeletal System" (F. G. Evans, ed.), pp. 81–135. Thomas, Springfield, Illinois.

Den Hartog, J. P. (1961a). "Mechanics." Dover, New York.

Den Hartog, J. P. (1961b). "Strength of Materials." Dover, New York.

Engels, W. L. (1938). Tongue musculature of passerine birds. *Auk* **55**, 642–650.

°Evans, H. E. (1969). Anatomy of the Budgerigar. *In* "Diseases of Cage and Aviary Birds." pp. 45–112. Lea & Febiger, Philadelphia, Pennsylvania.

Feng, T. P. (1932). The effect of length on the resting metabolism of muscle. *J. Physiol. (London)* **74**, 441–454.

°Fiedler, W. (1951). Beiträge zur Morphologie der Kiefermuskulatur der Oscines. *Zool. Jahrb., Abt. Anat. Ontog. Tiere* **71**, 235–288.

°Fisher, H. I. (1946). Adaptations and comparative anatomy of the locomotor apparatus of New World vultures. *Amer. Midl. Natur.* **35**, 545–727.

°Fisher, H. I. (1955a). Avian anatomy, 1925–1950, and some suggested problems. *In* "Recent Studies in Avian Biology" (A. Wolfson, ed.), pp. 57–104. Univ. of Illinois Press, Urbana.

Fisher, H. I. (1955b). Some aspects of the kinetics in the jaws of birds. *Wilson Bull.* **67**, 175–188.

Fisher, H. I. (1958). The "hatching muscle" in the chick. *Auk* **75**, 391–399.

°Fisher, H. I., and Goodman, D. C. (1955). The myology of the Whooping Crane, *Grus americana*. *Ill. Biol. Monogr.* **24**, 1–127.

Fitzgerald, T. C. (1970). "The Coturnix Quail/Anatomy and Histology." Univ. of Iowa Press, Ames.

Fourie, S. (1955). A contribution to the cranial morphology of *Nyctisyrigmus pectoralis pectoralis* with special reference to the palate and cranial kinesis. *Ann. Univ. Stellenbosch., Ser. A* **31**, 179–215.

°Fürbringer, M. (1888). "Untersuchungen zur Morphologie und Systematik der Vögel, Lugeich ein Beitrag zur Anatomie der Stütz und Bewegungsorgane," Vols. I and II. T. J. van Holkema, Amsterdam.

°Gadow, H. [and E. Selenka]. (1891–1893). Vögel. *Bronn's Klassen* **6**, Div. 4, Parts I and II.

°Gans, C., and Bock, W. J. (1965). The functional significance of muscle architecture—a theoretical analysis. *Ergeb. Anat. Entwicklungsgesch.* **38**, 115–142.

°George, J. C., and Berger, A. J. (1966). "Avian Myology." Academic Press, New York.

George, W. G. (1962). The classification of the Olive Warbler, *Peucedramus taeniatus*. *Amer. Mus. Nov.* **2103**, 1–41.

Ginsborg, B. L. (1960). Spontaneous activity in muscle fibres of the chick. *J. Physiol. (London)* **150**, 707–717.

Goodman, D. C., and Fisher, H. I. (1962). "Functional Morphology of the Feeding Apparatus in Waterfowl. Aves: Anatidae." Southern Illinois Univ. Press, Carbondale.

Goslow, G. E., Jr. (1972). Adaptive mechanisms of the raptor pelvic limb. *Auk* **89**, 47–64.

Gray, J. (1956). Muscular activity during locomotion. *Brit. Med. Bull.* **12**, 203–209.

°Gray, J. (1968). "Animal Locomotion." The World Naturalist. Weidenfeld & Nicolson. London.

Gringer, I., and George, J. C. (1969). An electron microscopic study of the pigeon breast muscle. *Can. J. Zool.* **47**, 517–523.

Harvey, E. B., Kaiser, H. E., and Rosenberg, L. E. (1968). An atlas of the domestic turkey (*Meleagris gallopavo*). Myology and osteology. *U.S. Ato. Energy Comm. 1123*, **TID-UC-48**, 1–247.

Hennig, W. (1966). "Phylogenetic Systematics." Univ. of Illinois Press, Urbana.

Hikida, R. S., and Bock, W. J. (1970). The structure of pigeon muscle and its changes due to tenotomy. *J. Exp. Zool.* **175**, 343–356.

Hikida, R. S., and Bock, W. J. (1971). Innervation of the avian latissimus dorsi anterior muscle. *Amer. J. Anat.* **130**, 269–280.

Hikida, R. S., and Bock, W. J. (1972). Effect of denervation on pigeon slow skeletal muscle. *Z. Zellforsch. Mikrosk. Anat.* **128**, 1–18.

°Hill, A. V. (1964). "Trails and Trials in Physiology." Arnold, London.

°Hofer, H. (1950). Zur Morphologie der Kiefermuskulatur der Vögel. *Zool. Jahrb., Abt. Anat. Ontog. Tiere* **70**, 427–600.

°Howard, H. (1929). The avifauna of Emeryville shellmound. *Univ. Calif., Berkeley, Publ. Zool.* **32**, 301–394.

°Hudson, G. E. (1937). Studies on the muscles of the pelvic appendage in birds. *Amer. Midl. Natur.* **18**, 1–108.

Hudson, G. E. (1948). Studies on the muscles of the pelvic appendage in birds. II. the heterogeneous order Falconiformes. *Amer. Midl. Natur.* **39**, 102–127.

°Hudson, G. E., and Lanzillotti, P. J. (1955). Gross anatomy of the wing muscles in the family Corvidae. *Amer. Midl. Natur.* **53**, 1–44.

°Hudson, G. E., and Lanzillotti, P. J. (1964). Muscles of the pectoral limb in galliform birds. *Amer. Midl. Natur.* **71**, 1–113.

*Hudson, G. E., Lanzillotti, P. J., and Edwards, G. D. (1959). Muscles of the pelvic limb in galliform birds. *Amer. Midl. Natur.* **61**, 1–67.
Hudson, G. E., *et al.* (1965). Ontogeny of the supernumerary sesamoids in the leg muscles of the Ring-necked Pheasant. *Auk* **82**, 427–437.
Hudson, G. E., *et al.* (1966). A numerical analysis of the modifications of the appendicular muscles in various genera of gallinaceous birds. *Amer. Midl. Natur.* **76**, 1–73.
Hudson, G. E., *et al.* (1969). A numerical study of the wing and leg muscles of Lari and Alcae. *Ibis* **111**, 459–524.
*Jollie, M. T. (1957). The head skeleton of the chicken and remarks on the anatomy of this region in other birds. *J. Morphol.* **100**, 389–436.
Kaplan, N. O., and Cahn, R. D. (1962). Lactic dehydrogenases and muscular dystrophy in the chicken. *Proc. Nat. Acad. Sci. U.S.* **48**, 2123–2130.
Kaupp, B. F. (1918). "Anatomy of the Domestic Fowl." Saunders, Philadelphia, Pennsylvania.
Kear, J., and Burton, P. J. K. (1971). The food and feeding apparatus of the Blue Duck *Hymenolaimus*. *Ibis.* **113**, 483–493.
Klemm, R. D. (1969). Comparative myology of the hind limb of procellariiform birds. *S. Ill. Univ. Monogr., Sci. Ser.* **2**, 1–269.
*Koch, J. C. (1917). The laws of bone architecture. *Amer. J. Anat.* **21**, 177–298.
*Krüger, P. (1952). "Tetanus und Tonus der quergestreiften Skelettmuskeln der Wirbeltiere und des Menschen." Akad. Verlagsges., Leipzig.
*Kummer, B. (1959a). "Bauprinzipien des Saugerskeletes." Thieme, Stuttgart.
*Kummer, B. (1959b). Biomechanic des Saugetierskelets. *In* "Handbuch der Zoologie," Vol. 8, Part 6, No. 2, pp. 1–80.
Kummer, B. (1962). Funktioneller Bau und funktionelle Anpassung des Knochens. *Anat. Anz.* **110**, 261–293.
*Kummer, B. (1966). Photoelastic studies on the functional structure of bone. *Folia Biotheor.*, **6**, 31–40.
Lack, D. (1965). Evolutionary ecology. *J. Ecol.* **53**, 237–245.
*Lakjer, T. (1926). "Studien über die Trigeminus-versorgte Kaumuskulatur der Sauropsiden." C. A. Reitzel, Copenhagen.
Lebedinsky, N. G. (1921). Zur Syndesmologie der Vögel. *Anat. Anz.* **54**, 8–15.
Lee, S. Y. (1971). A histochemical study of twitch and tonus fibers. *J. Morphol.* **133**, 253–272.
*Leiber, A. (1907). Vergleichende Anatomie der Spechtzunge. *Zoologica (Stuttgart)* **51**.
Levinson, I. J. (1961). "Introduction to Mechanics." Prentice-Hall, Englewood Cliffs, New Jersey.
Levinson, I. J. (1963). "Mechanics of Materials." Prentice-Hall, Englewood Cliffs, New Jersey.
Manger Cats-Kuenen, C. S. W. (1961). Casque and bill of *Rhinoplax vigil* (Forst.) in connection with the architecture of the skull. *Verh. Kon. Ned. Akad. Wetensch., Afd. Natuurk.* [2] **53**, 1–51.
Mayr, E. (1962). Accident or design, the paradox of evolution. *In* "The Evolution of Living Organisms" (G. W. Leeper, ed.), pp. 1–14. Melbourne Univ. Press, Victoria, Australia.
Miller, A. H. (1937). Structural modifications in the Hawaiian Goose (*Nesochen sandvicensis*). A study in adaptive evolution. *Univ. Calif., Berkeley, Publ. Zool.* **42**, 1–79.
Mivart, St. G. (1874). On the axial skeleton of the ostrich (*Struthio camelus*). *Trans. Zool. Soc. London* **8**, 385.

Mivart, St. G. (1879). On the axial skeleton of the Pelecanidae. *Trans. Zool. Soc. London* **10**, 315–378.

Moller, W. (1930). Über die Schnabel und Zungenmechanik blütenbesuchender Vögel. I. *Biol. Gen.* **6**, 651–726.

Moller, W. (1931). Über die Schnabel und Zungenmechanik blütenbesuchender Vögel. II. *Biol. Gen.* **7**, 99–154.

Mollier, G. (1937). Beziehungen zwischen Form und Funktion der Sehnen im Muskel-Sehnen-Knochen-System. *Morphol. Jahrb.* **79**, 161–199.

Mudge, G. P. (1903). On the myology of the tongue of parrots, with a classification of the order, based upon the structure of the tongue. *Trans. Zool. Soc. London* **16**, 211–278.

*Murray, P. D. F. (1936). "Bones. A Study of the Development and Structure of the Vertebrate Skeleton." Cambridge Univ. Press, London and New York.

Murray, P. D. F., and Drachman, D. B. (1969). The role of movement in the development of joints and related structures: The head and neck in the chick embryo. *J. Embryol. Exp. Morphol.* **22**, 349–371.

*Pauwels, F. (1965). "Gesammelte Abhandlungen zur funktionelle Anatomie des Bewegungsapparates." Springer-Verlag, Berlin and New York.

Pfuhl, W. (1937). Die gefiederten Muskeln, ihre Form und ihre Wirkungsweise. *Z. Anat. Entwicklungsgesch.* **106**, 749–769.

Popper, K. R. (1959). "The Logic of Scientific Discovery." Hutchinson, London.

Raikow, R. J. (1970). Evolution of diving adaptations in the stifftail ducks. *Univ. Calif., Berkeley, Publ. Zool.* **94**, (vi), 52.

Richards, L. P., and Bock, W. J. (1973). Functional anatomy and adaptive evolution of the feeding apparatus of the Hawaiian honeycreeper genus *Loxops* (Drepanididae). *A.O.U. Ornithol. Monogr.* **15**, (i–x), 1–173.

*Schildmacher, H. (1931). Untersuchungen über die Funktion der Herbstschen Körperchen. *J. Ornithol.* **79**, 374–415.

Schoonees, J. (1963). Some aspects of the cranial morphology of *Colius indicus*. *Ann. Univ. Stellenbosch*, Ser. A **38**, 216–246.

*Schumacher, G. H. (1961). "Funktionelle Morphologie der Kaumuskulatur." Fisher, Jena.

Sisson, S., et al. (1938). "The Anatomy of the Domesticated Animals. Saunders, Philadelphia, Pennsylvania.

Spring, L. W. (1965). Climbing and pecking adaptations in some North American woodpeckers. *Condor* **67**, 457–488.

Spring, L. (1971). A comparison of functional and morphological adaptations in the Common Murre (*Uria aalge*) and Thick-Billed Murre (*Uria lomvia*). *Condor* **73**, 1–27.

Starck, D. (1940). Beobachtungen an der Trigeminusmuskulatur der Nashornvögel. *Morphol. Jahrb.* **84**, 585–623.

*Starck, D. (1959). Neuere Ergebnisse der vergleichenden Anatomie und ihre Bedeutung für die Taxonomie erläutert an der Trigeminus-Muskulatur der Vögel. *J. Ornithol.* **100**, 47–59.

*Starck, D., and Barnikol, A. (1954). Beiträge zur Morphologie der Trigeminusmuskulatur der Vögel, besonders der Accipitres, Cathartidae, Striges und Anseres. *Morphol. Jahrb.* **94**, 1–64.

*Steinbacher, G. (1935). Funktionell-anatomische Untersuchungen an Vogelfüssen mit Wendezehen und Rückzehen. *J. Ornithol.* **83**, 214–282.

°Stolpe, M. (1932). Physiologisch-anatomische Untersuchungen über die hintere Extremität der Vögel. *J. Ornithol.* **80,** 161–247.
°Sy, M. (1936). Funktionell-anatomische Untersuchungen am Vogelflügel. *J. Ornithol.* **84,** 199–296.
vanden Berge, J. C. (1970). A comparative study of the appendicular musculature of the order Ciconiiformes. *Amer. Midl. Natur.* **84,** 289–364.
von Wahlert, G. (1965). The role of ecological factors in the origin of higher levels of organization. *Syst. Zool.* **14,** 288–300.
Wilson, E. O. (1965). A consistency test for phylogenies based upon contemporaneous species. *Syst. Zool.* **14,** 214–220.
Zusi, R. L. (1962). Structural adaptations of the head and neck in the Black Skimmer, *Rynchops nigra* Linnaeus. *Publ. Nuttall Ornithol. Club* **3,** (viii), pp. 1–101.
Zusi, R. L., and Storer, R. W. (1969). Osteology and myology of the head and neck of the pied-billed grebes (*Podilymbus*). *Misc. Publ. Mus. Zool., Univ. Mich.* **139,** 1–49.

Chapter 4

THERMAL AND CALORIC RELATIONS OF BIRDS

William A. Calder and James R. King

I.	Introduction	260
II.	A Simplified Heat Exchange Model	263
	A. Allometric Analysis	264
	B. Symbols and Units	266
	C. The Metabolism–Temperature Graph	266
	D. Standard Metabolic Rate	269
	E. Critical Temperatures	272
	F. Heat Transfer Coefficient	273
	G. Critical Temperatures and Extrapolations to Zero Metabolism	276
	H. Body Temperature	279
	I. Reconciliation of Empirical and Theoretical Treatments	284
III.	Basic Principles of Heat Transfer	287
	A. Energy Transfer by Thermal Radiation	287
	B. Net Radiation Exchange	293
	C. Heat Transfer by Conduction	294
	D. Heat Transfer by Convection	298
	E. Heat Transfer by Evaporation	302
	F. Interactions among Modes of Heat Transfer	307
	G. Résumé and Reiteration	308
IV.	Physiological Responses to Heat and Cold	309
	A. Ranges of Response	309
	B. Heat Production	310
	C. Heat Conservation	319
	D. Heat Storage in Hyperthermia	322
	E. Heat Dissipated by Evaporation	326

V.	Hypothermia	343
	A. Energetic Basis of Hypothermia	343
	B. Terminology	344
	C. The Hypothermic State	345
	D. Influence of Body Size on Torpor Entry and Arousal	350
VI.	Integration of Thermal and Caloric Responses	354
	A. Short-Term Regulation	354
	B. Long-Term Regulation	374
VII.	Evolutionary Aspects of Thermoregulation and Energy Metabolism	386
	A. Components of Adaptation	387
	B. Ecogeographic Rules	389
	References	393

I. Introduction

In this chapter we focus attention on the ways in which wild species of birds adjust to their caloric and thermal environments. Terrestrial animals must function thermodynamically within a "climate space" that consists of a minimum of four factors, including radiation, humidity, air velocity, and air temperature (Porter and Gates, 1969; Gates, 1970). In spite of large variations in these environmental factors, adult birds and mammals characteristically maintain the energy content and temperature of their bodies within narrow limits, with only temporary or adaptive lapses into obesity, hyperthermia, or hypothermia in various species. These forms of homeostasis are made possible by behavioral, physiological, and in some cases morphological modes of adjustment that we subsequently consider in detail. In doing so it is inevitable that we repeatedly compare the responses of birds to their caloric and thermal environments with the better known analogous processes in mammals. At the outset it is therefore appropriate to emphasize the ways in which the two classes of homeotherms are alike as well as different, and to examine some antiquated dogmas related to this subject.

In their descent through separate reptilian lines (Brodkorb, 1971), the birds and mammals constitute a remarkable case of parallel or convergent evolution of metabolism and thermoregulation, both qualitatively and quantitatively. Contrary to widespread belief, birds do not characteristically function at a higher metabolic intensity than mammals. The basal or standard metabolic rate of nonpasserine birds is indistinguishable from that of mammals (Lasiewski and Dawson,

1967). Although it is true that the standard metabolic rate of passerines (comprising about 60% of avian species) appears to be 30 to 70% greater than that of mammals in the range of overlap in the data (100–500 gm), the difference between passerines and nonpasserines is no greater than that found between taxa of mammals (Pearson, 1948; MacMillen and Nelson, 1969; T. J. Dawson and Hulbert, 1970), insects (Kayser and Heusner, 1964), or among various orders of nonpasserine birds (Zar, 1969). Thus, there has been some radiation of weight-relative metabolic intensity among taxa of both birds and mammals, but on the average both homeotherm classes have evolved to about the same quantitative level.

It is likewise clear that the energy cost of activities above the basal level is in general no greater in birds than in mammals, and may be less on the average. The energy cost of locomotion at comparable speeds is less for flying animals than for nonflying animals of the same weight (Tucker, 1970), although terrestrial locomotion may cost more calorically in bipedal birds than in quadrupedal mammals (Taylor et al., 1971) under comparable conditions. The energy expenditure of swimming in birds is no greater, and may be less, than the power requirement for transport in hydrodynamically similar vessels (Prange and Schmidt-Nielsen, 1970).

The body temperatures are also remarkably similar in birds and mammals (on the average, in view of the apparent range of evolutionary options) and are regulated in both groups at about the highest point compatible with protein stability (Section II). In view of the present-day similarity of caloric and thermal relationships in birds and mammals, and probably their common heritage in reptilian behavioral thermoregulation (Rodbard, 1948, Hammel et al., 1967; Cabanac et al., 1967), it is no surprise that their mechanisms of physiological thermoregulation are fundamentally alike (Section VI). The general thermodynamic similarity of the two homeotherm classes may represent convergence toward an optimal compromise between the conflicting requirements for water conservation and energy conservation (Section II,H).

The relatively minor differences that exist between birds and mammals as thermodynamic systems appear to have evolved in conjunction with flight and plumage, and with a sensorium emphasizing vision and hearing in birds, contrasted with olfaction and hearing in mammals. For instance, perhaps related to the mechanical properties of feathers and to the loss of the forelimbs for manipulative actions, birds have not evolved any truly fossorial forms, unlike small mammals, and thus do not characteristically evade environmental extremes (with some notable exceptions—Section VI, Table XV) by burrowing or hole-seeking. As a substitution, their powers of flight enable birds to evade environmental extremes by migration or, in resident forms, by relatively long-distance searches for favorable microenvironments.

Because of their dependence on vision, birds are predominantly daytime feeders (with the obvious exception of most owls and caprimulgids). Therefore, unlike the majority of mammals of comparable size, they must be active during at least part of the most stressful segment of the 24-hour cycle in hot environments. Furthermore, by resting during the colder (nocturnal) part of the 24-hour cycle in cold environments birds are unable to exploit the waste heat of muscular exercise when it would be most valuable for thermoregulation.

Physiologically, birds may be contrasted with mammals in the modes of thermoregulatory calorigenesis. Muscular activity, especially shivering, is a major source of heat in cool or cold environments in both classes of homeotherms; but among adults it is apparently only the mammals that are typically able to produce extra heat by nonshivering calorigenesis. In general, birds appear to lack this capacity (Chaffee and Roberts, 1971; see also Section VI,A,3).

Finally, it is frequently mentioned in elementary textbooks that birds lack sweat glands, and it is implied that this imposes unique constraints on avian patterns of evaporative heat transfer. Although the comparative anatomy and physiology of mammalian sweat glands are very imperfectly known, it is nevertheless clear that thermoregulatory sweating is absent or only weakly developed in a large majority of mammalian taxa, and that many species have only insignificant numbers of eccrine glands (Bligh and Allen, 1970; Adams, 1971). The familiar examples of mammals that depend heavily on thermoregulatory sweating (man, pigs, large ungulates) are the exceptions rather than the rule. The similarities between birds and mammals in this thermoregulatory attribute, as in the many others mentioned above, are more conspicuous than the differences. Furthermore, there is the possibility that transcutaneous evaporation of water in birds is a functional equivalent of sweating in mammals (Section IV,E). Birds and mammals have converged on similar levels of caloric and thermal homeostasis and share many basic modes of thermoregulation that differ mainly in physiological detail and behavioral alternatives.

The proliferation of studies of avian thermoregulation in the past two decades can probably be traced largely to the stimulating ideas of Scholander *et al.* (1950a,b,c) and their model of homeothermy. Indeed, most of the contemporary literature on avian thermoregulation features the Scholander model, and so our review logically begins by considering a bird in the simple laboratory environment to which this model can be applied. While this is an unnatural situation and its results are of limited value in extrapolations to the wild state, it does allow comparative evaluations of insulation, evaporative heat loss, body temperature, and other variables in the steady state, uncomplicated by activity.

We are aware that the Scholander (electrical analog or so-called Newtonian cooling) model is subject to criticism because it oversimplifies the basic physics of heat exchange (Gates, 1969, p. 126; Strunk, 1971, 1973; Tracy, 1972, 1973; Bakken and Gates, 1973, 1974) as well as for other reasons (Kleiber, 1961, 1972a,b, 1973; King, 1964; Tracy, 1972). Nevertheless, when coupled with adequate recognition of the complexities of real systems (Sections IV and VI), we find that the Scholander model remains as a convenient tool and as a frame of reference for comparative purposes that is unsurpassed by any extant alternative model.

Following an introduction of the Scholander model we will review the principles of heat transfer, emphasizing their application to avian energetics. We will then examine physiological and behavioral thermoregulatory responses and their integration, and will conclude with a brief account of some aspects of their evolution. Our treatments of these topics are often selective rather than comprehensive. The recent publication of substantial reviews of avian bioenergetics (Shilov, 1968; Dawson and Hudson, 1970; cognate chapters in the present volumes) makes it pointless for us to treat at length several subjects (e.g., ontogeny of thermoregulation, evolution of homeothermy, energetics of flight and migration) that might otherwise demand extensive review.

II. A Simplified Heat Exchange Model

The second law of thermodynamics states that energy transformations are less than 100% efficient. The chemical energy of the food intake is transformed, for instance, into the kinetic energy of circulating blood, the potential energy of electrical and osmotic gradients across membranes, and the chemical energy of tissue growth, egg production, and stored nutrients. Each transformation is accompanied by a loss of energy in the form of heat. Eventually, for an organism in a steady state (no change of mass, composition, or temperature), all of the metabolizable energy intake is degraded to heat. The loss of this waste product of metabolism is controlled as required to maintain a body temperature independent of environmental fluctuations (Clark, 1948).

A constant body temperature maintained through thermoregulation is one outstanding feature of the constant *milieu intérieur* of birds. It is possible only if there is a steady-state balance between heat income and heat loss (DuBois, 1937). The income is the sum of metabolic heat production, radiation, and in very hot climates, heat loading by conduction and convection. Heat loss occurs via radiation, conduction, convection, and evaporation. Natural environments are quite heterogeneous, the bird simultaneously facing one air temperature, another substrate temperature, several radiant temperatures, and fluctuating wind velocity. Thus, a quantitative account of thermal balance is very complex. Before considering the processes of heat exchange within such an environment, we should examine the more simplified situation of the resting bird in the thermally homogeneous surroundings of a laboratory experiment. This uniform environmental temperature, i.e., temperature of chamber walls the same as the air temperature, will be referred to as the ambient temperature, for which there is no true equivalent in nature.

A. Allometric Analysis

> I believe that the most useful path for physiology and medicine to follow now is to seek to discover new facts instead of trying to reduce to equations the facts which science already possesses. This does not mean that I condemn the application of mathematics to biological phenomena, because the science will later be established by this alone . . .
>
> —Claude Bernard (1865)

This was unquestionably a wise opinion in 1865, and the "facts" accumulated in the ensuing century have perhaps brought us close to the latter situation.

Since the contributions of Scholander *et al.* (1950a,b,c), data on metabolic rates have been published for at least 38 species of passerine birds and 34 species of nonpasserine birds. There are more than 8600 species of birds, some three-fifths of which are passerines (Storer, 1971; Brodkorb, 1971). We may ask at what point has a sample been obtained for a given physiological variable that is adequate to support reliable generalizations about the class Aves? Given limitations of time and funding, should one measure the heat transfer coefficient in the seventy-third species of bird, or begin to ask new questions? It is possible that an empirical equation for data already available could predict, very closely, the results of long hours of measurements on species 73. Also, applying the method of allometric cancellation (Stahl, 1962, 1963, 1967), more significant questions might emerge. By reducing the information in prose and tabular form to manipulable equations, we can evaluate quantitatively the possible evolutionary patterns, the constraints that nature places on them, and the economics of living organisms. This type of evaluation might help to prevent avian energetics from becoming one of those "other areas which suffered from an indigestion of facts, while data were collected without reference to problems" (Levins, 1968).

Body size is by far the major determinant of quantitative levels in metabolic variables. Consequently, weight-dependent variables can be properly compared only with reference to a common scale factor. Other evolutionary determinants of metabolic variables, such as phylogeny, morphology, behavior, or ecological constraints are evidently slight compared with body size. The most useful of the empirical expressions of variables (Y) (such as metabolic rate, insulation, and related factors) is as a power function of body mass (m):

$$Y = am^b$$

which is calculated after logarithmic transformation:

$$\log Y = \log a + b \log m$$

where a is a proportionality constant (see Lasiewski and Dawson, 1969, for a discussion of the legitimacy of this log transformation). Power functions or allometric equations of this type have been derived for standard metabolism, thermal conductance, and evaporative water loss, among other factors.

Allometric analysis has become a powerful and popular tool for the synthesis of overall patterns in heat exchange and the supporting functions of metabolic heat production and metabolic oxygen delivery [see Lasiewski (1972) for allometric analysis of avian respiration]. In addition to providing an expedient general model for avian thermoregulation, allometric equations are highly useful for prediction, and as a baseline for comparison of species that survive in unusual environments. They also provide insight into the problems and advantages of evolving small or large extremes of body size. We will rely extensively upon the allometric approach in this review. Expressions of statistical variability for previously unpublished allometric equations appear in Table XVII, at the end of this chapter.

B. Symbols and Units

To approach the physical basis of thermoregulation quantitatively and concisely, it is expedient that we utilize standardized symbols for physical and physiological quantities. We have attempted to follow the symbols and subscripts for thermoregulatory quantities proposed by Gagge *et al.* (1969) with some minor departures for clarity or our publisher's policy. We have utilized National Bureau of Standards (1968) policy in retaining units (e.g., grams, calories) that are customary in comparative physiology instead of translation to the new International System equivalents (newtons, joules). The symbols used in this chapter are as follows:

H = heat energy, made up of components:
  H_m = metabolic heat energy
  H_e = evaporation heat energy
  H_r = radiation heat energy
  H_k = conduction heat energy
  H_c = convection heat energy
  H_s = stored heat energy
$\dot{H}_b$ = standard, or basal, metabolic rate
$\dot{H}$ = dH/dt, or rate of heat loss or gain
T_b = body temperature
T_r = rectal or cloacal temperature
T_{hyp} = hypothalamic temperature
T_s = skin temperature
T_a = ambient (environmental) temperature
h = heat transfer coefficient
h_e = evaporative heat transfer coefficient
h_d = dry (nonevaporative) heat transfer coefficient
P = pressure
m = body mass in grams
M = body mass in kilograms
L = linear dimension
V = volume
A = area

C. The Metabolism–Temperature Graph

Heat flows only from a higher to a lower temperature. The rate of heat loss is a linear function of the temperature difference between an object and it surroundings if the insulation is held constant. This has been erroneously associated with Newton's law of cooling, which actually describes the rate of decrease of temperature change of a cooling object. However, it follows from the equation

$$\Delta H = c \, \Delta T \, m$$

where c is the specific heat, that the time derivatives of both sides of the equation are

$$\frac{dH}{dt} = c \, \frac{dT}{dt} \, m$$

Note that Newton's dT/dt, being an instantaneous rate, also involves no significant ΔT, so the biologist's misinterpretation of Newton's law is a reasonable approximation.

For the living bird, there would be no heat loss via radiation, conduction, or convection if the environmental or ambient temperature (T_a) were the same as the body temperature (T_b), say 40°C. Hence for thermoregulation, no metabolic heat would be required. In actual metabolic measurements in a resting bird there is no zero metabolism, but a minimal basal, or standard, metabolic rate ($\dot{H}_b$). This $\dot{H}_b$ represents the minimal cost of maintenance functions. The thermoneutral range is that T_a range in which $\dot{H}_b$ provides sufficient by-product heat to maintain an essentially constant body temperature. The thermoneutral range is bounded by the lower and upper critical temperature (T_{lc} and T_{uc}). Below the T_{lc}, metabolism must be increased to offset losses to a cooler environment.

The relationship of metabolic rate to T_a is usually approximated satisfactorily by a straight line.

$$\dot{H} = h(T_b - T_a) \qquad (1)$$

where h is a heat transfer coefficient, usually in calories per gram-hour-degree Centigrade (Burton, 1934). [This coefficient has subsequently been referred to as "thermal conductance" by many authors. Because this includes radiative and convective as well as conductive fluxes, we join Birkebak (1966) and Gagge et al. (1969) in reverting to the prior term.] The nonevaporative or dry heat losses are summarized by the Scholander model (Fig. 1) (Scholander et al., 1950a,b,c). If the heat loss is linearly proportional to $(T_b - T_a)$, the heat production requirement would theoretically be zero at $T_a = T_b$, thus providing a check on how well the model applies to a specific bird or how well the bird fits the model. Above T_{uc}, progressively

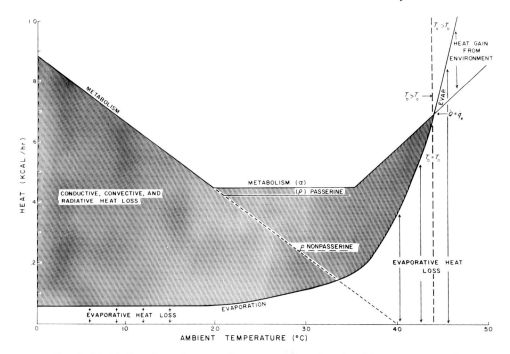

FIG. 1. Metabolic rate and evaporation rate as a function of ambient temperature for a hypothetical 32 gm passerine bird resting in a darkened metabolic chamber. α- and ρ-metabolism in thermoneutrality were predicted for the normal activity and rest periods, respectively, in a circadian cycle, from equations of Aschoff and Pohl (see text). Metabolic slopes below the thermoneutral range were predicted from the heat transfer coefficient for the 32 gm size, assuming that heat loss is proportional to the difference between a constant body temperature of 40°C and ambient temperature (Scholander model). The cross-hatching represents the composite of conductive, convective, and radiative heat losses, since heat loss must equal heat production if T_b is constant. At progressively higher T_a, the gradients for these losses become smaller. For thermoregulation, no metabolism is required at $T_a = T_b$, since there is no gradient for heat loss. This is the basis for the extrapolation of the subneutral metabolism slope to 0 kcal/hour. We will refer to the T_a at the end of this extrapolation as T_x (Section II,G). Because the production of heat does not go below the basal level, evaporation must dissipate a larger proportion of metabolic heat, aided by hyperthermia which tends to preserve a gradient for conductive loss. At 43°C, air temperature and hyperthermic body temperature are equal, so the only avenue for heat loss is by evaporation (see theoretical consideration of this oversimplification in Seymour, 1972). Above $T_a = 43°C$, heat flows inward, the heat gain being equal to evaporative heat loss minus metabolic heat production. For comparative purposes, (see p. 279), the standard metabolism, during the rest phase only, is shown for a 32 gm nonpasserine.

greater reliance is placed upon evaporative cooling. The metabolism is increased by the muscular effort of the necessary panting and the postural changes which facilitate heat loss, and by the Q_{10} effect of hyperthermia.

The shape of the metabolism–temperature curve is greatly influenced by bird size. The smaller the bird, the greater is its surface/volume ratio. Since heat loss occurs via the surface, but production is a function of mass or volume, the heat production per unit mass must be more intense as size decreases. The amount of insulative plumage a bird can transport is obviously limited by body size and has been quantified by Turček (1966; see also Brody, 1945, pp. 619f).

$$\text{plumage weight} = 0.09\, m^{0.95} \quad (2)$$

Since the weight of plumage is approximately a linear function of body weight [Eq. (2)], it may be shown on dimensional grounds alone that $P/A \propto m^{0.33}$, since $P \propto m^{1.0}$ and $A \propto m^{0.67}$, where P is plumage weight and A is body surface area. This prediction is approximated by Brody's (1945) empirical analysis of the data of Wetmore (1936) for small birds (3–120 gm) (see also Hutt and Hall, 1938).

$$\text{For males:} \quad P/A = 0.0060\, m^{0.32} \quad (3)$$

$$\text{For females:} \quad P/A = 0.0075\, m^{0.26} \quad (4)$$

where P/A is in grams per square centimeter. Thus, large birds have proportionately more plumage per unit surface area than small birds.

The discrepancies in exponents between theoretical and empirical estimates ($m^{0.33}$ versus $m^{0.32}$ and $m^{0.26}$) no doubt result from random variation, inaccuracy in estimating surface area, and in the fact that the empirical exponent for plumage weight as a function of body weight actually lies between 0.9 and 1.0 in most groups of birds excepting domestic fowl (Brody, 1945; Turček, 1966).

The manifestations of these scale effects seen in the metabolism curve are that a smaller bird will have a higher thermoneutral metabolism over a narrower thermoneutral range, and a steeper regression slope when exposed to temperatures below thermoneutrality. Numerous metabolic studies have provided sufficient data for allometric analysis of the relationship of metabolism and heat transfer coefficients to body size.

D. Standard Metabolic Rate

The standard, or basal, metabolic rate is the minimum rate of heat production in a resting bird in a thermoneutral environment, while not digesting or absorbing food (King and Farner, 1961). This quantity provides a reference basis for comparing birds with respect to size, phylogeny, state of acclimatization, and so on.

Empirical relationships of metabolic rate and body weight in birds have been derived by Brody and Proctor (1932) and King and Farner (1961), with additions and clarifications by Lasiewski and Dawson (1967) and Aschoff and Pohl (1970a,b). King and Farner (1961) had noted that species weighing less than about 100 gm did not conform very well with their general equation for all birds. Lasiewski and Dawson (1967) recognized that most of these smaller birds were passerines, and that their metabolic rates ($\dot{H}$) were significantly higher than those of nonpasserines, although having the same exponential relationship to body mass:

$$\text{Passerines:} \quad \dot{H}_b = 129 \, M^{0.724} \tag{5}$$

$$\text{Nonpasserines:} \quad \dot{H}_b = 78.3 \, M^{0.723} \tag{6}$$

The metabolic rate of passerines is thus estimated as $129/78.3 = 1.65$ times that of nonpasserines of the same weights. Zar (1968, 1969) further fragmented the data by orders and families and reported statistically significant differences in the elevations of the lines (metabolic rate per $kg^{0.75}$) among several orders. Metabolic rates for Falconiformes, Strigiformes, and Galliformes are lower than predicted from the nonpasserine Eq. (6), while Ciconiiformes, Columbiformes, Piciformes (data of E. Braun, personal communication), Anseriformes, and Apodiformes have higher values than predicted. The basic exponential relationship, $kg^{0.75}$, did not differ appreciably when the small size ranges represented are taken into account. The strength of allometric analysis lies in encompassing several orders of magnitude, such as hummingbird to ostrich, in which a size increase of four orders of magnitude minimizes the effects of errors of even 100% in metabolic measurements (see Kleiber, 1961, p. 207; Schmidt-Nielsen, 1970; Lasiewski and Calder, 1971). Such a range is not possible if regression lines are derived for each family or order separately, and the analysis thus loses power when the data are fragmented.

Lasiewski and Dawson (1967) considered that differences in standard metabolic rate between night and day in their data were insignificant, and they therefore combined data for both phases of the daily cycle. Aschoff and Pohl (1970a,b) have reemphasized the large differences in metabolic rates of resting birds between day and night, or between the activity (α) and resting (ρ) phases of the daily cycle. From a selection of data from the literature plus new data of their own, they analyzed separately the values for the activity and rest periods, with results shown in Fig. 2 and Table I. The equations at the left in Table I are the results of a reanalysis by Aschoff and Pohl (1970a) of the data selected by Lasiewski and Dawson (1967).

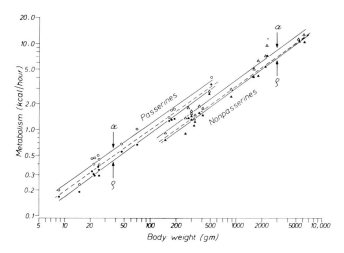

FIG. 2. The relationship of metabolism and body sizes in birds, showing differences between the Passeriformes and other birds, and between the normally active (α) and resting (ρ) portions of the circadian cycle. (From Aschoff and Pohl, 1970b, with permission.) The dashed lines correspond to Eqs. (5) and (6) derived by Lasiewski and Dawson (1967).

TABLE I
AVIAN STANDARD METABOLIC RATE IN RELATION TO BODY WEIGHT
AND PHASE OF THE DAILY CYCLE

	Rate[a] (number of species)			
	Lasiewski–Dawson[a]		Aschoff–Pohl	
Activity phase				
Passerines	$132.0 M^{0.73}$	(30)	$140.9 M^{0.704}$	(14)
Nonpasserines	$86.2 M^{0.72}$	(24)	$91.0 M^{0.729}$	(17)
Rest phase				
Passerines	$83.2 M^{0.613}$	(16)	$114.8 M^{0.726}$	(14)
Nonpasserines	$74.3 M^{0.76}$	(42)	$73.5 M^{0.734}$	(17)

[a] Metabolic rate in kilocalories per day, body weight in kilograms. Data of Lasiewski and Dawson (1967) reanalyzed by Aschoff and Pohl (1970a).

The Aschoff–Pohl equations confirm the Lasiewski–Dawson equations in revealing that metabolic intensity in passerines is 50–60% greater than in nonpasserines in both activity and rest periods of the daily cycle; but in addition, the new equations show clearly that metabolic rate is 20–25% greater, on the average, in quiescent birds during the day (α) than during the night (ρ). The reanalysis of the Lasiewski–Dawson data (Table I) follows the same trend, but the significance of quantitative comparisons among groups is not clear, since different sets of species are represented in the rest and the activity periods.

However ubiquitous, the empirical exponent of about ¾ (or its reciprocal counterpart −¼ as in gram-specific metabolic rate) was regarded vaguely as an apparent compromise between surface, $M^{0.67}$, and volume, $M^{1.0}$. McMahon (1973) has derived the ¾ exponent from arguments based on stability and flexure. It appears in metabolic studies throughout the animal kingdom (Hemmingsen, 1960), in mammalian metabolism and respiration (Kleiber, 1961; Stahl, 1967), in feeding territories of herbivorous but not predatory birds (Schoener, 1968), and in avian respiration (Lasiewski and Calder, 1971). We suggest at this point that allometric studies might be more profitably directed toward aspects of avian biology other than thermoneutral metabolism. The standard metabolism at 0°C, for example, has quite different weight relationships, $M^{0.42}$ for passerines, $M^{0.53}$ for nonpasserines (Kendeigh, 1969), no doubt because the given subneutral (0°C) temperature is not the same displacement from the lower critical temperature for all species. The analysis at 0°C therefore combines basal differences, insulative differences, and differences in displacement from thermoneutrality so that its interpretation is understandably intricate (see Section II, I).

E. CRITICAL TEMPERATURES

We might expect that the lower critical temperature would be inversely related to body size and the thermoneutral range directly related to size. The larger bird has a more favorable surface/volume ratio and more insulation to manipulate before energy reserves need be tapped to offset heat losses. Such is theoretically the case for mammals (Morrison, 1960). Unfortunately, the lower critical temperature is an imprecise entity. Anything that would elevate metabolism above basal (psychic, circadian, circannual, or experimental artifact) would give an apparently lower value for T_{lc}. Variability in calculated regression slopes for metabolism below thermoneutrality would also affect the T_{lc} intercept. The temperature of acclimation or acclimatization may also influence T_{lc} (Dawson, 1958; Gelineo, 1964; Pohl, 1969). Differences between species of the same size which are adapted to different environments would also cause variability in the apparent

T_{lc}. For example, the Snow Bunting (*Plectrophenax nivalis*, 43 gm) has a T_{lc} of 9°C, while the Common Cardinal (*Cardinalis cardinalis*, 41 gm) has a T_{lc} of 18°C (Scholander *et al.*, 1950a,b,c; Dawson, 1958). The Gray Jay (*Perisoreus canadensis*, 72 gm) has a T_{lc} of 6°C, compared with 18°C for the Blue Jay (*Cyanocitta cristata*, 81 gm) (Veghte, 1964; Misch, 1960). The highly aquatic American Dipper (*Cinclus mexicanus*, 50 gm) of cold mountain streams has a T_{lc} of 11.5°C, 4.5°C below that of the nonaquatic Evening Grosbeak (*Coccothraustes vespertinus*, 55 gm) (Murrish, 1970; Dawson and Tordoff, 1959). It is perhaps a consequence of this that no clear allometric relationship for T_{lc} is possible at this time.

Thus, although common sense tells us that the smaller the bird, the smaller its thermal inertia, and hence the smaller its zone of thermoneutrality, the determination of the T_{lc} incorporates variabilities in basal metabolism, the cold-stress regression line, and any possible adaptation to the environment, so that it is difficult to show the expected correlation between bird size and T_{lc}.

F. Heat Transfer Coefficient

Exposed to temperatures below thermoneutrality, a bird must increase metabolic heat production by shivering, by nonshivering calorigenesis, or by overt muscular activity. The relationship between resting metabolism and temperature is expressed by the heat transfer coefficient, which is calculated as

$$h = \dot{H}/(T_b - T_a) \tag{7}$$

The coefficient h includes (and may conceal from the unwary) all of the specific transfer coefficients for the various routes of heat transfer (h_r, h_c, h_k, h_e), not all of which are linear. The numerical values and dimensions of these quantities depend not only on basic physical properties, but also on factors in an animal's environment and on the animal's geometric coupling to these factors. In the metabolism chamber, the environmental factors are comparatively homogeneous and the bird's relationship to them varies minimally. Precaution must be taken in experimental design so that the use of the total heat transfer coefficient, h, becomes a useful and manageable empirical expression for comparative purposes. Heat radiated from the bird can be reflected back to the bird by shiny chamber walls, thereby artificially reducing metabolic heat requirements (Porter, 1969). Air-flow rates must be high enough to preserve normal vapor pressure differences (Lasiewski *et al.*, 1966a).

The analytical power of the empirical heat transfer coefficient is improved if the component of evaporative heat transfer (h_e) is factored out by subtracting evaporative heat loss from heat production, leaving the components of dry heat transfer (h_d), as prescribed by King and Farner (1964):

$$h_d = \frac{\dot{H} - \dot{H}_e}{T_b - T_a} \tag{8}$$

Calculated with the difference in temperature from body core to the environment, this encompasses the heat flux through the resistance of core, superficial musculature, skin, feathers, and boundary air layer. (No such reliance on deep-core temperature is included when the regression line slope is used directly as an expression of the heat transfer coefficient; e.g., Lasiewski et al., 1964, 1967.) The temperature or potential drop is probably slight across the first three thermal resistances, compared with the resistances of the feathers and boundary air, but quantitative studies are seriously needed here. Until there are extensive data partitioning the heat loss path into its components, the customary use of the deep body (usually cloacal or proventricular) temperature to calculate a heat transfer coefficient for the total path, as in the Scholander model, is an acceptable first approximation.

Herreid and Kessel (1967) determined the heat transfer coefficient from cooling curves of carcasses of thirteen species of birds as follows (h in calories per gram-hour-degree centigrade):

$$h = 4.57\, m^{-0.52} \quad \text{with plumage intact} \tag{9}$$

$$h = 7.24\, m^{-0.44} \quad \text{with plumage removed} \tag{10}$$

Thus, the feathers appear to reduce the heat loss by 37% for a theoretical 1 gm bird, and by a larger percentage with increasing size. Herreid and Kessel obtained an equation similar to Eq. (9) when they added published values from an additional eighteen species (mostly from metabolic determinations):

$$h = 4.59\, m^{-0.54} \tag{11}$$

Lasiewski et al. (1967) obtained a quite similar expression from a regression analysis of thirty-five species, metabolically determined:

$$h = 4.08\, m^{-0.51} \tag{12}$$

It is interesting to note that the birds utilized by Herreid and Kessel (1967) for Eq. (9) were all collected near Fairbanks, Alaska. The heat transfer coefficients from these Alaskan carcasses were not lower than those of predominantly temperate-zone species used in the analysis by Lasiewski et al. (1967). If the data from two different methods of determination are properly comparable, this implies that insulation is optimized at a maximum practical value in lower latitudes and is not a factor for significant adaptive increase in or near arctic environments. A different conclusion is that of Drent and Stonehouse (1971), who point out a number of comparisons between birds of similar size but different habitats that suggest adaptive differences in insulation. Pitelka (1962) noted that arctic finches have dense down feathers where tropical forms have almost bare apteria.

Since the power equation for metabolic rate in passerine birds is distinctly greater than, but parallel to, that for nonpasserines, perhaps heat transfer coefficients should be analyzed separately for these groups also. Combining data from Herreid and Kessell (1967), Lasiewski et al. (1967), and eight species published since that time, converted to common units, we find the following.

$$\text{Passerines:} \quad h = 4.55\, m^{-0.54} \tag{13}$$

$$\text{Nonpasserines:} \quad h = 4.06\, m^{-0.54} \tag{14}$$

Logarithmic transformations and expressions of statistical variability appear in Table XVII at the end of this chapter.

Kleiber (1970) has argued that conductance (heat transfer coefficient) should be expressed in physically correct dimensions. In conformity with this argument, we convert h to $h' = \text{cal/hour cm}^2\,°C$. The surface area of birds is generally assumed to be

$$A = 10\, m^{0.67} \tag{15}$$

This relationship has been verified to be a very good estimator of the body surface area beneath the feathers of birds (Drent and Stonehouse, 1971). Leighton et al. (1966) found that surface area within a strain of domestic fowl was proportional to $M^{0.73}$ to $M^{0.75}$, rather than $M^{0.67}$. These power functions, however, included the additional variable of growth, spanning an age range of 1 day to 37 weeks, and we regard them as a special case. Therefore

$$h' = h\, m / 10\, m^{0.67} \tag{15a}$$

for passerine birds:

$$h' = 0.455 m^{-0.21} \qquad (13a)$$

for nonpasserine birds:

$$h' = 0.406 m^{-0.21} \qquad (14a)$$

A regression for eighty-three data pairs for all birds and corrected according to Eq. (5) for evaporation from Table 2 of Drent and Stonehouse (1971) gives lower values and a smaller effect of body size (shallower slope):

$$h' = 0.322 m^{-0.15} \qquad (14b)$$

Note that for use in the Scholander Eq. (1), these values must be multiplied by the surface area. Equations (13) and (14) do not differ significantly, so the insulative properties of passerine plumage do not seem to differ from those of other birds, in contrast to the differences in the underlying metabolism. These allometric equations for metabolism and heat transfer may be used together for extensive predictive purposes. For example, hypothetical metabolism curves, $\dot{H} = f(T_a)$, can be drawn based upon basal metabolism and the slope of the metabolism curve below T_{lc}, as suggested by Lasiewski et al. (1967). They demonstrated a strong similarity between the predicted curves to actual determinations of four species of small birds.

G. Critical Temperatures and Extrapolations to Zero Metabolism

The Scholander model includes no implicit predictions about the level of the upper critical temperature, but requires that the line relating metabolic rate to air temperature extrapolates on the abcissa to a value of T_a equivalent to T_b at $\dot{H} = 0$. We will call this extrapolated temperature, T_x (Fig. 1). The intersection of this line with mean standard metabolic rate arbitrarily fixes the lower critical temperature.

It is evident from a survey of the literature (King, 1964) that the extrapolated temperature, T_x, frequently falls above or below the predicted point (T_b), and in some cases may diverge very widely (e.g., in the domestic pigeon, $T_x = 61.9°C$ when $T_b = 39.7°C$ (Calder and Schmidt-Nielsen, 1967). However, the heat loss is not from the core temperature, but from the skin or feather temperature to the environmental temperature. During exposure to cold, the difference between body surface temperatures and deep-body temperatures is greater. It is also greater the larger the bird (Veghte and Herreid, 1965) (see Fig. 3). Thus, we might expect that the larger the bird, the

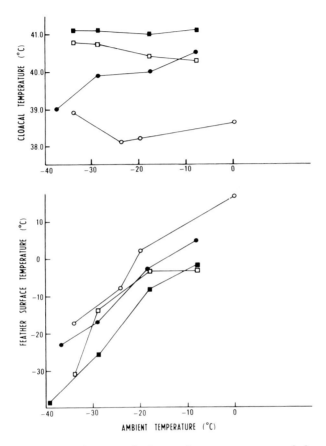

FIG. 3. A comparison of average feather surface temperatures and cloacal temperatures of four species of birds exposed to cold air temperatures. ○ = Black-capped Chickadee, ● = Gray Jay, □ = White-tailed Ptarmigan, ■ = Raven. (Reprinted from Veghte and Herreid, 1965, with permission.)

higher the T_a of the zero metabolism extrapolation (T_x). On the basis of data presently available, this does not seem to be the case. A least squares analysis of thirty-four species ranging from hummingbirds $(T_x = T_b)$ to pigeon (61.9°C versus 39.7°C) gives the relationship (see Table XVII)

$$T_x = 40.08\, m^{0.04} \qquad (16)$$

The slope of this curve is not significantly different from zero, so that 40.1°C can be regarded as a weight-independent constant ($m^0 = 1$).

This T_x satisfactorily approximates the resting T_b for most species of birds (King and Farner, 1961; Neumann et al., 1968; Dawson and Hudson, 1970). The conformity helps to verify the applicability of the model to birds as a group, but the divergences also emphasize the need for caution in the analysis or interpretation of individual sets of data in the context of the model.

The allometric reliability of the Scholander model in predicting realistic values of T_x allows an extension to predictions of the arbitrary T_{lc} as a function of body mass. At the lower critical temperature

$$\dot{H} = h\,(T_b - T_{lc}) \tag{1}$$

and the power functions can be substituted like individual values into the equation (see Stahl, 1967; Calder, 1968b; Lasiewski, 1972). Because of group-specific differences in standard metabolic rate (Table I) and slight differences in the heat-transfer coefficient [Eqs. (13 and 14)], passerines and nonpasserines are considered separately.

For passerines, the Aschoff–Pohl equation for metabolic rate in night-resting birds may be converted to units of calories, grams body mass, and hours and combined with Eq. (13) for the heat transfer coefficient to yield

$$31.7 m^{-0.274} = 4.55 m^{-0.54}(T_b - T_{lc}) \tag{17}$$

$$(T_b - T_{lc}) = 6.98 m^{0.266} \tag{18}$$

Similarly, for nonpasserines

$$19.2\,m^{-0.266} = 4.06\,m^{-0.54}(T_b - T_{lc}) \tag{19}$$

$$(T_b - T_{lc}) = 4.73\,m^{0.274} \tag{20}$$

If T_b is essentially independent of body size, as it seems to be when examined broadly through the avian class and the "noise" of fluctuations and measuring techniques, then the critical difference ($T_b - T_{lc}$; or the "critical gradient" of Scholander et al., 1950a) is strongly mass-dependent (Table II). The slopes for both passerines and nonpasserines, $m^{0.276}$ and $m^{0.274}$, are similar to that for mammals, $m^{0.25}$ (Morrison, 1960).

If we are correct in concluding that the insulative effectiveness and

TABLE II
PREDICTED CRITICAL GRADIENT ($T_b - T_{lc}$) AND LOWER
CRITICAL TEMPERATURE (T_{lc}) IN RELATION TO
BODY WEIGHT IN NONPASSERINE BIRDS

Body weight (gm)	$T_b - T_{lc}$ (°C)	T_{lc} (°C)[a]
1	4.7	35.3
10	8.9	31.1
100	16.7	23.3
1,000	31.4	8.6
10,000	59.0	−19.0
100,000	110.9	−70.9

[a] Assuming $T_b = 40°C$.

body temperatures of passerines and nonpasserines are essentially the same, but that these groups differ in standard metabolism, then we would expect a higher T_{lc} for a nonpasserine bird, on an equal-size basis. That is to say, that the line from $T_a = T_b$ to $T_a = 0°C$ would be the same, but a lower H_b line would intercept with that regression at a higher T_a (see Fig. 1). This is what Eqs. (17) and (19) say, also.

H. BODY TEMPERATURE

The body of a homeotherm comprises a series of isotherms, which have been broadly subdivided as the deep-body or "core" temperature, and the "shell" temperature consisting of the skin, feather surface, and peripheral temperatures. The core temperature is the most stable, being independent of a wide T_a range, and is usually recorded as T_b from proventricular or cloacal probes or quick-registering mercury thermometers. Especially in studies of the thermal problems of flight and the physiology of shivering, deep pectoral temperatures are determined from surgically implanted thermocouples or thermistors. The difference ($T_b - T_a$) diminishes with distance from the core, and the skin temperature (T_s) approaches T_a, more so in larger birds, with more insulation between skin and outer surface (see Fig. 3) as determined with infrared radiation thermometers or thermal scanning (Fig. 4).

It is possible to establish "standard" conditions for obtaining comparative data for various species (King and Farner, 1961), but these conditions are arbitrary, they have only an enigmatic relationship to the functional status of the animal in a natural environment, and it is questionable that they are really "standard" when large-bodied and

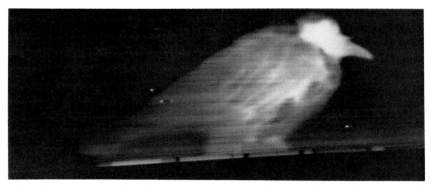

FIG. 4. Distribution of surface temperature on a Common Raven (*Corvus corax*) exposed to cold, as surveyed by an infrared scanning radiometer. The highest temperatures are shown in lightest tones. Per unit of surface, the greatest heat efflux is thus from the head and bill. (Photograph courtesy of Lt. Col. James H. Veghte.)

small-bodied birds are compared. Because of the slight thermal inertia but very intense metabolism of small birds, handling and cloacal or proventricular probing ("grab and jab") for temperature measurements can cause a rapid increase from the prehandling state (Fig. 5) (see also Dawson, 1954, p. 88). There is in addition some preliminary evidence that cloacal ("core") temperature as measured by probe in two species of birds is about 0.3° to 0.9°C lower than the simultaneous intraperitoneal temperature measured by an indwelling telemeter (Coulombe, 1970; Southwick, 1971).

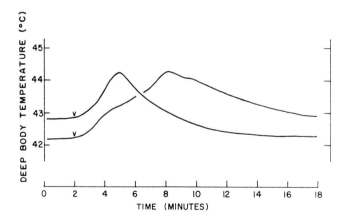

FIG. 5. Time course of telemetered intraperitoneal temperature in two *Zonotrichia leucophrys gambelii* following about 5 seconds of disturbance by observer (indicated by arrowheads); $T_a = 28°C$. (From Southwick, 1971.)

Thus, it appears that the range of variation of body temperature as measured by conventional methods in hand-held birds is potentially at least as great as the apparent range of variation among species. To this source of variation must be added the errors resulting from comparisons of individuals measured in different phases of the daily metabolic cycle. Just as there is a difference in diurnal and nocturnal metabolic rates (King and Farner, 1961; Aschoff and Pohl, 1970a,b), there is also a pronounced cycle in body temperatures, with diurnal temperatures 1°–3°C higher than diurnal temperatures in normothermic birds (e.g., see Dawson, 1954, 1958; Veghte, 1964; MacMillen and Trost, 1967; Aschoff and Pohl, 1970a,b; Coulombe, 1970; also Tables VIII, IX, and X in King and Farner, 1961). Small sunbirds, which experience large diurnal T_a oscillations in the mountains of Kenya, experience T_b cycles of 5° to 17°C without entering a true torpor (Cheke, 1971). In birds that become torpid at night, the extent of the cycling, is, of course, much more pronounced (see Section V). Generalizations about body temperatures must therefore be based upon data selected carefully with definite criteria as to the state of the birds to which the generalization applies, however arbitrary these criteria may be.

When measured with minimal disturbance during the night or the resting portion of the circadian cycle, T_b for birds appears to be nearly independent of body size, averaging about $40° \pm 1.5°C$, and thus slightly higher than the "typical" T_b of mammals. The T_b values for small, resting passerines are scarcely higher than the Ostrich's 39.2°C. [See Table VIII in King and Farner (1961), Table 59, part II, in Altman (1968), and Table I of Dawson and Hudson (1970). We will not attempt to retabulate these extensive collections.] Furthermore, normal body temperatures of desert and arctic birds do not appear to have been adaptively modified from the T_b range of birds in temperate climates (Dawson and Schmidt-Nielsen, 1964; Irving and Krog, 1954). Orders of birds that seem to have slightly lower temperatures are flightless or engage in hypothermic dormancy (Dawson and Hudson, 1970) or are water birds and hence associated with a cooler and more conductive medium. The penguins, ratites, and Procellariiformes do have resting T_b values slightly lower than other birds (Boyd and Sladen, 1971; Warham, 1971). On the other hand, the lower T_b values of penguins may be due only to the site of measurement, for gullet and stomach temperatures are higher in the Adélie Penguin (*Pygoscelis adeliae*) and the Humboldt Penguin (*Spheniscus humboldti*) (Goldsmith and Sladen, 1961; Drent and Stonehouse, 1971). If gullet and stomach temperature is taken as a more representative core T_b, then penguins are quite similar to other birds (38° to 40°C). These are phylogenetic variations in the class Aves, as contrasted to the hypothermia and hyperthermia exhibited by some birds, which are considered to be physiological adjustments away from the normothermic state.

Having documented the generalization that the normal T_b of a resting bird is approximately 40°C, we now seek an explanation for the natural selection of such a high T_b. This is maintained near the maximum tolerable by the steady-state turnover of body protein (Fig. 6), only 4.5° to 7°C below lethal limits (Morowitz, 1968; Dawson and

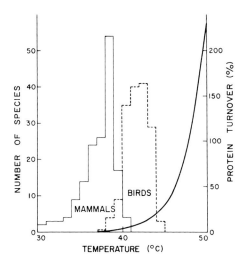

FIG. 6. Regulated body temperature in birds and mammals in relation to percent of body protein turnover per day. (From Morowitz, 1968, with permission.)

Schmidt-Nielsen, 1964). In intense activity or adaptive hyperthermia, the margin between vital and lethal is even less. On the other hand, the energetic cost of homeothermy is proportional to the difference $(T_b - T_a)$. The mean annual temperatures of even the tropics are only 25° to 27°C (Finch and Trewartha, 1949; Sellers, 1965). Thus, the mean temperature difference between bird and tropical environment is approximately 14°C. The average global air temperature at sea level is 15°C, and the earth's surface temperature is 13° to 14°C (Sellers, 1965; Geiger, 1966). Birds thus face, on the average, a temperature difference $(T_b - T_a)$ of 26°C.

From information on other animals, it would appear that the birds probably had a wide option for natural selection of T_b. The range of temperature tolerance for the protoplasm of normal, active animals extends over slightly more than 50°C, from that of supercooled fish in polar waters of $-1°C$ to $-2°C$ to desert arthropods and reptiles with high lethal temperatures extending to 50°C in some species (Gordon, 1964; DeVries and Wohlschlag, 1969; Cloudsley-Thompson, 1964; Schmidt-Nielsen and Dawson, 1964). Even within individuals of poikilothermic species, there can be thermal acclimation over a considerable range, accomplished by such mechanisms as isozyme induction (for review, see Fry and Hochachka, 1970). So the coevolution of thermoregulation and enzyme function would have been theoretically possible at other levels in the 50°C range, especially for temperatures displaced farther from the heat-death point.

The immediate needs for survival of the individual are food and water. Any adaptation that conserves energy or water facilitates the balancing of gain versus loss. The curves that relate metabolic rate and evaporation rate to ambient temperature have qualitatively very different shapes (Fig. 1), so that simultaneous optimization of the economy of both energy and water is impossible. To maintain a steady T_b, heat loss and heat gain must be equal. The high rates of metabolic heat production of homeotherms must be equalled by heat dissipation. We can conceive of two alternatives: (1) regulation at a lower T_b would be metabolically economical or (2) regulation at a high T_b would assist water economy.

Compare the economics of two hypothetical birds with $T_b = 40°C$ and 26°C (Fig. 7) both with the same basal metabolism and insulative properties predicted for 32 gm body mass, and both fitting the Scholander model. The vertical differences between metabolic curves below $T_a = 20°C$ (cross-hatched) show the energy savings of the hypothetical 26°C T_b compared to a 40°C bird, a maximum of 68% of basal metabolism for this reduction in heat loss potential. If, for this energetic economy, T_b were regulated close to mean air temperature, the small gradient for heat loss would necessitate high evaporation rates (Fig. 7). The vertical differences between evaporation curves (stippled) are proportional to the water savings of a 40°C bird— approximately 1.5 gm/hour when $T_a = 33.5°C$ and increasing to a maximum at higher T_a.

Even more dramatically during heat stress, the need for evaporative cooling is minimized by having a high T_b. This preserves a $T_b > T_a$ for nonevaporative heat loss. In extreme heat stress, where T_a has overtaken and surpassed T_b, the difference, and therefore the heat gain from the environment, is reduced by having T_b as high as it can be tolerated and safely regulated.

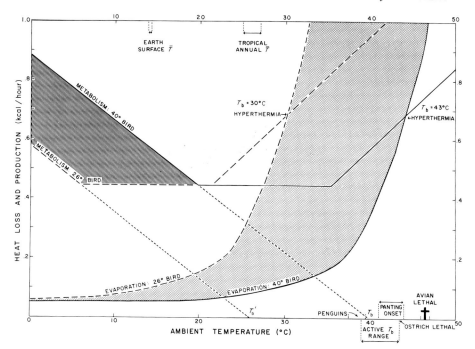

FIG. 7. A comparison of metabolic and evaporative costs of maintaining $T_b = 40°C$ and a hypothetical $T_b' = 26°C$. Metabolism and evaporation of the bird with $T_b = 40°C$ (solid lines and dotted extrapolation to $T_b = 40°C$) are those shown in Fig. 1. With the same basal metabolism and heat transfer coefficients, but regulation of $T_b' = 26°C$, the curves (broken lines) are shifted to the left. Note that a T_b' of 26°C corresponds to the mean annual T_a of tropical latitudes, or 12°C above mean surface temperature of the earth. The energetic savings of regulating $T_b = 26°C$ versus 40°C are represented by the cross-hatched difference between metabolic curves, while the water savings of regulating $T_b = 40°C$ versus 26°C are equal to the vertical difference between evaporation curves (stippled) and their extrapolations beyond the upper margin, divided by 0.58 kcal per gram of water.

The evolution of terrestriality made possible the exploitation of new food sources which permitted higher metabolic rates. However, this necessitated exposure to the threat of desiccation. Because the homeotherms have high metabolic rates, they must have high ventilation rates, a serious source of water loss via exhaled humidified air. We suggest, therefore, that by regulating high T_b values, natural selection opted for water economy instead of energy economy.

I. Reconciliation of Empirical and Theoretical Treatments

The validity of the Scholander model for large-scale comparisons can be examined by simultaneous solutions of two empirical allometric expressions, independently derived, and by substituting these expressions for variables in equations based on the model. The allometric terms on the two sides of the resulting equations should agree as to both coefficient and exponent, thus reconciling empirical and theoretical requirements.

Kendeigh (1969) has calculated regression lines relating body weight in a selection of avian species to resting metabolic rate at $T_a = 0°C$ and in the thermoneutral zone. He gives the following equations (in kcal/gm day).

Passerines

At $T_a = 0°C$: $\dot{H} = 4.769\, m^{0.417}$ (21)

At thermoneutrality: $\dot{H} = 0.882\, m^{0.693}$ (22)

Nonpasserines

At $T_a = 0°C$: $\dot{H} = 3.342\, m^{0.526}$ (23)

At thermoneutrality: $\dot{H} = 0.513\, m^{0.717}$ (24)

Realizing the uncertainty of extrapolations beyond the data (in the case of nonpasserines), we can nevertheless expect that the lines for metabolic rate at $T_a = 0°C$ and at thermoneutrality would intersect, if the Scholander model is suitable, at the body mass of a bird in which $T_{lc} = 0°C$. Thus, for passerine birds

$$4.769\, m^{0.417} = 0.882\, m^{0.693} \quad (25)$$

$$\text{Body mass} = 452.5 \text{ gm} \quad (26)$$

Equation (18) predicts that a bird of this size would have a critical gradient $(\Delta T = T_b - T_{lc})$ equal to:

$$T_b - T_{lc} = 6.98(452.5)^{0.266} = 35.5°C \quad (27)$$

This shows reasonably good agreement with the expected critical gradient (40°C) in view of the many possible sources of variation in these separate approaches.

The analogous operation predicting body weight for nonpasserines is

$$3.342\, m^{0.526} = 0.513\, m^{0.717} \quad (28)$$

$$\text{Body mass} = 18{,}204 \text{ gm} \quad (29)$$

For this bird, Eq. (20) predicts the critical gradient as:

$$T_b - T_{lc} = 4.73(18{,}204)^{0.27} = 66.4°C \tag{30}$$

The expected critical gradient is again 40°C. This larger disagreement is not surprising, since the intersection of Kendeigh's equations is considerably beyond the range of the data. Equations (23), (24), and/or (20) require further refinement.

A second examination of the extent to which the model accommodates empirical data is based on the fact that our Eqs. (13) and (14) multiplied by Eq. (16) (using T_x as T_b) should equal Eqs. (21) and (23) when converted to the same units. The results are shown in Table III. Again, considering the opportunities for variability, the slight numerical disagreement is not particularly surprising. Stahl (1967) regards residual mass exponents as probably not significant if less than 0.08. The residual mass exponents shown in Table III are 0.04 and −0.07. The coefficients of the equations differ by only 9% and 17% for passerines and nonpasserines, respectively.

TABLE III
RECONCILIATION OF EMPIRICAL AND THEORETICAL EQUATIONS ACCORDING TO THE SCHOLANDER MODEL

Group	Empirical metabolic rate[a] (cal/gm hour)	Predicted metabolic rate[b] (cal/gm hour)	Predicted/empirical[c]
Passerines	$198.7\,m^{-0.58}$	$181.9\,m^{-0.54}$	$0.91\,m^{0.04}$
Nonpasserines	$139.3\,m^{-0.47}$	$162.5\,m^{-0.54}$	$1.17\,m^{-0.07}$

[a] At $T_a = 0°C$; from Kendeigh (1969), with units converted.
[b] At $T_a = 0°C$, for passerines, from Eqs. (13) and (16): cal/gm hour $(4.55\,m^{-0.54})(40° - 0°)$; for nonpasserines, from Eqs. (14) and (16): cal/gm hour $(4.06\,m^{-0.54})(40° - 0°)$.
[c] Exponents are residual mass exponents.

We conclude from these results that current allometric expressions of empirical data, while in need of improvement in some cases, are reasonably good predictors of avian function, and that the Scholander model provides an adequate comparative framework for data obtained in a simplified laboratory environment. Other discussions of heat loss from an allometric viewpoint have been given by McNab (1970), Yarbrough (1971), Drent and Stonehouse (1971), Calder and King (1972), Calder (1972), and Kendeigh (1972). We turn attention next to a more exacting appraisal of the complexities of heat transfer.

III. Basic Principles of Heat Transfer

We have examined in the preceding section the relation of metabolic rate and body temperature to ambient temperature in a laboratory situation, unburdened by environmental complexity. We must now attempt to bridge the gap between the laboratory and the components of real microenvironments. This brief account is based mainly on viewpoints and treatments presented by Kleiber (1961), Birkebak (1966), Gates (1962), Porter and Gates (1969), and Gebhart (1971), to which the reader may refer for full details.

The relatively stable internal temperature of a bird represents a steady state between heat input and heat output. The routes of heat gain and loss can be summarized as:

$$H_m \pm H_r \pm H_c \pm H_k - H_e = H_s = 0.83m\,(\Delta T_b) \qquad (31)$$

or metabolic calorigenesis plus or minus net gains or losses via radiation, convection, and conduction, minus evaporative heat loss, equals storage heat (if any), which is specific heat (Hart, 1951) times body weight times change in T_b. Note that the storage heat is zero when T_b is constant; but if the body is becoming hyperthermic in heat stress or exercise, or hypothermic as in the development of torpor, the storage term must be taken into account.

The metabolic heat (H_m) and evaporative heat loss (H_e), although influenced by air temperature and vapor pressure, are internal components. The transfer of heat by the remaining routes — radiation, convection, and conduction — may be either inward or outward depending on the organism's surface properties in relation to ambient conditions. These relationships are embodied in several physical coefficients that define the functional interface between an organism and its physical environment. The terms H_r, H_c, H_k, and H_e in the heat-balance equation are functions of the basic dimensions of the "climate space" of Porter and Gates (1969). The coefficients and the functional relationships now to be discussed thus assume immediately an ecological significance.

A. Energy Transfer by Thermal Radiation

Energy is transmitted through space as electromagnetic radiation. The electromagnetic spectrum includes wavelengths from about $10^{-6}\,\mu$ to $10^8\,\mu$. The segment from 0.4 μ to about 100 μ is called thermal radiation and is subdivided into visible (0.4–0.7 μ) and invisible or near-red and infrared (>0.7 μ) parts on the basis of human perception. Organisms are coupled to their radiant energy environments through the emission properties of their surfaces and through the absorption properties of these surfaces in relation to the characteristics of incident radiation. The fundamental relationships are embodied in several classic physical laws that can be applied as limiting statements for living systems. For more detailed discussion of these principles we refer to Gates (1962), Reifsnyder and Lull (1965), Birkebak (1966), and Lowry (1969).

Emission of Thermal Radiation

Solids and liquids at any temperature above absolute zero emit radiation in proportion to the fourth power of their absolute temperature, as defined by the Stefan–Boltzmann equation:

$$\dot{H}_r = \epsilon \sigma T^4 \qquad (32)$$

where σ is the Stefan–Boltzmann proportionality constant, with a numerical value varying with units chosen and the source, e.g. 8.13×10^{-14} kcal/(cm² minute °K⁴) or 1.36×10^{-15} kcal/(cm² second °K⁴), and ϵ is the emissivity, a surface property of the object or substance. Thermal radiation of a given wavelength (monochromatic) falling on an object is reflected, absorbed, and/or transmitted. Since birds are opaque, the transmissivity is zero, and the incident radiation equals the fraction reflected plus the fraction absorbed, none being transmitted. A corollary of Kirchoff's law is that a good absorber is a good emitter *at the same wavelength*. Thus for one wavelength, ϵ = absorptivity = emissivity = 1 − reflectivity. A nonreflective black body is a perfect absorber and therefore a perfect emitter in the spectral range of its nonreflectivity.

The reflectivity refers to specific monochromatic radiation while the term "albedo" is used to designate the reflected portion of a wide spectral range, such as the visible portion of the solar spectrum. It is most commonly used with respect to environmental surfaces. Albedos of representative natural surfaces are found in Table IV.

It is important to note the relationships between: (1) the spectral distribution of radiation and the temperature of the radiator, and (2) the spectral distribution and the ϵ values of radiator and receiver. Wien's law states that the wavelength (λ) (in microns) of peak emissive power is related to the absolute temperature of the radiator as follows.

$$\lambda_{max} = 2897/T \qquad (33)$$

The lower the temperature of a radiator, the longer the wavelength

TABLE IV
ALBEDOS OF REPRESENTATIVE NATURAL
SURFACES WITH SUN AT ZENITH[a]

Surface	Albedo
Snow, fresh surface	75–95
Sand dune, dry	35–45
Sand dune, wet	20–30
Soil, dark	5–15
Soil, gray, or dry clay	10–20
Chaparral	15–20
Meadow, green	10–20
Tundra	15–20

[a] From Sellers, 1965; Geiger, 1966.

or lower the frequency of the peak radiation, and the spectral range of the radiated energy is shifted with this peak.

The plotting of energy or energy-related variables as a function of wavelength, as is customary with biologists, results in distortion of spectral energy distributions (Gates, 1962; Wald, 1965; Lowry, 1969). This is because energy is related directly to frequency, the energy of a photon being equal to Planck's constant (6.62×10^{-27} erg seconds) times the frequency. The frequency is equal to the speed of light divided by the wavelength. The traditional plot of spectral intensity as a function of λ makes the peak intensity of solar radiation appear as green light (0.47 μ), when the true peak actually occurs in the near infrared at 1.0 μ, with approximately 50% of solar radiation falling in the infrared, 25% in the visible red to yellow-green, and 25% in the green through ultraviolet.

To avoid the distortion of a λ plot, we will follow the recommendations of Gates (1962) and Wald (1965) and designate regions of the spectrum by wave numbers (λ^{-1}) (λ in millimeters), but to aid in this transition each value will be followed by the more familiar λ (in microns). Wien's law thus can be stated in terms of the wave number of maximum radiant energy as

$$\lambda^{-1} = T/5.099 \tag{33a}$$

while the frequency that divides the spectrum into two equal-energy halves is

$$\lambda^{-1}_{1/2} = T/4.110 \tag{33b}$$

(modified from Reifsnyder and Lull, 1965).

The reflectivity and the emissivity (= absorptance) of bird plumages bear a much more complex relationship to the spectral range, although measurements are in their infancy. These measurements are of two basic kinds. Broad spectrum measurements have provided information on mean solar reflectance (Table V) and total infrared

TABLE V
MEAN SOLAR REFLECTANCE (ρ) OF BIRD SKINS OF REPRESENTATIVE SPECIES[a]

Species	Reflectance		
	Wing	Breast	Mid-dorsal
Canada Goose, Branta canadensis	0.15	0.35	–
Mourning Dove, Zenaida macroura	0.30	0.39	–
Purple Martin, Progne subis	0.26	0.20	–
Red-eyed Vireo, Vireo olivaceus	0.29	0.54	–
Common Grackle, Quiscalus quiscula	0.25	0.21	–
Common Cardinal, Cardinalis cardinalis	0.23	0.40	0.24
Glaucous-winged Gull, Larus glaucescens	–	–	0.52
Bobwhite Quail, Colinus virginianus	–	–	0.22
European Starling, Sturnus vulgaris	–	–	0.15
Eastern Meadowlark, Sturnella magna	–	–	0.22

[a] From Birkebak (1966); mid-dorsal data from Porter and Gates (1969).

emissivity (Table VI). More recently, spectrophotometric determinations have described reflectance as a function of specific wavelengths in the visible and near-infrared ranges (Fig. 8).

From the broad-spectral measurements, we can generalize that the common birds reflect approximately one-fourth of incident solar radiation. Dark plumages, such as those of the European Starling (*Sturnus vulgaris*) and the wings of the Canada Goose (*Branta*

TABLE VI
TOTAL INFRARED EMISSIVITY (ϵ) OF REPRESENTATIVE ANIMAL SURFACES

Species	Conditions	Emissivity	Reference
Willow Ptarmigan (*Lagopus lagopus*)	Plumage, $T_s = -20°C$	0.98	Hammel, 1956
Red Fox (*Vulpes fulva*)	Dorsal fur, $T_s = -20°C$	0.98	Hammel, 1956
Gray Squirrel (*Sciurus carolinensis*)	Dorsal fur, $T_s = 38°C$	0.99	Birkebak, 1966
Human	Bare skin, $T_s \approx 32°C$	≥ 0.99	Hardy, 1939; Buchmüller, 1961; Mitchell, 1970

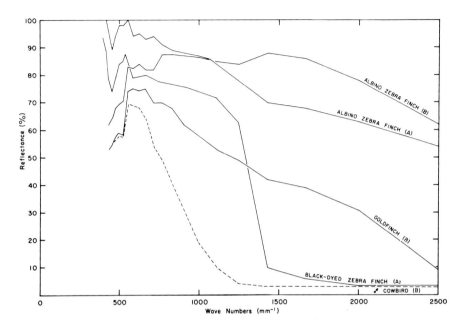

FIG. 8. Reflectance as percentage of incident radiation as a function of wave number. The only information for bird plumage in the infrared below 400 mm^{-1} ($\lambda = 2.5$ μ), which would constitute the emission from bodies of 0°–40°C according to Eq. (33a), indicates reflectance of 0–2% (see Table VI). (A) Data from Heppner (1970) and personal communication); (B) Data from Lustick (1971 and personal communication).

canadensis), reflect only 15%, while the Glaucous-winged Gull (*Larus glaucescens*) reflects 52% of the solar radiation (Birkebak, 1966; Porter and Gates, 1969).

There is much less information on the total infrared behavior of avian plumage, but measurements from fowl and ptarmigan further the assumptions from mammals that animal surfaces behave as black bodies in the infrared, with emissivity essentially 1.0 (Table VI).

Spectrophotometric studies give a more complex picture, with reflectances of 3–15% from dark-colored birds and 9–88% in light-colored to albino birds at 2500 mm^{-1} (0.4 μ) and 1429 mm^{-1} (0.7 μ), respectively, in the visible spectrum (Heppner, 1970; Lustick, 1969, 1971). The iridescent gorgets of hummingbirds have a very low reflectance over most of the visible spectrum, but are very highly reflective in the narrow range of their iridescence (Greenewalt, 1960).

In the near-infrared to 400 mm^{-1} (2.5 μ), several bird plumages examined so far are highly reflective (50–100%). Hummingbird gorgets are highly absorptive in this range. Black male Brown-headed Cowbirds (*Molothrus ater*) reflect 4% at 1250 mm^{-1} (0.8 μ) with a steady increase in reflectance to 54% at 714 mm^{-1} (1.4 μ; Greenewalt, 1960; Heppner, 1970, and personal communication; Lustick, 1971, and personal communication). A tail feather from *Cyanocitta stelleri* reflected maximally ($\sim$43%) at 400–500 mm^{-1} (2–2.5 μ) but < 10% was reflected at 72–180 mm^{-1} (5.5–14 μ) (Gates et al., 1966).

The biological significance of these relationships is not clear. In cold climates, it would be desirable to be highly absorbent in the solar spectrum, but to be a poor emitter in the infrared (at bird surface temperatures) to reduce radiative losses, particularly at night. In a hot sunny climate, a light coloration would seem desirable to reject some of the solar heat load. However, skin temperature of the Brown-necked Raven (*Corvus ruficollis*) of the Negev desert is no higher than that of the pale Desert Partridge (*Ammoperdix heyi*). Light is absorbed by the outer surface of the dark raven feathers, from which the heat is lost by convection (Marder, 1973; see also Øritsland, 1970; Davis and Birkebak, in press). What is the value, if any, of being highly reflective and consequently a poor emitter in the 400–1250 mm^{-1} near infrared? Perhaps this serves only to reject a part of the solar heat load in a portion of the spectrum not associated with visual concealment or display functions. The low T_b and T_s of a bird would not result in a significant radiation at these wavelengths, so this cannot be considered a heat-conservation adaptation. However, the emissivity has not been measured at wave numbers less than 400 mm^{-1} for these small birds. Radiation from a 32°C surface has a wave number range of about 25–250 mm^{-1} (converted from Hammel, 1956). From Eqs. (33a) and (33b), wave numbers of peak radiation intensity for 0°–30°C surfaces are 53.5–59.4 mm^{-1}, while median wave numbers for this temperature range are 66.4–73.2 mm^{-1}.

An organism is coupled to its radiation environment not only directly through its own absorptance but also indirectly through the reflectance of nearby objects. Examples of the reflectance of the solar spectrum by environmental surfaces (albedo) are shown in Table IV. The reflection of solar radiation from objects near an organism may multiply the total solar radiation load several-fold. Porter (1969) analyzes the metabolic consequences of emission and reflection of infrared radiation from the walls of metabolism chambers. The geometry of analogous ralationships in real microenvironments is extremely intricate, but an instructive theoretical treatment for a limiting case in lizards is presented by Norris (1967).

Organisms are coupled to the thermal radiation in the solar spectrum through total absorptance, but more specifically through the spectral distribution of absorptance at each wavelength (monochromatic absorptance); i.e., through the extent to which their absorption spectrum "matches" the solar spectrum. Total and monochromatic absorptance (reflectance, actually, but $\alpha = 1 - \rho$) can be measured over representative plumage segments by a reflectance spectrophotometer. The difference in the areas beneath the spectral curves for incident and reflected radiation is proportional to the radiant energy absorbed by the organism. This difference may be estimated by a form of numerical integration (Norris, 1967) or by graphic methods.

The effects of insolation on feeding behavior and food consumption in birds have been examined by Morton (1967a,b). Hamilton and Heppner (1967a,b), Lustick (1969, 1971), Heppner (1970), and Ohmart and Lasiewski (1971) have studied the metabolic and thermoregulatory consequences of insolation, particularly in relation to plumage or skin color. These investigations showed that insolation has important effects on both physiological and behavioral components of thermoregulation in birds, but the projection of these findings from the laboratory to the natural environment is still in a primitive stage of development. Note that plumage color is not a very reliable guide to reflectance of the solar spectrum, since only part of this spectrum is visible as reflected light (color). Very light-colored birds are only weakly coupled to visible solar radiation (reflectance is high). Black or dark-colored birds are more tightly coupled (reflectance is low) but the reflectance increases in the near infrared. Variation of color (e.g., black versus brown or various shades of brown) between these extremes are not necessarily accompanied by appreciable variations in mean reflectance (Lustick, 1970).

B. Net Radiation Exchange

Up to this point we have regarded birds as isolated objects that were either emitting or receiving radiation. Obviously, they both receive and emit concurrently under real conditions, and the net radiant energy load depends on the difference in intensity between absorbed and reradiated fluxes.

The net radiation exchange per unit area and time between two diffusely radiating black or gray bodies is described by a modification of the Stefan–Boltzmann law

$$H_r(\text{net}) = \epsilon_1 \epsilon_2 \sigma (T_1^4 - T_2^4) \tag{34}$$

where ϵ_1 and ϵ_2 are the emission coefficients of the two surfaces. The net rate of energy transfer is thus determined by the surface properties of the objects and the difference in the fourth powers of their absolute temperatures. Kleiber (1961; p. 136) shows that T^4 can be replaced by T and a new proportionality constant without much loss of accuracy if the temperature difference is relatively small (linear radiation equation). The net radiation exchange is not affected by the temperature of the air, except as it modifies surface temperatures.

In practice, the application of Eq. (34) to real organisms and their environments is greatly complicated by geometric relationships not taken into account by this version of the Stefan–Boltzmann law. In engineering practice, these relationships are subsumed within a "shape factor" or "angle factor" (Birkebak, 1966; Gebhart, 1971), which defines the extent to which two emitting surfaces intercept radiation from each other. Shape factors are obviously complex functions of the configuration and angle of the surfaces and of the distance between them. The rigorous analysis of even simple biological systems through the application of shape factors seems too complicated to be worthwhile, and the use of the net radiation law is thus confined to geometric models of animals (cylinders, spheres) and their environments. In this role, however, the equation is useful in helping to define boundary conditions and in suggesting the magnitude of heat loading and loss by radiation in various environments (see Porter and Gates, 1969).

C. Heat Transfer by Conduction

Internal convection is a function of blood flow, and cardiac output is probably proportional to a fractional power of body mass—assume $m^{3/4}$ for purposes of the present argument. The path length for thermal conduction is proportional to linear dimensions, or $m^{1/3}$. Assuming that the tissue conductivity is independent of body weight, the ratio of internal conduction to convection, $m^{1/3}/m^{3/4} = m^{-5/12}$, indicates that internal conduction becomes relatively more significant with decreasing body size. Regardless of its significance internally, heat conduction is a major link across the bloodless parts of the integument, across the boundary layer of still air against the integument, and between an animal's surface and the surrounding air or water.

The basic equation for steady-state heat conduction across a flat plate of area A is given by the integration of Fourier's law.

$$\dot{H}_k = (h_k A/L)(T_2 - T_1) \tag{35}$$

where $\dot{H}_k$ = heat transfer per unit time and area, A = the area of the plate, L = the length of the conduction path, and T_2 and T_1 are the temperatures in the direction of conduction. The quantity h_k is the conductivity coefficient, with units of heat per unit distance, time, and temperature. Its numerical value varies with the nature of the substance and its state.

The extension of Fourier's law to cylinders and spheres is more realistic for biological subjects. The relationships were first used in a biological context by van Dilla et al. (1968) in modeling the insulation properties of gloves, sleeves, and legs in arctic clothing. Their report may be consulted for additional details and for the derivation of the equations.

In a cylindrical shell of length L in which the radius to the outside surface is r_o and the radius to the inside surface is r_i, the heat transfer by conduction in a steady state is given by

$$\dot{H}_{k\ cyl} = \frac{2\pi L h_k (T_i - T_o)}{\ln(r_o/r_i)} \qquad (36)$$

The effect of curvature on conductive heat transfer thus becomes accentuated logarithmically with decreasing size (radius of curvature) and is further exaggerated in animals because the thickness of insulation $(r_o - r_i)$ also diminishes with decreasing size [Eqs. (6), (13), and (14)].

The analogous equation for a sphere is

$$\dot{H}_{k\ sph} = \frac{4\pi r_o r_i h_k (T_i - T_o)}{r_o - r_i} \qquad (37)$$

The thermal and caloric problems associated with small size (large surface/volume ratio) are accentuated by the effects of body geometry. These geometric relationships must therefore be kept in mind in assessing the effects of body size and insulation thickness on conductive heat transfer. A further source of variability is the amount and kind of body surface exposed or retracted in postural (behavioral) thermoregulation, which also affects convective heat loss (see Fig. 9, also Sections III,D and VI).

The conductance interface between an organism and its environment is the boundary layer—a film of still air (or water) against the skin. Heat is transported from the skin by conduction into the boundary layer and thence transferred from the limits of this layer by radiation and convection. The thickness of the boundary layer is a function of wind velocity and the state of erection of the fur and feathers. Even in normal swimming birds, a boundary layer of air is retained within

FIG. 9. Reduction of exposed surface area by retraction of extremities of the White Pelican (*Pelecanus erythrorhynchos*). See Fig. 19 for contrasting use of the pelican's oral surfaces in heat dissipation.

the feathers (Rutschke, 1960; Rijke, 1970). Since the conductivity of water is about 24-fold greater than that of air at normal environmental temperatures, the layer of air is an important component of insulation in aquatic homeotherms. Nye (1964) found that *Anas platyrhynchos* ducklings cannot maintain normal body temperature when swimming in a detergent solution that penetrates to the skin. Hartung (1967) reported that "heat-loss rate" (approximately conductance) in air was increased up to threefold in adult *Anas platyrhynchos* and *Anas rubripes* coated with oil that disrupted the arrangement of the feathers. These ducks were not studied while in water, but it is predictable that heat loss by conduction (and convection) would be augmented manyfold above that in air.

An important avenue of heat exchange is conduction to and from the ground or nest. We find no measurements of thermal conductances of nesting material or of changes in the overall conductance of birds whose ventral insulation is compressed against the substrate in repose, or whose brood-patches are exposed. Kendeigh (1963) has calculated the heat loss to eggs during incubation by the House Wren (*Troglodytes aedon*), and several authors provide data on egg and nest temperatures (e.g., Huggins, 1941; Irving and Krog, 1956; Spellerberg, 1969; Orr, 1970). Jones (1971) reviews briefly the thermal relationships of the incubation patch. There would be many technical problems in obtaining the steady-state heat flux and temperature measurements necessary to partition conductive losses in these situations, but this is an aspect of thermal biology that should be investigated (Drent, 1972, 1973).

D. Heat Transfer by Convection

Radiation and conduction are processes of energy exchange that do not involve concurrent transport of material. Convection, on the other hand, depends on conduction of heat into a moving stream of fluid and its transport away from the source. The force for the movement of fluid may be simply the bouyancy (reduction in density) resulting from heating. In this case the transfer is called free convection or natural convection. When the organism moves through the fluid, or when the force for fluid movement is remote, as in a wind or a stream of water or blood, the transfer is called forced convection. Heat transfer by free convection, the simplest case, can be described by the equation

$$\dot{H}_c = h_c A (T_s - T_a) \qquad (38)$$

where T_s is the surface temperature of the animal or object, and T_a is the temperature of free air beyond the boundary layer. This equation is analogous to the preceding equations for the net flux of radiation (Eq. 34) and transfer by conduction (Eq. 35). A coefficient, h_c, defines the functional interface between an organism and its environment, in which the driving force for energy exchange is a temperature difference. Unlike h_r and h_k, the h_c coefficient varies widely and depends strongly on geometric factors (surface curvature, shape and size of animal, orientation in stream vector), on laminar or turbulent flow, and on the density, viscosity, and thermal conductivity of the convection fluid (see Gates, 1962, p. 94ff). The convection coefficient thus conceals great complexity that impedes meaningful quantification and the rigorous application of convection theory to birds.

For instance, in free convection in air, it can be shown by dimensional analysis (Gates, 1962) that h_c for a smooth cylinder is a function of the thermal gradient and the diameter (D) of the cylinder:

$$h_c = 6 \times 10^{-3} [(T_s - T_a)/D]^{1/4} \tag{39}$$

In forced convection, the coefficient (in cal/cm² minute °C) is a function of air velocity (v) and diameter (W. P. Porter, personal communication).

$$h_c = 6.17 \times 10^{-3} (v^{1/2}/D^{2/3}) \tag{40}$$

Birds in flight move rapidly through the air, and so forced convection is potentially of greater significance to them than it is to earth-bound homeotherms. The heat loss per unit area by free convection from a smooth cylinder is approximated by

$$\frac{\dot{H}_c}{A} = 6.00 \times 10^{-3} \left(\frac{T_s - T_a}{D}\right)^{1/4} (T_s - T_a) \tag{41}$$

where terms are defined as previously. For a "bird" (smooth cylinder) at rest in still air, the importance of heat loss per unit area by free convection is slightly affected by body size and more strongly affected by thermal gradient. A halving of body diameter, for instance, is accompanied by about 1.2-fold increase of heat loss per unit area. A doubling of the thermal gradient causes about a 2.5-fold increase of heat loss per unit area. In comparison with heat loss by radiation, the losses by free convection are relatively small. In a hypothetical cylindrical bird with a diameter of 5 cm, for example, the heat loss by free convection when $T_s = 30°C$ and $T_s - T_a = 10°C$ is 0.0714 cal/cm² minute [from Eq. (41)]. By radiation, the net concurrent heat loss estimated by Eq. (34) in an environment of surfaces equilibrated to air temperature is 0.086 cal/cm² minute, similar to the loss by free convection.

Heat loss from the hypothetical cylindrical bird by forced convection is approximated by substitution of Eq. (40) for h_c in Eq. (38).

$$\frac{\dot{H}_c}{A} = 6.17 \times 10^{-3} \frac{v^{1/2}}{D^{2/3}} (T_s - T_a) \tag{42}$$

where v = velocity in centimeters per second and D = diameter in centimeters. This will serve to indicate the relative magnitude of

TABLE VII

COMPARISON OF CONVECTIVE AND NET RADIATIVE HEAT LOSSES FROM A HYPOTHETICAL CYLINDRICAL "BIRD"[a]

Air velocity (cm/second)	Temperature (°C)	Calculation	Heat loss (cal/cm² minute)	
0	$T_a = 20$	$6 \times 10^{-3} (10/5)^{1/4} (10°)$	0.0714	} 4×
0	0	$6 \times 10^{-3} (30/5)^{1/4} (30°)$	0.284	
223 (5 miles/hour)	20	By Eq. (40)	0.315	} 3×
223	0	By Eq. (40)	0.945	
894 (20 miles/hour)	20	By Eq. (40)	0.630	} 3×
894	0	By Eq. (40)	1.893	
Radiation	$\bar{T}_r = -273$	$8.13 \times 10^{-11} (303)^4$	= 0.685	
	$\bar{T}_r = 20$	$8.13 \times 10^{-11} (303^4 - 293^4)$	= 0.086	
	$\bar{T}_r = 0$	$8.13 \times 10^{-11} (303^4 - 273^4)$	= 0.234	

[a] Body diameter = 5 cm; body surface temperature = 30°C (303°K).

convection in heat budgets of birds under different conditions. The results of calculations for a hypothetical cylindrical "bird" of 5 cm body diameter appear in Table VII. Note that we have, for simplicity's sake, held the T_s constant at 30°C even at $T_a = 0$°C. Such would not be the case for a real bird (see Fig. 3).

There have been very few studies of the effects of forced convection on birds. Porter and Gates (1969) estimated the contribution of air velocity to the "climate space" (the limits of tolerance within which an animal must operate) in the Common Cardinal (*Cardinalis cardinalis*) and the Zebra Finch (*Poephila guttata*). Gessaman (1972) studied the effect of wind velocity on heat loss in the Snowy Owl (*Nyctea scandiaca*) and found that the metabolic rate at any given air temperature was a function of the square root of air velocity (v) (in meters per second) and temperature:

$$\dot{H}_{m\;v} = k v^{1/2} + \dot{H}_{m\;v=0} \qquad (43)$$

where $\dot{H}_{m\;v}$ = metabolic rate at any given air velocity in cubic centimeters of oxygen per gram-hour, $\dot{H}_{m\;v=0}$ = metabolic rate at negligible air velocity, and $k = 0.121 - 0.009\;T_a$.

A study of the effects of wind velocity on body temperature, heart rate, and respiratory rate in domestic fowl showed that birds in high air velocities (2500 cm/second) were able to maintain body temperature 1°–2°C lower at $T_a = 40$°C, and with considerably less cardiopulmonary stress, than birds at the same air temperature but with $v = 100$ cm/second (Siegel and Drury, 1968a,b).

Various kinds of avian behavior can probably be explained as instances of exploitation or avoidance of heat transfer by forced convection. The common sight of colonial seabirds with bills all oriented into the wind (e.g., Hesse *et al.*, 1937, p. 487) may reflect an adaptation to avoid ruffling the feathers and thus for preserving the boundary layer. To expedite convective heat transfer, Masked Boobies (*Sula dactylatra*) elevate the scapular feathers and expose the underlying skin while at the same time orienting the body so that exposed skin is shielded from direct radiation by the shadow of the head and neck. Concurrently, the wings may be drooped to expose the thinly feathered axillary region to air flow (Bartholomew, 1966). Finally, a bird may adjust its coupling to forced convection by moving in relation to wind profile in the microhabitat. This behavior would assume greatest importance in avoiding excessive heat loss in cold environments. The distribution of wind speed above various surfaces and around various objects is not accurately reflected by the standard meteorological measurements, and human observers are apt to overlook the range of options available to small animals. A thorough treatment of wind profiles in various microhabitats is presented by Geiger (1966).

Heat loss by convection may have a much greater impact on aquatic birds in many circumstances than upon terrestrial birds. Since the thermal conductivity of water is about 24-fold greater than that of air at the same temperature, the convective heat-transfer coefficients for air and water may show even greater differences (Birkebak, 1966, p. 278). These differences will be reduced by the fact that the thermal gradient across the boundary layer may be less in water than in air, and because the heat flow occurs over a series of thermal resistances (body core + shell + boundary layer) and a change in the medium changes only one part of this series (Calder, 1969). However, virtually no data exist for the feather–water interface in aquatic birds to aid in estimating convective heat flow. The experiments of Steen and Steen (1965) provide some initial insights by showing that the rate of heat loss through the legs and feet of the Great Black-backed Gull (*Larus marinus*) and the Gray Heron (*Ardea cinerea*) in water is about four times that in air at the same temperature. The velocity of the fluids, however, was not specified.

E. Heat Transfer by Evaporation

Heat loss by evaporation differs fundamentally from the dry heat loss processes of radiation, conduction, and convection, since it entails material as well as energy loss to the environment. Except in severe heat stress where $T_a \geq T_b$, this evaporation with the associated heat transfer is a liability in thermal and osmotic homeostasis. The water loss is an inevitable consequence of the need for gas exchanges to occur across a moist surface. The problem is accentuated for small birds, since the disparity between evaporative water loss and metabolic water production is inversely related to body weight (Bartholomew and Cade, 1963; Dawson and Bartholomew, 1968; Crawford and Lasiewski, 1968). The biological implications of evaporative heat transfer are treated in Section IV,E. We consider the basic physical relationships here. There are several excellent discussions of the physics of evaporation as related to biological systems (Thornthwaite, 1940; Hardy, 1968; Schmidt-Nielsen, 1964; Lowry, 1969).

The evaporation of water at the respiratory or cutaneous surfaces of a bird involves the change of state from liquid to vapor, the isothermal expansion of the vapor against external vapor pressure, and the cooling of the vapor to the temperature of air. These three processes extract heat from the boundary layer, which in a steady state is replaced by conduction of heat from underlying structures. The total heat extracted per gram of water vapor transferred to ambient air by this three-stage process is called the latent heat of vaporization. Its value is strongly dependent on the temperature of liquid water, and to a much lesser extent on the water vapor pressure and temperature of ambient air (Hardy, 1968). Exact values of the latent heat of vaporization can be obtained from standard handbooks, but for most biological applications in whole animals, the value 580 cal/gm ($T = 30°C$) is accepted without incurring serious error. Thus

$$\dot{H}_e = 580 \dot{E} \tag{44}$$

The rate of evaporation ($\dot{E}$) from a moist surface is proportional to the surface area (A) from which the evaporation is occurring and to the vapor pressure gradient ($P_s - P_a/x$) between that surface and the boundary layer of air over the surface.

$$\dot{E} \propto A(P_s - P_a)/x \tag{45}$$

where P_s equals vapor pressure of evaporating surface A, P_a equals ambient vapor pressure, and x equals distance across the boundary air layer. The value of P_a is readily obtained from the tabular values of vapor pressure corresponding to measured dew-point temperature or can be computed from measurements of relative humidity (R), where

$$R = 100 \ (P_a/P_s') \tag{46}$$

and P_s' is the vapor pressure of saturated air at prevailing air temperature.

It is not uncommon in biological applications to express atmospheric humidity in terms of relative humidity [Eq. (46)] or in terms of the vapor-pressure deficit (P_D), where

$$P_D = P_s' - P_a \tag{47}$$

The use of these indices of humidity can be misleading or irrelevant, and the P_D in particular should be used cautiously, or preferably not at all (Thornthwaite, 1940). Some examples will illustrate the point.

Air at 30°C and $R = 60\%$, for instance, has $P_a = 19.1$ mm Hg, compared with $P_s' = 31.8$ mm Hg (100% relative humidity) (see Fig. 10). It thus has a P_D of $31.8 - 19.1 = 12.7$ mm Hg. If a bird exposed to $T_a = 30°C$ at 60% relative humidity has a 33°C evaporating surface, then $P_a = 37.7$ mm Hg and $P_s - P_a = 37.7 - 19.1 = 18.6$ mm Hg. Note that it is this gradient, and not 60% relative humidity or 12.7 mm Hg vapor-pressure deficit, that quantitatively describes the potential for evaporation. The use of the vapor-pressure deficit as an index of evaporative potential would lead to the conclusion that no evaporation could occur into saturated air ($P_D = 0$) at, say 10°C. For a bird inhaling and warming this air, however, there would still be a gradient for evaporation from the respiratory passages, and the added moisture would condense as a fog upon exhalation.

Similar distorted perspectives may result from use of relative humidity as an index of vapor pressure or drying power. If air at 30°C and 60% relative humidity is heated to 40°C without adding or removing water vapor, it will have a lower relative humidity (35%) but the same P_a as at 30°C. Since P_a has not changed, the potential for evapora-

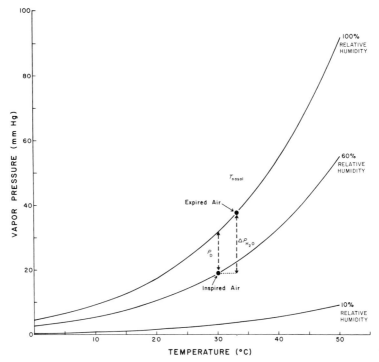

FIG. 10. Vapor pressure as a function of temperature for 100%, 60%, and 10% relative humidity. If a bird were inhaling air at 30°C and 60% relative humidity, the nasal surfaces might be cooled to 33°C by evaporation in saturating the inspired air. Expired air would then exist at 33°C saturated. The vapor pressure difference limiting the respiratory water loss would be as indicated, not by the vapor pressure deficit (P_D) of the inspired air. (Modified from Thornthwaite, 1940.)

tion will remain unchanged (even though the relative humidity is lower) when the air is inhaled over a moist respiratory surface. This conclusion would not be reached by the observation of relative humidity alone. Relative humidity and vapor-pressure deficit may be adequate for characterizing the physical environment, but they are deceptive when applied to the interactions of animal and environment with respect to water loss, and their use for this purpose is to be avoided. The relationships among psychrometric variables are summarized by Chambers (1970) in a very useful chart.

Until Lasiewski et al. (1966a,b) revealed that low airflow rates could limit evaporative cooling because of humidity buildup in experimental chambers, it was concluded that evaporative heat transfer in nearly all species of birds was limited to a maximum of about 50% of metabolic heat production (King and Farner, 1961, 1964; Dawson and Schmidt-Nielsen, 1964). Data from several species now show clearly that the capacity for evaporative heat loss can exceed resting metabolic calorigenesis (e.g., Bartholomew et al., 1962; Calder and King, 1963; Lasiewski et al., 1966a,b; Calder and Schmidt-Nielsen, 1966, 1967) (see Section IV). Nevertheless, there have been no comprehensive studies of a bird's evaporative heat transfer as a function of ambient water vapor pressure, and this aspect of the energy budget is much in need of further investigation before predictive models can be developed. At this stage, we can discern the effects of environmental humidity only through preliminary or fragmentary investigations.

The relative humidity in the chamber is calculated as follows, assuming complete mixing of air within.

$$R = \frac{\text{evaporation rate/air flow rate}}{\text{milligrams of water/volume saturated air}} \quad (48)$$

The evaporation rate is determined by water collection over a period of time. The water content of saturated air in the denominator is a function of temperature and may be found in steam or vapor density tables (List, 1951; Weast and Selby, 1971).

Air circulation or ventilation will increase the evaporation rate from a bird indirectly by removing humid air and maintaining a steeper vapor pressure gradient across a thinner layer of boundary air. The total evaporation rate is composed of the rates of respiratory and cutaneous evaporation. If the ambient temperature and humidity are held essentially constant the respiratory evaporation ($\dot{E}_r$) should be proportional to the ventilation.

TABLE VIII
HEAT LOSS RELATIONSHIPS[a]

Avenue	Heat loss (cal/minute)	=	Transfer coefficient	×	Area	×	Heat loss potential[b]
Radiation	$\dot{H}_r$		$\sigma \epsilon_s \epsilon_e$		A		$(T_s^4 - T_e^4)$
Conduction							
Flat plate	$\dot{H}_k$		h_k/L		A		$(T_2 - T_1)$
Cylindrical shell	$\dot{H}_k$		$h_k \dfrac{2\pi L}{\ln(r_0/r_i)}$				$(T_i - T_o)$
Hollow sphere	$\dot{H}_k$		$h_k \dfrac{4\pi r_0 r_i}{r_0 - r_i}$				$(T_i - T_o)$
Free convection							
Smooth cylinder	$\dot{H}_c$		$k_1 \left(\dfrac{T_s - T_a}{D}\right)^{1/4}$		A		$(T_s - T_a)$
Forced Convection							
Smooth cylinder	$\dot{H}_c$		$k_2 v^{1/2}/D^{2/3}$ $V^{1/2}$		A		$(T_s - T_a)$
Metabolic demand of Snowy Owl[c]	$\dot{H}_{m,v} - \dot{H}_{m,v=0} = \dot{H}_c$				$(0.97 - 0.072 \, T_a) \times 10^{-2}$		

[a] σ = Stefan–Boltzmann constant, 8.13×10^{-11} cal/cm^2 minute °K^4; ϵ_s, ϵ_e = emissivities of body surface and environmental object with which radiation exchange is occurring; L = linear path length; D = diameter; r_i = radius to inside surface; r_0 = radius to outside surface; T_s = skin temperature; T_e = surface temperature of environmental radiator; $\dot{H}_{m,v}$, $\dot{H}_{m,v=0}$ = metabolic rate at air velocities v and 0; k_1 = 6.0×10^{-3}; $k_2 = 6.17 \times 10^{-3}$ (Gates, 1962); $k_2 = 5.0 \times 10^{-3}$ (Lowry, 1969)).

[b] T for radiation exchange in °K = °C + 273°.

[c] Units converted from equation by Gessaman (1972).

$$\dot{E}_r \propto \dot{V}_e \propto m^{0.75} \tag{49}$$

where $\dot{V}_e$ is the minute respiratory ventilation. On the other hand, the cutaneous evaporation ($\dot{E}_s$) should be proportional to surface area.

$$\dot{E}_s \propto m^{0.67} \tag{50}$$

The results of regression analyses of available data do not conform to these expectations (Crawford and Lasiewski, 1968). This may be related to unrecognized variables in measuring techniques, or it could be influenced by selection of data from approximately the same T_a, rather than the same displacement below thermoneutrality.

F. Interactions among Modes of Heat Transfer

The foregoing account of basic physical principles has treated each mode of heat transfer separately. These are summarized in Table VIII. It is obvious, however, that the routes of heat exchange are functioning concurrently and that variation in the intensity of flux by one mode will affect the intensity of another. For instance, an increase of wind velocity or of the rate of percutaneous evaporation of water may reduce the surface temperature of the body, thus reducing also the infrared emission. The reduction of surface temperature, furthermore, decreases the thermal gradient for conduction across the boundary layer and decreases the gradient for convective heat transfer (when $T_b < T_a$). Forced convection prevents the buildup of water vapor above the evaporative surfaces, and thus enhances evaporative heat transfer.

The absorption of radiant energy can be used as a metabolic heat-sparing mechanism during exposure to cold air temperatures. This has been demonstrated with artificial sunlight (color temperature 2500°K) during indirect calorimetric studies of black dyed Zebra Finches (Hamilton and Heppner, 1967a; Heppner, 1970), Brown-headed Cowbirds (Lustick, 1969, 1970), Greater Roadrunners (*Geococcyx californianus*) (Ohmart and Lasiewski, 1971) and the White-crowned Sparrow (*Zonotrichia leucophrys*) (DeJong, 1971) (see Fig. 11).

It is extremely difficult to obtain quantitative estimates of the effects of interactions like these on the partitioning of heat flux, and their analysis stands as a major challenge to theoretical and experimental ingenuity. The empirical description of the simultaneous effect of wind velocity and temperature on metabolism of the Snowy

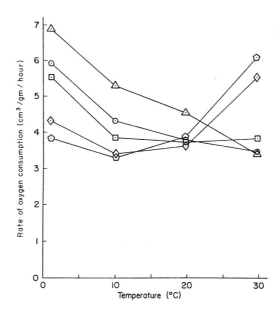

FIG. 11. Rate of oxygen consumption in White-crowned Sparrows (*Zonotrichia leucophrys gambelii*) exposed to different intensities of radiation from an infrared lamp at four temperatures. Radiation intensity (cal/cm^2/minute): triangle = 0.0, circle = 0.25, square = 0.5, diamond = 1.0, pentagon = 1.5. Mean values for 8–10 individuals are shown at each point. (From DeJong, 1971.)

Owl (Table VIII) is a step in this direction, analogous to the windchill factors and table given by Gates (1972). A knowledge of these interactions is required for a rigorous understanding of birds as thermal systems in natural environments, as is also a simultaneous knowledge of the microclimatology of these environments (Section VI). At present we have to rely on rather gross methods of approximation (Section IV).

G. Résumé and Reiteration

We have emphasized the ways in which birds are coupled to the major components (radiation, air temperature, humidity, and wind velocity) of their physical environment through corresponding physical coefficients and energy gradients (Table VIII). The analysis of these interfaces is extremely intricate and at the present time requires considerable simplification in practical applications. Several forms of simplification leading to models of variable predictive utility have attempted to cope with the problem (e.g., King and Farner, 1961; Birkebak *et al.*, 1965; Birkebak, 1966; Porter and Gates, 1969).

IV. Physiological Responses to Heat and Cold

A. RANGES OF RESPONSE

As we have seen from the discussion of the Scholander model (Section II), the range of environmental temperatures in which a bird can thermoregulate may be divided into three subranges. At one extreme, heat production and conservation are of prime importance. At the other, evaporative cooling becomes mandatory to preserve the system from heat damage. In all of these, there is at least a minimal involvement of metabolic heat generation, heat conservation, and evaporative heat transfer, not all of which are simultaneously advantageous. For example, in cold exposure, some evaporation occurs as a consequence of respiration, an undesirable loss of heat and water. In warm environments with a small outward heat flow gradient (T_a slightly below T_b), insulation is not desirable but can be reduced only to a minimum determined by the feathers and boundary layer. In severe heat stress, metabolic heat is disadvantageous, but is an unavoidable by-product of the inefficiency of metabolic processes. Thus, theoretically we may outline the thermal subranges and conceivable physiological responses as follows.

1. Cold stress
   a. Maximum heat conservation: ptiloerection, retracting extremities within plumage, peripheral vasoconstriction, countercurrent heat exchange (legs, nasal passages)
   b. Maximum water conservation: nasal counter-current serves both heat and water conservation
   c. Heat production augmented as necessary to effect adjustment
2. Thermoneutrality
   a. Heat conservation varied: manipulation of insulation and venous return rates, exposure of extremities
   b. Water conservation varied: a passive consequence of changes in circulation, exposure of lightly feathered but water-permeable surfaces
3. Heat stress
   a. Heat gain minimized: insulative and cardiovascular responses qualitatively approaching those of cold stress, as an unfavorable gradient exists, but in reversed direction; hyperthermia reduces the magnitude of the unfavorable gradient
   b. Metabolic heat production increased: an undesirable but necessary consequence of hyperthermia (Q_{10} effect) and increased effort in evaporative cooling effort (muscular work of panting and/or gular flutter)

c. Evaporative water loss augmented: as required to prevent rapid increase of body temperature, temperature conservation taking emergency priority over water conservation and acid–base regulation

Within this theoretical framework, we will now examine the processes of heat production, heat conservation, and evaporation individually.

B. Heat Production

1. Sources of Endogenous Heat

The conventional terminology that has been developed to classify heat production in homeotherms is not entirely satisfactory and requires some clarification. It is common to speak of shivering and nonshivering modes of calorigenesis. The former is unequivocal and can be regarded as regulatory calorigenesis. Nonshivering calorigenesis includes an obligatory fraction expressed as the basal or resting metabolic rate, and (in some species of mammals at least) a regulatory fraction that can be varied in accord with the demands of thermoregulation. Some writers seem to confine the source of regulatory nonshivering calorigenesis to visceral metabolism, excluding the voluntary muscles, but it is not clear why the definition should not also include the heat production of resting skeletal muscle. Additional sources of endogenous heat include voluntary muscular activity and, potentially, the calorigenic effects of feeding, digestion, and absorption.

The basal heat is an inevitable by-product of other energy transformations in the bird, as exercise heat is a waste product of locomotion and other movements. Depending upon the thermal circumstances, both sources of by-product heat can be advantageous, neutral, or disadvantageous. When cold stress is sufficient to render such by-product heat advantageous but insufficient, thermal balance requires an increase and uncoupling of some of these calorigenic processes to make them 100% inefficient in their primary nonthermal function or 100% heat-yielding. Shivering is the principle means for such augmentation, wherein the muscles are twitched without gross movement of the whole muscles. Nonshivering thermogenesis is a less common biochemical parallel in mammals in which the secretion of substances such as catecholamines increases the catabolism of energy reserves without anabolic objectives.

2. Heat Increment of Feeding

The basal metabolic rate is an important measurement for many purposes in comparative physiology, but is merely an abstraction when applied to free-living birds unless they are resting at night in a thermoneutral environment without food. By far, the greater majority of birds feed intermittently through the day and may be passing food from the crop to the stomach in the intervals between feeding, and in some species (e.g., *Lagopus* spp.) throughout the winter night. Heat production even in quiescent birds will therefore be augmented by the heat increment of feeding (also called the calorigenic effect, the specific dynamic effect, or the specific dynamic action of the ration). This added heat production is believed to result from the energy cost of assimilation. Our knowledge of its role in birds is confined to domesticated species. Because the heat increment of feeding is expressed in several different ways by different investigators (e.g., as a fraction of gross energy intake, of basal heat production, or of metabolizable energy) it is virtually impossible to provide a quantitative synthesis of the existing information. It is nevertheless evident that the magnitude of the heat increment of feeding is greater for proteins than for carbohydrates and fats, that it is directly proportional to the plane of nutrition, and that it is influenced by prior nutritional history of the animal (King and Farner, 1961; Whittow, 1965b; Romijn and Vreugdenhil, 1969), and perhaps by environmental temperature and other factors.

In an ecological context, the obligatory heat increment of feeding is of interest as a potential substitute for regulatory calorigenesis in cold surroundings. Hart (1963) rightly points out that the widespread assumption that the calorigenic effect of food reduces thermoregulatory expenditure in cold is not uniformly correct in the case of domestic mammals. In some cases or species, the substitution is only partial or is not evident at all. Adequate evidence on this subject in birds is elusive, and only the data of Berman and Snapir (1965) for three breeds of chickens appear to address the question (indirectly) over a short range of air temperature. The data are complicated by seasonal variations, but it is possible to reduce this source of error by expressing them as the ratio of metabolic rates in resting (but fed) and fasting birds. The relationship (Fig. 12) shows that the heat increment is about 20% of the resting metabolic rate at $T_a = 30°C$ but diminishes toward zero at the lower critical temperature (about 18°C in domestic fowl of comparable body weight). This is the trend to be

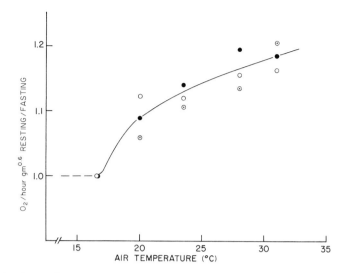

FIG. 12. Ratio of mean oxygen consumption in resting (but fed) and fasting chickens (three breeds indicated by symbols) in relation to air temperature. Body weight (in grams) is raised to the 0.6 power (empirically determined) to correct for differences in body weight among breeds. (Data from Berman and Snapir, 1965.)

expected if the heat increment of feeding substitutes for thermoregulatory calorigenesis. The subject requires further experimental examination, and general statements about the ecological significance of the heat increment of feeding in birds are not yet appropriate.

3. Heat of Muscular Exercise

Recent progress in the physiology of flight permits some quantitative examination of the thermal aspects of activity in birds. The metabolic rate has been measured in flight for several birds. The ratios of flight/standard metabolic rate range from 2.1 to 14.5 for passerines and from 3.0 to 14.5 for nonpasserines (see Table 1 of Farner, 1970). The mean values are 7.3 for the passerines and 9.6 for nonpasserines, if each species listed by Farner is given equal weight and values for the Laughing Gull (*Larus atricilla*) and the Purple Martin (*Progne subis*) are added (Tucker, 1970; Utter and LeFebvre, 1970). If these are analyzed together with values from Teal (1969) and Berger et al. (1970), higher ratios are obtained, but the nonpasserines are still characteristically higher than passerines. The differences between passerines and nonpasserines are apparently due to the differences in the reference level, since the standard metabolism of passerines is nearly 1.7 times that of nonpasserines with the same body mass. The lowest ratios in both groups are from birds that feed in forward flight, the swallows and swifts. The highest are from a finch and a hovering hummingbird.

If the flight metabolism is 25% efficient, this means that the other 75% is converted to heat and is thus potentially available for thermoregulation in a cold environment, or is a disposal problem limiting flight in a warm environment (see Kleiber, 1961; Tucker, 1968). A careful and enlightening analysis has been provided for the heat budget of the Budgerigar (*Melopsittacus undulatus*) from wind-tunnel measurements (Tucker, 1968).

The resting Budgerigar has a T_{lc} of 34°C. Homeostatic adjustment to 20°C is accomplished by a 108% increase of metabolic heat production above the basal level (Greenwald et al., 1967). In contrast, the flying Budgerigar produces the same amount of metabolic heat at 20°, 30°, and 37°C. The adjustment in flight is thus one of heat dissipation rather than production. While the exertion of flight produces more heat than at rest, the movement of flight constitutes a forced convection and probably compresses the feathers closer to the skin. The thinly feathered axillae beneath the wings are exposed and thereby facilitate heat loss during flight (Eliassen, 1962; Tucker, 1968). The metabolic demands of flight require a more rapid circulation (Hart and Roy, 1966; Berger et al., 1970). These factors result in a reduced insulation or a greater heat transfer coefficient (h). The value of h during flight compared with the value at rest in the Budgerigar is increased by 3.1 times at $T_a = 20°C$ and by 5 times at $T_a = 30°C$. In the Laughing Gull (*Larus atricilla*), the value of h during flight is 5.8 times the value at rest at $T_a = 30°C$ (Tucker, 1972).

A different pattern for disposition of metabolic heat in exercise is seen in oxygen consumption rates ($\dot{V}_{O_2}$) of Chaffinches (*Fringilla coelebs*). The relationship of $\dot{V}_{O_2}$ to T_a for five different levels of activity are essentially parallel in the 5°–20°C range, but higher for higher activity (Pohl, 1969). The nearly parallel appearance of these curves suggests no consistent increases in h. The $\dot{V}_{O_2}$ in the highest activity level at $T_a = 20°C$ is 57% higher than $\dot{V}_{O_2}$ at rest in $T_a = 5°C$. It is difficult to explain why $\dot{V}_{O_2}$ in the highest activity level increases by approximately the same amount when T_a is lowered from 20° to 5°C as it is at rest. At this point, we can only conclude that the heat production of exercise may not adequately substitute for that of shivering at low temperatures, since much of this increase above the resting level is lost via the forced convection of the exercise. This is an aspect of subneutral thermoregulation that deserves further attention.

Heat flow across heat flow discs placed subcutaneously over the pectoral muscles of pigeons increased sixfold from resting to flying in air at $T_a = 6°-17.5°C$ (Hart and Roy, 1967). At $T_a = 20°C$, 85% of the Budgerigar's heat production is dissipated by conduction, convection, and radiation. The remaining 15% is eliminated by evaporation. This is slightly more than the 9–10% evaporative heat transfer at 20°C in the resting bird and very similar to the 13% for pigeons flying at 20°C (Tucker's calculation from LeFebvre, 1964). In brief flights (≤ 18 seconds), the Black Duck (*Anas rubripes*) loses 19% of total heat production via evaporation from respiratory surfaces, independent of T_a in the range $-16°$ to $+19°C$ (Berger *et al.*, 1971).

Small size is advantageous for heat dissipation during flight, because of the greater surface/volume ratio and higher thermal conductance. Flying in wind tunnels in air temperatures of 25°–29°C, pigeons (350–400 gm) increased their body temperatures from 41.3° to 44.1°C and it had not stabilized after 10 minutes of flight, while Budgerigars (33.5 gm) heated from 41.1° to 42.1°C in three minutes of flight, and the body temperature then stabilized for the remainder of 7-minute flights. The birds were not, however, flying at the same speeds, but at 55 km/hour for the pigeons and 35 km/hour for the Budgerigars (Aulie, 1971).

Flying in air at $T_a = 30°C$, the Budgerigar can still dissipate 82% of its heat by nonevaporative means, the smaller heat loss potential being compensated for by an increase in heat transfer to 5.0 cal/gm hour °C. The 18% evaporative heat loss is very similar to that of the resting bird (17% at 30°C), but the absolute amount of water loss is of course some twelve times higher in flight. In air at $T_a = 37°C$, the gradient for dry heat loss is too small, and the heat transfer apparently cannot be increased beyond that at 30°C, even though the extension of the unfeathered portions of the legs beyond the feathers does not occur below 36°–37°C (Tucker, 1968).

This places a heavy demand on evaporative cooling, 47% of the heat production leaving thereby, contrasted to 28% in the resting bird exposed to 37°C. In terms of the absolute rate, the flying bird at 37°C increases its evaporation to 7½ to 9 times the rate when resting at 37°C. In a 13 minute flight (the longest time that one Budgerigar could fly at this high temperature) the evaporative rate of 37.3 mg/minute was inadequate to prevent a rise in body temperature. Consequently, 13% of the heat production was stored, or enough to raise the body temperature 2.8°C.

Unfortunately, there are no other published accounts of heat balance of birds in sustained flight. As pointed out by Tucker, the pigeons studied by Hart and Roy (1966, 1967) did not remain airborne beyond a transient 3–14 seconds, during which oxygen debt was still being accumulated. The results of Tucker's pioneering studies help to explain such common observations as reduced activity of birds in midday heat. Nocturnal migration is customarily explained as an adaptation to permit feeding by day and (or) to avoid predators (Thompson, 1964, pp. 469–535). However, Tucker's data indicate another advantage. The metabolic cost of flight is the same when T_a is 20° and 30°C. The water cost of ventilation to sustain this is 22½% greater in air at $T_a = 30°C$ than in air at $T_a = 20°C$. Thus, migration in the cooler night air, and radiating heat to the sky rather than submitting to a solar heat load, reduces the discrepancy between metabolic water production and evaporative plus fecal water losses. For a calculation of water balance during migratory flight, see Tucker (1968).

4. Nonexercise Augmentation of Calorigenesis

We have noted above that calorigenesis in muscular exercise is the result of inefficiency in converting metabolic energy into external work, perhaps 25% accountable as work and the other 75% as heat which can be utilized for thermoregulation (Kleiber, 1961; Tucker, 1968). We also noted that the heat transfer in flight is increased to some three to six times that at resting, owing to exposure of less insulated parts of the bird and to forced convection, external and internal. Suppose that a bird is subjected to cold stress and has no need or desire to go anywhere. The conversion of food energy to heat would be 100%, and this would escape more slowly because the boundary layer of air would not be disturbed by the convectional effect of movement. This heat production could be accomplished in two ways. One is shivering, the generation of heat by muscular contractions in an asynchronous pattern that does not result in gross movement of whole muscles. The other is to uncouple anabolic processes and stimulate calorigenesis chemically (nonshivering thermogenesis). Either case may be viewed as regulatory release of cellular machinery or muscle fibers from their primary function in order to balance heat loss.

Shivering is the major, if not the only, means of increasing heat production in the inactive bird (Section VI,A,3,b). Shivering has been closely correlated with the development of homeothermy, with pectoral heating, with degree of cold stress, and with rate of oxygen consumption below thermoneutrality.

The correlation between the appearance of muscle tremors and development of homeothermy was demonstrated in the House Wren and the Black-capped Chickadee (*Parus atricapillus*). Odum (1942) found that muscle tremors began in the House Wren when 3–6 days old, and were restricted to short periods. It is in this stage that body temperature can be elevated 0.5°–3.0°C above 21°C ambient, such calorigenesis preceding slightly the appearance of feathers. Tremors and appreciable elevation of T_b came at age 4 days in the chickadee. When 12 days old, the wren could maintain tremors continuously and $T_b = 15°C$ above the 21°C T_a. Precocial birds, which possess thermoregulatory powers along with the other manifestations of precocity soon after hatching, also lead in muscle tremor development. This is detected in the Ring-necked Pheasant (*Phasianus colchicus*) egg, unopened at the ninth day of incubation (Odum, 1942). Heat production by the embryo actually begins to raise egg temperature above incubation temperature in the domestic fowl at about this time, and at the end of incubation the duck embryo can elevate egg temperature 4°C above the 37.5°C environment of the incubator (Romanoff, 1941; Kashkin, 1961). An excellent discussion of the ontogeny of thermoregulation has been provided recently by Dawson and Hudson (1970).

The pectoral muscles are the largest component of body mass in flying birds. From extensive tabulations, the pectoral muscles are seen to average 15% of body weight, with 25% of body weight or more being contributed by the pectorals of strong flapping-flyers such as pigeons and doves (Hartmann, 1961). Birds lack the calorigenic brown fat depots of mammals (Johnston, 1972). Thus, the pectorals are the best organs for augmenting calorigenesis, a function that has been described in detail by electromyography.

Steen and Enger (1957) recorded electrical activity from needle electrodes in the pectoral muscles of the pigeon. Pigeons exposed to temperatures above 13°C had electromyograms of 15–18 Hz and low amplitude. At $-22°$ to $-24°C$, the frequency was more than doubled, while the amplitude increased tenfold. At intermediate temperatures (8°, $-6°C$) the electrical response was also intermediate. The electrical activity was accompanied by an increase in pectoral temperature, while abdominal temperatures decreased in the $-22°C$ exposure, a 1°C difference between pectoral and abdominal temperatures could develop in 5 minutes, and the difference was sometimes as great as 2°C, with the pectoral higher.

The peak-to-peak voltage developed by pectoral muscles was essentially a linear function of air temperature in the pigeon, increasing from about 50 μV at 30°C to about 450 μV at $-15°C$ (Hart, 1962) (see Fig. 13). West (1965) also found a linear correlation of electromyographic activity and oxygen consumption in four species of passerines (Fig. 14). This linearity is consistent with the hypothesis that there was no increment of nonshivering calorigenesis over a

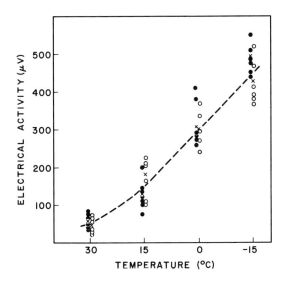

FIG. 13. The effect of environmental temperature on electrical activity (shivering) of the pectoral muscles of pigeons. Seasonal differences were not significant; ● = summer, ○ = winter, × = seasonal means. An average of values for the lower critical temperature of the pigeon from five studies is 24°C (range 14°–36°C) (see King and Farner, 1961). (From Hart, 1962, with permission.)

wide range of air temperature, or with the much less probable hypothesis that nonshivering calorigenesis contributed the same proportion of total heat production throughout the temperature range studied. Electromyograms have shown that much of the activity during shivering occurs at frequencies higher than 100 Hz in the Common Grackle (*Quiscalus quiscula*) and the Evening Grosbeak, with differences apparently related to species or body size in the frequency–temperature relationships.

Additional evidence of the presence or absence of nonshivering calorigenesis in adult birds is available from pharmacological investigations. The metabolic response to cold was abolished when Hart (1962) paralyzed the muscles of a pigeon with curare. Administration of noradrenaline following curare did not cause an increase of the pigeon's oxygen consumption, and there was therefore no adrenergic stimulus of nonshivering heat production analogous to that seen in some species of cold-acclimated adult mammals (Carlson, 1966; Jansky et al., 1969). Chaffee et al. (1963) subsequently showed in adult chickens that doses of noradrenaline up to threefold greater than effective doses for white rats had no calorigenic effect. Hissa and Palokangas (1970) reported that both adrenaline and noradren-

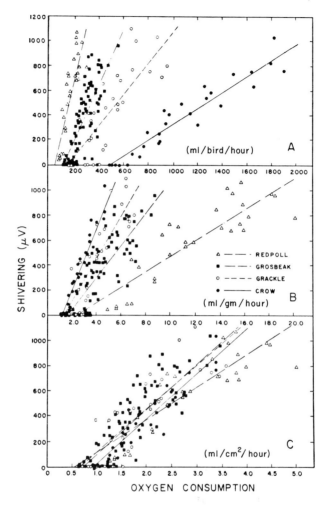

FIG. 14. Correlations of the electrical activity of shivering with simultaneous metabolic determinations in four passerine species, expressed as total oxygen consumption (A), oxygen consumption per gram body mass (B), and oxygen consumption per square centimeter of estimated body surface (C) (the least precise expression). (From West, 1965, with permission.)

aline failed to enhance oxygen consumption following intraperitoneal injection into cold-acclimated adult Great Tits (*Parus major*). On the contrary, both hormones induced statistically significant decreases in oxygen consumption that were inversely related to dose level. Brown fat, which is associated with nonshivering calorigenesis in mammals, could not be found in carcasses of eight species of birds which are known to become hypothermic (Johnston, 1971).

In short, there is no adequate evidence that regulatory nonshivering calorigenesis occurs in birds or is involved in their short-term thermoregulation. It is obvious that neither the obligatory nor the regulatory component of nonshivering heat production is augmented by catecholamines. Additional arguments concerning the existence of regulatory nonshivering heat production in immature birds are developed in Section VI.

C. Heat Conservation

Heat conservation mechanisms can be classified as follows:

>   Physical and physiological
>   Insulation and ptiloerection
>   Vasomotor control
>   Countercurrent heat conservation
>   Postural
>   Retraction of extremities
>   Bill under scapular feathers
>   Microhabitat selection (see Section VI)
>   Windbreaks
>   Radiation screening
>   Huddling

We have discussed previously (Section II) the measurement of the thermal properties of the plumage as expressed by the heat transfer coefficient (h). Insulation is resistance to heat transfer. Thus, the insulative value is inversely related to h. In addition to the physical amount of plumage, this physical resistance to heat flow can be varied by ptiloerection. Such "fluffing" has long been qualitatively appreciated as a means for increasing the thickness of the insulating coat (Moore, 1945).

Infrared scanning thermographs demonstrate dramatically the significance of the plumage in reducing surface temperatures and therefore heat loss, and also reveal the thermal heterogeneity of the body shell and the heat loss of the extremities (Fig. 4 in Veghte and Herreid, 1965). Further technical details of the method are given by Watmough et al. (1970).

The act of feather-fluffing, or ptiloerection, has, however, remained unquantified until two recent studies. Crawford and Schmidt-Nielsen (1967) measured a 7 cm increase in insulative thickness resulting from ptiloerection by the Ostrich (Struthio camelus). This was, however, in response to extreme heat stress, curiously observed at T_a from 25° to 51°C, when the advantage of greater insulation in heat stress would not occur unless $T_a > T_b$ (i.e., $T_a > 40°C$). McFarland and Baher (1968) established a scoring system for evaluating feather posture of the Ring Dove with 0 = "sleeked," 1 = "normal," and 2 = "raised." The sum from six body regions thus provided a quantitative feather index. This correlates well with T_a (Fig. 15), which also

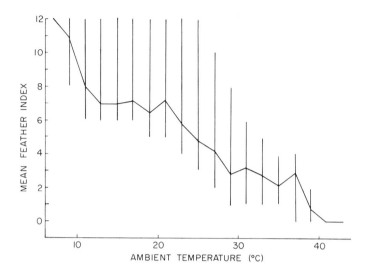

FIG. 15. The effect of ambient temperature on the feather index (an index of ptilomotor elevation) in the Ring Dove. (From McFarland and Baher, 1968, with permission.)

suggests strongly that conductance is not minimized at T_{lc} (see conductance values in King and Farner, 1964; Calder and Schmidt-Nielsen, 1967). Use of this promising technique, extended to other species in conjunction with metabolic determinations of T_{lc} and calculations of h, should be very fruitful. The feather index of the Ring Dove has been used with standard thermoregulation equations such as Eq. (1) in an analysis of feather position. Following a 20°C step increase in T_a, the time constants for exponential decreases in feather index, toward sleeking of the feathers, were progressively shorter with longer periods of water deprivation, which seems desirable for water economy. The opposite was true for food deprivation, which increased the time constant. Measurements of hypothalamic temperatures led the experimenters (McFarland and Budgell, 1970) to hypothesize that the positions of the body feathers were adjusted in response to changes in hypothalamic temperatures via a first-order feedback control system.

In addition to ptiloerection as a mode of heat conservation, the blood circulation to the skin and extremities can be reduced by vasoconstriction. This is established in mammalian thermoregulation

(Day, 1968; Schmidt-Nielsen et al., 1965; Aschoff and Pohl, 1970b; Walther et al., 1971) and can safely be assumed to occur in birds, though it does not seem to have been measured except in the extremities (Jones and Johansen, 1972).

Countercurrent heat exchange is an important heat conserving device, as pointed out originally by Claude Bernard (cited by Irving and Krog, 1955). Temperatures (T_{sc}) recorded from subcutaneous thermocouples in the leg of a Glaucous-winged Gull exposed to $T_a = 16°C$, declined from 37.8° to 15.1°C before the arterial blood passes below the feathered femoral portion of the leg. At the metatarsal joint, the T_{sc} was 7°C and in the webbing of the feet $T_{sc} = 0.0°$–4.9°C (Irving and Krog, 1955). Similar countercurrent heat conservation occurs in the American Wood-Stork (*Mycteria americana*) (Kahl, 1963) and in the Ring-necked Pheasant (Ederstrom and Brumleve, 1964).

Measurement of heat losses from the leg of a Gray Heron (*Ardea cinerea*) and of a Great Black-backed Gull (*Larus marinus*) show that this is a regulated conservation–dissipation mechanism (Steen and Steen, 1965). The countercurrent heat exchange from one leg immersed in water (initially 4°C and slowly increasing) amounted to less than 3% of total heat production at $T_a < 10°C$. The total amount of heat loss from one immersed leg was constant from $T_a = -10°C$ to $T_a = +20°C$. Above 20°C, heat loss increased linearly so that at $T_a = 30°C$ it was four times that at 20°C, accounting for 40% of the lower total heat production. Apparently this variability is accomplished by changes in rate of blood flow. The Gray Heron dissipated 10% of heat production at $T_a = 20°C$ and 60% at $T_a = 35°C$. The importance of this mode of heat dissipation in flight was noted by Steen and Steen (1965) and by Tucker (1968) in connection with the leg-extension observed in Budgerigars flying at $T_a = 36°$ to 37°C but not at lower temperatures. The thermoregulatory role of peripheral circulation in the legs in thermoregulation is discussed further by Jones and Johansen (1972).

As will be discussed in Section IV,E, countercurrent heat conservation in birds is not limited to the legs, but is also of considerable significance in the respiratory process, where the flow is alternating in the same channel as contrasted to the continuous flow in closely apposed vessels in the countercurrent exchange system of the legs.

In concluding this section on heat conservation, we must point out the lack of information on skin temperatures and the effects of forced convection upon T_{skin}. The one exception is a study on Great Black-backed Gulls (Eliassen, 1962, 1963). When at rest with their wings folded in a wind of 12 m/second (43 km/hour), the skin temperature of the gulls was the same in summer air (15°–19°C) as in winter air (−7° to −14°C). Only the temperature of the web of the wing decreased from 25° to 10°C in winter air. When the gulls were induced to glide for long periods, in the same wind tunnel conditions, the T_{skin} of the wings decreased by 3½° to 6°C in the winter air.

The Snowy Owl has a very low heat transfer coefficient ($h = 0.05$ cal/gm hour °C) in still air. When the owl is exposed to a 26.9 km/hour wind, the T_{lc} increases 11°C, and h increases to 2.7 times that in still air (Gessaman, 1972), which is the same magnitude of increase that the Budgerigar had when flying 35 km/hour (Tucker, 1968). The oxygen consumption of the Snowy Owl increases linearly with the square root of wind speed in air of $T_a = -20°C$ and $-30°C$ (Gessaman, 1972).

Just as convection increases the metabolic cost of thermoregulation in cold air, the movement of hot air over the bird, or of the bird through hot air, increases the problems of thermoregulation. Body temperatures of domestic fowl were positively correlated with air velocity (0.4–9 km/hour) when T_a exceeded T_b (Siegel and Drury, 1968a), which is reasonably explained in the same way that forced convection decreases the boundary layer and thereby facilitates heat transfer to cold air; only the direction of the gradient is reversed. This augmentation of heat influx by relative movement between air and bird is diminished or even prevented by hyperthermia, as has been measured for the Common Rhea (*Rhea americana*) during treadmill running with a simulated relative air speed (Taylor et al., 1971) (see next section).

These investigations of the effects of forced convection on skin temperature and metabolic rates have been major contributions to bridging the large gap between laboratory metabolism chambers and the natural situation. We hope that there will be more such studies on other species, and call attention to thermal studies of wind effects on White-tailed Deer (Moen, 1968; Stevens and Moen, 1970) for stimulating lines of approach.

D. Heat Storage in Hyperthermia

Heat-stressed birds characteristically become hyperthermic, the body temperature being regulated 2°–4°C above that of unstressed birds at rest. In the largest living bird, the Ostrich, hyperthermia occurs only when the birds are dehydrated (Crawford and Schmidt-Nielsen, 1967). The possible advantages of hyperthermia are three:

(1) If the air temperature approaches normal body temperature, a regulated elevation of the body temperature preserves a small gradient for heat loss via nonevaporative means (conduction, convection, and radiation). (2) If the air temperature is greater than the body temperature, heat will be gained from the environment in proportion to the temperature difference $(T_a - T_b)$. The hyperthermic T_b reduces this difference and therefore the rate of heat gain. (3) Heat is stored for dissipation later in the daily cycle when the environment cools (Schmidt-Nielsen, 1964; King and Farner, 1961; Dawson and Bartholomew, 1968; Dawson and Hudson, 1970).

The advantage of heat storage in water economy is dramatic and well-documented for the camel, which also undergoes a mild hypothermia at night. The water that would have been required for evaporative elimination of the heat stored between nocturnal hypothermia and diurnal hyperthermia (W) may be estimated as

$$W = \frac{H_s}{L} = \frac{c \cdot m\, \Delta T_b}{L} \qquad (51)$$

where H_s = stored heat, c = specific heat (calories per gram-degree Centigrade), ΔT_b = extent of hyperthermia, m = body mass, and L = latent heat of evaporation. The specific heat of animal tissues of average fatness is approximately 0.8 (Hart, 1951; Schmidt-Nielsen, 1964; Minard, 1970). For the camel, whose 260 kg body may have a T_b ranging from 34.5°C at night to 40.7°C in the heat of the day, this water savings through heat storage is 2.23 liters per day (Schmidt-Nielsen, 1964). The Ostrich does not drop T_b at night, but does have a daily amplitude of 4.2°C, which is equivalent to the saving of 0.58 liters of water.

Hyperthermic heat storage thus results in water economy, as a generalization for all birds. However, as was the case for evaporative heat transfer, quantitative comparisons are best related to metabolic rate in order to scale for body size (Dawson and Bartholomew, 1968; Dawson and Hudson, 1970). Alternatively, this water economy could be expressed as a fraction of water intake. However, the data on water intake are variable because of behavioral polydipsia in caged birds, complicated by the fact that water gains are a total from three sources (drinking water, free water in food, and oxidation water); furthermore, the available data represent a limited size range. Metabolic water production is directly proportional to metabolic heat production, so the latter is quite acceptable as a basis for comparison. A hypothermia of 2.5° to 4.0°C seems typical for most species that have been reported as capable of regulating T_b when $T_a = 44°–45°C$. For a 3°C hyperthermia, the amount of heat storage is

$$0.8 \text{ cal/gm °C} \times 3.0°C = 2.4 \text{ cal/gm} \qquad (52)$$

which should be multiplied by body mass (m) to estimate total heat storage. The significance of this, expressed as a fraction of one hour's metabolic heat production, will depend upon the metabolic equation used. Modifying the Lasiewski-Dawson (1967) equations [our Eq. (5) and (6)] to units of cal/gm hour gives the following.

For nonpasserines: $\quad \dfrac{2.4\,m}{22.1\,m^{0.72}} = 0.109\,m^{0.28} \quad$ in percent $\dot{H}_m \quad (53)$

For passerines: $\quad \dfrac{2.4\,m}{36.2\,m^{0.72}} = 0.066\,m^{0.28} \quad$ in percent $\dot{H}_m \quad (54)$

The exponent of these ratios ($m^{0.28}$) indicates that the significance of heat storage per se increases with body size, or conversely that it is proportionately less in the water economy of smaller birds. Thus a 10 gm nonpasserine bird stores the equivalent of 21% of an hour's basal heat production, while a 1 kg bird stores 75% of its hourly basal heat during the development of a 3°C hyperthermia. Both would save considerably more than the hour's metabolic water in the process (Table IX).

The effect of hyperthermia on the thermal gradient is very important to all birds, but especially to smaller birds. Heat gain from a hotter environment will follow Eq. (1), with a change in sign because of reversal of the thermal gradient direction:

$$\dot{H}_{in} = h(T_a - T_b) \qquad (1a)$$

If the T_a were 46°C, raising T_b from a normal 40°C to 43°C would cut heat gain in half.

Heat gain from the environment, related to an hour's resting metabolic heat of a passerine bird, for example, is, from Eq. (5) and (13)

$$\text{cal/hour} = \dfrac{4.55\,m^{-0.54} \times m^{1.0} \times (T_a - T_b)}{36.2\,m^{0.72}} = 0.126\,m^{-0.27}(T_a - T_b) \qquad (55)$$

Thus, the importance of hyperthermia in retarding heat gain is inversely related to body size.

TABLE IX
Significance of Heat Storage as a Function of Body Size

Body mass (gm)	Percent of 1 hour metabolic heat for 3°C increase[a]		Water savings (all birds)(ml/hour)	Metabolic water[c] (basal ml/hr)		Water savings/ metabolic water	
	Nonpasserines ($0.109 m^{0.28}$)	Passerines ($0.066 m^{0.28}$)		Nonpasserines	Passerines	Nonpasserines	Passerines
3	14.8	9.0	0.016	0.00593	0.00971	2.70	1.65
10	20.8	12.6	0.055	0.01416	0.02320	3.88	2.37
100	39.6	24.0	0.55	0.07481	0.12243	7.35	4.47
1,000	75.4	45.7	5.50	0.39534	0.65514	13.91	8.45
10,000	143.7	—	55.00	2.08917	—	26.32	—
100,000	273.8[b]	—	550.00	11.04009	—	49.82	—

[a] See text.
[b] A 3°C rise in T_b would take more than five hours of basal metabolism in a 45°C environment. This heat storage would be approximately one-half of basal heat production for the total period.
[c] Assuming RQ = 0.8 on a mixed fat and carbohydrate substrate.

Note that the exponent of this regression is quantitatively the inverse of that for heat storage [Eq. (52)]. Hyperthermia is advantageous to all birds, but the importance of the water economy attendant on hyperthermia is directly related to body size ($m^{0.28}$) while the significance of the reduced gradient for heat loading is inversely related to body size ($m^{-0.27}$).

Because of the technical problems involved in the acquisition of metabolic data from moving birds, our quantitative awareness of heat storage in avian thermoregulation has been limited mostly to the resting state, as has been the foregoing discussion. Notable recent exceptions are studies of flying Budgerigars (Tucker, 1968) and running Common Rheas (Taylor et al., 1971). Flying 35 km/hour in 37°C air, the Budgerigar stored 13% of its heat production. The eventual overheating from an increase of 0.2°C/minute limited the flight to less than 20 minutes, the T_b rising to 44°C. A rhea stored 57% of the heat produced while running 10 km/hour for 20 minutes in 35°C air (14% relative humidity). In air of 43°C and 8% relative humidity, 73% of the metabolic heat was stored in the same span of time at the same speed. Although the bird was hyperthermic, the T_b was usually below T_a, so that there was a continuous small heat gain from the environment. At the end of one run, however, T_b had risen to 46.4°C with apparently no ill effects. During an hour (20 minutes running plus 40 minutes recovery) the hyperthermia provided a gradient for nonevaporative heat loss amounting to over one-fourth of the heat increment due to the running. For comparison, a resting rhea, which did not become hyperthermic when exposed to air of 43°C, gained heat from the environment equivalent to 60% of its resting metabolic heat production.

E. Heat Dissipated by Evaporation

With regard to evaporative heat loss, the range of air temperatures tolerated by birds may be subdivided arbitrarily as follows: (1) cool temperatures at which water loss should be minimized, both to reduce heat loss and as an adaptation to terrestriality; (2) an intermediate temperature range wherein evaporation is gradually increased as dry heat losses are proportionately reduced with smaller thermal gradients; (3) warm to hot temperatures at which evaporation must be actively increased to dispose of metabolic and exogenous heat loads. In the cool and intermediate situations, ventilation is presumably regulated by requirements for exchanges of oxygen and carbon dioxide (for reviews, see Fedde, 1970; Lasiewski, 1972). When heat stress pushes T_b above the panting threshold, gas exchange and blood acid–base controls are overridden by thermoregulatory imperatives.

One common and useful expression of the thermal significance of evaporation ($\dot{H}_e$) is as a percentage of resting metabolic heat production ($\dot{H}_m$): $100(\dot{H}_e/\dot{H}_m)$. This ratio can be used for quantitative comparison of the proportion of heat dissipation as a function of ambient temperature. It is apparently independent of body size. Evaporative water loss occurs via respiratory and cutaneous routes (see Table X). Respiratory minute volume and metabolic rate are both proportional to $M^{3/4}$ (Lasiewski and Dawson, 1967; Lasiewski and Calder, 1971). Cutaneous water loss would be expected to be proportional to surface area ($m^{0.67}$), but at 25°C is a function of $m^{\sim 0.6}$ (Crawford and Lasiewski, 1968). Thus, the division of evaporative heat by metabolic heat production is only a first approximation of a mass-independent expression.

The relationship of ($\dot{H}_e/\dot{H}_m$) to T_a is shown in Fig. 16, representing data selected from twenty species ranging in size from 6 to 100,000 gm. The selection of data was based upon Table V of Dawson and Hudson (1970) with some substitutions, additions, and reinterpretations which are not tabulated here in order to conserve space. Criteria for selection were: birds with *ad libitum* water intake, low humidity exposure, and approximate agreement of T_a for $\dot{H}_e = \dot{H}_m$ with $T_a = T_b$.

TABLE X
RESPIRATORY WATER LOSS

Species	T_a (°C)	Respiratory evaporation (% of total evaporation)	Reference
Struthio camelus	To 35	>98	Schmidt-Nielsen et al., 1969
Coturnix chinensis	25	41.5	Bernstein, 1971a
	30	55.1	Bernstein, 1971b
	35	41.9	Bernstein, 1971a
Columba livia	Neutral ($T_b = 40$°C)	16.2	Smith, 1969
	>45 ($T_b = 43$°C)	33.5	Smith, 1969
	>45 ($T_b = 46$°C)	71.6	Smith, 1969
Melopsittacus undulatus	30	41.1	Bernstein, 1971b
Phalaenoptilus nuttallii	35	48.7	Lasiewski et al., 1971
Geococcyx californianus	35	49.0	Lasiewski et al., 1971
Ploceus cucullatus	30	48.9	Bernstein, 1971b
Poephila guttata	25	57.5	Lee and Schmidt-Nielsen, 1971
	30	37.1	Bernstein, 1971b

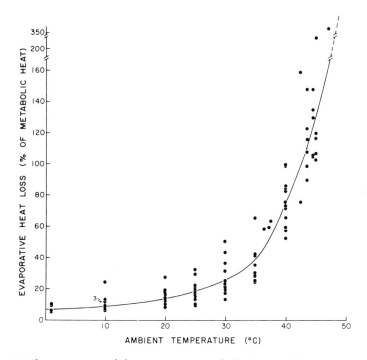

FIG. 16. The proportion of the concurrent metabolic heat production that is dissipated by evaporation as a function of chamber air temperature. Analysis with HP-9100B electronic calculator system, with separate curves for $0° < T_a \leq 38°C$ and $40° \leq T_a \leq 47°C$ faired together.

Strictly speaking, this is not true, for as Seymour (1972) points out, during evaporative cooling, the evaporative surfaces (T_{es}) must be cooler than the body core temperature. If $T_{core} = T_a$ and $T_{es} < T_{core}$, then there exists a gradient for heat influx from the environment to the evaporating surfaces. And, since there is a measurable component of cutaneous evaporation, some heat could flow from the air to the skin even when the core was isothermic to the air. As a first approximation, however, the correspondence of $\dot{H}_e = \dot{H}_m$ at $T_a = T_b$ is valid, because "the difference is within the limits of accuracy of the present methods of measurement." The ratio increases exponentially with increasing temperature in birds exposed to T_a up to 38°C, the relationship being

$$100(\dot{H}_e/\dot{H}_m) = 5 + 1.48\ e^{0.087 T_a} \tag{56}$$

where $e = 2.718$ (base of natural logarithms); $p < 0.001$. This curve suggests a relationship to the curves for vapor density of saturated air or vapor pressure as functions of temperature (see Fig. 11).

A change in inflection of the $\dot{H}_e/\dot{H}_m$ curve is seen in the T_a range of 35°–40°C for many species. A curve for eighteen species in the T_a range of 40°–47°C intersects the $0° < T_a < 38°C$ curve at 37°C ($p < 0.001$) (see Fig. 16). This inflection is undoubtedly related to the transition in respiratory pattern and conductance associated with facilitation of heat dissipation to compensate for reduced or unfavorable thermal gradients.

By comparing four very different kinds of birds of the same body size Lasiewski and Seymour (1972) demonstrated clearly that the thermoregulatory effectiveness of the evaporative heat dissipation is determined not by the evaporation rates alone, but also by the relationship of the evaporation to heat production. Thus, in order to regulate T_b during heat stress, a weaver bird with the high metabolism characteristic of passerines (See Table I) must evaporate considerably more than nonpasserine doves and quail, and the Poorwill (*Phalaenoptilus nuttallii*) is the best of the four in this regard, with the low resting metabolic rate of a caprimulgid (Bartholomew *et al.*, 1962; Lasiewski and Dawson, 1964; Lasiewski and Seymour, 1972).

1. Cutaneous Water Loss

The absence of sweat glands and the presence of plumage that impedes cutaneous convection currents had perpetuated the assumption that all significant evaporation is respiratory. This has been shown to be the case for the Ostrich (Schmidt-Nielsen *et al.*, 1969), but several recent studies indicate that this is not true for smaller birds (Table X of Smith, 1969; Smith and Suthers, 1969; Bernstein, 1971a,b; van Kampen, 1971; Lee and Schmidt-Nielsen, 1971; Lasiewski *et al.*, 1971; Lasiewski, 1972) (Section III,D). Except in the Ostrich, respiratory evaporation constitutes only 16–58% of the total evaporation in approximate thermoneutrality. The contribution of respiratory evaporation increases in heat stress, as measured by Smith (1969) in pigeons and van Kampen (1971) in chickens, and as is obvious from observation of panting birds. Partitioning evaporation into cutaneous (plus plumage) and respiratory (plus head) components has produced differing results. For example, at $T_a = 30°C$, the Zebra Finch lost only 37% via respiration (Bernstein, 1971b), while at 25°C, Lee and Schmidt-Nielsen (1971) accounted for 57% via respiration, the latter in good agreement with respiratory considerations. Unlike respiratory evaporation, cutaneous evaporation may be associated with a considerable time lag while the vapor moves through the feathers. This could delay the attainment of a steady state for measurement. The ability of feathers to carry liquid water has been documented in the unspecialized Cape Sparrow (*Passer melanurus*) as well as the specialized feathers of the Namaqua Sandgrouse (*Pterocles namaqua*) (Cade and Maclean, 1967). Adsorbed water already present in the feathers could be released upon transfer of the birds to dry air for measurement of water loss, giving exaggerated values for the cutaneous (plus plumage) water loss. Therefore, the equilibration period before measurements might be crucial. Taking this into account, Lee and Schmidt-Nielsen (1971) exposed their Zebra Finches for a minimum of 2½ hours in dry circulating air before measurements. They also used a smaller airflow, which could have reduced the vapor pressure difference (see Lasiewski *et al.*, 1966a). Either of these factors may account for their lower values for the cutaneous proportion and for total evaporation than those obtained in other studies (Calder, 1964; Cade *et al.*, 1965; Bernstein, 1971a,b).

Thus, in birds, we find a parallel to the cutaneous evaporation of reptiles. While the common-sense assumption had been that the reptilian integument was also practically impermeable to water, measurements showed that 52–88% of the evaporated water escaped across the scaled surfaces (Bentley and Schmidt-Nielsen, 1966; Schmidt-Nielsen and Bentley, 1966; Prange and Schmidt-Nielsen, 1969). The effect of the plumage in trapping a boundary layer of air may seem an improvement in retarding evaporation, but the proportion of cutaneous loss is still a major factor in the water and heat balance of birds; and, in absolute values, the birds are losing much more water per day across the skin than are the reptiles. For example, a 124 gm *Iguana iguana* loses 0.72 gm of water per day through the skin when exposed to $T_a = 23°C$. A 124 gm nonpasserine bird might lose 42–84% of its 6.71 gm evaporative total, or 2.8–5.6 gm of water per day via the skin. This loss, four- to eight-fold higher in birds than in lizards, probably reflects the thinner, more permeable skin of birds and the higher vapor pressure at their higher surface temperature.

Urohidrosis is a specialized form of evaporative heat loss. When the American Wood-Stork (*Mycteria americana*) is subjected to $T_a > 45°C$, the frequency of excretions increased in parallel to panting, and the excretions were directed at the legs. Artificial wetting of the legs in this environment (46.5°–50.5°C) accomplished a 1°C drop in T_b (Kahl, 1963). The Turkey Vulture (*Cathartes aura*) also excretes on its legs when T_a rises into this range (Hatch, 1970), and excreting on the legs is listed for twelve other species of storks and the Black Vulture

(*Coragyps atratus*), but without measurements of the thermal significance (Kahl, 1963).

2. Respiratory Evaporation

Both the amount of ventilation and probably its route can be varied, so that respiratory evaporation is vital as a variable effector of thermal homeostasis (see Lasiewski, 1972; Section IV). For terrestrial animals exposed to low temperatures, neither augmented heat loss by evaporation nor the water loss is desirable. In breathing air at 15°C and 25% relative humidity, a Cactus Wren (*Campylorhynchus brunneicapillus*) loses 7.8 cal/liter to warm the air, and another 33.5 cal by evaporation as it is saturated. Of the total, 75% of this heat (81% of the warming, 74% of the latent heat, and therefore 74% of the water loss) is recovered by countercurrent exchange before this air is returned to a 15°C environment in expiration (Fig. 17). This saving is 16.1% of the estimated metabolic heat production during the period. In

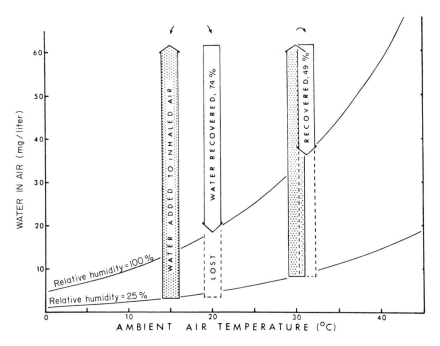

FIG. 17. Estimated water recovery by condensation during expiration in the Cactus Wren. Calculations were based upon 1 liter of air (standard temperature and pressure) inhaled 25% saturated (relative humidity) at 15°C and 30°C (shaded bars) and exhaled saturated at nasal temperatures from Fig. 18, curve 5. (From Schmidt-Nielsen *et al.*, 1970b, with permission.)

30°C air of 25% relative humidity, the heat is reabsorbed and water condenses when the exhaled air passes over progressively cooler respiratory surfaces going from the lungs to the external nares. Such countercurrent exchange probably is the case in all birds. The exhaled air temperatures that establish it have been measured in seven species of four orders, an 18–3000 gm size range (Fig. 18) (Schmidt-Nielsen et al., 1970b).

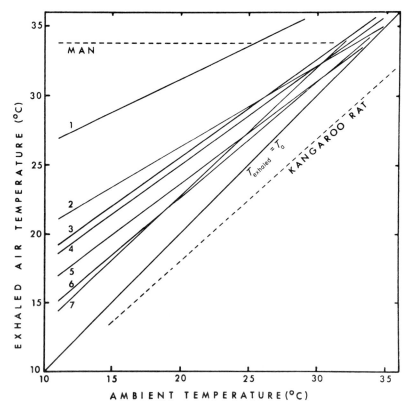

FIG. 18. Temperatures of exhaled air as a function of ambient temperature, from Schmidt-Nielsen et al. (1970b) with permission. Note that all birds represented, including the Cactus Wren, exhale air that is warmer than that from the desert-dwelling Kangaroo Rat. Thus, the birds regain somewhat less water by condensation than does the Kangaroo Rat. (1) Domestic Duck, (2) House Sparrow, (3) Budgerigar, (4) pigeon, (5) Cactus Wren, (6) Carolina Wren, (7) Hermit Thrush.

Perhaps preadaptively, the recovery of heat from warm, humidified air is much less when the exchanges are with outside air at $T_a = 30°C$, 49% of the water being reclaimed with only 52% of the heat loss during inspiration. At progressively higher temperatures, increases in ventilation become necessary for adequate elimination of heat.

3. Panting and Gular Flutter

High temperatures greatly reduce or even reverse the gradient for heat loss by nonevaporative means. Evaporation via skin and normal breathing may become inadequate to dissipate the metabolic heat, let alone any gain from the surroundings. The countercurrent heat exchange in the upper respiratory passages conserves water as well as heat (Schmidt-Nielsen et al., 1970b). When this heat conservation is no longer desirable and higher evaporation rates are useful, however, the countercurrent exchange can be reduced by returning the air by another route, e.g., orally. This has not been demonstrated in birds, but is the case in the panting dog (Schmidt-Nielsen et al., 1970a).

Until heat stress is applied, respiration is controlled by the respiratory center in the medulla (Lasiewski, 1972). With increasing body (and therefore brain) temperatures, the tidal volume appears to be increased [Calder and Schmidt-Nielsen (1966); however, as pointed out by Smith (1969), the tidal volumes estimated by those authors are probably too high]. If the temperatures increase still further, the hypothalamic panting center takes over the control of respiration (see Section VI,A,1).

The panting threshold is reached at body temperatures between 41° and 44°C (King and Farner, 1964). The onset of panting may be an abrupt (step) or gradual increase from normal respiratory frequency (Lasiewski et al., 1966b). Comparison of allometric expressions for the two frequencies shows that the panting rates are 15.7–26.5 times resting rates (26.5 in doves) (W. A. Calder and R. C. Lasiewski, unpublished observation). Evaporative cooling is also augmented by gular flutter in several taxa, the hyoid apparatus and associated musculature moving the moist gular region (Fig. 19). Three patterns of enhanced evaporation have been noted (Lasiewski et al., 1966a; Calder and Schmidt-Nielsen, 1968; Bartholomew et al., 1968) (see also Section IV,A and Chapter 5, Volume II of this treatise, pp. 318–325): (1) panting only (e.g., Passeriformes, Cathartidae); (2) panting with synchronous gular flutter (e.g., Columbiformes, Galliformes, Anseriformes); (3) panting with independent gular flutter (e.g., Phalacrocoracidae, Pelecanidae).

Evolution of the capacity for enhanced evaporative heat loss must have been influenced by several factors, among which one might think first of mechanical efficiency and the primary functions of the structures "borrowed" for evaporation. Evaporation occurs passively along a vapor pressure gradient. Enhancement of this function at the evaporative surface can occur only by maintaining the steepest possible gradient through forced convection. The necessary convection could probably be achieved more economically by moving the

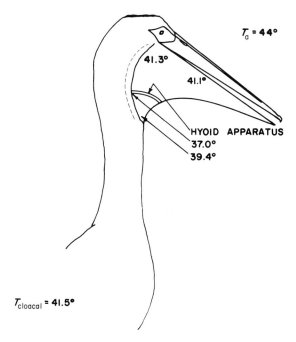

FIG. 19. Mean surface temperatures (°C) in the pouch and gular region of White Pelicans (*Pelecanus erythrorhynchos*) during panting in a 44°C room. Gular flutter is accomplished by contraction of muscles attached to the hyoid apparatus. The trachea opens to the buccal cavity via the glottis, located approximately where the arrow meets the hyoid. (From Calder and Schmidt-Nielsen, 1968, with permission.)

respiratory air than by moving the whole body. Probably less effort yet is needed to move the gular pouch. Both cutaneous and gular routes would obviate any acid–base disturbance that might accompany hyperventilation.

Any effort involved in heat dissipation by ventilation is accompanied by the heat production of that effort. The larger this heat production, the less effective is the heat dissipation. The efficiency can therefore be expressed as

$$\text{Efficiency} = \frac{\text{heat loss} - \text{heat of evaporatory effort}}{\text{heat loss}} \qquad (57)$$

The minimization of this thermoregulatory effort is obviously desirable in the interests of water economy, energy economy, and in permitting survival of severe threats to thermal homeostasis. Quantitative measurements are difficult, and consequently few, because the elevation in metabolism above thermoneutrality reflects not only panting effort but also the increased metabolic reaction rates at elevated tissue temperatures (Q_{10} effect) [see Eq. (72)] plus any restlessness and irritability evoked by the heat stress.

The work of panting must overcome three kinds of resisting forces in any oscillating system, such as in respiratory pumping. These forces are (1) frictional resistance to air and thoracic movement, (2) inertance or reactance to acceleration of the sum of air volume and the mass of the thoracoabdominal system, and (3) elastic reactance to stretching of structures and compression of volume. At the natural (resonant or harmonic) frequency of a system, the inertance and elastic reactances are of equal magnitude but opposite phase, so they cancel each other. Thus, the only work necessary for movement is to overcome frictional resistance. This represents a minimal work load (Hull and Long, 1961) and minimal heat production.

Consequently, the natural frequency is the most efficient one for panting, as has been elegantly demonstrated in the dog (Crawford, 1962) and the pigeon (Crawford and Kampe, 1971). The pigeon has a resting thermoneutral respiratory rate of 29/minute and a panting rate of 650/minute, the transition between these being abrupt as T_b attains the panting threshold as a consequence of step increase in T_a (Calder and Schmidt-Nielsen, 1966, 1967). Smith (1969) found a more gradual increase in panting frequency of pigeons in a whole-body plethysmograph, but this may have been an artifact of gular constriction in the collar of the apparatus. However, a gradual rise in T_a also produced a gradual transition from resting respiration to synchronized panting and gular flutter, from approximately 590/minute at $T_b = 42.7°C$ to 705/minute at $T_b = 43°C$ (Weathers, 1972).

The natural frequency (f) of a resonating system is related to its mass (M) as follows

$$f = \frac{(S/M)^{1/2}}{2\pi} \tag{58}$$

where S is an elastic constant (the reciprocal of compliance). Thus, if panting occurs at the natural frequency, panting frequency should be proportional to a negative fractional power of body mass. This was found to be the case in allometric analysis, although the particular exponents differ quantitatively from what might be expected from the above relationship (W. A. Calder and R. C. Lasiewski, unpublished observation) (see Fig. 20).

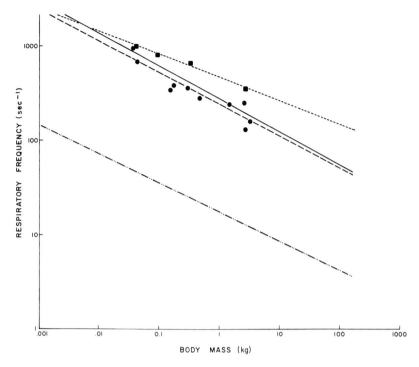

FIG. 20. Respiratory frequencies during panting compared with normal respiration in absence of heat stress as a function of body mass of birds. The regression line for doves and pigeons is similar to that of panting mammals. Panting frequencies of the other birds are considerably lower than their mammalian counterparts, as are normal resting respiratory frequencies. Key: *Panting:* doves (■)(----), $456M^{-0.25}$; nondoves (●)(---), $236\ M^{-0.34}$; all (———), $269\ M^{-0.35}$. *Resting:* nonpasserines (—··—), $17.2M^{-0.31}$. (From W. A. Calder and R. C. Lasiewski, unpublished observation.)

For nonpasserine birds: $f_{pant} = 296.4 M^{-0.35}$ (59)

$f_{rest} = 17.2 M^{-0.31}$ (60)

For doves: $f_{pant} = 455.9 M^{-0.25}$ (61)

For nondoves: $f_{pant} = 235.8 M^{-0.34}$ (62)

For nonpasserines: $f_{pant}/f_{rest} = 15.7 M^{-0.04}$ (63)

For mammals: $f_{pant}/f_{rest} = 8.8 M^{0.05}$ (64)

The similar exponents for the body mass dependence for both panting and resting birds strongly suggest an underlying relationship to mechanical efficiency which would be afforded by oscillation at a natural resonant frequency. Thus the observations of pigeons by Crawford and Kampe (1971) can tentatively be assumed to apply for birds in general.

There are major differences in panting frequency between birds of similar sizes such as the Greater Roadrunner (285 gm; $f_{\text{pant}} = 356$/minute) and the pigeon (315 gm; $f_{\text{pant}} = 650$/minute), and on an allometric basis between doves and nondoves (Calder and Schmidt-Nielsen, 1967; W. A. Calder and R. C. Lasiewski, unpublished observation). Since resting respiration and panting seem to have different natural frequencies for mechanical efficiency, there must be an adjustment in the elastic component to accomplish the shift from low to high natural frequency. This shift could be one that matches the thoracoabdominal system to the hyoid–gular system in those birds with synchronous gular flutter during panting. The pigeons and doves have much smaller gular proportions than the roadrunner. If the gular region is analogous to a vibrating string, its natural frequency would be

$$f = \frac{(s/m)^{1/2}}{2L} \tag{65}$$

where $L =$ length, $s =$ tension, and $m =$ linear density. This could explain why these birds of similar size and f_{rest} have such different f_{pant} (Calder and Schmidt-Nielsen, 1967).

4. Maximum Evaporation Rates

As indicated in Fig. 1, when the T_a exceeds T_b, there is a gradient for heat from the environment. The only route for dissipating this heat gain and the metabolic heat is via evaporation. Values for the high evaporation rates required in such heat stress are summarized in Table XI. While such conditions ($T_a > T_b$) exist only on deserts and as extremes for relatively short periods, the heat burden is similar in magnitude to the sum of metabolic heat and solar radiation in somewhat lower air temperatures, so evaporation rates like these may occur frequently in nature.

Analyzed allometrically, these maximum evaporation rates may be summarized as:

$$\dot{E} = 258.6 M^{0.80} \tag{66}$$

That is, a 1 kg bird is predicted to evaporate 258.6 mg water per minute. This may be compared with the Crawford–Lasiewski (1968)

TABLE XI

MAXIMUM RECORDED RATES OF EVAPORATION OF BIRDS EXPOSED TO
AMBIENT TEMPERATURES GREATER THAN BODY TEMPERATURES

Species	Body mass (kg)	Evaporation (mg/minute)	Reference
Poephila guttata (Zebra Finch)	0.0117	10.20	Calder, 1964
Carpodacus mexicanus (House Finch)	0.0199	11.50	R. C. Lasiewski, personal communication
Pyrrhuloxia sinuata (Pyrrhuloxia)	0.0315	17.27	Hinds and Calder, 1973
Cardinalis cardinalis (Common Cardinal)	0.0402	18.82	Hinds and Calder, 1973
Coturnix chinensis (Painted Quail)	0.0409	19.10	R. C. Lasiewski, personal communication
Scardafella inca (Inca Dove)	0.0415	10.80	Macmillen and Trost, 1967
Phalaenoptilus nuttallii (Poorwill)	0.0495	33.40	Lasiewski, 1969
Eurostopodus guttatus (Spotted Nightjar)	0.0875	58.33	Dawson and Fisher, 1969
Podargus ocellatus (Little Papuan Frogmouth)	0.145	48.33	Lasiewski et al., 1970
Geococcyx californianus (Greater Roadrunner)	0.2847	64.24	Calder and Schmidt-Nielsen, 1967
Columba livia (Domestic Pigeon)	0.3146	93.59	Calder and Schmidt-Nielsen, 1967
Columbia livia (Domestic Pigeon)	0.359	106.80	Calder and Schmidt-Nielsen, 1966
Struthio camelus (Ostrich)	88.00	11,000.00	Schmidt-Nielsen et al., 1969

equation for all birds at 25°C (the units changed to $\dot{E}$ in milligrams per minute):

$$\dot{E} = 17.1 M^{0.585} \tag{67}$$

Thus, exposed to $T_a > T_b$, a 1 kg bird would evaporate 15.3 times as much water as he would when exposed to 25°C. Crawford and Lasiewski noted that their exponent, 0.585, did not fit either a body surface (0.67) or ventilation (0.75) function. Of the two, the body surface exponent is closer. It is interesting that the exponent for maximum evaporation comes close to that for a function of ventilation. If this is not just coincidental, it suggests that augmentation of evaporation at higher temperatures is more closely reliant on ventilation than on cutaneous evaporation.

5. Hypocapnia in Panting

When evaporation via the skin, resting respiration, and gular flutter are inadequate to cope with heat stress, necessary augmentation is accomplished by hyperventilating the respiratory system (Fig. 21).

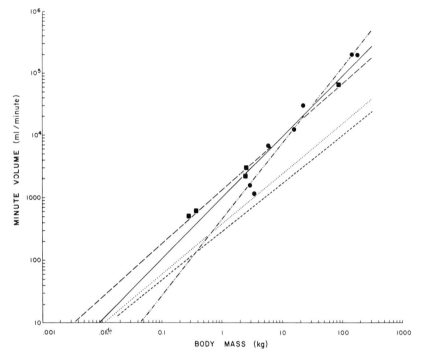

FIG. 21. Minute ventilation volumes, comparing resting birds and mammals with panting animals, as a function of body size. In birds, the panting volumes are more than four times normal ventilation volumes. Key: Birds: resting (----), $284 M^{0.77}$; panting (■)(– – –), $1312 M^{0.85}$. Mammals: resting (·····), $379 M^{0.80}$; panting (●) (–··–), $444 M^{1.22}$. Both: panting (———), $1011 M^{0.97}$. (W. A. Calder and R. C. Lasiewski, unpublished observation.)

The mesencephalic panting center must drive the respiration more intensely than gas exchange control by the medullary centers would dictate. As a consequence, abnormal amounts of carbon dioxide are washed out of the lungs and blood, leading to respiratory alkalosis (Fig. 22) and a need for metabolic or renal compensation, or both, to return blood pH to normal (Linsley and Burger, 1964; Calder and Schmidt-Nielsen, 1966, 1967). This also occurs in panting dogs and cattle (see review by Richards, 1970b), but the Ostrich does not incur alkalosis from panting (Schmidt-Nielsen et al., 1969). Thus, for birds other than the Ostrich, the sharing of the same effector system for gas

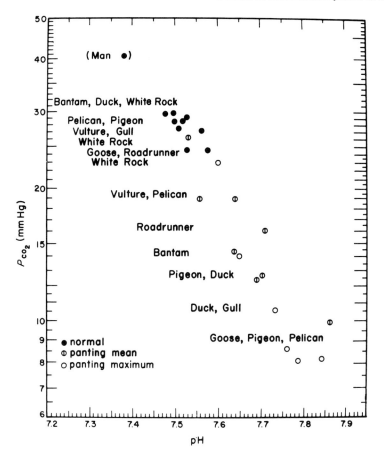

FIG. 22. Arterial pH and carbon dioxide tensions in panting birds (open circles) compared with values (filled circles) at rest in $T_a = 20°-25°C$. [From Calder and Schmidt-Nielsen (1968) with permission.]

exchange and heat dissipation results in a conflict in which the former is disturbed. This occurs despite a reduction to less than half the normal tidal volume, increasing respiratory dead space to minimize carbon dioxide washout.

6. Heat Transfer Coefficients during Heat Stress

At lower temperatures, the heat transfer coefficient can be estimated either from the slope of the regression of metabolic rate on T_a or from Eq. (1). The slope method is based upon the assumption that the total metabolic level at each temperature is determined entirely by heat balance requirements (no work) and that the body core temperature is independent of environmental temperature.

At high temperatures, these assumptions do not hold, so the slope of metabolism above thermoneutrality is a composite of metabolic increases due to the Q_{10} effect of hyperthermia and the increment of physical effort of panting plus the higher conductance of exposed oral surfaces and modifications in ptiloerection (Dawson and Fisher, 1969).

Thus, in the higher T_a ranges, the heat transfer coefficient h can be calculated only according to King and Farner's (1964) Eq. (10).

$$h = \frac{\dot{H}_m - \dot{H}_e}{T_s - T_a} \tag{68}$$

or in absence of skin temperature (T_s) data, h for body core to ambient difference:

$$h = \frac{\dot{H}_m - \dot{H}_e}{T_b - T_a} \tag{68a}$$

It must be noted that as $T_a \to T_b$, $\dot{H}_m \to \dot{H}_e$. Ideally, both measurements of $(\dot{H}_m - \dot{H}_e)$ and $(T_b - T_a)$ approach zero simultaneously (Fig. 1), but these will not coincide exactly due to experimental errors and variability. This means that calculated h values may arithmetically approach 0 or ∞ at $T_a \approx T_b$; it must be noted that such values are, of course, meaningless.

Here again we see the importance of the criterion that $\dot{H} \approx \dot{H}_e$ when $T_a \approx T_b$. The number of studies that have been carried to these high temperatures and show such agreement are few in a sizable literature on metabolism and thermoregulation of birds. Values of h (in calories per gram/hour/degree centigrade) calculated according to King and Farner (1964) relate to body size as follows (Table XII; see Table XVII for S_{yx}, S_b).

$$0° \leq T_a \leq 11.1°C \quad h = 5.88\,m^{-0.626} \tag{69}$$

$$43.8° \leq T_a \leq 46.5°C \quad h = 15.31\,m^{-0.611} \tag{70}$$

This confirms the observation that h is often deleteriously greater at high T_a values than at low T_a values (see Table XII for examples), which would result in greater heat gain from the environment, to be offset in turn by evaporative water loss (Dawson and Fisher, 1969). Such could be explained as the combined effects of postural exposure or behavioral extension of extremities, feather position, increased ventilation, and exposure of the evaporatively cooled buccopharyngeal

TABLE XII
A Comparison of Calculated Heat Transfer Coefficients from Cold
($0°C \leq T_a \leq 11.1°C$) and Heat ($T_a > T_b$) Stress

Species	Body mass (kg)	h (cal/gm hour °C) In cold	h (cal/gm hour °C) In heat	h ratio (heat/cold)	Reference
Poephila guttata (Zebra Finch)	0.0117	1.142	8.099	7.1	Calder, 1964
Pyrrhuloxia sinuata (Pyrrhuloxia)	0.032	0.701	0.510	0.73	Hinds and Calder, 1973
Phalaenoptilus nuttallii (Poorwill)	0.040	0.411	1.317	3.2	Bartholomew et al., 1962
Cardinalis cardinalis (Common Cardinal)	0.041	0.702	0.761	1.08	Hinds and Calder, 1973
Chordeiles minor (Common Nighthawk)	0.073	0.472	1.785	3.8	Lasiewski and Dawson, 1964
Speotyto cunicularia (Burrowing Owl)	0.147	0.203	0.773	3.8	Coulombe, 1970
Geococcyx californianus (Greater Roadrunner)	0.285	0.206	1.023	5.0	Calder and Schmidt-Nielsen, 1967
Columba livia (Domestic Pigeon)	0.315	0.194	0.596	3.1	Calder and Schmidt-Nielsen, 1967
Struthio camelus (Ostrich)	100.0	0.004	0.011	2.9	Crawford and Schmidt-Nielsen, 1967
Mean ratio:				3.4	
Ratio of allometric equations [(70)/(69)]:				$2.6 m^{0.015}$	

surfaces. Studies of thermoregulation at $T_a > T_b$, and therefore this analysis, have been limited to desert birds and the ubiquitous pigeon, birds most exposed to natural selection for protection and efficient thermal relationship at high T_a values. Of these, only the Pyrrhuloxia (*Pyrrhuloxia sinuata*) and the Common Cardinal seem to have minimized heat gain when the gradient is reversed to one causing heat influx (Hinds and Calder, 1973).

V. Hypothermia

A. Energetic Basis of Hypothermia

The regulation of a high T_b within narrow limits becomes energetically more expensive with increasing difference between T_b and T_a. When the light intensity is inadequate for efficient feeding or when low T_a renders poikilothermic prey inactive and unavailable, the bird enters a negative energy balance, drawing upon stored reserves which, being finite, limit the endurance of fasting.

$$t_{fast} = \frac{H_{crop} + H_{tissues}}{h(T_b - T_a)} - \frac{H_{search}}{\dot{H}_{flight}} \tag{71}$$

where t_{fast} = maximum tolerable duration of fasting (hours); $H_{crop} + H_{tissues}$ = total energy stores; H_{search} = energy reserved as an investment for the next successful feeding; $\dot{H}_{flight}$ = metabolic rate when searching for and replenishing the energy supply; other symbols as in Eq. (1). Crop volume and nectar concentration for tropical hummingbirds have recently been measured as a basis for estimating metabolic endurance (Hainsworth and Wolf, 1971). From this it is seen that with given reserves (H) and rates of consumption [$\dot{H}_m$ and $h(T_b - T_a)$], the components that could be varied to extend the endurance of fasting are T_a and T_b. The selection of a warmer T_a is one solution which we will discuss in Section VI. The other alternative is to lower the T_b.

Hypothermia would not only decrease the potential for heat loss to the environment, but the metabolism would be less than basal, affording additional energetic economy through the Q_{10} effect. The Q_{10} is an expression of the effect of temperature on physiological rates and ultimately on the underlying biochemical reaction rates. It is calculated on the basis of 10°C temperature difference but may be calculated for any temperature span as follows.

$$Q_{10} = (R_2/R_1)^{10/(T_2 - T_1)} \tag{72}$$

where R_1 and R_2 are rates at temperatures T_1 and T_2, respectively. For many physiological processes, Q_{10} values are approximately 2, but the metabolism of torpid hummingbirds is affected twice as much by temperatures, with Q_{10} values of 4 or more (Lasiewski, 1963; Lasiewski and Lasiewski, 1967). This means that a 10°C decrease of body temperature in torpid birds causes oxygen consumption to decrease to one-quarter or less of its value at the higher temperature. The small thermal gradient and much reduced heat production owing to the Q_{10} effect during hypothermia thus makes possible a substantial savings of energy. In his pioneering study in behavioral energetics, Pearson (1954) calculated a 24-hour budget of 7.55 kcal for a male Anna's Hummingbird (*Calypte anna*) which utilized the economy of nocturnal torpidity, as compared with 10.32 kcal in continuous normothermy, a savings of 27%.

B. Terminology

Hypothermia may be a brief or prolonged, regulated or passive abandonment of thermoregulation, and it may be a slight departure from normal T_b or a profound drop resulting in sluggishness and loss of coordination. These variations on the general theme of hypothermia may be distinguished by the following terms:

Hypothermia — Body temperature below the normal (normothermia) for resting birds. This may be only a few degrees (e.g., T_b = 34°C in the Turkey Vulture; Heath, 1962) or profound (T_b = 8.8°C in the Anna's Hummingbird; Bartholomew et al., 1957)

Poikilothermia — Body temperature variable, tending toward conformity with the environment; a passive nonregulated process

Heliothermia — Utilization of solar radiation for body warming, a form of ectothermia (heat from the environment rather than internal metabolism)

Regulated hypothermia — Body temperature reduced below normal but regulated at some lower limit (see Hainsworth and Wolf, 1970; Wolf and Hainsworth, 1972b)

Torpor — Profound hypothermia in which heart, respiratory, and metabolic rates are greatly depressed, coordination is essentially absent, and response to external stimulation diminished or absent (Jaeger, 1948, 1949; Bartholomew et al., 1957; Pearson, 1960)

Hibernation — Torpor over a long period of food shortage in the winter

Blurring these distinctions are cyclical and arrhythmic fluctuations in T_b of essentially normothermic birds, while superimposed upon the classification are the inevitable effects of body size, as in many other aspects of avian thermal biology discussed previously.

C. THE HYPOTHERMIC STATE

1. Distribution

As has been the case for other phenomena of avian physiology, the entry of hummingbirds into a torpid state was noted by aviculturalists before physiologists (Huxley et al., 1939). Those authors reported that colies (*Colius*) and swallows also became torpid (see also Bartholomew and Trost, 1970; Lasiewski and Thompson, 1966; Serventy, 1970), and they pointed out the similarity between the regular torpidity of hummingbirds and mammalian hibernation. In fact, the Poorwill (*Phalaenoptilus nuttallii*) does enter true hibernation in the winter when the insect supply is inadequate to support normothermy (Jaeger, 1948, 1949).

The majority of bird species that are now known to utilize hypothermia are in the related orders Apodiformes and Caprimulgiformes, but some degree of hypothermia is found in one or more species of at least six other orders. Dawson and Hudson (1970) provide an excellent tabulation of the phylogenetic variety and thermal depths of hypothermia in birds. To this may now be added the reports of irregular depressions of 0.5°–8.4°C lasting 30–188 minutes recorded from Snowy Owls in Alaska (Gessaman and Folk, 1969), of hypothermia in Black-capped Chickadees (Budd, 1972), and of torpidity in the White-backed Swallow (*Cheramoeca leucosternum;* Serventy, 1970) and the Violet-green Swallow (*Tachycineta thalassina;* Lasiewski and Thompson, 1966). Birds that are able to enter and arouse from torpor do not always become torpid at night. Even when the T_a is as low as 12°C, if in good condition and well fed, hummingbirds often maintain normothermia all night (Lasiewski, 1963).

2. Body Temperatures during Hypothermia

The lowest T_b values in hypothermia listed by Dawson and Hudson (1970) range from 4.8° to 34.0°C. The Turkey Vulture has both the largest body size and highest hypothermic T_b, this hypothermia not producing torpor (Heath, 1962). Similar nontorpid, shallow hypothermia is described for two cuculids, the Smooth-billed Ani (*Crotophaga ani*) and the Greater Roadrunner (Warren, 1960; Ohmart and Lasiewski, 1971). Lasiewski (1964) considered T_b values of 31.2°–36.3°C, reported as "torpid" or "hibernating" for hummingbirds by previous authors, to be indicative of "deep sleep" but still homeothermic. Newly captured passerines became hypothermic ($T_b = 30°$–38°C) when exposed to nocturnal temperatures of −5° and −10°C (Steen, 1958). Larger birds (geese, gulls, owls, magpie) had nocturnal

T_b depressions of 0.9,–4.0°C at $T_a = -20°C$ (Irving, 1955), while sunbirds (*Nectarinia*) exhibit amplitudes of 5°–17°C in T_b in high mountains (2700–4500 m) where T_a cycles from intense heat to freezing at night (Cheke, 1971). The relatively small *Parus major* and *P. atricapillus* also experience a hypothermia of 8–12°C during cold exposure in winter (Steen, 1958; Budd, 1972). As indicated in Section II,H, diurnal cycles of normothermic T_b are well known (Dawson, 1954; Dawson and Hudson, 1970; Aschoff and Pohl, 1970a,b; Binkley et al., 1971).

Considered together, these facts suggest no really clear distinction between nocturnal hypothermia and the basic daily cycle in T_b of Pearson's (1960) "obligate homeotherms." Such an impression is reinforced by the rather low temperatures reported for coordination and flight in arousing birds. White-throated Swifts (*Aeronautes saxatalis*) could move with effective coordination with $T_b = 25.8°C$ and had normal activity at $T_b = 35°C$ (Bartholomew et al., 1957). Normal activity with thermogenic "wing-buzzing" appeared in Anna's Hummingbird at 35.5°C. The Poorwill could fly at 34°C (Bartholomew et al., 1957). Austin and Bradley (1969) reported good flight in Poorwills when T_b reached 27.4°C.

3. Physiology and Behavior in Hypothermia

While the distinction between diurnal cycling of T_b and hypothermia is blurred, torpidity has been regarded as qualitatively different from homeothermy. As Dawson and Hudson (1970) have pointed out, there is little quantitative information from birds in natural torpidity. In the laboratory, body temperatures apparently can become too low to allow spontaneous arousal. Lasiewski (1963) did not find spontaneous arousal with $T_b = 12°C$ or lower in temperate-zone hummingbirds. Wolf and Hainsworth (1972a) have shown that hummingbirds of tropical highlands (*Eugenes fulgens* and *Panterpe insignis*) possess a regulated torpor with spontaneous arousal from T_b values of 10°–12°C. A Poorwill aroused spontaneously from a T_b of 6°C (Ligon, 1970). The related European Nightjar (*Caprimulgus europaeus*) shows two types of torpor in the laboratory. One is a light torpor of short duration, $T_b > 15°C$, which occurs in day and night, between crepuscular activity periods. Arousal is spontaneous and rapid, even with $T_a < 10°C$ (0°–10°C). The other was a deep torpor, $T_b < 15°C$, from which there was no spontaneous arousal (Peiponen, 1965, 1966, 1970). In torpor, the metabolic rates bear a direct exponential relationship to T_b, with Q_{10} values of 1.5–5.6 (mostly 4.1–4.3) for small hummingbirds (Lasiewski, 1964; Lasiewski and Lasiewski, 1967; Hainsworth and Wolf, 1970).

The metabolic cost of fasting is thus reduced by hypothermia considerably below that if normothermia is maintained, but is still measureable. Young European Swifts (*Apus apus*) which were fasted to enter terminal torpor lost an average of 52.8% of the prefasting weight. Weight losses of organs (percent of initial weights) were: fat, 100%; liver, 66%; pectoral muscles, 59%; lungs, 49%; heart, 41%; alimentary canal, 36% (Koskimes, 1948, 1950). Note, however, that it is not possible to distinguish between weight loss by catabolism and weight loss by dehydration in these data, both food and water having been denied the subjects.

Heart rates are also directly related to T_b, although the heart rate does not decrease as much as the oxygen consumption for a given drop in T_a (Lasiewski, 1964). Atrioventricular dissociation developed in Costa's Hummingbird (*Calypte costae*) and the Black-chinned Hummingbird (*Archilochus alexandri*) at T_b of 11.5°C and 7°C, respectively, but this was soon followed by death (Lasiewski, 1964). No major qualitative difference occurred in cardiac function between normothermic and torpid Poorwills, but only a prolongation of the interval between atrial and ventricular excitations (Bartholomew *et al.*, 1962). The breathing of hummingbirds during torpor was irregular, with periods of apnea (Lasiewski, 1964).

Thus, torpidity seems to be mainly a quantitative slowing of physiological rates. The qualitative differences between torpidity and diurnal T_b cycling are irregular breathing, the impairment of coordination, and diminished response to external stimuli which accompany deep torpor (Jaeger, 1948, 1949; Bartholomew *et al.*, 1957; Lasiewski and Thompson, 1966).

Recent studies of regulation during torpor in three species of hummingbirds suggest that the torpid state is more akin to homeothermy than has been generally assumed. The linear correlation between T_b and T_a in torpor of temperate-zone hummingbirds has been generally and quite logically regarded as "abandonment" of homeothermy (Lasiewski, 1964). Similarly, T_b decreases almost linearly with decreases of T_a down to 18°C for *Eulampis jugularis* and down to approximately 12°C for *Eugenes fulgens* and *Panterpe insignis*. This seems to be a poikilothermic transition between regulation in the active or normothermic state and regulation in the torpid or hypothermic state (see Figs. 23 and 24). However, there is in torpidity or hypothermia a second lower critical temperature ($T_{lc\ hypo}$) which is 12°–18°C below the T_{lc} of the nontorpid state. When these three species of hummingbirds are exposed to T_a below this $T_{lc\ hypo}$, metabolism is increased to prevent further cooling. Of particular interest are the regression lines for oxygen consumption below the $T_{lc\ hypo}$

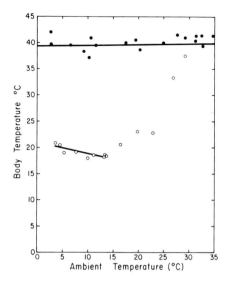

FIG. 23. Body temperature as a function of ambient temperature in the tropical hummingbird *Eulampis jugularis*. The solid circles and upper line is from nontorpid, normothermic birds. The open circles and lower line show body temperatures during torpor, which are regulated during exposure below approximately 12°C ambient. (Reprinted by courtesy of Hainsworth and Wolf, 1970; copyright 1970 by the American Association for the Advancement of Science.)

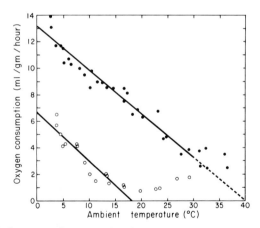

FIG. 24. Metabolism as a function of ambient temperature in a tropical hummingbird. *Eulampis jugularis*. The closed circles and upper line are from nontorpid birds. Open circles and lower regression line obtained during torpor. Note that the regression line in torpor extrapolates approximately to the body temperature in torpor shown in the previous figure. Thus, body temperature is regulated during torpor in accordance with the Scholander model, with the heat transfer coefficient or insulation approximately the same as during normal, nontorpid regulation. (Figure reproduced with permission from Hainsworth and Wolf, 1970; copyright 1970, American Association for the Advancement of Science.)

which extrapolate approximately to the regulated hypothermic T_b for all three species if metabolism were zero, i.e., $T_b = T_x$ (Hainsworth and Wolf, 1970; Wolf and Hainsworth, 1972a) (see Fig. 24).

Thus, in regulated torpor, the Scholander model seems to describe the heat exchange very well (Table XIII) and the existence of regula-

TABLE XIII
REGULATED HYPOTHERMIA IN SMALL BIRDS[a]

Species	Body weight (gm)	Regulated T_b in torpor or hypothermia (°C)	Scholander extrapolated T_x (°C)	Mean minimum T_a (°C)
Panterpe insignis[a]	4.9–6.5	10–12	12	4–6
Eugenes fulgens[a]	8.0–9.5	10–12	11	4–6
Eulampis jugularis[a,b]	7.0–10.9	18–20	18	16–19
Acanthis flammea[c]	10–12	32–34	33	(−)24
Parus major[c]	17–20	29–31	29	(−)24
Passer montanus[c]	22–24	28–31	29	(−)24
Passer domesticus[c]	26–28	35–37	34	(−)24
Fringilla montifringilla[c]	26–28	38–40	38	(−)24
Carduelis chloris[c]	29–32	33–35	33	(−)24

[a] Table adapted from Wolf and Hainsworth (1972a).
[b] Hainsworth and Wolf (1970).
[c] Steen (1958) and personal communication; for newly caught birds only.

tion during torpor is clearly substantiated. A similar response for hibernating bats has been proposed by Henshaw (1968). Similar agreement was also found in the less profound hypothermia of newly caught passerine birds exposed to nocturnal cold stress in the subarctic by Steen (1957). Wolf and Hainsworth (1972a) correlated the regulated body temperatures in torpor with the environmental temperature, there being no apparent body-size relationship in these three species of hummingbirds. It would be interesting to learn how the regulated T_b is related to habitat over a wide range of hummingbird species. Are North American hummingbirds of the temperate zone poikilothermic conformers, incapable of arousing from temperatures below 13°C, while tropical hummingbirds can survive lower temperatures at which they regulate? Female Calliope Hummingbirds (*Stellula calliope*) maintain normothermy during incubation and brooding during exposure to 0°C in northwestern Wyoming (Calder, 1971), but what do the males do at night? Anna's Hummingbird survives occasional winter cold snaps to −9°C in southeastern Arizona. Thus, the phenomenon of regulated hypothermia may be more widespread than is presently evident from laboratory studies, and we are again acutely aware of the paucity of data from wild torpid birds. Records of hypothermia (n = 12) in nests of the Broad-tailed Hummingbird (*Selasphorus platycercus*) were correlated with reduced feeding opportunity, and nest temperatures stabilized at 6.5°C or higher as if regulated (Calder and Booser, 1973; Calder, 1974).

4. Torpor Entry and Arousal

As we have indicated above, hypothermia is a significant way to conserve energy. Cooling from a normal to a hypothermic state results from a rate of heat loss that exceeds heat production. This may be accomplished by decreasing (minimizing) heat production, by increasing heat loss (increased h), or by a combination of these. Once the time has come to suspend activity and enter torpor, the sooner the T_b decreases, the more energy is conserved, and the longer the bird can remain at the most economical level of fuel consumption consistent with protection from thermal damage. On the other hand, given the time or need for activity, rapid warming is advantageous. Thus, maximal heat transfer and minimal heat production are best for economy in cooling, while the opposite obtains for rewarming.

There have been no detailed analyses of heat exchange during the torpor cycle in birds, such as Tucker (1965) has made for the California Pocket Mouse (*Perognathus californicus*), and understanding of heat exchange in the avian torpor cycle is incomplete. We can speculate that the bird, like the pocket mouse, probably reduces metabolism almost to the minimum at a given T_b, while h is increased above minimum to facilitate cooling. The entry into torpor approximates a Newtonian cooling curve, i.e., $\log(T_b - T_a)$ is linearly related to time. Arousal from torpor is more rapid than entry into torpor (Pearson, 1960; Lasiewski and Lasiewski, 1967; Lasiewski *et al.*, 1967).

D. Influence of Body Size on Torpor Entry and Arousal

Hypothermia has been observed in birds ranging in body mass from 3 to 2230 gm (Dawson and Hudson, 1970), the heaviest of which weighs less than the largest hibernating mammals, marmots (*Marmota* sp.; Benedict and Lee, 1938) and the echidna (*Tachyglossus aculeatus*; Augee and Ealey, 1968). Thus the hypothermia of torpor and hibernation appears to occur only in smaller birds, whose greater metabolic intensities make it more important to them. In this subsection we examine the allometric aspects of hypothermia.

The ratio of metabolic rates in torpor to rates of resting normothermic birds ($T_a = 15°–16°C$) is inversely related to body mass: 1:54–1:59 in hummingbirds and 1:13 in the larger (75 gm) Common Nighthawk, *Chordeiles minor* (Lasiewski and Lasiewski, 1967). Of course, part of the saving is used in the arousal process, but only 1/85 of the total 24 hour metabolic expenditure of a 4 gm hummingbird is incurred going from 10° to 40°C T_b (Pearson, 1960).

It has been noted that entry into torpor with consequent energetic economy is expedited by reducing heat production and increasing heat loss. While these responses have not been quantified in the physiology of any species, the influence of body size on rates of entry into and arousal from torpor is well documented.

The smaller the bird, the higher the heat transfer coefficient and the ratio of surface to volume, and the lower the total heat capacity of the body. The hummingbird can therefore cool more rapidly than the caprimulgid; i.e., the rate of cooling is inversely related to body mass (see Fig. 25) (Lasiewski and Lasiewski, 1967; Lasiewski et al., 1967).

The smaller bird has a more intense metabolism (calories per gram-hour) and a smaller mass to warm up, so that warming rate is also inversely related to body size (Bartholomew et al., 1957; Lasiewski and Lasiewski, 1967; Lasiewski et al., 1967). The rate of warming of twenty-four species of birds and mammals was summarized by Heinrich and Bartholomew (1971) (see Fig. 26).

$$°C/\text{minute} = 2.03\,m^{-0.40} \tag{73}$$

where $m =$ body mass in the range of 3.2–220 gm; T_a and initial T_b were between 20° and 25°C. Extrapolated, this predicts that a 2230 gm Turkey Vulture could warm 0.093°C/minute compared with about 1.3°C/minute for a 3.2 gm hummingbird (Lasiewski and Lasiewski, 1967), and with a mean rate of 0.05°C/minute actually recorded from a Turkey Vulture (Heath, 1962). Heath also reported a warming from 34° to 37.5°C in 3 minutes attributed to effective use of shivering when the bird was awakened by probing. This, however, must represent internal heat transfer or local calorigenesis only, for if the entire body warmed at that rate, assuming a specific heat of 0.8 cal/gm °C, the metabolic heat production would equal 21 times the standard metabolism, not counting any loss by conduction, convection, radiation, or evaporation. Pearson (1960) calculated that a 200 kg bear would need the equivalent of its total energy budget for 24 hours to warm from 10° to 37°C, contrasted with a hummingbird's expenditure of 1/85 of

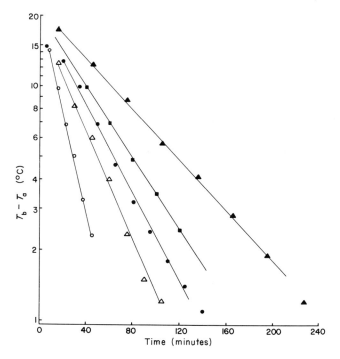

FIG. 25. Rates of body cooling during entry into torpor in five species of birds. Semilogarithmic plotting transforms the exponential curves into straight lines, the negative slopes of which are inversely related to body size. A comparable line for the Turkey Vulture's hypothermia would extend from ΔT of 23° to 20°C over the same time period. Key: O = *Archilochus* (4.0 gm), ▲ = *Eugenes* (6.8 gm), ● = *Lampornis* (8.5 gm), ■ = *Patagona* (21 gm), △ = *Phalaenoptilus* (40 gm). (Reprinted from Lasiewski and Lasiewski (*Auk* **84**, 1967) with permission.)

the daily budget (17 minutes at the average rate) in warming from 10° to 40°C. Lasiewski and Lasiewski (1967) calculated that a bird weighing 80–100 gm would take 12 hours to cool to 20°C and rewarm to normothermy utilizing endogenous heat only (see Table XIV). Thus, larger birds are limited by their body mass to relatively shallow fluctuations in T_b, rather than profound torpor, and/or to utilization of supplementary heating from solar radiation. Both shallow hypothermia and sunning behavior, but not deep torpor, are reported for the Smooth-billed Ani (Warren, 1960), the Greater Roadrunner (Ohmart and Lasiewski, 1971), and the Turkey Vulture (Heath, 1962).

In summary, the hummingbirds can rapidly cool and rewarm; because of their high metabolic intensity, it is very economical for them to become torpid for short nocturnal periods. Birds of intermediate size, e.g., the Poorwill, can survive longer periods of cold and food

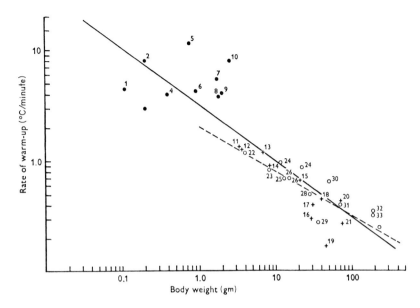

FIG. 26. Rates of rewarming during arousal from torpor as a function of body weight (W). Ambient and initial body temperatures were between 20° and 25°C. The larger the animal, the slower it warms, related to the lower metabolism per gram of tissue. The solid line indicates rates for insects, birds, and mammals (°C/minute = $3.22W^{-0.51}$); the broken line indicates rates for birds and mammals only (°C/minute = $2.03W^{-0.40}$). (Reprinted from Heinrich and Bartholomew, 1971, with permission.)

TABLE XIV
TIME REQUIRED FOR COOLING TO $T_b \approx 20°C$ AND ENDOTHERMIC REWARMING AT TORPOR ENTRY AND AROUSAL RATES FOR SOME BIRDS ($T_a = 20°C$)[a]

Species	Body mass (gm)	Approximate entry and arousal time (minutes)
Archilochus alexandri	4	75
Eugenes fulgens	6.8	126
Lampornis clemenciae	8.5	160
Phalaenoptilus nuttallii	40	<285
Hypothetical	80–100	720

[a]Calculations of Lasiewski and Lasiewski (1967).

shortage by hibernating (Jaeger, 1949; Bartholomew et al., 1957). Still larger birds, with lower metabolism per gram of body mass, lose heat at modest rates and cool slowly, so that even hibernation may not be practical. However, in combination with heliothermia, limited nocturnal hypothermia becomes practical for some larger birds such as the Turkey Vulture and the roadrunner.

VI. Integration of Thermal and Caloric Responses

The modes of thermal and caloric response described in the preceding sections are controlled and integrated in reaction to minute-to-minute variations in environmental conditions (thermoregulation in the usual sense) and in response to seasonal variation (acclimatization). It is convenient to treat these topics separately, although it should be kept in mind that adjustments on either time scale have elements in common. We proceed from the hypothesis that the predominant regulatory systems have evolved so that core body temperature is kept within narrow limits at a minimum average energy cost to the animal. Exceptionally, a few species of birds have developed a capacity for regulated hypothermia in conditions in which the energy drain of normothermia is potentially lethal (Section V).

A. Short-Term Regulation

1. Central Neural Control

Insufficient information is available from birds to allow more than a superficial characterization of their thermoregulatory systems. At present it appears that thermoregulation in birds does not differ in any important way from the much better known analogous processes in mammals (Bligh, 1966; Hammel, 1968; Hardy et al., 1970). Briefly stated, and ignoring apparent differences among species, the mammalian thermoregulatory system is integrated through nuclei of the spinal cord and ventral preoptic hypothalamus. It is thought that an area of the anterior hypothalamus functions as a heat loss center regulating panting, sweating, and vasodilation, while an area of the posterior hypothalamus serves as a heat maintenance center, regulating shivering and (in the newborn calf and the cold-acclimated adult mammal) nonshivering calorigenesis. Sensory information is derived from thermosensitive neurons of the hypothalamus and spinal cord as well as from thermoceptors in the skin and perhaps elsewhere in the body. While the hypothesis may still be controversial, it appears probable that hypothalamic temperature is tightly regulated and is adjusted through reference to a set point of neural activity that itself may be modulated by afferent input from spinal and peripheral receptors. It is against the foregoing comparative background that what is known of the avian thermoregulatory apparatus may now be summarized (see also Whittow, 1965a; Dawson and Hudson, 1970).

a. Central Components and Their Organization. Thermoregulatory functions of the avian brain have been explored through the ablation of various structures (by transection or by stereotaxic placement of lesions) and by stereotaxic exploration with stimulating electrodes or with thermodes allowing localized stimulus or cooling and heating of selected structures. The results of these methods of investigation are in general agreement, revealing an anatomical arrangement and function of thermoregulatory centers analogous to those of mammals. They confirm and extend the earlier studies by von Saalfeld (1936) and of Rogers and co-workers (e.g., Rogers, 1919, 1928; Rogers and Wheat, 1921; Rogers and Lackey, 1923).

Kanematsu et al. (1967) found in the chicken that bilateral electrolytic lesions in the hypothalamus anteroventral to the anterior commissure resulted in a persistent or a definite transient hyperthermia (45.1°–46.0°C in nine birds, compared with 42.2°C in sham-operated controls). In contrast, birds with bilateral lesions ventrocaudal to the anterior commissure lost the ability to maintain normal rectal temperature when placed in a cold environment. These birds showed persistent hypothermia or poikilothermia, the rectal temperature changing with each change of air temperature. These results are consistent with the proposal of an anterior hypothalamic heat loss center and a posterior hypothalamic heat maintenance center in mammals (Benzinger, 1970).

Similarly, bilateral lesions in the region of the anterior commissure in chickens interfere with thermoregulation (Lepkovsky et al., 1968). In House Sparrows (*Passer domesticus*), bilateral lesions in the anterior hypothalamus and preoptic area likewise prevent the birds from maintaining normal body temperature when $T_a = 10°–40°C$; low body temperature was associated with low air temperature. Lesions slightly rostrad in the anterior forebrain and preoptic area had only slight effects on thermoregulation in this species (Mills and Heath, 1972).

The results of the selective ablations in chickens and sparrows thus confirm the results of the pioneering experiments of Rogers (1919) in domestic pigeons. His pigeons, decerebrated by transection of the neuraxis at the level of the cerebral peduncles, were essentially normothermic at $T_a = 16°–40°C$ and possessed functional shivering, panting, and ptilomotor reflexes. Cauterization of the thalamus (presumably including the hypothalamus) abolished these reflexes and rendered the pigeons poikilothermic.

The primary thermoregulatory nuclei of the anterior hypothalamus in birds (chickens and pigeons) function together with a mesencephalic panting center (von Saalfeld, 1936; Sinha, 1959). Recent stereotaxic exploration with stimulating electrodes in chickens and pigeons has shown that the panting center is located in a discrete part of the dorsal mesencephalon corresponding to the nucleus mesencephalicus lateralis pars dorsalis (Richards, 1971a), a region that Richards suggests may be anatomically homologous as well as functionally analogous to the mammalian pneumotaxic center.

The mesencephalic panting center is essential for the polypneic response to thermal stress in the pigeon and chicken (Richards, 1971a), and probably in all avian species. Its relationship to the hypothalamic thermoregulatory centers is not completely clear. Electrical stimulation of the pigeon hypothalamus at the level of the anterior commissure induced panting that ceased immediately with cessation of stimulation; at slightly more rostral levels (n. preopticus medialis et lateralis), polypnea persisted after the cessation of stimulus, the duration of the after effect being proportional to the strength and duration of stimulation (Åckerman et al., 1960). Richards (1971a) reports similar results from the pigeon, but could not elicit panting by hypothalamic stimulation in the chicken. Transection of the brainstem behind the hypothalamus in this species did not always abolish panting in response to hyperthermia, although it raised the threshold. This increase of threshold may explain the fact that some of the chickens studied by Lepkovsky et al. (1968) and Kanematsu et al. (1967) failed to respond by panting in environments that produced this response in control birds. The difference between pigeons and chickens in the relationship between the hypothalamus and the panting center may be only quantitative, resulting from different levels of facilitation (greater in the chicken than the pigeon) of the panting center by impulses descending from the hypothalamus.

b. *Central Neurotransmitter Substances in Thermoregulation.* Feldberg (1965, 1970) proposed the hypothesis for mammals that hypothalamic thermoregulatory neurons are affected by noradrenaline and 5-hydroxytryptamine, probably produced as neurotransmitters by hypothalamic presynaptic neurons in the afferent pathways. Release of noradrenaline is believed to result in cutaneous vasodilation, decreased muscle tone, polypnea, and a decrease of body temperature; release of 5-hydroxytryptamine is believed to result in cutaneous vasoconstriction, increased muscle tone and shivering, and increased body temperature. This is the pattern observed in cats, dogs, and monkeys in response to microinjections of the amines into the hypothalamus or the third ventricle, but other mammalian species show opposite or different patterns of response (Feldberg, 1970).

Similar studies among birds have been made only with chickens (Marley and Stephenson, 1970), which appear to respond like cats, dogs, and monkeys. Injection of 0.05–0.2 μmoles of α-methylnoradrenaline into the hypothalamus (by way of a chronically implanted microcannula) of chickens 14–21 days of age consistently reduced body temperature by as much as 6°C and concurrently reduced oxygen consumption (Marley and Stephenson, 1970). Noradrenaline produced the same effects. The sensitivity of the hypothalamus to α-methylnoradrenaline was greatest in the region above the optic chiasma; injections of this amine into the cerebral tissue (neostriatum intermediale) or into the brain stem at the level of the trochlear nerve had no effect. Previous studies had shown identical results from large doses administered intravenously or intraperitoneally (Allen and Marley, 1967; Allen et al., 1970), and it is surmised that these effects were exerted as the amines penetrated the hypothalamus. A sharp decline in body temperature and oxygen consumption and a transient reduction in shivering resulted from systemic administration of α-methylnoradrenaline in spite of a strong peripheral vasoconstriction that would reduce heat loss (Allen et al., 1970). In adult chickens, in which the blood–brain barrier is mature, intravenous injection of catecholamines failed to induce hypothermia, although microinjection into the hypothalamus did so (Grunden and Marley, 1970). There is more than a tenfold reduction between the ages of 1 and 30 days in the rate at which tritium-labeled noradrenaline accumulates in the brain of chickens (Spooner et al., 1966).

The thermoregulatory effects of intravenous doses of noradrenaline in adult chickens in other experiments by Allen and Marley (1967) were slight and erratic among individuals even at very high dose levels (to about 3200 μg/kg). In contrast, Freeman (1970a) observed significant decreases of body temperature in Japanese Quail 6–9 weeks old (sexually mature) within 1 hour after intraperitoneal injection of adrenaline, noradrenaline, and 5-hydroxytryptamine (300 μg/kg).

The absence of thermoregulatory effects, at least in adult chickens, of catecholamines administered systemically suggests that these substances are not acting directly on peripheral receptors. Further evidence supporting this viewpoint was obtained from chickens 10–16 days old in which the brainstem was transected completely, caudal to the respiratory center. Intravenous administration of α-methylnoradrenaline (to 40 μmoles/kg) was ineffective in inducing hypothermia in four of the five experimental birds, and in the fifth a dose of 20 μmoles/kg reduced the body temperature by only 1.5°C (Allen et al., 1970). Plainly, an intact neuraxis is required for the full expression of noradrenaline-induced hypothermia in young chickens.

Presumably, efferent pathways must be intact to convey signals from the amine-activated (or suppressed) hypothalamic centers, although a modulating role of sensory input cannot be discounted (Section VI,A,2). The thermoregulatory effects of noradrenaline and α-methylnoradrenaline were prevented or substantially reduced by pretreatment of the chicks with phenoxybenzamine (which blocks transmission at sympathetic α-adrenoceptors) administered either intravenously or in the hypothalamus. The β-blocking agent propranolol also prevented the hypothermic effects of the catecholamines, which was an unexpected result suggesting that the hypothalamic thermoregulatory nuclei contain catecholamine-sensitive neurons with the characteristics of both α and β receptors (Marley and Stephenson, 1970).

Earlier experiments by Allen and Marley (1967) had shown that intravenous administration of α-methyltryptamine (a long-lasting functional analog of 5-hydroxytryptamine) to chickens 12–15 days old increased body temperature (e.g., 1.1°C) and oxygen consumption (e.g., 45%), and approximately doubled electromyographic activity. Intrahypothalamic effects of tryptamine and its derivatives have not yet been reported for birds, but the intravenous effects conform with those found in dogs, cats, and monkeys (Feldberg, 1970), suggesting provisionally that similar mechanisms occur in the two classes of homeotherms.

Several other investigators have explored the potential thermoregulatory role of catecholamines in chickens (Freeman, 1966, 1967, 1970a; Wekstein and Zolman, 1968, 1969); but the interpretation of the results of these investigations is confounded by differences in the maturity of the experimental birds, in the dosages of the hormones and the routes of administration, by problems of distinguishing between central and peripheral effects, and by other discrepancies of technique. A useful review of the data is not possible in the space available here. Certain aspects will be considered in Section VI,3,b.

2. Sensory Input

There is indirect evidence that information about the thermal status of the body is derived from hypothalamic, spinal, and cutaneous receptors in birds, but direct measurement of potentials in thermosensitive receptors or afferent tracts has been accomplished only in the sensory fields of the glossopharyngeal (Kitchell *et al.*, 1959) and trigeminal (Necker, 1972) nerves in the tongue or beak of chickens and pigeons. Neural responses to thermal stimuli could be invoked only from the edges and "inside" of the beak in the case of pigeons, but not from the rhamphotheca and cere (Necker, 1972). It seems likely that these receptors function mainly in sampling the temperature of food and water, and their role, if any, in general thermoregulation remains unclear. By analogy with what is known of mammals (Bligh, 1966; Hensel, 1970), it may be surmised that birds possess widespread cutaneous thermoceptors as well as thermosensitive neurons in the brain and spinal cord.

Heating the anterior hypothalamus of the House Sparrow to 44°–45°C by means of miniature thermodes, stereotaxically placed, induced reduction of oxygen consumption and cloacal temperature. Conversely, cooling of the thermode to 18°–21°C induced an increase of oxygen consumption and of cloacal temperature (Mills and Heath, 1970). The effective hypothalamic locus for these responses was centered in the preoptic region dorsal and anterior to the n. preopticus anterior (Fig. 27). The responsive region was bordered dorsally and rostrally by thermally unresponsive regions.

Correlations between hypothalamic temperature and thermo-

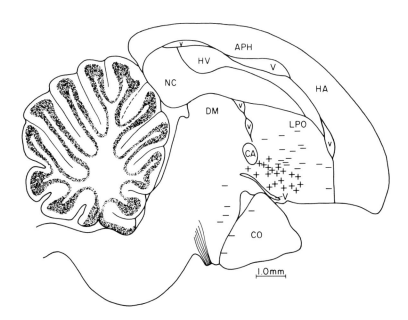

FIG. 27. Parasagittal section (about 0.5 mm from midline) of the brain of a House Sparrow (*Passer domesticus*), showing thermally sensitive (+) and insensitive (−) points for induction of thermoregulatory responses. Explorations were made by means of miniature thermodes. Salient hypothalamic landmarks: CO = optic chiasma; CA = anterior commissure. (From Mills and Heath, 1970, as modified by courtesy of J. E. Heath.)

regulatory responses in birds provide further evidence about the functions of central thermosensitive elements. Deep brain temperature in chickens kept in constant environmental conditions displays a diurnal cycle with a mean amplitude of 1.3°C per cycle (von Saint Paul and Aschoff, 1968), or 1.1°–2.2°C per cycle, in parallel with a cycle of deep body temperature (Scott et al., 1970; Scott and van Tienhoven, 1971). Minor waves of 15–30 minutes in duration were superimposed on the longer rhythm. At different depths in the brain down to 12 mm, the short oscillations were mainly in phase with each other (von Saint Paul and Aschoff, 1968; Scott and van Tienhoven, 1971). These and other data are all consistent with the hypothesis that temperature changes in the brain reflect mainly changes in blood flow; some data disagree with the alternative hypothesis that the temperature variations reflect changes in blood temperature of the intensity of tissue metabolism. These tentative conclusions suggest that brain temperature is dependent primarily on modulations of vasomotor tonus, a point that has significant implications for the interpretation of feedback loops in the thermoregulatory system.

The hypothalamic temperature (T_{hyp}) of unstressed chickens is consistently lower (by 0.9°–1.0°C on the average) than intraabdominal or rectal temperature (Scott and Van Tienhoven, 1971; Richards, 1970a, 1970d). The amplitude of the daily cycle of T_{hyp} is greater than that of T_b, and consequently the difference $T_b - T_{hyp}$ is greater at night, reaching 1.5°–2.0°C (Fig. 28). The difference diminishes progressively while air temperature increases to a level (40°C) causing panting (Richards, 1970a), but the difference is reversed (hypothalamus warmer) apparently only in anesthetized birds (Richards, 1970a).

Richards (1970a, 1970d) suggests that a hypothalamic temperature lower than rectal temperature can result from relatively cool blood from the skin of the head entering the cavernous sinus via the superficial ventral ophthalmic veins; the cavernous sinus itself surrounds the intercarotid anastomosis from which the hypothalamus is directly supplied with blood in the chicken (Richards, 1967), Japanese Quail (Sharp and Follett, 1969), and other species (Vitums et al., 1964, 1965).

In contrast with the condition found in chickens, the hypothalamic temperature in Japanese Quail reared and kept at $T_a = 21°C$ averaged 42.7°–42.8°C, or consistently 0.6°–0.7°C *more* than mean rectal temperature (Yousef et al., 1966). There were no significant differences among birds reared at $T_a = 21°C$ or 32°C or in daily photoperiods of L:D 8:16 and L:D 10:14. Exposure of quail reared at 32°C to a cold stress ($T_a = 5°C$) resulted in a sharp decrease of T_{hyp} to a level that

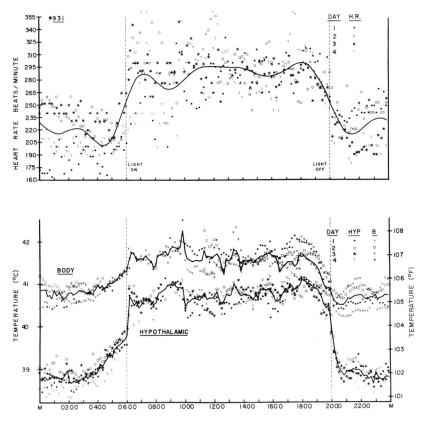

FIG. 28. Diurnal cycle of heart rate and of intraperitoneal (body) temperature and hypothalamic temperature measured simultaneously in unrestrained chickens. Air temperature 20°–24.3°C. (From Scott and van Tienhoven, 1971.)

remained below T_b until the birds were returned to a thermoneutral environment (Fig. 29). It is not clear whether the difference observed between quail and chickens is genuine species difference or an artifact of measuring technique. Scott and van Tienhoven (1971), Scott et al. (1970), and Richards (1970a, 1971b) recorded temperatures remotely from indwelling sensors in undisturbed chickens. Yousef et al. (1966) evidently handled the quail during measurements at 2-hour intervals.

It is evident that there are interactions between hypothalamic temperature and input from peripheral thermosensors in the control of thermoregulatory reactions. Exposure of chickens to air temperature increasing either gradually or abruptly from 21°C to 40°C induces panting only when both rectal and hypothalamic temperatures

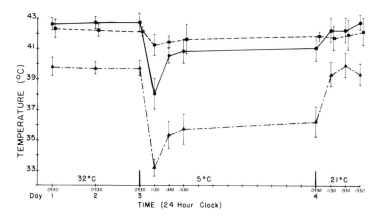

Fig. 29. Hypothalamic (●), rectal (■), and skin (▲) temperatures (mean ± SD) of hand-held Japanese Quail. Segments along the abcissa indicate the sequence of exposure to T_a = 32°, 5°, and 21°C. (From Yousef et al., 1966.)

are elevated above the normal range (Richards, 1970a). Heating the skin alone does not induce panting unless there is also an increase of central body temperature. Activation of cutaneous thermoceptors alone is therefore not a sufficient stimulus for panting in chickens. Randall (1943) reached this same conclusion for day-old chicks in which skin temperature was raised artificially to 45°C while the carotid blood was cooled by means of a collar through which cold water was circulated.

While an increase in cutaneous temperature alone is not sufficient to initiate panting in chickens, it is nevertheless evident from the experiments by Richards (1970a, 1971b) that there are interactions between central and peripheral thermal stimulation as well as between hypothalamic and extracranial deep body temperatures. The increment (above control levels) of both rectal and hypothalamic temperatures at the onset of panting was an inverse curvilinear function of air temperature, and presumably of skin temperature, in chickens exposed to infrared irradiation (Fig. 30), suggesting that cutaneous thermoceptors have at least an inhibitory influence on the central regulation of panting. Randall and Hiestand (1939) had noted earlier that panting in chickens was immediately inhibited by a blast of cold air, and Steen and Steen (1965) found that panting stopped immediately in a Great Black-backed Gull when the legs were irrigated with cold water.

An interaction between hypothalamic temperature and extracranial deep body temperature is suggested by several aspects of the investi-

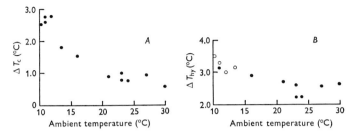

FIG. 30. The relationship between air temperature and the increment of colonic temperature (A) and hypothalamic temperature (B) in adult chickens at the onset of panting in response to infrared irradiation of the thorax and abdomen (A) and of the head (B). The open circles in B indicate the increment of hypothalamic temperature in birds in which panting could not be elicited. (From Richards, 1970a.)

gations by Richards (1970a, 1971b), but is illustrated best by his experiments with anesthetized chickens in which hypothalamic temperature and rectal temperature were controlled separately (Fig. 31). When hypothalamic temperature was held at 40°C, raising the rectal temperature to 46°C was ineffective in initiating panting; but when

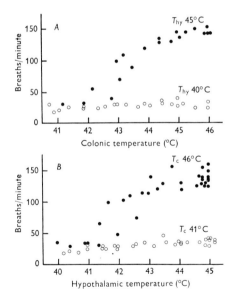

FIG. 31. The influence of hypothalamic temperature and of extracranial deep body temperature on the respiratory frequency of anesthetized chickens. (A) The effect of raising the deep body temperature at two controlled hypothalamic temperatures. (B) The effect of raising hypothalamic temperature at two controlled colonic temperatures. (From Richards, 1970a.)

hypothalamic temperature was maintained at about 45°C, respiratory frequency rose sharply when rectal temperature exceeded about 42.5°C (Fig. 31A). Data from the reciprocal experiment (T_r constant, T_{hyp} increasing), with similar results, are shown in Fig. 31B. These results imply that a facilitating effect from extracranial deep body receptors is necessary before panting will occur. Rautenberg (1969a,b) has provided strong evidence that thermosensitive structures lie within the spinal cord in pigeons. By heating or cooling thermodes that were chronically implanted in the peridural space he was able to produce the full range of thermoregulatory responses, including panting, peripheral vasomotion, ptilomotor reactions, shivering, and variations of oxygen consumption, without significant variations of brain temperature or rectal temperature. Additional investigations (Rautenberg *et al.*, 1972) in pigeons show that temperature signals generated by the hypothalamic region affect mainly ptilomotor and vasomotor reactions to a mild thermal load, while changes of spinal cord temperature invoke shivering and panting that counteract hypo- and hyperthermia in birds subjected to stronger thermal stress. There thus appears to be a hierarchical arrangement of regulatory responses in the neuraxis.

In the studies by Rautenberg (1969a,b) the threshold temperatures (peridural) for panting induced by heating the spinal cord thermode was inversely related to T_a and T_{skin} in a manner analogous to that in the chicken. The temperature threshold for the onset of shivering in further experiments with pigeons (Rautenberg, 1971) was correlated better with skin temperature than with any combination of skin and core temperature that was examined. The threshold (T_{skin}) for shivering increased from 34° to 41°C as the temperature of the spinal cord was reduced from 43° to 37°C. From these observations plus additional data on heat production, it was concluded that heat production is influenced by a proportional control mechanism and that the change of skin temperature is accompanied by an inverse change in the set point of the regulated body temperature (spinal cord temperature). Thus, for instance, if the spinal cord tends to cool, the change invokes a compensatory increase in heat production that is amplified by decreasing skin temperature. The gain of the control system ($\Delta BMR/\Delta T_{sp}$) is the same for all skin temperatures within the range examined.

Finally, in connection with sensory input to the brain stem, some mention must be made of the role of the vagus nerves in relation to panting (see also Lasiewski, 1972). There are apparently differences among species in the dependence of panting on vagal input. Bilateral vagotomy in the domestic duck, Japanese Quail, and chicken abolishes reflex panting in response to heat stress (Richards, 1968). Panting in these species is evidently dependent on afferent vagal drive or facilitation. In contrast, panting in the domestic pigeon is reduced in frequency but not abolished after bilateral vagotomy (von Saalfeld, 1936), and may be restored to normal by electrical stimulation of the vagus (Sinha, 1959). The control of panting in the pigeon, as in many mammalian species, is dominated by central controls (for review, see Richards, 1970b). The characteristics of the sense organs involved in generating the vagal facilitation of panting are essentially unknown at present (Richards, 1970b, p. 244).

3. Effector Responses

We have described elsewhere in this chapter several aspects of thermoregulatory responses, including shivering (Section IV,B,3), vasomotion in appendages (Section IV,C), ptilomotor reactions (Section IV,C), and various behavioral reactions (Section VI,A,3c), and we will confine attention in the present section to what little is known of the central control of these phenomena. A cybernetic model emphasizing ptilomotor control in the Ring Dove is presented by McFarland and Budgell (1970).

a. Vasomotor Reactions. Direct measurements of blood flow in superficial tissues of birds have been made only in the legs and feet (Jones and Johansen, 1972), and assessment of vasomotor reactions and variations in blood flow in other appendages and the skin of the body must depend on the reasonable assumption that variation of temperature in superficial structures reflects variation of blood flow (e.g., von Saint Paul and Aschoff, 1968; van Kampen, 1971). On this basis, Richards (1971b) estimated that the range of vasomotor control of heat loss from the feathered surfaces of chickens is insignificant, compared with the range of control in the legs, comb, and wattles. The rhythmic variations of heat loss from the comb and wattles of hyperthermic chickens (van Kampen, 1971) indicate a cycle of vasomotion with a period of about 5 minutes.

Knowledge of the thermoregulatory functions of the legs and feet of aquatic birds has been thoroughly reviewed by Jones and Johansen (1972) and need not be reiterated in detail here. Briefly, the evidence indicates that seven- to twentyfold increases in blood flow can occur in the legs and feet of Giant Fulmars (*Macronectes giganteus*), and presumably in other web-footed birds subjected to heat stress, compared with the flow in thermoneutral surroundings. The control of vasomotion is by a synergistic neurogenic mechanism involving both a cholinergic vasodilator and an adrenergic vasoconstrictor. Presumably the reflux is initiated by cutaneous receptors in the legs and/or feet, since it is induced by immersion of the legs of normothermic birds in ice water.

At least in the chicken and the pigeon, there is a parallel central control of thermoregulatory vasomotion in the legs, independent of reflexes initiated by cutaneous receptors. Richards (1968) found that toe temperature in chickens increased from about 20° to about 35°C ($T_a = 20°C$) when hypothalamic temperature was increased 3°C by infrared irradiation of the head. Rectal temperature and skin temperature on the back were stable during this period. Rautenberg (1969b) obtained similar results upon varying the temperature of the spinal cord in pigeons. The most prominent effects were observed on warming the cord (e.g., his Fig. 8), when foot temperature increased from 27° to 38°C while rectal, cranial, and skin (rump) temperatures actually decreased slightly. Less consistent or prominent effects (decreases of foot temperature) were initiated by cooling the cord. These data implicate extracranial thermosensitive elements in the control of peripheral vasomotion. The data of Murrish (1970) on leg temperature of the American Dipper are consistent with the concept of a centrally initiated vasodilation, but the experiments were not designed to discriminate clearly between central and reflex initiation.

Very little is known about thermoregulatory vasomotion in structures of birds other than the legs. The combs and wattles of chickens are significant avenues of regulated heat transfer (Whittow *et al.*, 1964) and appear to be affected by vasomotor controls like those of the legs (Richards, 1970a, 1971b; van Kampen, 1971). Richards suggests that vasomotion in the extremities (and presumably in the skin generally) is under a "fine reflex control." The reciprocal wavelets of temperature variation observed by von Saint Paul and Aschoff (1968) in skin temperature and hypothalamic temperature of chickens indeed suggest a sensitive calibration of vasomotor activity in skin, but the data do not discriminate between central and reflex controls.

b. Calorigenesis. We have noted already (Section IV,B,3) the evidence that shivering is the only mode of thermoregulatory heat production in fasting adult birds at rest, and that electromyographic activity tends to be an inverse linear function of air temperature (West, 1965; West *et al.*, 1968) and a direct function of oxygen consumption (Fig. 14). We summarize in this section the available data concerning the control of shivering.

As mentioned previously, Mills and Heath (1970) showed that

oxygen consumption by House Sparrows could be varied in the expected directions by heating or cooling the anterior hypothalamus by means of thermodes. Presumably, the variation of oxygen consumption reflects variation of muscle tonus and the intensity of shivering. Shivering can be initiated and terminated in pigeons by varying the temperature of the spinal cord, as shown by Rautenberg's (1969a) elegant experiments. As in the cases of panting and of peripheral vasomotion, this shows a role of extracranial thermoceptors in the control of heat production. Within the thermoneutral range ($T_a = 20°$–$30°C$) there was an inverse correlation between air temperature and the slowly increasing peridural temperature at which electromyographic evidence of shivering was first detected. This variation of the shivering threshold with variation of air temperature suggests a modifying influence from other thermoceptors, probably cutaneous (Rautenberg, 1971).

The upper frequency of shivering (electromyographic activity of pectoral muscle) was a direct function of air temperature in the Evening Grosbeak (~ 57 gm), so that the maximum power output tended to be concentrated at lower frequencies as air temperature decreased. Such was not the case of the Common Grackle (~ 125 gm), whose upper frequencies were essentially independent of air temperature through a range of $40°C$ and were about half the frequency observed in the grosbeak (West et al., 1968). This suggests that shivering is modulated by air temperature in the grosbeak but not in the grackle, but the differences of size and, potentially, resonant properties of muscle vibration in these two species make it impossible to state rigorous conclusions at the present time.

Although the current evidence suggests that adult birds do not employ regulatory nonshivering calorigenesis, it has been proposed that newly hatched precocial birds (chickens) depend initially on nonshivering calorigenesis in a manner analogous to that of at least some species of mammals (Freeman, 1966, 1967, 1970a; Wekstein and Zolman, 1968, 1969). This assumption may be premature and needs to be examined in relation to the empirical data. It appears that the concept originated by analogy with the mammalian pattern and from the observations of Randall (1943) and Freeman (1966) that the increase of oxygen consumption noted in neonate chickens exposed to cold was "accompanied by little or no shivering" (Freeman, 1966, p. 378). This is not particularly persuasive evidence of the absence of shivering. Electromyographic evidence of increased muscle tone and shivering may be detected without overt, visually detectable tremors (Hemingway, 1963, p. 399). Visible shivering with increase in the rate of body warming occurs in 1-day-old chickens recovering from anesthesia induced by sodium pentobarbital (Freeman, 1970a), which reveals that chickens of this age already possess functional efferent pathways. Odum (1942) had previously observed that brief tremors with a frequency matching those of muscle tremors could be detected in embryos of Ring-necked Pheasants as early as the ninth day of incubation. Kespaik and Davydov (1966) detected shivering (chiefly in the legs) electromyographically in Black-headed Gulls (*Larus ridibundus*) on the day of hatching. Allen et al. (1970, p. 677) consistently detected electromyographic activity in chickens 1–7 days old (exact age not specified) at $T_a = 31°C$. Upon transfer to $T_a = 16°C$, eight of eleven chicks in this age group showed an increase of mean oxygen consumption amounting to 444%, and an increase of 1280% in the integrated electromyogram (mean body temperature decreased 7.5°C); three of eleven chicks were not able to thermoregulate (mean body temperature decreased 18.9° from 39.9°C), and mean oxygen consumption decreased 8% although the integrated electromyographic activity increased by 124%. Older chicks (up to 21 days of age) were able to thermoregulate at 16°C, although imperfectly, with lesser increases of oxygen consumption and electromyographic activity.

Adrenaline enhances the contraction of chicken and mammalian white muscle, both directly and through actions on neuromuscular transmission. One or both of these effects may be reduced or abolished by drugs blocking the action of α- or β-adrenoceptors (for review, see Bowman and Nott, 1969). The β-active amine α-methylnoradrenaline injected intravenously into 8-day-old chickens is followed by an approximate doubling of electromyographic activity simultaneously with about a 40% *decrease* of oxygen consumption (Allen et al., 1970). This decrease of metabolic intensity can perhaps be accounted for in part by the Q_{10} effect of the concurrent 5°C decrease in body temperature. The effectiveness of shivering in maintaining body temperature is reduced by administration of the β-blocking drug propranolol in young lambs (Alexander and Williams, 1968). Freeman (1970b) invokes this response to account for the inability of cold-stressed adult Japanese quail to maintain normal body temperature after treatment by propranolol.

Transferred to the case of immature domestic fowl, this concept of catecholamine-modulated effectiveness of shivering may explain the inability of chicks to maintain normal body temperature after administration (intravenous) of propranolol (Wekstein and Zolman, 1968, 1969). These authors show that propranolol treatment also reduces oxygen consumption, but the Q_{10} effect of hypothermia (body temperature as much as 5.5°C below that of saline-injected controls) was not taken into account, and it is not clear how much of the reduction of oxygen consumption results from hypothermia and how much, if any, results from the alleged reduction of regulatory nonshivering calorigenesis. In addition, it should be noted that systemic administration of propranolol typically reduces heart rate and myocardial contractility, blocks the mobilization of metabolic substrates, and induces vasomotor responses that may influence heat transfer (for review, see Himms-Hagen, 1967; Kattus *et al.*, 1970), all of which are activities that may affect body temperature without primary involvement of "nonshivering thermogenesis." Finally, it has already been noted (Section VI,A,1b) that systemic injection of catecholamines or their blockers may have both hypothalamic and peripheral effects. That these effects have been intermingled in experimental situations is suggested by the enigmatic fact that both β-active amines (Allen *et al.*, 1970) and the β-blocker propranolol (Wekstein and Zolman, 1969) induce hypothermia when administered intravenously to young chicks.

In short, it seems from the foregoing arguments that it is not necessary to invoke for young birds a thermoregulatory nonshivering calorigenesis of the type associated with neonate mammals possessing brown adipose tissue—birds have no brown adipose tissue at any age (Freeman, 1967; Johnston, 1971). It has not yet been persuasively shown that shivering alone is not adequate to sustain body temperature within the range found in young birds, and the existence of regulatory nonshivering thermogenesis in birds of any age (with the possible exception of chickens; see Section VI,B,2a) must receive the Scottish verdict: not proved.

c. Behavioral Reactions. The capacity for selecting a favorable thermal environment is an important aspect of thermoregulation in free-living birds and involves many different behavior patterns. Voluntary selection of thermal environments is known in representatives of all classes of vertebrates (Hammel, 1968, p. 651), but evidently has been infrequently studied, quantitatively, among birds. Domestic geese from 3 to 21 days of age when given a choice of $T_a = 16°-19°$, $19°-23°$, and $23°-28°$C tended to congregate in the intermediate compartment and to avoid air temperatures above 23°C and, less clearly, below 19°C (Poczopko, 1967). Day-old chickens tested in a linear thermal gradient showed a narrow quasinormal distribution of preference with a mean at 41.3°C (Ogilvie, 1970). In this same investigation it was found that the mean temperature preferred by Japanese Quail decreased from 44.2°C at 1–2 days of age to 37.8°C at 6–8 days of age, and to 35.1°C at 13–15 days of age. Presumably this decline in preferred temperature reflects a maturation in the control of thermoregulatory processes.

Some attributes of the voluntary selection of thermal surroundings are revealed by experiments with Ring Doves trained to regulate their ambient temperature within narrow limits (32°–34°C) by pecking keys controlling an air-conditioning apparatus (Budgell, 1971). The evidence from trials involving step changes and sinusoidal oscillations of air temperature is consistent with the hypothesis that the responses of the doves to these changes are governed by a simple feedback loop involving brain temperature acting on an off–on threshold for the behavioral responses. Presumably a similar control system accounts for the selection of preferred ambient temperature in unrestrained birds.

In addition to the capacity for voluntary selection of their thermal regime, birds possess somatomotor reactions, besides panting and gular fluttering, that are externalized as patterns of "behavioral thermoregulation." The simpler of these include, for instance, wing-drooping (exposing the thinly feathered axillary and costal regions) in hot environments, and squatting on the legs in cold environments. Simple reactions such as these may be at least partly under reflex or involuntary control. Lesions of the anterior hypothalamus in adult chickens abolished the wing-drooping that is characteristic of normal adults subjected to heat stress (Lepkovsky *et al.*, 1968). Ostensibly identical postural reactions were induced in young chickens by cold stress and by intrahypothalamic or intravenous injection of α-methyl-noradrenaline (Allen and Marley, 1967; Marley and Stephenson, 1970). Similarly, administration of α-methyltryptamine induced wing-drooping like that seen in heat-stressed controls. The tryptamine response but not the noradrenaline response persisted in adult chickens.

Various relatively complex behavior patterns apparently responding to thermal stress are commonly observed in free-living birds. Although the thermoregulatory consequences of these reactions have rarely been quantified, it is nevertheless plausible on theoretical grounds that they are very significant in alleviating thermoregulatory stress. In an effort to focus attention on potentially fruitful lines of investigation, we summarize in Table XV some of the more prominent patterns of behavioral thermoregulation in relation to the elements of the general heat-balance equation. Although these are presented in

TABLE XV

SELECTED EXAMPLES OF BEHAVIORAL THERMOREGULATION TO ATTAIN HEAT BALANCE $(H_\mathrm{m} \pm H_\mathrm{r} + H_\mathrm{c} \pm H_\mathrm{k} - H_\mathrm{e} \pm H_\mathrm{s} = 0)$

Major effect	Behavior	Species	Habitat	Reference
Minimize H_m	Reduced activity in mid-day heat	*Campylorhynchus brunneicapillus*	Desert	Ricklefs and Hainsworth, 1968
		Geococcyx californianus	Desert	Calder, 1968a; see also Kavanau and Ramos, 1970
		Various (10 spp.)	Desert	Sopyev, 1968
		Various	Tropical	Medway, 1969
Reduce H_r load	Back to sun, shading gular area and feet to enhance their convective and conductive heat loss	*Sula dactylatra*	Tropical	Bartholomew, 1966
	Orientation to minimize surface area exposed to solar radiation during incubation	*Chordeiles minor*	Temperate	Weller, 1958
	Nesting site between cactus pads affording almost complete, continuous shade	*Lophortyx gambelii*	Desert	W. A. Calder, unpublished
	Roofed nest in exposed situations	*Campylorhynchus brunneicapillus*	Desert	Hensley, 1954; Ricklefs and Hainsworth, 1969
		Auriparus flaviceps	Desert	Hensley, 1954
	Nesting within interior of dense shrubs	*Polioptila melanura*	Desert	Hensley, 1954
	Nest orientation and location to avoid mid-day H_r	*Ammomanes deserti*	Desert	Orr, 1970
	Shade-seeking	Various	Desert	Madsen, 1930, cited by Schmidt-Nielsen, 1964

TABLE XV (continued)

Major effect	Behavior	Species	Habitat	Reference
Reduce H_r load and avoid high T_a	Nesting in Saguaro Cactus cavities	*Melanerpes uropygialis*	Desert	Hensley, 1954
		Myiarchus cinerascens	Desert	Hensley, 1954
		Myiarchus tyrannulus	Desert	Hensley, 1954
		Micrathene whitneyi	Desert	Ligon, 1968
Reduce nocturnal H_r loss	Roosting under eaves, roofs, in buildings, tunnels, and caves	*Leucosticte arctoa atrata* and *L. a. tephrocotis*	Temperate	French, 1959; King and Wales, 1964
	Individual roosting in cavities in trees	*Sitta carolinensis*	Temperate	Kilham, 1971
	Communal roosting in cavities or under cover	*Sitta pygmaea*	Temperate	Knorr, 1957
		Sialia sialis	Temperate	Frazier and Nolan, 1959
	Nesting under branches to shield from night sky	*Stellula calliope*	Montane	Calder, 1971
		Selasphorus platycercus	Montane	Calder, 1973
	Sunning behavior	*Zonotrichia leucophrys*	Temperate	Morton, 1967a
		Calypte anna	Desert (winter)	Calder, 1973
		Geococcyx californianus	Desert	Ohmart and Lasiewski, 1971
Increase H_r gain		*Molothrus ater*	Temperate (lab)	Lustick, 1969
		Various	—	Teager, 1967; Goodwin, 1967; Kennedy, 1969; King, 1970
Minimize or maximize H_c	Orientation of nest opening; avoid winds during cold (early nests); face winds during hot season	*Campylorhynchus brunneicapillus*	Desert	Ricklefs and Hainsworth, 1969
Reduce h, H_k	Retraction of extremities	Various	—	Alder, 1963
	Huddling of 15 to 20 birds	*Certhia brachydactyla*	Temperate	Löhrl, 1955

4. THERMAL AND CALORIC RELATIONS OF BIRDS

	Behavior	Species	Habitat	Reference
Decrease ΔT	Retraction of legs, locomotion on snow surface by means of wings	*Acanthis* spp.	Arctic	Johnson, 1957
	Roosting in pairs, fours	*Parus atricapillus*	Arctic	Johnson, 1957
	Exposure of thinly feathered axillary regions to facilitate heat loss	*Sturnus vulgaris*	Experimental	Brenner, 1965
		Sula dactylatra	Tropical	Bartholomew, 1966
	Selected T_a within thermoneutral range	*Streptopelia "risoria"*	Laboratory	Budgell, 1971
Influence ΔT, H_c, and/or H_r	Roosting in old rodent burrows, squirrel nests, cavities and crevices in trees, under moss	*Parus montanus*	Boreal	Zonov, 1967
	Roosting in nesting box	*Passer domesticus*	Temperate	Kendeigh, 1960
	Roosting and nesting in caves, and tunnels in high Andes	*Oreotrochilus estella* and others	Alpine	Pearson, 1953
	Spending night under snow cover	*Junco hyemalis*	Temperate	George, 1967
		Acanthis flammea and others cited	Arctic	Sulkava, 1969
		Various tetraonids	—	Irving, 1960; Volkov, 1968 Welty, 1962
	Feeding under snow cover	Various	Arctic	
		Acanthis	Arctic	Cade, 1953; Johnson, 1957
	Avoidance of heat stress in rock crevices	*Salpinctes obsoletus*	Desert	Smyth and Bartholomew, 1966
	Altitudinal movement to roost in caves at lower elevations	*Columba albitorques*	Montane	Boswall and Demment, 1970
	Soaring at higher altitudes to avoid heat stress	Large soaring birds	Desert	Madsen, 1930, cited by Schmidt-Nielsen, 1964
Increase H_r and H_k loss, reduce H_c gain	Sitting on heels with feet off ground in shade of body	Juvenile *Diomedea* spp.	Tropic	Howell and Bartholomew, 1961

the context of short-term adjustments, it is also evident that they may have consequences in long-term acclimatization (Section VI,B,2d).

B. Long-Term Regulation

In this section we consider compensatory adjustments of thermoregulation—acclimation and acclimatization—that occur on a timescale of weeks or months. Adjustments in response to a single variable, usually air temperature in experimental situations, constitute *acclimation*, while compensatory adjustments to concurrent variation of several factors, as in natural situations or outdoor cages, constitute *acclimatization* (Prosser, 1973). In addition to the physiological and behavioral components of response that characterize short-term regulation, long-term adjustments may include also morphological components such as variation in the weight of plumage or in the density of cutaneous capillaries. These definitions of acclimation and acclimatization refer to phenotypic events that are not to be confused with genetic adaptations.

The functional consequences of experimental acclimation and natural acclimatization are not usually identical in either birds or mammals (Hart, 1957, 1964), and the results of experimental acclimation must be extrapolated cautiously to the long-term thermoregulatory adjustments of free-living homeotherms. The recent literature on temperature acclimation and allied phenomena in birds has been summarized by Chaffee and Roberts (1971). In the present account we will, therefore, confine attention to a development of general perspectives illustrated by selected investigations. Because there have been relatively few studies of acclimatization to heat stress in birds, we necessarily develop our themes primarily in the context of acclimatization to cold, reserving for a terminal section (Section VI,B,4) a brief summary of adjustments to heat.

1. The Evidence of Acclimatization

Evidence of seasonal acclimatization or thermal acclimation is most easily obtained by measurement of the tolerance of birds to thermal stress, in terms either of mortality or the rate of weight loss. Extensive data are available only for cold tolerance. Pigeons, European Starlings, and Evening Grosbeaks exposed to extreme cold ($-48°$ to $-57°C$) in continuous light with ample food and water survive much longer if winter-acclimatized than if summer-acclimatized (Hart, 1962). Newly captured birds in this investigation survived a shorter time than those held previously in cages outdoors for 1 week or longer, a fact that Hart attributes to the inexperience of the new captives in finding food and water.

House Sparrows held in captivity outdoors in Illinois show a distinct annual cycle of cold tolerance when tested in conditions similar to those used by Hart (1962) but exposed to the photoperiod normal for the season (Barnett, 1970). The LD_{50} (air temperature at which 50% died within 6 days) for these birds varied from about $-25°C$ in midwinter to about $0°C$ in late summer, the onset of improved tolerance being coincident with the annual molt. In contrast, West (1960) found no seasonal variation in the ability of Tree Sparrows (*Spizella arborea*) to tolerate cold stress. The limits of tolerance were about $-28°C$ in both winter-acclimatized and summer-acclimatized birds. West correlates this lack of seasonal adjustment with the migratory habits of the birds, which allow them to exist in relatively uniform environmental conditions throughout the year (see Section VI,B,2,d).

An additional simple index of acclimation or acclimatization is the rate of overnight loss of body weight upon exposure to cold. This technique has not been widely used, and the only extensive data are those of Dolnik (1967) for fifteen species of passerines exposed in cages (without food) at night to normal air temperature at Rybatschi (about 55° N latitude). Selected data are shown in Table XVI and reveal large-scale differences in the rate of nocturnal weight loss (approximately proportional to energy expenditure) that appear to be correlated with the winter ranges of the species in question. The rate of weight loss in migrant species typically wintering in middle latitudes is about twice as great as that in resident species or short-distance migrants wintering at high latitudes. These differences among species may result from genetic adaptations in addition to differences in the capacity for acclimatization, but they nevertheless suggest, in an ecologically significant context, the range of adaptability to be expected among small-bodied passerine birds.

2. Modes of Acclimatization

The seasonal differences in weight loss and mortality from thermal stress that have just been described can potentially be explained by metabolic, insulative, and (in free-living birds) nutritional and behavioral forms of adjustment. The range of options available through metabolic and insulative acclimatization is illustrated by Fig. 32.

a. Metabolic Adjustment. Metabolic adjustment by birds to cold stress under natural or seminatural conditions evidently most commonly involves an increase in the capacity for heat production, lead-

TABLE XVI
WEIGHT LOSS IN SELECTED SPECIES KEPT OVERNIGHT IN CAGES WITHOUT FOOD[a]

Species	Mean annual body weight (gm)	Condition during winter			Condition during summer molt	
		Evening fat reserve (gm)	Nocturnal weight loss		Evening fat reserve (gm)	Nocturnal weight loss (gm/hour)
			(gm/hour m^2)	(gm/hour)		
Winter latitude 45°N						
Carpodacus erythrinus	24.4	0.8	22.1	0.17	0.5	0.25
Emberiza hortulana	24.4	0.7	17.8	0.15	0.2	0.26
Winter latitude about 55–60°N						
Fringilla coelebs	23.4	1.0	11.0	0.09	0.3	0.22
F. montifringilla	25.3	1.1	10.4	0.09	0.4	0.22
Winter latitude about 62–66°N						
Carduelis spinus	14.8	2.5	9.9	0.06	0.7	0.13
Acanthis flammea	18.0	5.0	7.2	0.05	—	—

[a] Selected data from Dolnik (1967). Captive birds retained and studied at the Biological Station at Rybatschi (about 55°N).

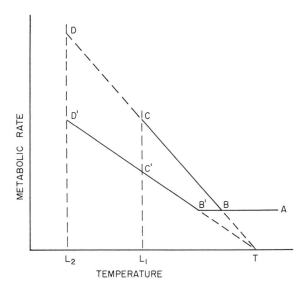

FIG. 32. Metabolic versus insulative acclimatization. L_1 and L_2 represent low temperature limits of animals acclimatized to warm and cold climates, respectively. Extension of limits from L_1 to L_2 may be carried out through insulative or metabolic adjustment. Insulative acclimatization entails a decrease of the critical temperature from B to B' and change in slope from BC to BC'D', so that the energy expenditure is not greater at L_2 (D') than at L_1 (C). Metabolic acclimatization entails an extension of the line BC to BCD, with greater energy expenditure at L_2 (D) than at L_1 (C). (From Hart, 1957.)

ing to a reduction in the lower lethal temperature (the extension of the line CD in Fig. 32), as shown for a variety of species and experimental conditions (for review, see Hart, 1957, 1962; West, 1962; also Davis, 1955). The acclimation of small birds to cold indoors commonly results in substantial increases of basal metabolic rate (elevation of the line ABB' in Fig. 32), and theoretically (but usually not actually) to a decrease in the lower critical temperature (Gelineo, 1955, 1964; Hart, 1957). In captive birds exposed to seminatural conditions, however, seasonal changes of basal metabolic rate correlated inversely with air temperature are usually absent or slight (Hart, 1957; Dawson, 1958; Veghte, 1964; Gavrilov et al., 1970; Hissa and Palokangas, 1970; Pohl, 1971). The basal metabolic rate of Willow Ptarmigan (*Lagopus lagopus*) confined in cages outdoors is slightly *lower* in winter than in summer (West, 1972). The only evidence for a large-scale seasonal cycle of basal metabolic rate was observed in *Passer domesticus* (in males, an amplitude of 74% above the midsummer minimum) by Miller (1939); but, as already noted by Hart (1964), there is no reason to suppose that a similar cycle occurs in free-living birds. The low-amplitude annual cycles of basal metabolism observed in chickens—e.g., winter means about 12–15% above summer means (Winchester, 1940)—are inversely correlated with air temperature but directly correlated with egg production, and it is not clear what fraction of the annual change, if any, results from thermal acclimatization and what fraction is associated with the demands of ovogenesis.

The evidence about the physiological mechanisms allowing increase of thermoregulatory heat production upon acclimatization to cold is conflicting for various species. Among mammals, some (but not all) species respond to experimental cold stress by a gradual decrease of electromyographic activity (for review, see Hart, 1964) and by a gradual increase of nonshivering calorigenesis mediated (in some, but again not in all species) by catecholamines (Jansky et al., 1969). The modulations of shivering and nonshivering calorigenesis are of course reciprocal modes of acclimatization (or acclimation), and it is not possible to invoke one response without involving compensatory change in the other.

Among birds, with respect first to the responses of shivering to chronic cold stress, West (1965) observed no differences in the electromyographic activity of groups of Evening Grosbeaks acclimated for 3 weeks to air temperatures of 18°, 10°, and −15°C. Rautenberg (1969a) found that the thresholds (rectal temperature, skin temperature) at which shivering began were the same in pigeons acclimated to 10° and to 29°C for several months. These data suggest that there was no replacement of shivering by nonshivering calorigenesis upon chronic cold exposure in these species. This observation is supported by the fact that Chaffee et al. (1963) were not able to induce increased oxygen consumption by injection of L-noradrenaline (up to three times the dose producing a 200% increase in rats) into anesthetized chickens acclimated to either 23° or 1°C for 3 months, a result resembling the previous findings of Hart (1962) with winter-acclimatized pigeons. These results of course do not prove that regulatory nonshivering calorigenesis is absent or negligible in birds, since the relationship between electromyographic activity and muscular heat production is probably labile, and nonshivering calorigenesis may be controlled in some species principally by substances other than noradrenaline (Jansky et al., 1969). Barnett (1970) shows that the oxygen consumption of brain and muscle excised from acclimated or acclimatized *Passer domesticus* is usually but not always significantly greater *in vitro* when obtained from birds exposed to cold. The oxygen consumption of excised liver showed the opposite trend. Elevated rates (compared with controls) of oxygen consumption in excised brain and in heart muscle as well as liver were found in cold-acclimated *Passer domesticus, Columba livia,* and *Pyrrhula pyrrhula* in experiments mentioned by Slonim (1971). The significance of these adjustments at the tissue level for the acclimatization and nonshivering calorigenesis in the intact bird is obscure, but should nevertheless not be overlooked.

Very long periods of experimental acclimation may result in responses that are not evident in the shorter experiments just described. For chickens acclimated to 21° and 5°C for up to 9 months (11 months of age), El Halawani et al. (1970) present evidence indicating that nonshivering calorigenesis gradually replaces shivering in thermoregulatory heat production. After 2 months of treatment, the acclimated birds (5°C) were not different from the controls (21°C) in oxygen consumption, electromyographic activity, or response to tyramine (which depletes peripheral adrenergic fibers of catecholamines). Between 2 and 5 months of acclimation, however, the oxygen consumption of the cold-acclimated group increased significantly above that of the control group, while mean electromyographic activity decreased to about one-tenth of its level at 2 months and remained at this reduced level until the end of the experiment at 9 months. The electromyographic activity of the control group decreased by about half during the same period, presumably as a consequence of age. Tyramine treatment significantly depressed oxygen consumption in birds acclimated to 5°C, but usually not in the control birds. At the same time, this drug increased the mean electromyographic activity fourfold or more in the acclimated birds, but reduced electromyographic activity in the controls. The pattern of results in this investigation is consistent with the interpretation that peripheral catecholamines are responsible for the "nonshivering" heat production invoked by long-term acclimation (see also Lin and Sturkie, 1968). Because detectable levels of shivering persisted in acclimated birds, it is not possible to say conclusively that muscular heat production was replaced fully by visceral heat production, since it is plausible that catecholamines exert their effect on nonshivering calorigenesis in part by modulating the relationship between electrogenesis and calorigenesis in skeletal muscle. Slonim (1971) refers to experiments in mammals in which cold-acclimation invoked a threefold to fourfold increase of heat production per unit electromyographic activity in muscles studied *in situ*. Critical evidence on this subject in birds requires experiments involving inactivation of the shivering reflex, for instance by curare treatment (Hart, 1962).

The ecological significance of the acclimation in the chicken revealed in the investigation by El Halawani et al. (1970), even assuming that this process mimics acclimatization in free-living birds, is not entirely clear. It required 3 to 4 months to invoke a perceptible response to 5°C. This is an adjustment that is not sensitive enough, temporally, to be significant to a 2 kg bird like a chicken in a natural environment. The equivalent adjustment time for a smaller bird cannot be predicted exactly; but assuming as a first approximation that metabolic time is inversely proportional to the body weight raised to the one-fourth power (Kleiber, 1961, p. 364), the time required for a 25 gm bird to attain the state of acclimation reached by a 2 kg chicken in 3 months would be about 1 month. A regulatory inertia of this magnitude is of dubious significance to wild species in adjusting to seasonal fluctuations of temperature.

b. Insulative Adjustment. The existence of insulative adjustment to chronic thermal stress is indicated by variation to the slope of the line BCD in Fig. 32. An increase of insulation, for instance, in response to cold stress would result in decrease of slope BC'D', with a concurrent decrease of the lower critical temperature if the basal metabolic rate remained stable. Significant seasonal adjustments of insulation (conductance) occur in some species exposed in captivity to seminatural conditions (e.g., *Perisoreus canadensis*, Veghte, 1964; *Parus major*, Hissa and Palokangas, 1970; Gavrilov et al., 1970; *Lagopus lagopus*, West, 1972; domestic pigeons, Riddle et al., 1934) but not in others (e.g., *Passer domesticus, Sturnus vulgaris, Coccothraustes vespertinus,* feral pigeons, Hart, 1962; *Cardinalis cardinalis*, Dawson, 1958; *Sitta canadensis*, Mugaas and Templeton, 1970).

Seasonal variability of insulation, when it occurs, may be attributed to adjustments of both plumage insulation and tissue insulation. One index of plumage insulation is the weight of feathers. This varies greatly through the year as a consequence of wear and replacement, the latter mainly at the time of postnuptial molt. In *Passer domesticus*, for instance, the weight of body feathers varies from 0.90 gm/bird at the end of the breeding season in August to 1.53 gm/bird at the end of molt, a 70% increase (Barnett, 1970). The weight of body plumage in this species decreases slightly during the winter, and rapidly after the onset of the breeding season in March. As noted previously, these changes are paralleled by changes of cold resistance in the same birds. For other nonmigratory species, it can be estimated that the annual alternation between worn and fresh plumage entails an increase of weight above the worn condition amounting to about 45% in *Cyano-*

citta stelleri (total plumage, Pitelka, 1958) and about 53% in *Pyrrhula pyrrhula* (body plumage only, Newton, 1968). In the migratory Tree Sparrow (*Spizella arborea*), West (1960) observed an analogous increase of about 25%.

In addition to the obligatory variations of plumage weight attending the cycle of wear and replacement in all species of birds, the effectiveness of plumage insulation is also determined by its thickness, and therefore in part by ptilomotor control. Although the data of Wallgren (1954, p. 55) suggest that there may be important differences among species in the strength of ptilomotor responses to cold, it is not known if this mechanism is susceptible to seasonal acclimatization.

The insulation (conductivity) of superficial tissues is also potentially variable and susceptible to seasonal acclimatization. The estimates made by Veghte (1964) for *Perisoreus canadensis* suggest that tissue insulation may approximately triple in this species between summer and winter. The changes that make this possible have not been directly investigated in birds, but probably involve chronic vasomotor adjustments and added insulation supplied by subcutaneous fat, which commonly increases in winter in birds of high latitudes (King and Farner, 1966; King, 1972) and may nearly ensheath the body (King and Farner, 1965). Although it is sometimes stated (e.g., Porter and Gates, 1969) that subcutaneous fat confers no special insulative advantage because its conductivity, as a dead tissue, is about the same as that of muscle, the fact remains that skin temperature and heat loss are lower in fat men than in lean men (Baker and Daniels, 1956; Cannon and Keatinge, 1960; see also Bryden, 1964, 1968). The problem awaits experimental examination in small animals.

c. Nutritional Adjustment. The physiological consequences of variation in the composition or caloric density of the diet are not commonly treated as an aspect of acclimatization in wild species of animals, but they nevertheless logically belong in this category of adjustment. There are only a few data available for natural species of birds, but they are adequate to indicate that variability of nutrients may have significant effects on the capacity of birds to withstand thermal stress. For instance, the level of protein in isocaloric diets fed to captive *Passer domesticus* (3, 5, and 9% protein) and *Spizella arborea* (2, 4, 8, and 16% protein) significantly affected survival in a standard cold test (Martin, 1968, 1970). *Passer domesticus* survived longest when fed a diet containing 5% protein, and *Spizella arborea* survived best on 8% protein. It is noteworthy that North American experimenters commonly feed granivorous birds high-protein rations

(15–20% protein), which significantly impair cold resistance in the species noted above.

The caloric density of nutrients may also have a significant effect on tolerance of thermal stress. Brooks (1968) estimated that substitution of birch seeds (the natural food) in place of an experimental ration of lower caloric density would reduce the lower lethal temperature by about 20°C in redpolls (*Acanthis* spp.), assuming that the digestibility of the two foods was the same. It is not certain, however, that birds habitually select foods of maximum caloric density under natural conditions. *Spizella arborea* wintering in Illinois, for instance, preferentially utilize seeds of low caloric density compared with those available (West, 1967). Rock Ptarmigan (*Lagopus mutus*) in Iceland, in contrast, appear to feed preferentially on foods of high nutritive value (Gardasson and Moss, 1970). Emlen (1966) provides an instructive model for the interpretation of the significance of variations in preference.

In relation to heat resistance, Robinson and Lee (1947) have shown that the heat tolerance of adult chickens is inversely related to the plane of nutrition (total caloric intake), but is unaffected by the level of dietary protein. Persons *et al.* (1967) also showed that various levels of dietary protein had no effect on the survival of the young chickens exposed to heat stress, although the addition of fat to the diet significantly reduced survival.

d. Behavioral Adjustments. Facultative adjustments of behavior are probably more significant in the seasonal thermoregulatory acclimatization of wild homeotherms than has been generally realized heretofore, and may in many cases augment or replace physiological modes of adjustment. We have described in Section VI,A,3,c some behavioral forms of short-term thermoregulation, and we confine attention now to seasonal behavioral plasticity that may be regarded as energy sparing. Many of the same behavioral adjustments noted in relation to short-term regulation may of course have supplementary roles in seasonal acclimatization, and the distinction is therefore blurred and arbitrary. Of necessity, this account consists merely of miscellaneous observations, the significance of which cannot be satisfactorily assessed until better data become available on the microhabitats of the species in question.

The most conspicuous forms of seasonal thermoregulatory behavior are known from cold regions or seasons, in which birds indulge in several modes of shelter seeking. Burrowing in snow affords protection from low air temperature, wind, and from radiation to the cold sky. This is a habit long known in tetraonids (for review, see Irving, 1960; Volkov, 1968; Sulkava, 1969), but also reported for the Snow Bunting (Welty, 1962, p. 127), for the Common Redpoll (*Acanthis flammea*) (Sulkava, 1969; see also Cade, 1953), and for the Willow Tit (*Parus montanus*) (Zonov, 1967). In the case of *Acanthis*, subnival burrows as long as 73 cm and as much as 11 cm below the surface were found. Judging from the accumulation of droppings, Sulkava (1969) estimated that some burrows were used only between feeding sessions during the day, while others were used throughout the night. From the results of a 3-year study near Lake Baikal, Zonov (1967) reports that *Parus montanus* habitually excavate burrows in the snow, occupying the same burrows throughout the winter. He further reports that a 20 cm passage can be excavated in 10–15 seconds and describes the movements of head, wings, and feet by which this is accomplished.

Another form of shelter seeking in winter is roosting in holes in trees and in other protected places, as shown by nuthatches (*Sitta* spp.) (Knorr, 1957; Kilham, 1971), Eastern Bluebirds (*Sialia sialis*) (Frazier and Nolan, 1959), rosy-finches (*Leucosticte* spp.) (King and Wales, 1964), parids (*Parus* spp.) (Kluyver, 1957; Zonov, 1967), and other species (Busse and Olech, 1968). Often this involves communal roosting. Knorr (1957) estimated that more than 100 *Sitta pygmaea* occupied a single cavity in a pine tree, a density that may occasionally lead to suffocation of some individuals. Frazier and Nolan (1959) found as many as fourteen Eastern Bluebirds roosting in a peculiar head-downward huddle in a single cavity, and the authors suggest from the sluggishness of the birds in the morning that they may develop a nocturnal hypothermia. Busse and Olech (1968) found that the percentage of available nest boxes occupied by four species of small songbirds near Warsaw in winter was in inverse linear function of air temperature at 15:00.

Communal roosting or huddling may also occur in open roosts at night in the winter in a variety of species (Tucker, in Welty, 1962, p. 127; Löhrl, 1955; Armstrong, 1955). The huddling of small birds may, as in groups of European Starlings (Brenner, 1965), ameliorate energy expenditure at night as a result of the reduction of surface-to-volume ratio.

Still another type of behavioral acclimatization to winter is the development of communal food-gathering and food storage in special caches. Food storage for use in winter is known from several species, but is developed to an apparently unusual degree among corvids (Dow, 1965; Turček and Kelso, 1968; Balda and Bateman, 1971).

Short-term food storage is also possible in the crop and/or esophageal diverticula of wild birds in which these structures are adequately developed. This form of winter adaptation should perhaps be regarded as a nutritional rather than as a behavioral adjustment, insofar as it involves internal storage of energy supplementing that stored as fat and glycogen. Many species of birds are able to store small quantities of food in the crop, aiding overnight survival even in relatively mild climates (e.g., *Emberiza citrinella*, Evans, 1969a; *Loxia* spp., Dawson and Tordoff, 1964); but the pattern is best known in arctic birds such as *Acanthis flammea* and *A. hornemanni* (Brooks, 1968) and *Lagopus lagopus* (Irving et al., 1967). Neither of these forms in central Alaska accumulates unusual amounts of fat in winter (King and Farner, 1966; West and Meng, 1968), and it appears that stored food is a vital overnight energy resource. In *Lagopus* sp. it is evident that behavioral modes of thermoregulation coupled with very good surface insulation (Johnson, 1968) allow the birds to exist in winter at about the same level of energy intake required in summer (West, 1968).

Finally, it is not far fetched to regard migration as a special case of thermoregulatory acclimatization, or as a substitute for it, since the seasonal movements take the birds away from regions in which they actually or potentially could not maintain energy balance in the winter (e.g., Seibert, 1949; Wallgren, 1954; West, 1962; Zimmerman, 1965). It may be significant in this context that Gelineo (1964) reports that migratory species of birds respond more slowly in experimental temperature acclimation than do nonmigratory species. The correlation between seasonal movement and trophic conditions is especially conspicuous in partial migrants or irruptive migrants that are specialized for feeding predominantly on a single crop, such as birch seeds or conifer seeds (e.g., *Acanthis flammea*, Evans, 1966, 1969b; Eriksson, 1970; and *Loxia* and *Nucifraga*, for review, see Newton, 1970; Schüz, 1971, pp. 203 and 210). Among regular migrants, it is known that at least one subspecies, *Zonotrichia leucophrys gambelii*, has only a slight capacity for thermoregulatory acclimation to various long-continued thermal or photoperiodic regimes (Southwick, 1971), suggesting that it has substituted migration for thermal acclimatization, and has lost (or never evolved) a significant capacity for the latter.

3. Patterns of Acclimatization

Since thermoregulatory acclimatization involves a subtle interplay of metabolic, insulative, nutritional, and behavioral variables, it is reasonable to expect that different species, and perhaps even different geographic populations of a given species, have not exploited identical patterns of adjustment to thermal and trophic stress, and do not depend on identical environmental signals to bring the adjustments into play (Hart, 1957). This may explain the conflicting results that have already been noted in various investigations of acclimatization in birds, although the effects of differences in experimental methods can by no means be discounted. Progress beyond the present morass of apparently conflicting information toward a better understanding of thermoregulatory acclimatization will be aided by the development of standardized experimental methods that ask first the question: Does this species (or better, this population) possess an ecologically significant capacity for physiological acclimatization? The answer will determine the subsequent procedures for the exploration of physiological processes or of nutritional and behavioral alternatives.

4. Acclimatization to Heat Stress

Thermoregulatory acclimatization to warm seasons or regions consists basically of the reversal of the adjustments in response to cold, but may be amplified in cases of severe, prolonged heat stress. The adjustments in this case are complicated by the concurrent problems of minimizing water loss, maximizing heat loss, and (in very hot environments) reducing heat uptake. The resulting responses are thus a compromise between osmoregulatory and thermoregulatory requirements.

It appears that there have been no extensive quantitative studies of seasonal variation of mortality from heat stress. In chickens maintained in natural conditions of air temperature and photoperiod and tested at $T_a = 40.6°C$ and relative humidity = 70%, the mean survival time for a "summer group" was 190 minutes, compared with 127 minutes for a "winter group" (Weiss and Borbely, 1957), indicating a significant ($P < 0.01$) acclimatization.

Acclimation of chickens to warm environments typically reduces the basal metabolic rate (e.g., Huston et al., 1962). Similar results were obtained for *Emberiza citrinella* and *E. hortulana* by Wallgren (1954) (for other species, see Gelineo, 1964) kept in stable warm environments. Wallgren also found, however, that *Emberiza* spp. showed no change of basal metabolic rate if the birds were exposed for 8 hours per day to room temperature instead of continuously to higher air temperatures. This makes it questionable, as in other cases previously mentioned, that the results of typical acclimation experiments are meaningful in analyzing the adjustments of free-living birds.

The rate at which body temperature increases during acute heat stress is typically reduced by acclimation of chickens to warm environments (e.g., Hutchinson and Sykes, 1953; Weiss et al., 1963). Summer-acclimatized *Cardinalis cardinalis* become hyperthermic at higher air temperatures than winter-acclimatized birds (Dawson, 1958). Relatively little is known about the mechanisms that allow this improved heat tolerance. Acclimation to heat reduces the respiratory rate, respiratory minute volume, and other cardiopulmonary variables in unstressed warm-acclimated chickens (Hillerman and Wilson, 1955; Weiss et al., 1963). Acute heat stress, however, induced similar cardiopulmonary responses in both warm-acclimated and control birds, and Weiss et al. (1963) estimated that the slight differences observed in respiratory variables were insufficient to explain the slower development of debilitating hyperthermia in the acclimated birds. DeShazer et al. (1970) showed by direct calorimetry that the evaporative heat loss of hens acclimated to 35°C for 11 weeks was actually about one-third that of controls acclimated to 24°C for the same period, when both groups were tested at 35°C. This was due to a tripling of the conductivity of the body shell in the heat-acclimated hens compared with the control hens. These results suggest a pattern of reciprocal adjustments involving conservation of water while enhancing heat transfer by chronic vasodilation in superficial tissues. Such results, however, are not uniformly obtained. Hutchinson and Sykes (1953), for instance, found that evaporative heat loss did not differ significantly between heat-acclimated and control chickens.

Such apparent contradictions among the results of investigations of chickens may originate from differences among breeds, or even among genetic lines within breeds (for review, see Whittow, 1965a, p. 214; also Kamar and Khalifa, 1964; Wilson et al., 1966; Persons et al., 1967; Gilbreath and Ko, 1970). This complication, plus the peculiarities of energy metabolism and thermoregulation that have been entrained by the selection of domestic fowl for continuous egg production, make it questionable that data about the capacities of chickens for thermal acclimation have a general relevance for the responses of nondomesticated species of birds.

VII. Evolutionary Aspects of Thermoregulation and Energy Metabolism

Having described the range of adjustments of thermoregulatory variables in individual birds (Sections V and VI), we now turn to the subject of genotypic radiation of these variables among populations in response to the selection pressures of climatic regions, habitats, or modes of life. In a classic paper, Bartholomew (1958) pointed out: (1) "small physiological differences between closely related species need not be adaptive," and (2) "the evolution of terrestrial vertebrates has been characterized first, by increasingly effective homeostatic mechanisms and second, by increasingly variable and effective behavior." Behavioral adjustments may supplement or replace physiological or morphological adjustments in thermoregulation on an evolutionary time scale as well as within the life span of individual birds. One must not rely exclusively on comparative physiology to explain thermoregulatory adaptations.

A. COMPONENTS OF ADAPTATION

The main thermoregulatory variables upon which natural selection might act to induce adjustments to specific environments include the mean level and lability of body temperature, the basal and maximum metabolic rates, the conductance of the body shell, and the sensitivity of these traits to acclimatization. The level of body temperature is nonadaptive for different climatic regions, as documented in Section II,H (see also King and Farner, 1961; Dawson and Hudson, 1970), although lability of body temperature (facultative hypothermia, Section V) has evolved in various phylogenetic lines concurrently with small body size and/or dependence on unstable food supplies. We confine attention in the following discussion to metabolic rate, conductance, and allied traits.

Little is known of the susceptibility of avian thermoregulatory traits to selection. Investigations of domesticated species of birds show that slight differences of mean body temperature or intolerance to heat stress can be produced by artificial selection over a few generations (Persons *et al.*, 1967; Gilbreath and Ko, 1970). It is of course apparent from differences in basal metabolic rate between passerines and nonpasserines (Section II,D) and perhaps among the other avian taxa (Zar, 1969) that there has been significant phylogenetic radiation in this variable that could conceivably be "preadaptive" for various habitats; but apparently there have been no controlled investigations of the genetic plasticity of metabolic rate and insulation in relation to thermal selection in any avian species.

Scholander *et al.* (1950c) concluded that basal metabolic rate is not significantly adaptive to tropical and arctic regions, and that adaptation of insulation is far more important. Bartholomew (1972, p. 59) reasserted this viewpoint, but added that a reduction of basal metabolic rate, while irrelevant to Arctic-dwelling species, might have selective significance for birds exposed chronically to heat stress. Information on these subjects is commonly sought by comparing thermoregulatory variables in closely related species or populations presumably adapted to different environments (e.g., Scholander et al., 1950a; Winkel, 1951; Dawson, 1954; Wallgren, 1954; Saxena, 1957; Koskimies and Lahti, 1964; Hudson and Kimzey, 1966; Dolnik, 1967; Blem, 1969; Rising, 1969; Hinds and Calder, 1973). As noted in particular by Dawson (1954) and by Hudson and Kimzey (1966), it is not always possible to show conclusively that observed differences are genetically determined. Apparent differences among experimental populations may result merely from differences in the states of acclimatization (or acclimation), without necessarily indicating fixed genetic traits. This artifact (evidently controlled in most or all of the investigations mentioned above) must be eliminated by methods of cross-acclimatization, or at least by exposing the experimental populations to the same conditions of acclimatization until a steady state is reached. Natural selection may act on the limits and rates of acclimatization as well as on the mean level of thermoregulatory traits, and so the design of fully controlled experiments becomes very intricate.

In comparing pairs or groups of species from different environments, Scholander et al. (1950c) and Wallgren (1954) found no differences in basal metabolic rate (see also Dawson, 1954; Rising, 1969) but conspicuous differences in conductance that are correlated adaptively with climate. The weight of the contour plumage in adult Blackcaps (*Sylvia atricapilla*) averages 21% greater than in the slightly heavier Garden Warbler (*S. borin*). This is interpreted (Berthold and Berthold, 1971) as an adaptation of insulation to the cold (European) winter quarters of the Blackcap and the warmer (North African) winter quarters of the Garden Warbler.

The data of Dolnik (1967) (see Table XVI, this chapter) integrate metabolic and insulative attributes, but indicate differences among species that can be interpreted as reflecting genetic traits correlated with cold stress in the natural environments of the species in question. Winkel (1951) and Saxena (1957) claim that basal metabolic rate increases northward in the middle European and African species that they studied, or was greater in species of temperate zones compared with those of tropic and subtropic zones; but it is evident in retrospect that the observed differences are small, are no greater than those commonly found between species within zones, and were not often examined for statistical significance. Although Saxena (1957) asserts that the observed differences of basal metabolic rate are genetically fixed, his own data for *Euplectes hordaceus*, for instance, show that acclimatization of this tropical-zone ploceid increases its basal metabolic rate to the level found in middle European finches of similar body weight.

It is of interest that a small but appreciable climate-correlated radiation of thermoregulatory traits has occurred in North American *Passer domesticus* in the century since the introduction of the parent stock. Hudson and Kimzey (1966) found no significant differences in temperate-zone populations of *P. domesticus* from Colorado, Michigan, and New York, but birds from a population in the hot–humid coastal plains of Texas showed significantly lower basal metabolic rate (16% below the next highest), greater evaporative heat loss, and apparently greater insulation, although the latter was only marginally different from the Michigan ($P > 0.8$) and the Colorado and New York ($P > 0.9$) populations. In contrast, Blem (1969) found a decrease of conductance northward in populations of *P. domesticus* from eleven localities extending from southern Florida to northern Manitoba. This was evidently attributable to tissue insulation (subcutaneous fat) rather than to plumage, and hence may be an aspect of nutritional acclimatization (Section VI,A,2,c), genetically determined. Further evidence for adaptive radiation of thermoregulatory traits in *P. domesticus* is presented by Blem (1969).

Finally, it should be mentioned that thermoregulatory adaptations of young birds as well as of their parents may influence the habitat distribution of a species, an effect that is apt to be particularly conspicuous in forms producing precocial young. This introduces an obvious complication in rationalizing avian distribution in terms of physiological constraints. The only large-scale investigation of this subject appears to be that of Koskimies and Lahti (1964), who showed that 1-day-old ducklings of several species are much more cold-resistant than day-old chicks of gulls (*Larus* spp.) and several species of gallinanceous birds. Among the ducks, interpretation of the basis of the differences in cold-resistance is complicated by differences in body size among species; but it is nevertheless clear that cold-hardiness is broadly correlated with climatic conditions in the natural habitats.

B. ECOGEOGRAPHIC RULES

In many but not all taxa of homeotherms (and some poikilotherms), there are clines of increasing body size (Bergmann's rule) and decreasing length of appendages (Allen's rule) that are correlated with increasing latitude or altitude. These clines have been traditionally interpreted in a thermoregulatory context as resulting from the selective advantage of a smaller surface-to-volume ratio in colder regions, and conversely. Comparative physiologists have not always agreed with this interpretation, noting the questionable thermoregulatory advantage of the small differences of body size that are usually observed (Scholander, 1955; Irving, 1957), although LeFebvre and Raveling (1967) have recently estimated that the size differences (rather substantial) between two forms of the Canada Goose (*Branta canadensis maxima* and *B. c. parvipes*) entails thermoregulatory advantages allowing the larger form to winter far north of the smaller one. The literature on this subject has been summarized by Mayr (1956), King and Farner (1961), James (1968, 1970), and others, and does not need to be repeated here. Some recent developments, however, merit notice.

In a variety of avian species, body size is correlated with climatic and geographic variables with a precision that was previously unsuspected (e.g., Fig. 33). Wing length (shown to be an adequate index

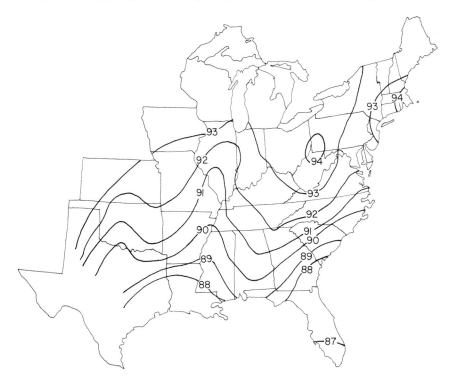

FIG. 33. Isometric lines (isophenes) of wing length in the Downy Woodpecker, *Dendrocopos pubescens*. (From James, 1970.)

of body size) is correlated significantly with a variety of seasonal temperatures (wet bulb and dry bulb) in eight North American species (James, 1970). In general, the correlations (all negative) were highest between wing length and wet-bulb temperatures or indexes including a component of humidity as well as air temperature (Fig. 34). In the

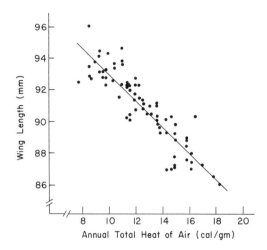

FIG. 34. The relationship ($r = -0.90$) between wing length and an index of air temperature and humidity in various populations of the Downy Woodpecker, *Dendrocopos pubescens*. (Modified from James, 1970.)

Red-winged Blackbird (*Agelaius phoeniceus*), Power (1969) found significant partial regression coefficients correlating wing length and several geographic and climatic variables whose influence is modified by body size. These included correlations between wing length and July wet-bulb temperature (negative in both sexes) and wing length and latitude (negative in males, and positive but not significant at $P = 0.05$ in females). He speculates that the opposite latitudinal influences on males (black) and females (brown) may result from differential effects of insolation.

In *Passer domesticus*, a multivariate analysis of fifteen environmental variables in relation to indexes of skeletal size and form showed that Principal Component I (an index of general body size) has a significant negative regression on all measures of winter temperature in both sexes, but has no significant relationship to summer temperatures (Johnston and Selander, 1971). Principal Component II (an index of the relationship between core body size and the length of appendages) showed significant positive regressions on measures

TABLE XVII
LOGARITHMIC TRANSFORMATIONS AND STATISTICAL VARIABILITY IN ALLOMETRIC EQUATIONS

Equation number	Logarithmic form	S_{yx}	S_b	95% confidence interval
13	$\log h = 0.6577 - 0.5363 \log m$	0.0925	0.03945	4.227–4.891 cal/gm hour °C
14	$\log h = 0.6087 - 0.5409 \log m$	0.2123	0.03606	3.435–4.804 cal/gm hour °C
16	$\log T_x = 1.6029 + 0.0377 \log m$	0.0750	0.0209	37.869–42.420°C
66	$\log \dot{M}_e = 2.1427 + 0.8005 \log M$	0.1348	0.0397	214.0–312.6 mg/minute
69	$\log h = 0.7703 - 0.6263 \log m$	0.0955	0.0294	4.954–7.008 cal/gm hour °C
70	$\log h = 1.1848 - 0.6110 \log m$	0.3232	0.0994	8.513–27.520 cal/gm hour °C

of summer temperature, and generally also on winter temperature, in both sexes. The authors conclude that Principal Component I conforms with Bergmann's rule, and Principal Component II with Allen's rule.

Certain aspects of the evidence available to date merit special emphasis. First, the general relationship between indexes of body size and of macroenvironmental temperature and humidity, while not proof of cause and effect, nevertheless is strong evidence that selection for body size commonly acts through the thermoregulatory consequences of size. The analysis by Kendeigh (1969) (see Eqs. 21 and 23 in Section II) supplies quantitative insight into the increasing caloric cost of thermoregulation at 0°C as body size decreases among species.

Second, although the clinal variations in size are small and their thermoregulatory significance has been questioned (Scholander, 1955; King and Farner, 1961), it has been pointed out by Mayr (1956, 1970) that selective advantages are independent and additive, and that the greater quantitative importance of other thermoregulatory adjustments does not preclude adaptations of body size.

And finally, the quantitative variability among species in the response of body size to variation of temperature and humidity suggests that the evolution of body size is influenced not only by selection for bioenergetic traits, but also by other selection pressures. Particularly perceptive analyses of some of these alternatives, including such factors as method of locomotion, feeding habits, and behavioral traits, are presented by Hamilton (1961), Rosenzweig (1968), Kendeigh (1969), and McNab (1971).

ACKNOWLEDGMENTS

This analysis was supported in part by National Science Foundation grant GB-13249 (WAC) and by a grant from the Graduate School Development Fund, Washington State University (JRK). The late Dr. Robert C. Lasiewski made a major contribution of ideas for allometric analysis of thermoregulatory panting, the data for which were obtained by Robert Epting and Roger Seymour. We are indebted to Peggy Himes for great patience and meticulous care in typing several drafts and the final manuscript and preparing the reference list.

REFERENCES

Adams, T. (1971). Carnivores. *In* "Comparative Physiology of Thermoregulation" (G. C. Whittow, ed.), Vol. 2, pp. 151–189. Academic Press, New York.

Åkerman, B., Andersson, B., Fabricius, E., and Svensson, L. (1960). Observations on central regulation of body temperature and of food and water intake in the pigeon (*Columbia livia*). *Acta Physiol. Scand.* **50**, 328–336.

Alder, L. P. (1963). Alternative leg positions of birds in cold weather. *Brit. Birds* **56**, 219–220.

Alexander, G., and Williams, D. (1968). Shivering and non-shivering thermogenesis during summit metabolism in young lambs. *J. Physiol. (London)* **198**, 215–276.
Allen, D. J., and Marley, E. (1967). Effect of sympathomimetic and allied amines on temperature and oxygen consumption in chickens. *Brit. J. Pharmacol. Chemother.* **31**, 290–312.
Allen, D. J., Garg, K. N., and Marley, E. (1970). Mode of action of α-methylnoradrenalin on temperature and oxygen consumption in young chickens. *Brit. J. Pharmacol.* **38**, 667–687.
Altman, P., ed. (1968). "Metabolism." Biol. Handb. Ser., Fed. Amer. Soc. Exp. Biol., Bethesda, Maryland.
Armstrong, E. A. (1955). "The Wren." Collins, London.
Aschoff, J., and Pohl, H. (1970a). Rhythmic variations in energy metabolism. *Fed. Proc., Fed. Amer. Soc. Exp. Biol.* **29**, 1541–1552.
Aschoff, J., and Pohl, H. (1970b). Der Ruheumsatz von Vögeln als Funktion der Tageszeit und der Körpergrösse. *J. Ornithol.* **111**, 38–47.
Augee, M. L., and Ealey, E. H. M. (1968). Torpor in the echidna, *Tachyglossus aculeatus. J. Mammal.* **49**, 446–454.
Aulie, A. (1971). Body temperatures in pigeons and budgerigars during sustained flight. *Comp. Biochem. Physiol.* **38**, 173–176.
Austin, G. T., and Bradley, W. G. (1969). Additional responses of the Poor-will to low temperatures. *Auk* **86**, 717–725.
Baker, P. T., and Daniels, F., Jr. (1956). Relationship between skinfold thickness and body cooling for two hours at 15°C. *J. Appl. Physiol.* **8**, 409–416.
Bakken, G. S., and Gates, D. M. (1974). On heat loss in a Newtonian animal. *J. Theor. Biol.* (in press).
Bakken, G. S., and Gates, D. M. (1974). Linearized heat transfer relations in biology. *Science* (in press).
Balda, R. P., and Bateman, G. C. (1971). Flocking and annual cycle of the Piñon Jay, *Gymnorhinus cyanocephalus. Condor* **73**, 287–302.
Barnett, L. B. (1970). Seasonal changes in the temperature acclimatization of the House Sparrow, *Passer domesticus. Comp. Biochem. Physiol.* **33**, 559–578.
Bartholomew, G. A. (1958). The role of physiology in the distribution of terrestrial vertebrates. *In* "Zoogeography," Publ. No. 51, pp. 81–95. Amer. Ass. Advan. Sci., Washington, D.C.
Bartholomew, G. A. (1966). The role of behavior in the temperature regulation of the Masked Booby. *Condor* **68**, 523–535.
Bartholomew, G. A. (1972). Energy metabolism. *In* "Animal Physiology: Principles and Adaptations" (M. S. Gordon *et al.*, eds.), 2nd ed., pp. 44–72. Macmillan, New York.
Bartholomew, G. A., and Cade, T. J. (1963). The water economy of land birds. *Auk* **80**, 504–539.
Bartholomew, G. A., and Trost, C. H. (1970). Temperature regulation in the Speckled Mousebird, *Colius striatus. Condor* **72**, 141–146.
Bartholomew, G. A., Howell, T. R., and Cade, T. J. (1957). Torpidity in the White-throated Swift, Anna Hummingbird and Poor-will. *Condor* **59**, 145–155.
Bartholomew, G. A., Hudson, J. W., and Howell, T. R. (1962). Body temperature, oxygen consumption, evaporative water loss, and heart rate in the Poor-will. *Condor* **64**, 117–125.
Bartholomew, G. A., Lasiewski, R. C., and Crawford, E. C., Jr. (1968). Patterns of panting and gular flutter in cormorants, pelicans, owls, and doves. *Condor* **70**, 31–34.

Benedict, F. G., and Lee, R. C. (1938). Hibernation and marmot physiology. *Carnegie Inst. of Wash., Publ.* **497**.

Bentley, P. J., and Schmidt-Nielsen, K. (1966). Cutaneous water loss in reptiles. *Science* **151**, 1547-1549.

Benzinger, T. H. (1970). Peripheral cold reception and central warm reception, sensory mechanisms of behavioral and autonomic thermostasis. *In* "Physiological and Behavioral Temperature Regulation" (J. D. Hardy, A. P. Gagge, and J. A. J. Stolwijk, eds.), pp. 831-855. Thomas, Springfield, Illinois.

Berger, M., Hart, J. S., and Roy, O. Z. (1970). Respiration, oxygen consumption and heart rate in some birds during rest and flight. *Z. Vergl. Physiol.* **66**, 201-214.

Berger, M., Hart, J. S., and Roy. O. Z. (1971). Respiratory water and heat loss of the Black Duck during flight at different ambient temperatures. *Can. J. Zool.* **49**, 767-774.

Berman, A., and Snapir, N. (1965). The relation of fasting and resting metabolic rates to heat tolerance in the domestic fowl. *Brit. Poultry Sci.* **6**, 207-216.

Bernard, C. (1865). "An Introduction to the Study of Experimental Medicine" (republication by Dover, New York, 1957) pp. 129-134.

Bernstein, M. H. (1971a). Cutaneous and respiratory evaporation in the Painted Quail, *Excalfactoria chinensis*, during ontogeny of thermoregulation. *Comp. Biochem. Physiol. A* **38**, 611-617.

Bernstein, M. H. (1971b). Cutaneous water loss in small birds. *Condor* **73**, 468-469.

Bernstein, M. H. (1971c). Vascular adjustments, and their control, influencing heat loss in pigeon feet. *Amer. Zool.* **11**, Abstr. No. 259, 671.

Berthold, P., and Berthold, H. (1971). Über jahreszeitliche Änderungen der Kleinfederquantität in Beziehung zum Winterquartier bei *Sylvia atricapilla* und *S. borin*. *Vogelwarte* **26**, 160-164.

Binkley, S., Kluth, E., and Menaker, M. (1971). Pineal function in sparrows: Circadian rhythms and body temperature. *Science* **174**, 311-314.

Birkebak, R. C. (1966). Heat transfer in biological systems. *Int. Rev. Gen. Exp. Zool.* **2**, 269-344.

Birkebak, R. C., Cremers, C. J., and LeFebvre, E. A. (1965). Thermal modeling applied to animal systems. *J. Heat Transfer* **88**, 125-130.

Blem, C. R. (1969). Geographical variation in the bioenergetics of the House Sparrow, *Passer domesticus*. Ph.D. Dissertation, University of Illinois, Urbana.

Bligh, J. (1966). The thermosensitivity of the hypothalamus and thermoregulation in mammals. *Biol. Rev. Cambridge Phil. Soc.* **41**, 317-367.

Bligh, J., and Allen, T. E. (1970). A comparative consideration of the modes of evaporative heat loss from animals. *In* "Physiological and Behavioral Temperature Regulation" (J. D. Hardy, A. P. Gagge, and J. A. J. Stolwijk, eds.), pp. 97-107. Thomas, Springfield, Illinois.

Boswall, J., and Demment, M. (1970). The daily altitudinal movement of the White-collared Pigeon *Columba albitorques* in the High Simien, Ethiopia. *Bull. Brit. Ornithol. Club.* **90**, 105-107.

Bowman, W. C., and Nott, M. W. (1969). Actions of sympathomimetic amines and their antagonists on skeletal muscle. *Pharmacol. Rev.* **21**, 27-72.

Boyd, J. D., and Sladen, W. J. L. (1971). Telemetry studies of the internal body temperatures of Adélie and Emperor Penguins at Cape Crozier, Ross Island, Antarctica. *Auk* **88**, 366-380.

Brenner, F. J. (1965). Metabolism and survival time of grouped Starlings at various temperatures. *Wilson Bull.* **77**, 388-395.

Brodkorb, P. (1971). Origin and evolution of birds. *In* "Avian Biology" (D. S. Farner and J. R. King, eds.), Vol. 1, pp. 19-55. Academic Press, New York.

Brody, S. (1945). "Bioenergetics and Growth." Von Nostrand-Reinhold, Princeton, New Jersey.

Brody, S., and Proctor, R. C. (1932). Growth and development with special reference to domestic animals. XXIII. Relation between basal metabolism and mature body weight in different species of mammals and birds. *Mo. Agr. Exp. Sta., Res. Bull.* **166**, 89-101.

Brooks, W. S. (1968). Comparative adaptations of the Alaskan redpolls to the arctic environment. *Wilson Bull.* **80**, 253-280.

Bryden, M. M. (1964). Insulating capacity of the subcutaneous fat of the Southern Elephant Seal. *Nature (London)* **203**, 1299-1300.

Bryden, M. M. (1968). Growth and function of the subcutaneous fat of the elephant seal. *Nature (London)* **220**, 597-599.

Buchmüller, K. (1961). Über die ultrarote Emission, Reflexion und Durchlässigkeit der lebenden menschlichen Haut im Spektralbereich $\lambda \approx 3\text{-}15$ μm. *Pfluegers Arch. Gesamte Physiol.* **272**, 360-371.

Budd, S. M. (1972). Thermoregulation in Black-capped Chickadees (*Parus atricapillus*). *Amer. Zool.* **12** (3), Abstr. No. 402.

Budgell, P. (1971). Behavioural thermoregulation in the Barbary Dove (*Streptopelia risoria*). *Anim. Behav.* **19**, 524-531.

Burton, A. C. (1934). The application of the theory of heat flow to the study of energy metabolism. *J. Nutr.* **7**, 497-533.

Busse, P., and Olech, B. (1968). Some problems relating to retirement of birds to nest boxes for the night. *Acta Ornithol.* **11**, 1-26 (translated from Polish).

Cabanac, M., Hammel, H. T., and Hardy, J. D. (1967). *Tiliqua scincoides:* Temperature-sensitive units in lizard brain. *Science* **158**, 1050-1051.

Cade, T. J. (1953). Sub-nival feeding of the Redpoll in interior Alaska: A possible adaptation to the northern winter. *Condor* **55**, 43-44.

Cade, T. J., and Maclean, G. L. (1967). Transport of water by adult sandgrouse to their young. *Condor* **69**, 323-343.

Cade, T. J., Tobin, C. A., and Gold, A. (1965). Water economy and metabolism of two estrildine finches. *Physiol. Zool.* **38**, 9-33.

Calder, W. A. (1964). Gaseous metabolism and water relations of the Zebra Finch, *Taeniopygia castanotis*. *Physiol. Zool.* **37**, 400-413.

Calder, W. A. (1968a). The diurnal activity of the Roadrunner, *Geococcyx californianus*. *Condor* **70**, 84-85.

Calder, W. A. (1968b). Respiratory and heart rates of birds at rest. *Condor* **70**, 358-365.

Calder, W. A. (1969). Temperature relations and underwater endurance of the smallest homeothermic diver, the Water Shrew. *Comp. Biochem. Physiol.* **30**, 1075-1082.

Calder, W. A. (1971). Temperature relationships and nesting of the Calliope Hummingbird. *Condor* **73**, 314-321.

Calder, W. A. (1972). Heat loss from small birds: Analogy with Ohm's law and a reexamination of the "Newtonian model." *Comp. Biochem. Physiol. A* **43**, 13-20.

Calder, W. A. (1974). Factors in the energy budget of mountain hummingbirds. *In* "Perspectives in Biophysical Ecology" (D. Gates, ed.). Ecological Studies, Analysis and Synthesis. Springer-Verlag, New York (in press).

Calder, W. A., and J. Booser (1973). Hypothermia of Broad-tailed Hummingbirds during incubation in nature with ecological correlations. *Science* **180**, 751-753.

Calder, W. A., and King, J. R. (1963). Evaporative cooling in the Zebra Finch. *Experientia* **19**, 603.

Calder, W. A., and King, J. R. (1972). Body weight and the energetics of temperature regulation: A re-examination. *J. Exp. Biol.* **56**, 775–780.
Calder, W. A., and Schmidt-Nielsen, K. (1966). Evaporative cooling and respiratory alkalosis in the pigeon. *Proc. Nat. Acad. Sci. U.S.* **55**, 750–756.
Calder, W. A., and Schmidt-Nielsen, K. (1967). Temperature regulation and evaporation in the pigeon and roadrunner. *Amer. J. Physiol.* **213**, 883–889.
Calder, W. A., and Schmidt-Nielsen, K. (1968). Panting and blood carbon dioxide in birds. *Amer. J. Physiol.* **215**, 477–482.
Cannon, P., and Keatinge, W. R. (1960). The metabolic rate and heat loss of fat and thin men in heat balance in cold and warm water. *J. Physiol. (London)* **154**, 329–344.
Carlson, L. D. (1966). The role of catecholamines in cold adaptation. *Pharmacol. Rev.* **18**, 291–301.
Chaffee, R. R. J., and Roberts, J. C. (1971). Temperature acclimation in birds and mammals. *Annu. Rev. Physiol.* **33**, 155–202.
Chaffee, R. R. J., Mayhew, W. W., Drebin, M., and Cassuto, Y. (1963). Studies on thermogenesis in cold-acclimated birds. *Can. J. Biochem. Physiol.* **41**, 2215–2220.
Chambers, A. B. (1970). A psychrometric chart for physiological research. *J. Appl. Physiol.* **29**, 406–412.
Cheke, R. A. (1971). Temperature rhythms in African montane sunbirds. *Ibis* **113**, 500–506.
Clark, W. M. (1948). "Topics in Physical Chemistry," p. 369. Williams & Wilkins, Baltimore, Maryland.
Cloudsley-Thompson, J. L. (1964). Terrestrial animals in dry heat: Arthropods. *In* "Handbook of Physiology" (D. B. Dill, ed.), Sect. 4, pp. 451–465. Amer. Physiol. Soc., Washington, D.C.
Coulombe, N. H. (1970). Physiological and physical aspects of temperature regulation in the Burrowing Owl, *Speotyto cunicularia*. *Comp. Biochem. Physiol.* **35**, 307–337.
Crawford, E. C. (1962). Mechanical aspects of panting in dogs. *J. Appl. Physiol.* **17**, 249–251.
Crawford, E. C., and Kampe, G. (1971). Resonant panting in pigeons. *Comp. Biochem. Physiol.* **40**, 549–552.
Crawford, E. C., and Lasiewski, R. C. (1968). Oxygen consumption and respiratory evaporation of the Emu and Rhea. *Condor* **70**, 333–339.
Crawford, E. C., and Schmidt-Nielsen, K. (1967). Temperature regulation and evaporative cooling in the Ostrich. *Amer. J. Physiol.* **212**, 347–353.
Davis, E. A., Jr. (1955). Seasonal changes in the energy balance of the English Sparrow. *Auk* **72**, 385–411.
Davis, L. B., Jr., and Birkebak, R. C. (1974). On the transfer of energy in layers of fur. *Biophysical J.* (in press).
Dawson, T. J., and Hulbert, A. J. (1970). Standard metabolism, body temperature and surface areas of Australian marsupials. *Amer. J. Physiol.* **218**, 1233–1238.
Dawson, W. R. (1954). Temperature regulation and water requirements of the Brown and Abert Towhees, *Pipilo fuscus* and *Pipilo aberti*. *Univ. Calif., Berkeley, Publ. Zool.* **59**, 81–124.
Dawson, W. R. (1958). Relation of oxygen consumption and evaporative water loss to body temperature in the cardinal. *Physiol. Zool.* **31**, 37–48.
Dawson, W. R., and Bartholomew, G. A. (1968). Temperature regulation and water economy in desert birds. *In* "Desert Biology" Vol. 1 (G. W. Brown, Jr., ed.), pp. 357–394. Academic Press, New York.
Dawson, W. R., and Fisher, C. D. (1969). Responses to temperature by the Spotted Nightjar (*Eurostopodus guttatus*). *Condor* **71**, 49–53.

Dawson, W. R., and Hudson, J. W. (1970). Birds. *In* "Comparative Physiology of Thermoregulation" (G. C. Whittow, ed.), Vol. 1, pp. 223–310. Academic Press, New York.

Dawson, W. R., and Schmidt-Nielsen, K. (1964). Terrestrial animals in dry heat: Desert birds. *In* "Handbook of Physiology" (D. B. Dill, ed.), Sect. 4, pp. 481–492. Amer. Physiol. Soc., Washington, D.C.

Dawson, W. R., and Tordoff, H. B. (1959). Relation of oxygen consumption to temperature in the Evening Grosbeak. *Condor* **61**, 388–396.

Dawson, W. R., and Tordoff, H. B. (1964). Relation of oxygen consumption to temperature in the Red and White-winged Crossbills. *Auk* **81**, 26–35.

Day, R. (1968). Regional heat loss. *In* "Physiology of Heat Regulation and the Science of Clothing" (L. H. Newburgh, ed.), pp. 240–261. Hafner, New York (reprint of 1949 edition).

DeJong, A. A. (1971). The role of incident radiation in the energy economy of the White-crowned Sparrow, *Zonotrichia leucophrys gambelii*. Ph.D. Dissertation, Washington State University, Pullman.

DeShazer, J. A., Jordan, K. A., and Suggs, C. W. (1970). Effect of acclimation on partitioning of heat loss by the laying hen. *Trans. ASAE (Amer. Soc. Agr. Eng.)* **13**, 82–84.

DeVries, A. L., and Wohlschlag, D. E. (1969). Freezing resistance in some antarctic fishes. *Science* **163**, 1073–1075.

Dolnik, V. R. (1967). Bioenergetische Anpassungen der Vögel and die Überwinterung in verschiedene Breiten. *Falke* **14**, 305–306 and 347–349.

Dow, D. D. (1965). The role of saliva in food storage in the Gray Jay. *Auk* **82**, 139–154.

Drent, R. (1972). Adaptive aspects of the physiology of incubation. *Proc. Int. Ornithol. Congr., 15th, 1970*, pp. 255–280.

Drent, R. (1973). The natural history of incubation. *In* "Breeding Biology of Birds" (D. S. Farner, ed.), pp. 262–311. Natl. Acad. Sci., Washington, D.C.

Drent, R. H., and Stonehouse, B. (1971). Thermoregulatory responses of the Peruvian Penguin, *Spheniscus humboldti*. *Comp. Biochem. Physiol. A* **40**, 689–710.

DuBois, E. F. (1937). The mechanism of heat loss and temperature regulation. Lane Medical Lectures. *Stanford Univ. Publ., Univ. Ser., Med. Sci.* **3**, 313–407.

Ederstrom, H. E., and Brumleve, S. J. (1964). Temperature gradients in the legs of cold-acclimatized pheasants. *Amer. J. Physiol.* **207**, 457–459.

El Halawani, M. E.-S., Wilson, W. O., and Burger, R. E. (1970). Cold-acclimation and the role of catecholamines in body temperature regulation in male leghorns. *Poultry Sci.* **49**, 621–632.

Eliassen, E. (1962). Skin temperature of seagulls exposed to air currents of high speed and low temperatures. *Årbok Univ. Bergen, Mat.-Naturvitensk. Ser.* **17**, 1–10.

Eliassen, E. (1963). Preliminary results from new methods of investigating the physiology of birds during flight. *Ibis* **105**, 234–237.

Emlen, J. M. (1966). The role of time and energy in food preference. *Amer. Natur.* **100**, 611–617.

Eriksson, K. (1970). The autumn migration and wintering ecology of the siskin *Carduelis spinus*. *Ornis Fenn.* **47**, 52–68.

Evans, P. R. (1966). Autumn movements, moult and measurements of the Lesser Redpoll *Carduelis flammea cabaret*. *Ibis* **108**, 183–216.

Evans, P. R. (1969a). Winter fat deposition and overnight survival of Yellow Buntings (*Emberiza citrinella* L.). *J. Anim. Ecol.* **38**, 415–423.

Evans, P. R. (1969b). Ecological aspects of migration, and premigratory fat deposition in the Lesser Redpoll, *Carduelis flammea cabaret*. *Condor* **71**, 316–330.

Farner, D. S. (1970). Some glimpses of comparative avian physiology. *Fed. Proc., Fed. Amer. Soc. Exp. Biol.* **29,** 1649–1663.

Fedde, M. R. (1970). Peripheral control of avian respiration. *Fed. Proc., Fed. Amer. Soc. Exp. Biol.* **29,** 1664–1673.

Feldberg, W. A. (1965). A new concept of temperature control in the hypothalamus. *Proc. Roy. Soc. Med.* **58,** 395–404.

Feldberg, W. A. (1970). The monoamines of the hypothalamus as mediators of temperature responses. *In* "Physiological and Behavioral Temperature Regulation" (J. D. Hardy, A. P. Gagge, and A. J. A. Stolwijk, eds.), pp. 493–506. Thomas, Springfield, Illinois.

Finch, V. C., and Trewartha, G. T. (1949). "Elements of Geography: Physical and Cultural." McGraw-Hill, New York.

Frazier, A., and Nolan, V., Jr. (1959). Communal roosting by the Eastern Bluebird in winter. *Bird-Banding* **30,** 219–226.

Freeman, B. M. (1966). The effects of cold, noradrenalin and adrenalin upon the oxygen consumption and carbohydrate metabolism of the young fowl (*Gallus domesticus*). *Comp. Biochem. Physiol.* **18,** 369–382.

Freeman, B. M. (1967). Some effects of cold on the metabolism of the fowl during the perinatal period. *Comp. Biochem. Physiol.* **20,** 179–193.

Freeman, B. M. (1970a). Thermoregulatory mechanisms of the neonate fowl. *Comp. Biochem. Physiol.* **33,** 219–230.

Freeman, B. M. (1970b). Some aspects of thermoregulation in the adult Japanese Quail (*Coturnix coturnix japonica*). *Comp. Biochem. Physiol.* **34,** 871–881.

French, N. R. (1959). Life history of the Black Rosy Finch. *Auk* **76,** 159–180.

Fry, F. E. J., and Hochachka, P. W. (1970). Fish. *In* "Comparative Physiology of Thermoregulation" (G. C. Whittow, ed.), Vol. 1, pp. 79–134. Academic Press, New York.

Gagge, A. P., Hardy, J. D., and Rapp, G. M. (1969). Proposed standard system of symbols for thermal physiology. *J. Appl. Physiol.* **27,** 439–446.

Gardasson, A., and Moss, R. (1970). Selection of food by Icelandic ptarmigan in relation to its availability and nutritive value. *In* "Animal Populations in Relation to their Food Resources" (A. Watson, ed.), pp. 47–71. Blackwell, Oxford.

Gates, D. M. (1962). "Energy Exchange in the Biosphere." Harper, New York.

Gates, D. M. (1969). Climate and stability. *In* "Diversity and Stability in Ecological Systems" (G. M. Woodwell and H. H. Smith, eds.), pp. 115–127. Brookhaven Symposia in Biol. 22. Brookhaven Natl. Lab., Upton, N.Y.

Gates, D. M. (1970). Animal climates (Where animals must live). *Environ. Res.* **3,** 132–144.

Gates, D. M. (1972). "Man and His Environment: Climate." Harper, New York.

Gates, D. M., Keegan, J. J., and Weidner, V. R. (1966). Spectral reflectance and planetary reconnaissance. Scientific experiments for manned orbital flight. *Sci. Technol.* **4,** 71–86.

Gavrilov, V. M., Dolnik, V. R., and Keskpaik, J. (1970). The winter energy metabolism of the Chaffinch. *Izv. Akad. Nauk Est. SSR* **19,** 211–218 (In Russian).

Gebhart, B. (1971). "Heat Transfer," 2nd Ed. McGraw-Hill, New York.

Geiger, R. (1966). "The Climate Near the Ground." Harvard Univ. Press, Cambridge.

Gelineo, S. (1955). Température d'adaptation et production de chaleur chez les oiseaux de petite taille. *Arch. Sci. Physiol.* **9,** 225–243.

Gelineo, S. (1964). Organ systems in adaptation: The temperature regulating system. *In* "Handbook of Physiology" (O. B. Bill, ed.), Sect. 4, pp. 259–282. Amer. Physiol. Soc., Washington, D.C.

George, J. (1967). Long night of the snowbirds. *Nat. Wildl.* **5,** 14–17.
Gessaman, J. A. (1972). Bioenergetics of the Snowy Owl (*Nyctea scandiaca*). *Arctic Alp. Res.* **4,** 223–238.
Gessaman, J. A., and Folk, G. E., Jr. (1969). Body temperature and thermal conductance of the Snowy Owl. *Physiologist* **12,** 234.
Gilbreath, J. C., and Ko, R.-C. (1970). Sex differential for body temperature in Japanese Quail. *Poultry Sci.* **49,** 34–36.
Goldsmith, R., and Sladen, W. J. L. (1961). Temperature regulation of some antarctic penguins. *J. Physiol. (London)* **157,** 251–262.
Goodwin, D. (1967). Some possible function of sun-bathing in birds. *Brit. Birds* **60,** 363–364.
Gordon, M. S. (1964). Animals in aquatic environments: Fishes and amphibians. *In* "Handbook of Physiology" (D. B. Dill, ed.), Sect. 4, pp. 697–713. Amer. Physiol. Soc., Washington, D.C.
Greenewalt, C. H. (1960). Flight. *In* "Hummingbirds" pp. 203–238. Doubleday, Garden City, New York.
Greenwald, L., Stone, W. B., and Cade, T. J. (1967). Physiological adjustments of the Budgerigah (*Melopsittacus undulatus*) to dehydrating conditions. *Comp. Biochem. Physiol.* **22,** 91–100.
Grunden, L. R., and Marley, E. (1970). Effects of sympathomimetic amines injected into the third ventricle in adult chickens. *Neuropharmacology* **9,** 119–128.
Hainsworth, F. R., and Wolf, L. L. (1970). Regulation of oxygen consumption and body temperature during torpor in a hummingbird, *Eulampis jugularis*. *Science* **168,** 368–369.
Hainsworth, F. R., and Wolf, L. L. (1971). Crop volume and hummingbird energetics. *Amer. Zool.* **11,** Abstr. No. 258, 671.
Hamilton, T. H. (1961). The adaptive significance of interspecific difference trends of variation in wing length and body size among bird species. *Evolution* **15,** 180–195.
Hamilton, W. J., III, and Heppner, F. (1967a). Radiant solar energy and the function of black homeotherm pigmentation: An hypothesis. *Science* **155,** 196–197.
Hamilton, W. J., III, and Heppner, F. (1967b). Black pigmentation: Adaptation for concealment or heat conservation. *Science* **158,** 1340–1341.
Hammel, H. T. (1956). Infrared emissivities of some arctic fauna. *J. Mammal.* **37,** 375–381.
Hammel, H. T. (1968). Regulation of internal body temperature. *Annu. Rev. Physiol.* **30,** 641–710.
Hammel, H. T., Caldwell, F. T., Jr., and Abrams, R. M. (1967). Regulation of body temperature in the Blue-tongued Lizard. *Science* **156,** 1260–1262.
Hardy, J. D. (1939). The radiating power of human skin in the infrared. *Amer. J. Physiol.* **127,** 454–462.
Hardy, J. D. (1968). Heat transfer. *In* "Physiology of Heat Regulation and the Science of Clothing" (L. H. Newburgh, ed.), pp. 78–108. Hafner, New York (reprint of 1949 edition).
Hardy, J. D., Gagge, A. P., and Stolwijk, J. A. J., eds. (1970). "Physiological and Behavioral Temperature Regulation." Thomas, Springfield, Illinois.
Hart, J. S. (1951). Average body temperature in mice. *Science* **113,** 325–326.
Hart, J. S. (1957). Climatic and temperature induced changes in the energetics of homeotherms. *Rev. Can. Biol.* **16,** 133–174.
Hart, J. S. (1962). Seasonal acclimatization in four species of small wild birds. *Physiol. Zool.* **35,** 224–236.

Hart, J. S. (1963). Physiological responses to cold in nonhibernating homeotherms. In "Temperature: Its Measurement and Control in Science and Industry" (J. D. Hardy, ed.), Vol. III, Part 3, pp. 373–406. Van Nostrand-Reinhold, Princeton, New Jersey.

Hart, J. S. (1964). Geography and season: Mammals and birds. In "Handbook of Physiology" (D. B. Dill, ed.), Sect. 4, pp. 295–321. Amer. Physiol. Soc., Washington, D.C.

Hart, J. S., and Roy, O. Z. (1966). Respiratory and cardiac responses to flight in pigeons. *Physiol. Zool.* **39**, 291–306.

Hart, J. S., and Roy, O. Z. (1967). Temperature regulation during flight in pigeons. *Amer. J. Physiol.* **213**, 1311–1316.

Hartmann, F. A. (1961). Locomotor mechanisms of birds. *Smithson. Misc. Collect.* **143**, 1–91.

Hartung, R. (1967). Energy metabolism in oil-covered ducks. *J. Wildl. Manage.* **31**, 798–804.

Hatch, D. E. (1970). Energy conserving and heat dissipating mechanisms of the Turkey Vulture *Auk* **87**, 111–124.

Heath, J. E. (1962). Temperature fluctuation in the Turkey Vulture. *Condor* **64**, 234–235.

Heinrich, B., and Bartholomew, G. A. (1971). An analysis of pre-flight warm-up in the Sphinx Moth, *Manduca sexta. J. Exp. Biol.* **55**, 223–239.

Hemingway, A. (1963). Shivering. *Physiol. Rev.* **43**, 397–422.

Hemmingsen, A. M. (1960). Energy metabolism as related to body size and respiratory surfaces and its evolution. *Rep. Steno Mem. Hosp. Nord. Insulinlab.* **9**, Part II, 1–110.

Hensel, H. (1970). Temperature receptors in the skin. In "Physiological and Behavioral Temperature Regulation" (J. D. Hardy, A. P. Gagge, and J. A. J. Stolwijk, eds.), pp. 442–453. Thomas, Springfield, Illinois.

Henshaw, R. E. (1968). Thermoregulation during hibernation: Application of Newton's law of cooling. *J. Theor. Biol.* **20**, 79–90.

Hensley, M. M. (1954). Ecological relations of the breeding bird population of the desert biome in Arizona. *Ecol. Monogr.* **24**, 185–207.

Heppner, F. (1970). The metabolic significance of differential absorption of radiant energy by black and white birds. *Condor* **72**, 50–59.

Herreid, C. F., II., and Kessel, B. (1967). Thermal conductance in birds and mammals. *Comp. Biochem. Physiol.* **21**, 405–414.

Hesse, R., Allee, W. C., and Schmidt, K. P. (1937). "Ecological Animal Geography." Wiley, New York.

Hillerman, J. P., and Wilson, W. O. (1955). Acclimation of adult chickens to environmental temperature changes. *Amer. J. Physiol.* **180**, 591–595.

Himms-Hagen, J. (1967). Sympathetic regulation of metabolism. *Pharmacol. Rev.* **19**, 367–461.

Hinds, D. S., and Calder, W. A. (1973). Temperature regulation of the Pyrrhuloxia and the Arizona Cardinal. *Physiol. Zool.* **46**, 55–71.

Hissa, R., and Palokangas, R. (1970). Thermoregulation in the Titmouse (*Parus major* L.). *Comp. Biochem. Physiol.* **33**, 941–953.

Hoesch, W. (1959). Brutverhalten bei starker Sonneneinwirkung. *J. Ornithol.* **100**, 173–175.

Howell, T. R., and Bartholomew, G. A. (1961). Temperature regulation in Laysan and Black-footed Albatrosses. *Condor* **63**, 185–197.

Hudson, J. W., and Kimzey, S. L. (1966). Temperature regulation and metabolic rhythms in population of the House Sparrow, *Passer domesticus*. *Comp. Biochem. Physiol.* **17**, 203–217.

Huggins, A. (1941). Egg temperatures of wild birds under natural conditions. *Ecology* **22**, 148–157.

Hull, W. E., and Long, E. C. (1961). Respiratory impedance and volume flow at high frequency in dogs. *J. Appl. Physiol.* **16**, 439–443.

Huston, T. M., Cotton, T. E., and Cameron, J. L. (1962). The influence of high environmental temperature on the oxygen consumption of mature domestic fowl. *Poultry Sci.* **41**, 179–183.

Hutchinson, J. C. D., and Sykes, A. H. (1953). Physiological acclimatization of fowl to a hot, humid environment. *J. Agr. Sci.* **43**, 294–322.

Hutt, F. B., and Hall, L. (1938). Number of feathers and body size in passerine birds. *Auk* **55**, 651–657.

Huxley, J. S., Webb, C. S., and Best, A. T. (1939). Temporary poikilothermy in birds. *Nature (London)* **143**, 683–684.

Irving, L. (1955). Nocturnal decline in the temperature of birds in cold weather. *Condor* **57**, 362–365.

Irving, L. (1957). The usefulness of Scholander's views on adaptive insulation of animals. *Evolution* **11**, 257–259.

Irving, L. (1960). "Birds of Anaktuvuk Pass, Kobuk, and Old Crow." *U.S. Nat. Mus. Bull.* **217**. Smithsonian Institution, Washington, D.C.

Irving, L., and Krog, J. (1954). Body temperatures of arctic and subarctic birds and mammals. *J. Appl. Physiol.* **6**, 667–680.

Irving, L., and Krog, J. (1955). Temperature of skin in the Arctic as a regulator of heat. *J. Appl. Physiol.* **7**, 355–364.

Irving, L., and Krog, J. (1956). Temperature during the development of birds in arctic nests. *Physiol. Zool.* **29**, 195–205.

Irving, L., West, G. C., and Peyton, L. C. (1967). Winter feeding program of Alaska Willow Ptarmigan shown by crop contents. *Condor* **69**, 69–77.

Jaeger, E. C. (1948). Does the Poor-will "hibernate"? *Condor* **50**, 45–46.

Jaeger, E. C. (1949). Further observations on the hibernation of the Poor-will. *Condor* **51**, 105–109.

James, F. C. (1968). A more precise definition of Bergmann's Rule. *Amer. Zool.* **8**, 815–816.

James, F. C. (1970). Geographic size variation in birds and its relationship to climate. *Ecology* **51**, 365–390.

Jansky, L., Bartuňkova, R., Kočkova, J., Mejsnar, J., and Zeisberger, E. (1969). Interspecies differences in cold adaptation and nonshivering thermogenesis. *Fed. Proc., Fed. Amer. Soc. Exp. Biol.* **28**, 1053–1058.

Johnson, H. McC. (1957). "Winter Microclimates of Importance to Alaskan Small Mammals and Birds," AAL Tech. Rep. 57-2. Alaska Air Command, Arctic Aeromed. Lab., Fort Wainwright, Alaska.

Johnson, R. E. (1968). Temperature regulation in the White-tailed Ptarmigan, *Lagopus leucurus*. *Comp. Biochem. Physiol.* **24**, 1003–1014.

Johnston, D. W. (1971). The absence of brown adipose tissue in birds. *Comp. Biochem. Physiol. A* **40**, 1107–1108.

Johnston, R. F., and Selander, R. K. (1971). Evolution in the House Sparrow. II. Adaptive differentiation in North American populations. *Evolution* **25**, 1–28.

Jones, D. R., and Johansen, K. (1972). The blood vascular system of birds. *In* "Avian

Biology" (D. S. Farner and J. R. King, eds.), Vol. 2, pp. 157–285. Academic Press, New York.

Jones, R. E. (1971). The incubation patch of birds. *Biol. Rev. Cambridge Phil. Soc.* **46**, 315–339.

Kahl, M. P., Jr. (1963). Thermoregulation in the Wood Stork, with special reference to the role of the legs. *Physiol. Zool.* **36**, 141–151.

Kamar, G. A. R., and Khalifa, M. A. S. (1964). The effect of environmental conditions on body temperature of fowls. *Brit. Poultry Sci.* **5**, 235–244.

Kanematsu, S., Kii, M., Sonada, T., and Kato, Y. (1967). Effects of hypothalamic lesions on body temperature in the chicken. *Jap. J. Vet. Sci.* **29**, 95–104.

Kashkin, V. (1961). Heat exchange of bird eggs during incubation. *Biophysics (USSR)* **6**, 57–63.

Kattus, A. H., Ross, G., and Hall, V. E., eds. (1970). "Cardiovascular Beta Adrenergic Responses," UCLA Forum in Medical Sciences, No. 13. Univ. of California Press, Berkeley.

Kavanau, J. L., and Ramos, J. (1970). Roadrunners: Activity of captive individuals. *Science* **169**, 780–782.

Kayser, C., and Heusner, A. (1964). Etude comparative du métabolisme énergétique dans la série animale. *J. Physiol. (Paris)* **56**, 489–524.

Kendeigh, S. C. (1960). Energy of birds conserved by roosting in cavities. *Wilson Bull.* **73**, 140–147.

Kendeigh, S. C. (1963). Thermodynamics of incubation in the house wren, *Troglodytes aedon*. *Proc. Int. Ornithol. Congr., 13th, 1962* pp. 884–904.

Kendeigh, S. C. (1969). Tolerance of cold and Bergmann's Rule. *Auk* **86**, 13–25.

Kendeigh, S. C. (1972). Energy control of size limits in birds. *Amer. Natur.* **106**, 79–88.

Kennedy, R. J. (1969). Sunbathing behaviour of birds. *Brit. Birds* **62**, 249–258.

Keskpaik, J., and Davydov, A. (1966). Factors determining the cold-hardiness of *Larus ridibundus* L. on the first day after hatching. *Eesti NSV Tead. Akad. Toim., Biol.* 1966, No. 4, 485–491 (in Russian).

Kilham, L. (1971). Roosting habits of White-breasted Nuthatches. *Condor* **73**, 113–114.

King, B. (1970). Sunbathing by migrant and vagrant passerines. *Brit. Birds* **63**, 37–38.

King, J. R. (1964). Oxygen consumption and body temperature in relation to ambient temperature in the White-crowned Sparrow. *Comp. Biochem. Physiol.* **12**, 13–24.

King, J. R. (1972). Adaptive periodic fat storage by birds. *Proc. Int. Ornithol. Congr., 15th, 1970* pp. 200–217.

King, J. R., and Farner, D. S. (1961). Energy metabolism, thermoregulation and body temperature. *In* "Biology and Comparative Physiology of Birds" (A. J. Marshall, ed.), Vol. 2, pp. 215–288. Academic Press, New York.

King, J. R., and Farner, D. S. (1964). Terrestrial animals in humid heat: Birds. *In* "Handbook of Physiology" (D. B. Dill, ed.), Sect. 4, pp. 603–624. Amer. Physiol. Soc., Washington, D.C.

King, J. R., and Farner, D. S. (1965). Studies of fat deposition in migratory birds. *Ann. N.Y. Acad. Sci.* **131**, 422–440.

King, J. R., and Farner, D. S. (1966). The adaptive role of winter fattening in the White-crowned Sparrow with comments on its regulation. *Amer. Natur.* **100**, 403–418.

King, J. R., and Wales, E. E., Jr. (1964). Observations on migration, ecology, and population flux of wintering Rosy Finches. *Condor* **66**, 24–31.

Kitchell, R. L., Ström, L., and Zotterman, Y. (1959). Electrophysiological studies of thermal and taste reception in chickens and pigeons. *Acta Physiol. Scand.* **46**, 133–151.

Kleiber, M. (1961). "The Fire of Life." Wiley, New York.

Kleiber, M. (1970). Conductivity, conductance and transfer constant for animal heat. *Fed. Proc., Fed. Amer. Soc. Exp. Biol.* **29**, 660.

Kleiber, M. (1972a). Body size, conductance for animal heat flow and Newton's Law of cooling. *J. Theor. Biol.* **37**, 139–150.

Kleiber, M. (1972b). A new Newton's law of cooling? *Science* **178**, 1283–1285.

Kleiber, M. (1973). Technical comments: Perspectives of linear heat transfer. *Science* **181**, 186.

Kluyver, H. N. (1957). Roosting habits, sexual dominance and survival in the Great Tit. *Cold Spring Harbor Symp. Quant. Biol.* **22**, 281–285.

Knorr, O. A. (1957). Communal roosting of the Pigmy Nuthatch. *Condor* **59**, 398.

Koskimies, J. (1948). On temperature regulation and metabolism in the swift. *Micropus a. apus* L., during fasting. *Experientia* **4**, 274–276.

Koskimies, J. (1950). The life of the Swift, *Micropus apus*, (L.) in relation to the weather. *Ann. Acad. Sci. Fenn., Ser. A* **15**, 1–151.

Koskimies, J., and Lahti, L. (1964). Cold-hardiness of the newly hatched young in relation to ecology and distribution in ten species of European ducks. *Auk* **81**, 281–307.

Lasiewski, R. C. (1963). Oxygen consumption of torpid, resting, active and flying hummingbirds. *Physiol. Zool.* **36**, 122–140.

Lasiewski, R. C. (1964). Body temperature, heart and breathing rate, and evaporative water loss in hummingbirds. *Physiol. Zool.* **37**, 212–223.

Lasiewski, R. C. (1969). Physiological responses to heat stress in the Poorwill. *Amer. J. Physiol.* **217**, 1504–1509.

Lasiewski, R. C. (1972). Respiratory function in birds. *In* "Avian Biology" (D. S. Farner and J. R. King, eds.), Vol. 2, pp. 287–342. Academic Press, New York.

Lasiewski, R. C., and Calder, W. A. (1971). A preliminary allometric analysis of respiratory variables in resting birds. *Resp. Physiol.* **11**, 152–166.

Lasiewski, R. C., and Dawson, W. R. (1964). Physiological responses to temperature in the Common Nighthawk. *Condor* **66**, 477–490.

Lasiewski, R. C., and Dawson, W. R. (1967). A re-examination of the relation between standard metabolic rate and body weight in birds. *Condor* **69**, 13–23.

Lasiewski, R. C., and Dawson, W. R. (1969). Calculation and miscalculation of the equations relating avian standard metabolism to body weight. *Condor* **71**, 335–336.

Lasiewski, R. C., and Lasiewski, R. J. (1967). Physiological responses of the Blue-throated and Rivoli's Hummingbirds. *Auk* **84**, 34–48.

Lasiewski, R. C., and Seymour, R. S. (1972). Thermoregulatory responses to heat stress in four species of birds weighing approximately 40 grams. *Physiol. Zool.* **45**, 106–118.

Lasiewski, R. C., and Thompson, H. J. (1966). Field observation of torpidity in the Violet-green Swallow. *Condor* **68**, 102–103.

Lasiewski, R. C., Hubbard, S. H., and Moberly, W. R. (1964). Energetic relationships of a very small passerine bird. *Condor* **66**, 212–220.

Lasiewski, R. C., Acosta, A. L., and Bernstein, M. H. (1966a). Evaporative water loss in birds. I. Characteristics of the open flow method of determination, and their relation to estimates of thermoregulatory ability. *Comp. Biochem. Physiol.* **19**, 445–457.

Lasiewski, R. C., Acosta, A., and Bernstein, M. H. (1966b). Evaporative water loss in birds. II. A modified method for determination by direct weighing. *Comp. Biochem. Physiol.* **19**, 459–470.

Lasiewski, R. C., Weathers, W. W., and Bernstein, M. H. (1967). Physiological responses of the Giant Hummingbird, *Patagona gigas*. *Comp. Biochem. Physiol.* **23**, 797–813.

Lasiewski, R. C., Dawson, W. R., and Bartholomew, G. A. (1970). Temperature regulation in the Little Papuan Frogmouth, *Podargus ocellatus*. *Condor* **72**, 332–338.

Lasiewski, R. C., Bernstein, M. H., and Ohmart, R. D. (1971). Cutaneous water loss in the Roadrunner and Poor-will. *Condor* **73**, 470–472.

Lee, P., and Schmidt-Nielsen, K. (1971). Respiratory and cutaneous evaporation in the Zebra Finch: Effect on water balance. *Amer. J. Physiol.* **220**, 1598–1605.

LeFebvre, E. A. (1964). The use of D_2O^{18} for measuring energy metabolism in *Columba livia* at rest and in flight. *Auk* **81**, 403–416.

LeFebvre, E. A., and Raveling, D. G. (1967). Distribution of Canada Geese in winter as related to heat loss at varying environmental temperatures. *J. Wildl. Manage.* **31**, 538–546.

Leighton, A. T., Jr., Siegel, P. B., and Siegel, H. S. (1966). Body weight and surface area of chickens (*Gallus domesticus*). *Growth* **30**, 229–238.

Lepkovsky, S., Snapir, N., and Furuta, F. (1968). Temperature regulation and appetitive behaviour in chickens with hypothalamic lesions. *Physiol. & Behav.* **3**, 911–915.

Levins, R. (1968). "Evolution in Changing Environments." Princeton Univ. Press, Princeton, New Jersey.

Ligon, J. D. (1968). The biology of the Elf Owl, *Micrathene whitneyi*. *Univ. Mich. Mus. Zool., Misc. Publ.* **136**, 1–70.

Ligon, J. D. (1970). Still more responses of the Poor-will to low temperatures. *Condor* **72**, 496–497.

Lin, Y. C., and Sturkie, P. D. (1968). Effect of environmental temperatures on the catecholamines of chickens. *Amer. J. Physiol.* **214**, 237–240.

Linsley, J. G., and Burger, R. E. (1964). Respiratory and cardiovascular responses in the hyperthermic domestic cock. *Poultry Sci.* **43**, 291–305.

List, R. J. (1951). "Smithsonian Meteorological Tables," 6th revised ed., Smithson. Misc. Collect. No. 114. Smithsonian Institution, Washington, D.C.

Löhrl, H. (1955). Schlafengewöhnheiten der Baumläufer (*Certhia brachydactyla, C. familiaris*) und andere Kleinvögel in kalten Winternachten. *Vogelwarte* **18**, 99–104.

Lowry, W. P. (1969). "Weather and Life." Academic Press, New York.

Lustick, S. (1969). Bird energetics: Effects of artificial radiation. *Science* **163**, 387–390.

Lustick, S. (1970). Energetics and water regulation in the Cowbird (*Molothrus ater obscurus*). *Physiol. Zool.* **43**, 270–287.

Lustick, S. (1971). Plumage color and energetics. *Condor* **73**, 121–122.

McFarland, D. J., and Baher, E. (1968). Factors affecting feather posture in the Barbary Dove. *Anim. Behav.* **16**, 171–177.

McFarland, D. J., and Budgell, P. W. (1970). The thermoregulatory role of feather movements in the Barbary Dove (*Streptopelia risoria*). *Physiol. & Behav.* **5**, 763–771.

McMahon, T. (1973). Size and shape in biology. *Science* **179**, 1201–1204.

Macmillen, R. E., and Nelson, J. E. (1969). Bioenergetics and body size in dasyurid marsupials. *Amer. J. Physiol.* **217**, 1246–1251.

MacMillen, R. E., and Trost, C. H. (1967). Thermoregulation and water loss in the Inca Dove. *Comp. Biochem. Physiol.* **20**, 263–273.

McNab, B. K. (1970). Body weight and the energetics of temperature regulation. *J. Exp. Biol.* **53**, 329–348.

McNab, B. K. (1971). On the ecological significance of Bergmann's rule. *Ecology* **52**, 845–854.

Marder, J. (1973). Body temperature regulation in the Brown-necked Raven (*Corvus corax ruficollis*) – II. Thermal changes in the plumage of ravens exposed to solar radiations. *Comp. Biochem. Physiol. A.* **45**, 431–440.

Marley, E., and Stephenson, J. D. (1970). Effects of catecholamines infused into the brain of young chickens. *Brit. J. Pharmacol.* **40**, 639–658.

Martin, E. W. (1968). The effects of dietary protein on the energy and nitrogen balance of the Tree Sparrow (*Spizella arborea arborea*). *Physiol. Zool.* **41**, 313–331.

Martin, E. W. (1970). Tolerance to temperature stress in House Sparrows (*Passer domesticus*) and Tree Sparrows (*Spizella arborea*). *Int. Stud. Sparrows* **4**, 45–48.

Mayr, E. (1956). Geographical character gradients and climatic adaptation. *Evolution* **10**, 105–108.

Mayr, E. (1970). "Populations, Species, and Evolution." Harvard Univ. Press, Cambridge, Massachusetts.

Medway, Lord. (1969). The diurnal activity cycle among forest birds in Ulu Gombak. *Malay. Nature J.* **22**, 184–186.

Miller, D. S. (1939). A study of the physiology of the sparrow thyroid. *J. Exp. Zool.* **80**, 259–281.

Mills, S. H., and Heath, J. E. (1970). Thermoresponsiveness of the preoptic region of the brain in House Sparrows. *Science* **168**, 1008–1009.

Mills, S. H., and Heath, J. E. (1972). Anterior hypothalamic/preoptic lesions impair normal thermoregulation in House Sparrows. *Comp. Biochem. Physiol. A* **43**, 125–129.

Minard, D. (1970). Body heat content. *In* "Physiological and Behavioral Temperature Regulation" (J. D. Hardy, A. P. Gagge, and J. A. J. Stolwijk, eds.), pp. 345–357. Thomas, Springfield, Illinois.

Misch, M. S. (1960). Heat regulation in the Northern Blue Jay, *Cyanocitta cristata bromia* Oberholser. *Physiol. Zool.* **33**, 252–259.

Mitchell, D. (1970). Measurement of the thermal emissivity of human skin *in vivo*. *In* "Physiological and Behavioral Temperature Regulation" (J. D. Hardy, A. P. Gagge, and J. A. J. Stolwijk, eds.), pp. 25–33. Thomas, Springfield, Illinois.

Moen, A. N. (1968). The critical thermal environment: A new look at an old concept. *BioScience* **18**, 1041–1043.

Moore, A. D. (1945). Winter night habits of birds. *Wilson Bull.* **57**, 253–260.

Morowitz, H. J. (1968). "Energy Flow in Biology." Academic Press, New York.

Morrison, P. R. (1960). Some interrelations between weight and hibernation function. *Bull. Mus. Comp. Zool., Harvard Univ.* **124**, 75–91.

Morton, M. L. (1967a). The effects of insolation in the diurnal feeding pattern of White-crowned Sparrows (*Zonotrichia leucophrys gambelii*). *Ecology* **48**, 690–694.

Morton, M. L. (1967b). Diurnal feeding patterns in White-crowned Sparrows, *Zonotrichia leucophrys gambelii*. *Condor* **69**, 491–512.

Mugaas, J. N., and Templeton, J. R. (1970). Thermoregulation in the Red-breasted Nuthatch (*Sitta canadensis*). *Condor* **72**, 125–132.

Murrish, D. E. (1970). Responses to temperature in the Dipper, *Cinclus mexicanus*. *Comp. Biochem. Physiol.* **34**, 859–869.

National Bureau of Standards. (1969). NBS interprets policy on SI units. *Nat. Bur. Stand. (U.S.), News Bull.* **52**, No. 6.

Necker, R. (1972). Response of trigeminal ganglion neurons to thermal stimulation of the beak in pigeons. *J. Comp. Physiol.* **78**, 307–314.

Neumann, R., Hudson, J. W., and Hock, R. J. (1968). Body temperatures. *In* "Metabolism" (P. Altman, ed.), pp. 334–343. Biol. Handb. Ser., Fed. Amer. Soc. Exp. Biol., Bethesda, Maryland.

Newton, I. (1968). The temperatures, weights, and body composition of molting Bullfinches. *Condor* **70,** 323-332.

Newton, I. (1970). Irruptions of crossbills in Europe. *In* "Animal Populations in Relation to their Food Resources" (A. Watson, ed.), pp. 337-357. Blackwell, Oxford.

Norris, K. S. (1967). Color adaptation in desert reptiles and its thermal relationships. *In* "Lizard Ecology: A Symposium" (W. W. Milstead, ed.), pp. 162-229. Univ. of Missouri Press, Columbia.

Nye, P. (1964). Heat loss in wet ducklings and chicks. *Ibis* **106,** 189-197.

Odum, E. P. (1942). Muscle tremors and the development of temperature regulation in birds. *Amer. J. Physiol.* **136,** 618-622.

Ogilvie, D. M. (1970). Temperature selection in day-old chickens (*Gallus domesticus*) and young Japanese Quail (*C. coturnix japonicus*). *Can. J. Zool.* **48,** 1295-1298.

Ohmart, R. D., and Lasiewski, R. C. (1971). Roadrunners: Energy conservation by hypothermia and absorption of sunlight. *Science* **172,** 67-69.

Øritsland, N. A. (1970). Energetic significance of absorption of solar radiation in polar homeotherms. *In* "Antarctic Ecology" (M. W. Holdgate, ed.). Vol. 1, pp. 464-470. Academic Press, New York.

Orr, Y. (1970). Temperature measurements at the nest of the Desert Lark (*Ammomanes deserti deserti*). *Condor* **72,** 476-478.

Pearson, O. P. (1948). Metabolism of small mammals, with remarks on the lower limit of mammalian size. *Science* **108,** 44.

Pearson. O. P. (1953). Use of caves by hummingbirds and other species at high altitudes in Peru. *Condor* **55,** 17-20.

Pearson, O. P. (1954). The daily energy requirements of a wild Anna Hummingbird. *Condor* **56,** 317-322.

Pearson, O. P. (1960). Torpidity in birds. *Bull. Mus. Comp. Zool. Harvard Univ.* **124,** 93-103.

Peiponen, V. A. (1965). On hypothermia and torpidity in the Nightjar (*Caprimulgus europaeus* L.). *Ann. Acad. Sci. Fenn,* A4 **87,** 1-15.

Peiponen, V. A. (1966). The diurnal heterothermy of the Nightjar (*Caprimulgus europaeus* L.). *Ann. Acad. Sci. Fenn.,* A4 **101,** 1-35.

Peiponen, V. A. (1970). Body temperature fluctuations in the Nightjar (*Caprimulgus europaeus* L.) in light conditions of southern Finland. *Ann. Zool. Fenn.* **7,** 239-250.

Persons, J. N., Wilson, H. R., and Harms, R. H. (1967). Relationships of diet composition to survival time of chicks when subjected to high temperature. *Proc. Soc. Exp. Biol. Med.* **126,** 604-606.

Pitelka, F. A. (1958). Timing of molt in Steller Jays of the Queen Charlotte Islands, British Columbia. *Condor* **60,** 38-49.

Pitelka, F. A. (1962). Remarks on latitudinal trends in the density of down feathers. *In* "Comparative Physiology of Temperature Regulation" (J. P. Hannon and E. Viereck, eds.), Part 3, pp. 329-331. Arctic Aeromed. Lab., Fort Wainwright, Alaska.

Poczopko, P. (1967). Studies on the effect of environmental thermal conditions on geese. I. Thermal preference in goslings. *Acta Physiol. Pol.* **18,** 335-340.

Pohl, H. (1969). Some factors influencing the metabolic response to cold in birds. *Fed. Proc., Fed. Amer. Soc. Exp. Biol.* **28,** 1059-1064.

Pohl, H. (1971). Seasonal variation in metabolic functions of Bramblings. *Ibis* **113,** 185-193.

Porter, W. P. (1969). Thermal radiation in metabolic chambers. *Science* **166,** 115-117.

Porter, W. P., and Gates, D. M. (1969). Thermodynamic equilibria of animals with environment. *Ecol. Monogr.* **39,** 227-244.

Power, D. M. (1969). Evolutionary implications of wing and size variation in the Red-winged Blackbird in relation to geographic and climatic factors: A multiple regression analysis. *Syst. Zool.* **18**, 363–373.

Prange, H. D., and Schmidt-Nielsen, K. (1969). Evaporative water loss in snakes. *Comp. Biochem. Physiol.* **28**, 973–975.

Prange, H. D., and Schmidt-Nielsen, K. (1970). The metabolic cost of swimming in ducks. *J. Exp. Biol.* **53**, 763–777.

Prosser, C. L. (1973). "Comparative Animal Physiology." Saunders, Philadelphia, Pennsylvania.

Randall, W. C. (1943). Factors influencing the temperature regulation of birds. *Amer. J. Physiol.* **139**, 56–63.

Randall, W. C., and Hiestand, W. A. (1939). Panting and temperature regulation in the chicken. *Amer. J. Physiol.* **127**, 761–767.

Rautenberg, W. (1969a). Untersuchungen zur Temperaturregulation wärme- und kälteakklimatisierter Tauben Z. *Vergl. Physiol.* **62**, 221–234.

Rautenberg, W. (1969b). Die Bedeutung der zentralnervösen Thermosensitivität für die Temperaturregulation der Taube. *Z. Vergl. Physiol.* **62**, 235–266.

Rautenberg, W. (1971). The influence of skin temperature on the thermoregulatory system of pigeons. *J. Physiol. (Paris)* **68**, 396–398.

Rautenberg, W., Necker, R., and May, B. (1972). Thermoregulatory responses of the pigeon to changes of the brain and spinal cord temperatures. *Pfluegers Arch.* **388**, 31–42.

Reifsnyder, W. E., and Lull, H. W. (1965). Radiant energy in relation to forests. *U.S. Dep. Agr., Forest Serv., Tech. Bull.* **1344**.

Richards, S. E. (1967). Anatomy of the arteries of the head in domestic fowl. *J. Zool.* **152**, 221–234.

Richards, S. A. (1968). Vagal control of thermal panting in mammals and birds. *J. Physiol. (London)* **199**, 89–101.

Richards, S. A. (1970a). The role of hypothalamic temperature in the control of panting in the chicken exposed to heat. *J. Physiol. (London)* **211**, 341–358.

Richards, S. A. (1970b). The biology and comparative physiology of thermal panting. *Biol. Rev. Cambridge Phil. Soc.* **45**, 223–264.

Richards, S. A. (1970c). Physiology of thermal panting in birds. *Ann. Biol. Anim., Biochim., Biophys.* **10**, Suppl. Ser. 2, 151–168.

Richards, S. A. (1970d). Brain temperature and the cerebral circulation of the chicken. *Brain Res.* **23**, 265–268.

Richards, S. A. (1971a). Brain stem control of polypnoea in the chicken and pigeon. *Resp. Physiol.* **11**, 315–326.

Richards, S. A. (1971b). The significance of changes in the temperature of the skin and body core of the chicken in the regulation of heat loss. *J. Physiol. (London)* **216**, 1–10.

Ricklefs, R. E., and Hainsworth, F. R. (1968). Temperature dependent behavior of the Cactus Wren. *Ecology* **49**, 227–233.

Ricklefs, R. E., and Hainsworth, F. R. (1969). Temperature regulation in nestling Cactus Wrens: The nest environment. *Condor* **71**, 32–37.

Riddle, O., Smith, G. C., and Benedict, F. G. (1934). Seasonal and temperature factors and the determination in pigeons of percentage metabolism range per degree of temperature change. *Amer. J. Physiol.* **107**, 333–342.

Rijke, A. M. (1970). Wettability and phylogenetic development of feather structure in water birds. *J. Exp. Biol.* **52**, 469–479.

Rising, J. D. (1969). A comparison of metabolism and evaporative water loss of Baltimore and Bullock Orioles. *Comp. Biochem. Physiol.* **31**, 915–925.

Robinson, K. W., and Lee, D. H. K. (1947). The effect of the nutritional plane upon the reactions of animals to heat. *J. Anim. Sci.* **6**, 182–194.

Rodbard, S. (1948). Body temperature, blood pressure and hypothalamus. *Science* **108**, 413–415.

Rogers, F. T. (1919). Studies of the brain stem. I. Regulation of body temperature in the pigeon and its relation to certain cerebral lesions. *Amer. J. Physiol.* **49**, 271–283.

Rogers, F. T. (1928). Studies of the brain stem. XI. The effects of artificial stimulation and of traumatization of the avian thalamus. *Amer. J. Physiol.* **86**, 639–650.

Rogers, F. T., and Lackey, R. W. (1923). Studies of the brain stem. VII. The respiratory exchange and heat production after destruction of the body temperature-regulating centers of the thalamus. *Amer. J. Physiol.* **66**, 453–460.

Rogers, F. T., and Wheat, S. D. (1921). Studies on the brain stem. V. Carbon dioxide excretion after destruction of the optic thalamus and the reflex functions of the thalamus in body temperature regulation. *Amer. J. Physiol.* **57**, 218–227.

Romanoff, A. L. (1941). Development of homeothermy in birds. *Science* **94**, 218–219.

Romijn, C., and Vreugdenhil, E. L. (1969). Energy balance and heat regulation in the white leghorn fowl. *Neth. J. Vet. Sci.* **2**, 32–58.

Rosenzweig, M. L. (1968). The strategy of body size in mammalian carnivores. *Amer. Midl. Natur.* **80**, 299–315.

Rutschke, E. (1960). Untersuchungen über Wasserfestigkeit und Struktur des Gefieders von Schwimmvögeln. *Zool. Jahrb., Abt. Anat. Ontog. Tiere* **87**, 441–506.

Saxena, B. B. (1957). Unterschiede physiologischer Konstanten bei Finkenvögeln aus verschiedenen Klimazonen. *Z. Vergl. Physiol.* **40**, 376–396.

Schmidt-Nielsen, K. (1964). "Desert Animals: Physiological Problems of Heat and Water." Oxford Univ. Press (Clarendon), London and New York.

Schmidt-Nielsen, K. (1970). Energy metabolism, body size, and problems of scaling. *Fed. Proc., Fed. Amer. Soc. Exp. Biol.* **29**, 1524–1532.

Schmidt-Nielsen, K., and Bentley, P. J. (1966). Desert tortoise *Gopherus agassizii*: Cutaneous water loss. *Science* **154**, 911.

Schmidt-Nielsen, K., and Dawson, W. R. (1964). Terrestrial animals in dry heat: Desert birds. *In* "Handbook of Physiology" (D. B. Dill, ed.), Sect. 4, pp. 481–492. Amer. Physiol. Soc., Washington, D.C.

Schmidt-Nielsen, K., Dawson, T. J., Hammel, H. T., Hinds, D., and Jackson, D. C. (1965). The Jack Rabbit—a study in its desert survival. *Hvalradets Skr.* **48**, 125–142.

Schmidt-Nielsen, K., Bretz, W. L., Taylor, C. R. (1970a). Panting in dogs: Unidirectional air flow over evaporative surfaces. *Science* **169**, 1102–1104.

Schmidt-Nielsen, K., Kanwisher, J., Lasiewski, R. C., Cohn, J. E., and Bretz, W. L. (1969). Temperature regulation and respiration in the Ostrich. *Condor* **71**, 341–352.

Schmidt-Nielsen, K., Hainsworth, F. R., and Murrish, D. E. (1970b). Countercurrent heat exchange in the respiratory passages: Effect on water and heat balance. *Resp. Physiol.* **9**, 263–276.

Schoener, T. W. (1968). Sizes of feeding territories among birds. *Ecology* **49**, 123–141.

Scholander, P. F. (1955). Evolution of climatic adaptation in the homeotherms. *Evolution* **9**, 15–24.

Scholander, P. F., Walter, V., Hock, R., and Irving, L. (1950a). Body insulation of some arctic and tropical mammals and birds. *Biol. Bull.* **99**, 225–236.

Scholander, P. F., Hock, R., Walters, V., Johnson, F., and Irving, L. (1950b). Heat regulation in some arctic and tropical mammals and birds. *Biol. Bull.* **99**, 237–258.

Scholander, P. F., Hock, R., Walters, V., and Irving, L. (1950c). Adaptation to cold in arctic and tropical mammals and birds in relation to body temperature, insulation, and basal metabolic rate. *Biol. Bull.* **99**, 259–271.

Schüz, E. (1971). "Grundriss des Vogelzugskunde," 2nd ed. Parey, Berlin.
Scott, N. R., and van Tienhoven, A. (1971). Simultaneous measurement of hypothalamic and body temperatures and heart rate of poultry. *Trans. ASAE (Amer. Soc. Agr. Eng.)* 14, 1027–1033.
Scott, N. R., Johnson, A. T., and van Tienhoven, A. (1970). Measurement of hypothalamic temperature and heart rate of poultry. *Trans. ASAE (Amer. Soc. Agr. Eng.)* 13, 342–347.
Seibert, H. C. (1949). Differences between migrant and non-migrant birds in food and water intake at various temperatures and photoperiods. *Auk* 66, 128–153.
Sellers, W. D. (1965). "Physical Climatology." Univ. of Chicago Press, Chicago, Illinois.
Serventy, D. L. (1970). Torpidity in the White-backed Swallow. *Emu* 70, 27–28.
Seymour, R. S. (1972). Convective heat transfer in the respiratory systems of panting animals. *J. Theor. Biol.* 35, 119–127.
Sharp, P. J., and Follett, B. K. (1969). The blood supply to the pituitary and basal hypothalamus in the Japanese Quail (*Coturnix coturnix japonica*). *J. Anat.* 104, 227–232.
Shilov, I. A. (1968). "Thermoregulation in Birds." Moscow Univ. Press (in Russian).
Siegel, H. S., and Drury, L. N. (1968a). Physiological responses of chickens to variations in air temperature and velocity. *Poultry Sci.* 47, 1120–1127.
Siegel, H. S., and Drury, L. N. (1968b). Physiological responses to high lethal temperature and air velocity in young fowl. *Poultry Sci.* 47, 1230–1235.
Sinha, M. P. (1959). Observations on the organization of the panting centre in avian brain. *Int. Congr. Physiol. Sci., 21st, Symp. Spec. Lect., 1959* p. 24.
Slonim, A. D. (1971). Eco-physiological patterns of thermoregulation and the shivering thermogenesis. *J. Physiol. (Paris)* 68, 418–420.
Smith, R. M. (1969). Cardiovascular, respiratory, temperature and evaporative water loss responses of pigeons to varying degrees of heat stress. Ph.D. Dissertation, Indiana University, Bloomington.
Smith, R. M., and Suthers, R. (1969). Cutaneous water loss as a significant contribution to temperature regulation in heat stressed pigeons. *Physiologist* 12, 358.
Smyth, M., and Bartholomew, G. A. (1966). The water economy of the Black-throated Sparrow and the Rock Wren. *Condor* 68, 447–458.
Sopyev, O. (1968). On nestling feeding activity in the desert situation. *Ornitologiya* 9, 142–145 (in Russian).
Southwick, E. A. (1971). Effects of thermal acclimation and daylength on the cold-temperature physiology of the White-crowned Sparrow, *Zonotrichia leucophrys gambelii*. Ph.D. Dissertation, Washington State University, Pullman.
Spellerberg, I. F. (1969). Incubation temperatures and thermoregulation in the McCormick Skua. *Condor* 71, 59–67.
Spooner, C. R., Winters, W. D., and Mandell, A. J. (1966). DL-norepinephrine-7-H^3 uptake, water content, and thiocyanate space in the brain during maturation. *Fed. Proc., Fed. Amer. Soc. Exp. Biol.* 25, 451.
Stahl, W. R. (1962). Similarity and dimensional methods in biology. *Science* 137, 205–212.
Stahl, W. R. (1963). Similarity analysis of physiological systems. *Perspect. Biol. Med.* 6, 291–321.
Stahl, W. R. (1967). Scaling of respiratory variables in mammals. *J. Appl. Physiol.* 22, 453–460.
Steen, J. (1957). Food intake and oxygen consumption in pigeons at low temperatures. *Acta Physiol. Scand.* 39, 22–26.

Steen, J. (1958). Climatic adaptation in some small northern birds. *Ecology* **39**, 626–629.
Steen, J., and Enger, P. S. (1957). Muscular heat production in pigeons during exposure to cold. *Amer. J. Physiol.* **191**, 157–158.
Steen, J., and Steen, I. B. (1965). The importance of the legs in the thermoregulation of birds. *Acta Physiol. Scand.* **63**, 285–291.
Stevens, D., and Moen, A. N. (1970). Functional aspects of wind as an ecological and thermal force. *Trans. N. Amer. Wildl. Natur. Resour. Conf.* **35**, 106–114.
Storer, R. W. (1971). Classification of birds. *In* "Avian Biology" (D. S. Farner and J. R. King, eds.), Vol. 1, pp. 1–18. Academic Press, New York.
Strunk, T. H. (1971). Heat loss from a Newtonian animal. *J. Theor. Biol.* **33**, 35–61.
Strunk, T. H. (1973). Technical Comments: Perspectives of linear heat transfer. *Science* **181**, 184–185.
Sulkava, S. (1969). On small birds spending the night in the snow. *Aquilo, Ser. Zool.* **7**, 33–37.
Taylor, C. R., Dmi'el, R., and Fedak, M. (1971). Running in a large bird, the Rhea: Energetic cost and heat balance. *Fed. Proc., Fed. Amer. Soc. Exp. Biol.* **30**, Abstr. No. 716, 320.
Teager, C. W. (1967). Birds sun-bathing. *Brit. Birds* **60**, 361–363.
Teal, J. M. (1969). Direct measurement of CO_2 production during flight in small birds. *Zoologica (New York)* **54**, 17–23.
Thompson, A. L. (1964). "A New Dictionary of Birds." McGraw-Hill, New York.
Thornthwaite, C. W. (1940). Atmospheric moisture in relation to ecological problems. *Ecology* **21**, 17–28.
Tracy, C. R. (1972). Newton's law: Its application for expressing heat losses from homeotherms. *BioScience* **22**, 656–659.
Tracy, C. R. (1973). Technical Comments: Perspectives of linear heat transfer. *Science* **181**, 185–186.
Tucker, V. A. (1965). The regulation between torpor cycle and heat exchange in the California Pocket Mouse *Perognathus californicus*. *J. Cell. Comp. Physiol.* **65**, 405–414.
Tucker, V. A. (1968). Respiratory exchange and evaporative water loss in the flying Budgerigar. *J. Exp. Biol.* **48**, 67–87.
Tucker, V. A. (1970). Energetic cost of locomotion in animals. *Comp. Biochem. Physiol.* **34**, 841–846.
Tucker, V. A. (1972). Metabolism during flight in the Laughing Gull, *Larus atricilla*. *Amer. J. Physiol.* **222**, 237–245.
Turček, F. J. (1966). On plumage quantity in birds. *Ekol. Pol., Ser. A* **14**, 617–634.
Turček, F. J., and Kelso, L. (1968). Ecological aspects of food transportation and storage in the Corvidae. *Commun. Behav. Biol.* **1**, 277–297.
Utter, J. M., and LeFebvre, E. A. (1970). Energy expenditure for free flight by the Purple Martin (*Progne subis*). *Comp. Biochem. Physiol.* **35**, 713–719.
van Dilla, M., Day, R., and Siple, P. A. (1968). Special problems of hands. *In* "Physiology of Heat Regulation and the Science of Clothing" (L. H. Newburgh, ed.), pp. 374–388. Hafner, New York (reprint of 1949 edition).
van Kampen, M. (1971). Some aspects of thermoregulation in the White Leghorn fowl. *Int. J. Biometeorol.* **15**, 244–246.
Veghte, J. H. (1964). Thermal and metabolic responses of the Gray Jay to cold stress. *Physiol. Zool.* **37**, 316–328.
Veghte, J. H., and Herreid, C. F. (1965). Radiometric determination of feather insulation and metabolism of arctic birds. *Physiol. Zool.* **38**, 267–275.
Vitums, A., Mikami, S.-I., Oksche, A., and Farner, D. S. (1964). Vascularization of the

hypothalamo-hypophysial complex in the White-crowned Sparrow, *Zonotrichia leucophrys gambelii*. *Z. Zellforsch. Mikrosk. Anat.* **64**, 541–569.

Vitums, A., Mikami, S.-I., and Farner, D. S. (1965). Arterial blood supply to the brain of the White-crowned Sparrow *Zonotrichia leucophrys gambelii*. *Anat. Anz.* **116**, 309–326.

Volkov, N. I. (1968). An experimental study of thermal conditions in snow burrows of tetraonid birds. *Zool. Zh.* **47**, 283–285 (in Russian).

von Saalfeld, E. (1936). Untersuchungen über das Hacheln bei Tauben. *Z. Vergl. Physiol.* **23**, 727–743.

von Saint Paul, U., and Aschoff, J. (1968). Gehirntemperaturen bei Hühnern. *Pfluegers Arch. Gesamte Physiol.* **301**, 109–123.

Wald, G. (1965). Frequency or wavelength? *Science* **150**, 1239–1240.

Wallgren, H. (1954). Energy metabolism of two species of the genus Emberiza as correlated with distribution and migration. *Acta. Zool. Fenn.* **84**, 1–110.

Walther, O. E., Simon, E., and Jessen, C. (1971). Thermoregulatory adjustments of skin blood flow in chemically spinalized dogs. *Pfluegers Arch. Gesamte Physiol.* **322**, 323–335.

Warham, J. (1971). Body temperature of petrels. *Condor* **73**, 214–219.

Warren, J. W. (1960). Temperature fluctuation in the Smooth-billed Ani. *Condor* **62**, 293–294.

Watmough, D. J., Fowler, P. W., and Oliver, R. (1970). The thermal scanning of a curved isothermal surface: Implication for clinical thermography. *Phys. Med. & Biol.* **15**, 1–8.

Weast, R. C., and Selby, S. M. (1971). "Handbook of Chemistry and Physics." Chem. Rubber Publ. Co., Cleveland, Ohio.

Weathers, W. W. (1972). Thermal panting in domestic pigeons, *Columba livia*, and the Barn Owl, *Tyto alba*. *J. Comp. Physiol.* **79**, 79–84.

Weiss, H. S., and Borbely, E. (1957). Seasonal changes in the resistance of the hen to thermal stress. *Poultry Sci.* **36**, 1383–1384.

Weiss, H. S., Frankel, H., and Hollands, K. G. (1963). The effect of extended exposure to a hot environment on the response of the chicken to hyperthermia. *Can. J. Biochem. Physiol.* **41**, 805–815.

Wekstein, D. R., and Zolman, J. F. (1968). Sympathetic control of homeothermy in the young chick. *Amer. J. Physiol.* **214**, 908–912.

Wekstein, D. R., and Zolman, J. F. (1969). Ontogeny of heat production in chicks. *Fed. Proc., Fed. Amer. Soc. Exp. Biol.* **28**, 1023–1028.

Weller, M. W. (1958). Observations on the incubation behavior of a common nighthawk. *Auk* **75**, 48–59.

Welty, J. C. (1962). "The Life of Birds." Knopf, New York.

West, G. C. (1960). Seasonal variation in the energy balance of the Tree Sparrow in relation to migration. *Auk* **77**, 306–329.

West, G. C. (1962). Responses and adaptations of wild birds to environmental temperature. *In* "Comparative Physiology of Temperature Regulation" (J. P. Hannon and E. G. Viereck, eds.), Part 3, pp. 291–333. Arctic Aeromed. Lab., Fort Wainwright, Alaska.

West, G. C. (1965). Shivering and heat production in wild birds. *Physiol. Zool.* **38**, 111–120.

West, G. C. (1967). Nutrition of Tree Sparrows during winter in central Illinois. *Ecology* **48**, 58–67.

West, G. C. (1968). Bioenergetics of captive Willow Ptarmigan under natural conditions. *Ecology* **49**, 1035–1045.

West, G. C. (1972). Seasonal differences in resting metabolic rate of Alaska ptarmigan. *Comp. Biochem. Physiol. A* **42**, 867–876.

West, G. C., and Meng, M. S. (1968). Seasonal changes in body weight and fat and the relation of fatty acid composition to diet in the Willow Ptarmigan. *Wilson Bull.* **80**, 426–441.

West, G. C., Funk, E. R. R., and Hart, J. S. (1968). Power spectral density and probability analysis of electromyograms in shivering birds. *Can. J. Physiol. Pharmacol.* **46**, 703–706.

Wetmore, A. (1936). The number of contour feathers in birds. *Auk* **53**, 159.

Whittow, G. C. (1965a). Regulation of body temperature. *In* "Avian Physiology" (P. D. Sturkie, ed.), 2nd ed., pp. 186–238. Cornell Univ. Press, Ithaca, New York.

Whittow, G. C. (1965b). Energy metabolism. *In* "Avian Physiology" (P. D. Sturkie, ed.), 2nd ed., 239–271. Cornell Univ. Press, Ithaca, New York.

Whittow, G. C., Sturkie, P. D., and Stein, G., Jr. (1964). Cardiovascular changes associated with thermal polypnea in the chicken. *Amer. J. Physiol.* **207**, 1349–1353.

Wilson, H. R., Armas, A. E., Ross, I. J., Dorminey, R. W., and Wilcox, C. J. (1966). Familial differences of Single Comb White Leghorn Chickens in tolerance to high ambient temperature. *Poultry Sci.* **45**, 784–788.

Winchester, C. F. (1940). Seasonal metabolic and endocrine rhythms in the domestic fowl. *Mo., Agr. Exp. Sta., Res. Bull.* **315**, 1–56.

Winkel, K. (1951). Vergleichende Untersuchungen einiger physiologischer Konstanten bei Vögeln aus verschiedenen Klimazonen. Ein tiergeographisches Problem. *Zoologische Jahrb. Abt. Syst., Oekol. Geogr. Tiere* **80**, 256–276.

Wolf, L. L., and Hainsworth, F. R. (1972a). Environmental influence on regulated body temperature in torpid hummingbirds. *Comp. Biochem. Physiol. A* **41**, 167–173.

Wolf, L. L., and Hainsworth, F. R. (1972b). Time and energy budgets of territorial hummingbirds. *Ecology* **52**, 980–988.

Yarbrough, C. G. (1971). The influence of distribution and ecology on the thermoregulation of small birds. *Comp. Biochem. Physiol.* **39**, 235–266.

Yousef, M. K., McFarland, L. Z., and Wilson, W. O. (1966). Ambient temperature effects on hypothalamic, rectal and skin temperatures in Coturnix. *Life Sci.* **5**, 1887–1896.

Zar, J. H. (1968). Calculation and miscalculation of the allometric equation as a model in biological data. *BioScience* **18**, 1118–1120.

Zar, J. H. (1969). The use of the allometric model for avian standard metabolism-body weight relationships. *Comp. Biochem. Physiol.* **29**, 227–234.

Zimmerman, J. L. (1965). The bioenergetics of the Dickcissel, *Spiza americana. Physiol. Zool.* **38**, 370–389.

Zonov, G. B. (1967). On the winter roosting of Paridae in Cisbaikal. *Ornitologiya* **8**, 351–354 (in Russian).

Chapter 5

PHYSIOLOGY AND ENERGETICS OF FLIGHT

*M. Berger and J. S. Hart**

I.	Introduction	416
II.	Respiratory Mechanics	416
	A. Coordination with Wing Movements	416
	B. Frequency of Respiration, Tidal Volume, and Ventilation	422
III.	Respiratory Gas Exchange	426
	A. Oxygen Consumption and Utilization	426
	B. Carbon Dioxide Production and Respiratory Quotient	429
	C. Relation to Flight Speed and Angle	429
	D. Relation to Body Size	430
IV.	Circulation	434
	A. Relation of Heart Rate to Heart Weight and Body Weight	434
	B. Heart Weight and Body Weight in Birds and Mammals	438
	C. Oxygen Pulse and Minimum Cardiac Output	439
V.	Temperature Regulation	443
	A. Body Temperature	443
	B. Heat Production and Efficiency	447
	C. Heat Loss	449
VI.	Water Loss	455
VII.	Energy Turnover in Migratory Flights	459
	A. Material Balance Studies	459
	B. Flight Duration and Range	461
	C. Temperature and Altitude	464
	References	467

*Deceased.

I. Introduction

The history of the physiology of flight includes accounts of imaginary capabilities and legends. Strange birds with wondrous physiology have been described; they are able to fly because the air sacs are reservoirs for oxygen and even in long flights no respiratory movements are necessary (Baer, 1896); filled air sacs facilitate flying since the volume of the body is increased and specific gravity decreased; air sacs are also used for balancing (von Lucanus, 1929); respiration is synchronous with wing beats, and this can be assumed even in hummingbirds (Zimmer, 1935; Stolpe and Zimmer, 1959); flight work can be reduced by a head wind, and in the White Stork (*Ciconia ciconia*) it becomes zero at 10 m/second; with a fresh breeze even a man could fly by his own muscular strength (Lilienthal, 1889); birds have tremendous hearts that make up over 20% of body weight (which would be about 20 times as large as in mammals) (reported by Portmann, 1950); birds do not like to fly with a tail wind because their feathers become ruffled up (Prechtl, 1846).

Much work of a more basic nature has been carried out since the above-mentioned reports were published, but these studies have still raised as many problems as they have solved. Later generations of physiologists may regard the contents of this review as being naive, but we hope that it will be useful for a few years in a rapidly growing field.

II. Respiratory Mechanics

A. Coordination with Wing Movements

The early studies on avian flight physiology have been very largely concerned with the problem of the linkage between breathing and wing movements since it had long been assumed that increased mechanical efficiency would apply if the two actions were synchronous. Since the first studies indicated synchrony for pigeons (Marey, 1890) and were later confirmed for crows (*Corvus* sp.) (Groebbels, 1929, 1932; Zimmer, 1935), it was indeed assumed that a one-to-one coordination between breathing and wing action applied for all birds. Thus, early indications that the frequency of respiration (breaths per minute) in the Chaffinch (*Fringilla coelebs*) is lower than the wing rate (Fraenkel, 1934) were considered to be experimental artifacts (Zimmer, 1935). Weighing the different results and ideas, Salt and Zeuthen (1960) stated that "respiration and flight movements can hardly be synchronized at all times."

Almost all recent studies have shown that wing rates exceed respiration rates in birds and have emphasized the problem of coordination rather than synchrony of these activities. Wing rates exceeded respiration rates in ducks and gulls (Tomlinson, 1963), in the Budgerigar (*Melopsittacus undulatus*)(Tucker, 1968b), and in the Mallard (*Anas platyrhynchos*)(Lord et al., 1962), but the phase relationships and coordination could not be assessed from the methods used in the tests. The following account is based primarily on a comparative study of nine species of different sizes (waxbill to pheasant) in which details of the phase relationships have been defined (Berger et al., 1970a).

1. Types of Coordination

Simultaneous recordings of wing beats and respiratory air flow revealed a coordination in nine species (see review in Table I). The most common coordination is the 3:1 type (3 wing beats during the time of 1 breath) in which the beginning of inspiration occurs mostly at the end of upstroke and beginning of expiration at the end of downstroke (Fig. 1c). The 5:1 type (Fig. 1d) showed the same phase relationships. Different types were found not only within a species but also within one individual and even during one flight of a bird.

In recent publications, a synchrony was found only in pigeons (Hart and Roy, 1966) and during take-off in the Common Crow (*Corvus brachyrhynchos*). This is in agreement with the older studies, but the phase relationships were found to be contradictory.

While coordination between wing action and respiration in birds is a commonly observed phenomenon, it is not obligatory. Examples of an apparent absence of coordination were found in the Budgerigar (Tucker, 1968b) where the ratio of wing beats to respiration varied between 3.1 to 1 and 4.2 to 1, depending on flight speed. In a hummingbird, the Glittering-throated Emerald (*Amazilia fimbriata*) (M. Berger and J. S. Hart, unpublished), the ratio during hovering varied between 5.6 to 1 and 7.9 to 1. In most small birds, parts of some flights show no coordination, particularly during irregular or undulating horizontal flight when the breathing frequency is quite regular. However, even in undulating flyers such as the Black-rumped Waxbill, (*Estrilda troglodytes*) and in the Black-capped Chickadee (*Parus atricapillus*)(Fig. 1e), an interdependence was demonstrated with the beginning of inspiration occurring preferentially during the end of upstroke and end of downstroke of the wing.

2. Changes during Flight

A change in coordination during flight can be associated with change in wing frequency, change in respiration frequency, or both.

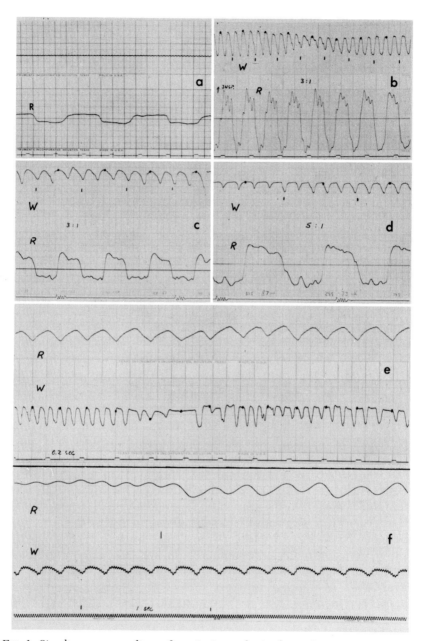

FIG. 1. Simultaneous recordings of respiration and wing beats. Instantaneous respiratory flow is shown in a, b, c, and d; respiration frequency in e and f. R = respiration, W = wing beats. Dots in the wing-beat curve mark beginning of inspiration. Graphs

Commonly, sudden changes in wing frequency (e.g., stops or increases during a rapid turn) cause decoordination. During take-off, wing rates are fairly high and the rate decreases as the bird reaches its normal speed: e.g., European Blackbird (*Turdus merula*); European Starling (*Sturnus vulgaris*); Hooded Crow (*Corvus corone cornix*) (Oehme, 1963, 1965); Brünnich's Murre (*Uria lomvia*)(Rüppell, 1969).

The decline in the rates of wing beat and respiration following the high rates at take-off were frequently associated with a relatively greater decrease in respiration rates and with an increase in coordination numbers. Thus, in the Common Crow (Fig. 1f), the ratio increased from 1:1 to 3:2, from 3:1 to 4:1 in the Evening Grosbeak (*Coccothraustes vespertinus*) and from 4:1 to 5:1 in the Black Duck (*Anas rubripes*). However, increases in respiration frequency during flight also occurred. In general, the transitions from one frequency to another were quite abrupt so that the phase relationships were interrupted only for a short time.

3. Factors Controlling Coordination Types

In order to interpret the coordination types in different species it is necessary to study the factors controlling wing beat and respiration rate. It has been known for a long time that wing rate decreases with increasing body weight (Greenewalt, 1960; Hertel, 1963). The wing rate depends also on the wing area. Birds with relatively small wings exhibit high frequencies compared with birds of the same weight but with large wings. A Mallard weighing 1 kg with a wing area of 1000 cm² has a wing rate of about 8 per second during cruising, whereas the Herring Gull (*Larus argentatus*) — weight (W) 1 kg, but wing area (A) 1700 cm² — has a wing rate of about 3 per second.

Combining the influence of size and relative wing area, it was found that the wing-loading index $W^{2/3}/A$ strongly influences the coordination type (Table I). Birds with low indexes have the lowest coordination numbers, while the highest coordination numbers were found in birds with the highest wing-loading indexes (i.e., with relatively small wings). The data do not show an influence of body weight alone;

---

a, b, e and f are originals; c and d are from Berger *et al.*, 1970a. (a) Evening Grosbeak, rest, time marks 0.2 seconds. (b) Evening Grosbeak, 3:1 coordination, time marks 0.2 second. Marked wing influence on tidal volume during take-off. (c) Black Duck, 3:1 coordination, time marks 1 second. (d) Black Duck, 5:1 coordination, time marks 1 second. (e) Black-capped Chickadee, irregular wing beats, beginning of inspiration mostly at the end of the upstroke, time marks 0.2 second. (f) Common Crow, end of take-off, change from 1:1 to 3:2 coordination, time marks 1/60 second.

TABLE I
WING-BEAT AND RESPIRATION RATES IN BIRDS

Species	Weight (W) (gm)	Wing area (A) (cm²)	$W^{2/3}/A$	Wing beats:breaths[a,b]							
				1:1	3:2	2:1	5:2	3:1	7:2	4:1	5:1
Larus delawarensis	427	1280	0.044	+				(+)			
Corvus brachyrhynchos	455	1310	0.045		+	+	+				
Parus atricapillus	10.3	65	0.073					+	+	(+)	
Columba livia	380	660	0.080	+							
Estrilda troglodytes	7.0	43	0.084					+		+	
Coccothraustes vespertinus	60	160	0.095					+	+	+	
Aix sponsa	600	700	0.10					+	+	+	
Anas rubripes	1030	820	0.12					+	+	+	
Phasianus colchicus	1440	1030	0.12					+		+	+
Coturnix coturnix	132	153	0.16							+	+

[a] Respiratory data from Berger *et al.* (1970a).
[b] + = coordination found quite regularly; (+) = coordination found only during short periods of flight.

5. PHYSIOLOGY AND ENERGETICS OF FLIGHT   421

a 3:1 and 4:1 coordination was established in the 7 gm waxbills as well as in the 1440 gm pheasants. The only mammal studied so far, the Spear-nosed Bat (*Phyllostomus hastatus*) showed also a 1:1 coordination (Suthers *et al.*, 1972). This is consistent with observations on birds, since bats have low wing-loading indexes.

4. Significance of Coordination and Phase Relationships

Groebbels (1929, 1932) found in pigeons and in Carrion Crows (*Corvus corone*), that inspiration is associated with upstroke, while Zimmer (1935) concluded the contrary from his experiments applying the same method to the same species. Hart and Roy (1966) showed in pigeons that inspiration occurred during the second half of upstroke and first half of downstroke, and Berger *et al.* (1970a) obtained similar results for Common Crows (*Corvus brachyrhynchos*). In most other types of coordination, the beginning of inspiration was also correlated with the middle or end of upstroke.

For anatomical reasons, respiratory and wing movement systems are essentially independent and controlled by different factors (Duncker, 1968a,b, 1971). In accordance with this, it has been demonstrated that they can work independently. Usually, however, there is a linkage, exhibited only in a few species by a synchronization (1:1 coordination), but in most species by a coordination of 2:1 to 5:1. The assumption that wing muscles as the largest muscles of the body, averaging 17% of body weight (Greenewalt, 1962), support respiration during flight has not been confirmed, since in most of the species so far studied respiration frequency is lower than wing frequency. In a 5:1 coordination, 2½ wing beats occur during the period of inspiration so that a possible advantage during one phase is offset in the opposite phase, yielding no advantage.

Under certain circumstances, the maximal muscle action during take-off tends to decrease the inspiratory flow rate (Fig. 1b), tidal volume, and ventilation (compared with levels found in horizontal flight) and probably contributes to accumulation of an oxygen debt during this period. A mechanical assistance to breathing from wing action in nonsynchronous coordination would occur only in a very short phase of the respiratory cycle during reversal of respiratory flow. The change from inspiration to expiration is associated in many cases with a phase of the wing cycle that markedly reduces inspiratory flow. In this sense, wing action controls respiration to some extent, but this rhythm cannot be considered to dominate the respiratory cycle. In general, the coordination between respiration and wing frequency appears to be centrally controlled (see von Holst, 1939).

B. Frequency of Respiration, Tidal Volume, and Ventilation

Respiration frequency in flying birds has been measured by a variety of methods. In the older studies, small pressure capsules or lamps were fastened to the bird and, from changes in pressure or in displacement, respiration movements were derived (Marey, 1890; Groebbels, 1929; 1932; Fraenkel, 1934; Zimmer, 1935). Tomlinson and McKinnon (1957) and Tomlinson (1963) used balloons that covered the entire beak but allowed no gas exchange. Myograms of respiratory and wing muscles were recorded by Wick (1964). Hart and Roy (1966) and Berger et al. (1970b) measured respiratory flow rates directly by using head masks. Temperature differences between inspired and expired air have been measured in a mask (Tucker, 1968b), at the nostril opening (Roy and Hart, 1966), or in the expired air stream using an open circuit device (Berger and Hart, 1972).

Tidal volume and ventilation in flight have been determined directly in pigeons (Hart and Roy, 1966), grosbeaks, gulls, and ducks (Berger et al., 1970b) by measuring integrated instantaneous inspiratory flow rates with a pressure transducer and calibrated resistance in a head mask. Indirect estimates on Budgerigars (Tucker, 1968b) and hummingbirds (Berger and Hart, 1972) were calculated from respiratory moisture loss assuming saturation of expired air at T_b (body temperature) which may be true only at high ambient temperatures (35° and 38°C).

The respiratory frequency of birds has been found to vary with flight speed, body weight, and other factors. Tucker (1968b) has apparently made the only observations on the effect of flight speed on frequency of respiration by showing for Budgerigars a maximum rate at low speed (270 per minute), a minimum (200 per minute) at intermediate speeds, and an increase again to 260 per minute at high speeds, correlated apparently with the changes in oxygen consumption with flight speed.

The correlation between resting frequencies of respiration and body weight in birds has been critically reviewed by Calder (1968). These rates decrease with increasing body weight (W) in proportion to $W^{-0.28}$ to $W^{-0.33}$, but the exponents are lower than those in resting mammals ($W^{-0.5}$ to $W^{-0.33}$).

During flight, respiration rate is about 3–19 times higher than in rest (Fig. 2). The increase appears largest in large birds, and therefore the decrease with increasing body weight is less pronounced than in rest (see also Fig. 5 for heart rate).

As found in resting birds and mammals, tidal volume (V_T) and ventilation ($\dot{V}$) increase during flight with increasing body weight (Table

5. PHYSIOLOGY AND ENERGETICS OF FLIGHT

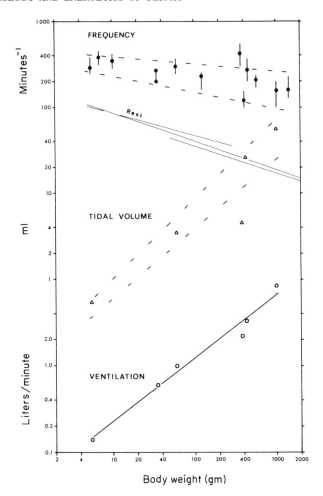

FIG. 2. Respiration frequency (minutes^{-1}), tidal volume (ml), and ventilation (liters/minute) during flight in relation to body weight in different species. Broken lines for frequency are assumed limits (excluding the pigeon), and with these, broken lines for tidal volume are estimated limits according to $V_T = \dot{V}/f$ using Eq. (1) for ventilation (solid line). Data are from Hart and Roy (1966), Tucker (1968b), Berger et al. (1970a), Berger and Hart (1972).

II). In the relationship $V_T = \dot{V}/f$, where f = frequency of respiration, variation of $\dot{V}$ was relatively smaller than that of f, but V_T was also highly variable, since f and V_T tended to show compensatory shifts (Fig. 3). The correlation between $\dot{V}$ and W (Fig. 2) for six species is given by the equation

$$\dot{V} = 42W^{0.73} \text{ ml/minute (W in gm)} \qquad (1)$$

TABLE II
RESPIRATORY CHARACTERISTICS OF FLYING BIRDS

Species	T_a (°C)	Body weight (gm)	Frequency (minutes^{-1})	Tidal volume[a] (ml)	Ventilation[a] (liters/minute)	Oxygen extracted (%)	References
Amazilia fimbriata	35	5.5	280	0.54	0.14	2.6	Berger and Hart (1972)
Melopsittacus undulatus	19–21	35	199	1.15	0.23	5.5	Tucker (1968b)
Coccothraustes vespertinus	36–37	35	—	—	0.72	1.8	Tucker (1968b)
Columba livia	22	60	294	3.5	1.0	3.3	Berger et al. (1970b)
	20[b]	380	487	4.5	2.2	3.4	LeFebvre (1964); Hart and Roy (1966)
Larus delawarensis	22	420	122	26	3.3	2.0	Berger et al. (1970b)
Anas rubripes	19	1000	158	56	8.6	2.8	Berger et al. (1970b)

[a]Volumes are STPD, except for *Melopsittacus*, where assumed as 42°C, saturated.
[b]Temperatures are assumed, records not published.

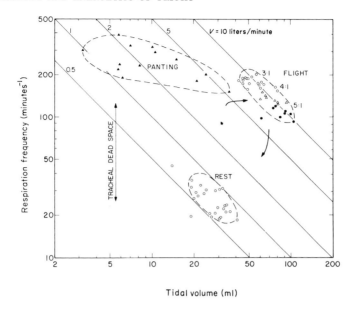

FIG. 3. Respiration frequency and tidal volume during rest, panting, and flight in the Black Duck. From Berger et al. (1970b).

With the upper and lower limits for frequency obtained from Fig. 2 ($f_{max} = 460W^{-0.083}$ per minute and $f_{min} = 360W^{-0.20}$ per minute), estimates for tidal volume range from $0.09\ W^{0.8}$ to $0.12W^{0.9}$ ml. This is in rough agreement with the average of $0.10\ W^{0.9}$ obtained from data of Table II (excluding the pigeon). These data, however, are not sufficient to reveal whether or not the increase in tidal volume of flying birds is proportional approximately to $W^{1.08}$ as found for resting birds (Lasiewski and Calder, 1971) and mammals (Stahl, 1967).

The increase of ventilation in flight varies between 12 times resting (Budgerigar at 35°C; Tucker, 1968b) and 20 times resting in the pigeon (Hart and Roy, 1966). The calculated ventilation in flight from the equation $V = 42W^{0.73}$ ml/minute (data in Table II) is approximately 20 times that calculated for resting birds of the same body weight (Lasiewski and Calder, 1971). Owing to the larger carbon dioxide production in flight, this increase is considerably greater than that which led to respiratory alkalosis in resting birds at high temperatures (Calder and Schmidt-Nielsen, 1966). The increase in ventilation was due mainly to increase in respiratory frequency in pigeons, whereas both frequency and tidal volume increased by a factor of 3 to 5 in Evening Grosbeaks and Black Ducks (Fig. 3).

III. Respiratory Gas Exchange

A. Oxygen Consumption and Utilization

The study of flight metabolism of birds, numerous estimates from aerodynamic theory, changes in body composition, changes in body weight, and measurement of gas exchange provide partial answers to the central question of the energy utilized for flight in nature. Estimates from changes in body weight and composition can be related directly to natural conditions, but they are relatively inaccurate. The most accurate estimates of flight metabolism are provided from measurements of oxygen consumption, but it is not known to what extent these exceed normal flight metabolism owing to very short flights or stimulation by the experimental conditions. Data in which oxygen consumption has been measured or calculated from carbon dioxide production or from change in body composition (as opposed to weight loss only) are given in Table III for eleven species.

The first measurements of the rate of oxygen consumption ($\dot{V}_{O_2}$) in flying birds were done by Pearson (1950), who measured the change in P_{O_2} in a closed system in which hummingbirds were hovering. This system was also used by Lasiewski (1962, 1963a,b), and one of his birds (Lasiewski, 1963a) allowed very accurate determination while hovering for 35 minutes (given in Table III). Also in hovering hummingbirds, $\dot{V}_{O_2}$ was measured by means of an open flow system (Hainsworth and Wolf, 1969; Berger and Hart, 1972). In the latter experiments, the birds drank from a small funnel, and at the same time all expired air was collected and analyzed. In the experiments of Tucker (1966, 1968b, 1969), the birds were trained to fly for long periods in a wind tunnel while $\dot{V}_{O_2}$ was determined by head masks from which expired air was drawn. Wind speed, flight angle, and temperature were varied. Berger et al. (1970b) measured ventilation and F_{O_2} (fractional content of oxygen) in three species by means of head masks. Telemetry was used to transmit the data in outdoor experiments with ducks. Since these flights lasted only 10–15 seconds, they may conform mainly to take-off conditions in which $\dot{V}_{O_2}$ might exceed that in true level flight.

Oxygen consumption calculated from carbon dioxide production of sixteen individual birds of thirteen species (Teal, 1969) from measurements of carbon dioxide production during short intermittent flights in an inflated plastic tube has not been included in Table III. The method does not distinguish between flight and post-flight metabolism and gives results about 100–200% above those by measurement of oxygen uptake (referred to flight time) compared with other tests

TABLE III
Oxygen Consumption of Birds in Flight[a]

Species	Weight (gm)	Flight conditions	$\dot{V}_{O_2}$ (ml/min)	Data source
Costa's Hummingbird *Calypte costae*	3.0	Hovering	2.1	Lasiewski (1963b)
Glittering-throated Emerald *Amazilia fimbriata*	5.5	Hovering	4.0	Berger and Hart (1972)
Purple-throated Carib *Eulampis jugularis*	8.3	Hovering	6.0	Hainsworth and Wolf (1969)
Chaffinch[a] *Fringilla coelebs*	22.5	Migratory flight	12	Dolnik et al. (1963)
Budgerigar *Melopsittacus undulatus*	35	Wind tunnel, horizontal, 35 km/hour	12	Tucker (1968b)
Purple Martin[a] *Progne subis*	51	Homing after dislocation	12	Utter and LeFebvre (1970)
Evening Grosbeak *Coccothraustes vespertinus*	60	Short flights	34	Berger et al. (1970b)
Laughing Gull *Larus atricilla*	350	Wind tunnel, horizontal, 30 km/hour	58	Tucker (1969)
Pigeon[a] *Columba livia*	384	Homing after dislocation	77	LeFebvre (1964)
Ring-billed Gull *Larus delawarensis*	410	Short flights	66	Berger et al. (1970b)
Black Duck *Anas rubripes*	1020	Short flights	237	Berger et al. (1970b)

[a] Value for $\dot{V}_{O_2}$ was calculated from carbon dioxide production or fat loss.

[Eq. (2)]. Oxygen consumption calculated from post-flight extrapolation (Keskpaik, 1968) yielded very low values (two to three times standard levels) in two swallows, *Hirundo rustica* and *Delichon urbica*, and has also been excluded from Table III.

Finally, data are included for free-flying birds calculated from fat loss in migrating Chaffinches (Dolnik *et al.*, 1963) and from fat loss and carbon dioxide production in homing pigeons (LeFebvre, 1964) using the $D_2^{18}O$ method (Lifson *et al.*, 1955) and in Purple Martins (*Progne subis*)(Utter and LeFebvre, 1970) using the same method. The errors of the method and the difficulties of assuring that the flight periods were uninterrupted have been pointed out (LeFebvre, 1964). Nevertheless, these data and results based on weight loss during migrations provide valuable reference for comparison with flight metabolism obtained under less natural conditions. We are excluding from this discussion several papers in which weight loss was used to estimate oxygen consumption or flight energy (discussed in Sections VI and VII,A).

The oxygen extraction during flight (Table II) was calculated from the ratio of oxygen consumed to minute volume (total ventilation) in pigeons (LeFebvre, 1964; Hart and Roy, 1966), Budgerigars (Tucker, 1968b), and hummingbirds (Berger and Hart, 1972) or directly as the difference in oxygen content between inspired and expired air for Evening Grosbeaks, Ring-billed Gulls (*Larus delawarensis*), and Black Ducks (Berger *et al.*, 1970b). The results in most of these species indicate that 2–3.5% of the inspired oxygen is extracted during flight compared with values up to 7% for resting birds obtained by Scharnke (1934, 1938), Zeuthen (1942), and Piiper *et al.* (1970). For flight, the mean value from Eqs. (1) and (2) yields an extraction of 2.4% with nearly independence of body weight ($W^{-0.01}$). In resting birds, Lasiewski and Calder (1971) provide an allometric analysis of ventilation and oxygen consumption for six species in which the oxygen extraction can be calculated as approximately 5%.

Tucker's (1968b) data at 20°C appear exceptional in showing an oxygen extraction during flight that is similar to that in resting birds, but the assumption of 42°–43°C for the temperature of expired air is not proved, and therefore the ventilation could be higher and oxygen extraction lower. Resting birds show definitely lower T_{ex} especially at low T_a (Schmidt-Nielsen *et al.*, 1970).

Thus it appears from the limited data that there may be a decrease in the oxygen extraction from rest to flight, suggesting that the ventilation increases more than that required to satisfy metabolic needs. This would signify either a decrease in the proportion of the ventilation perfusing the air capillaries (decreased effective ventilation) or a decrease in the extraction efficiency at the respiratory surfaces, which seems unlikely. From the limited data available, it appears that the efficiency of external respiration in mammals is also reduced during heavy exercise (dog, Young et al., 1959; man, Dejours et al., 1963). Probably the decrease in efficiency during flight is an adaptive adjustment that increases evaporative heat loss without hyperventilation of the respiratory surfaces, especially at high air temperatures.

B. CARBON DIOXIDE PRODUCTION AND RESPIRATORY QUOTIENT

Evidence obtained from long flights, especially without feeding, indicates that fat is the main source of energy (George and Berger, 1966; Drummond, 1967; Drummond and Black, 1960). In the pigeon (LeFebvre, 1964), measurements of carbon dioxide production (3.14 liters/hour) and fat loss (2.26 gm/hour, yielding 3.2 liters carbon dioxide per hour) indicated that all the carbon dioxide was derived from combustion of fat. This assumption has also been made in calculations of $\dot{V}_{O_2}$ from data of Dolnik et al. (1963) and Utter and LeFebvre (1970) in Table III. Tests of carbon dioxide production by Teal (1969) do not include estimates of the respiratory quotient (RQ = $\dot{V}_{CO_2}/\dot{V}_{O_2}$ assumed as 0.8).

Both oxygen consumption and carbon dioxide production have been reported for only two species during flight. In the Budgerigar (Tucker, 1968b) during level flight at 35 km/hour, the mean RQ in five experiments was 0.780 ± 0.032 (SD). With this figure, it was concluded that 72% of the energy for flight is derived from fat metabolism. During the first five minutes of flight, carbon dioxide production was lower (0.198 ml/gm/minute) than during the next 15 minutes (0.271 ml/gm/minute). This difference was associated with an oxygen debt during onset of flight.

The hummingbird *Amazilia fimbriata* (Berger and Hart, 1972) showed variations in RQ ranging from 0.66 to 0.85, with a small increase with rise in temperature in flights lasting up to 70 seconds. The lowest RQs were found at the beginning of flight and in the first flights of consecutive experiments, suggesting carbon dioxide retention in the body. It is difficult to decide whether or not the RQ represents the metabolic state of the body in these short flights.

C. RELATION TO FLIGHT SPEED AND ANGLE

Oxygen consumption in relation to flight speed and angle was studied in Budgerigars and Laughing Gulls (*Larus atricilla*) in wind

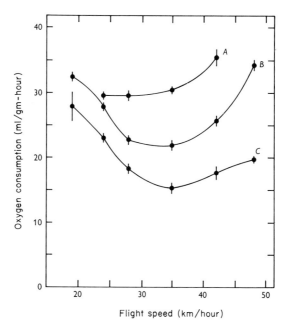

FIG. 4. Oxygen consumption in the Budgerigar during level (B), 5° ascending (A), and 5° descending flight (C) at different speeds. From Tucker (1968b).

tunnel experiments by Tucker (1968b, 1969). In Budgerigars (Fig. 4) at a given speed, more oxygen was consumed in ascending flight than in horizontal flight; in descending flight, consumption was less. A speed range was found in ascending, horizontal, and descending flight at which oxygen consumption was lowest, with significant increase at low and high speeds. In calculations of flight power of different species, Pennycuick (1968b) obtained similar curves with a minimum at certain speeds. Comparable results have been found for sinking speed of birds during gliding. In the Black Vulture (*Coragyps atratus*)(Raspet, 1960); Northern Fulmar (*Fulmarus glacialis*)(Pennycuick, 1960); and Laggar Falcon (*Falco jugger*)(Tucker and Parrott, 1970), sinking speed showed a minimum at a certain speed, as known also in gliding airplanes. Though sinking speed in gliding cannot be converted directly to power in flapping flight, it at least indicates change in aerodynamic performance which has to be compensated for by muscular work.

D. RELATION TO BODY SIZE

The correlation between oxygen consumption and body weight on an interspecific basis should be considered in regard to maximal power during brief bursts and continuous power (work output or oxygen consumption) in long flights.

While no conclusive data on maximal power are available, some of the data in Table III, which include take-off, can be considered as close to maximal effort. In theory, the maximal metabolism should increase with body weight as in rest, i.e., about with $W^{0.7}$ (Hemmingsen, 1960, for oxygen uptake; or Wilkie, 1959; Pennycuick, 1968b, for power available for flight).

Oxygen consumption in continuous flight has been measured for certain species during hovering or in long flights in a wind tunnel (Table III). We can expect that birds with high wing loading W/A or wing-loading index $W^{2/3}/A$ (which would result in a higher sinking speed compared with other birds of the same weight) would have a relatively high oxygen consumption. Because of insufficient data, such differences cannot be shown, or at least do not account for the variability observed. The least-square regression of all data of Table III is

$$\dot{V}_{O_2} = 1.01\ W^{0.72}\ \text{ml}\ O_2/\text{min (body weight in gm)} \qquad (2)$$

A similar regression was found by using only the data for sustained flight (five species). Experimental data for oxygen consumption during flight, therefore, show a dependency on body weight as found in resting birds (see King and Farner, 1961; Lasiewski and Dawson, 1967; Aschoff and Pohl, 1970). When calculated to a body weight of the House Sparrow (*Passer domesticus*) (35 gm), the values are about twice as great as the maximal potential metabolism calculated by Kendeigh (1969b). Values as derived from Eq. (2) can be considered likely even for the largest flying birds. Rheas (*Rhea americana*, weight 18–25 kg) running at high speeds (36 and 45 km/hour) are able to increase their oxygen consumption to about 12–16 times resting (Taylor *et al.*, 1971). The maximal value of 4.6 liters O_2/kg/hour is even slightly above the prediction of Eq. 2.

The power required for horizontal flight should be proportional to $W^{1.17}$ (von Helmholtz, 1874; Wilkie, 1959; Pennycuick, 1968b). If it is assumed that flight speed increases with $W^{1/6}$ and that thrust or drag is proportional to W (assuming a constant lift–drag ratio), the power required for flight would be $P = $ weight $\times$ speed/effective lift–drag ratio. At present, data on oxygen consumption do not show this relationship. If the calculations based on pigeons and gulls are applicable for long flights in other birds, then small birds should have a flight metabolism that would be only 2 times basal (at 10 gm body weight, 1 ml O_2/minute), and in large birds oxygen consumption would exceed values that are considered to be at the limits for muscular power. In a 10 kg swan, which is able to fly continuously by its own strength, extrapolation yields 3 liters O_2/minute (about 50 times basal). Though hovering is not comparable with forward flight, the difference between measured data of hovering hummingbirds (about 40 ml O_2/gm hour) and the theoretical value for horizontal flight in a 5–10 gm bird (6 ml O_2/gm hour) appears too large.

It has been suggested (Dolnik, 1969a, 1970) that variations in the difference between available power (= maximum power output in long flights) and required (theoretical) flight power was due to differences in aerodynamic quality. In large birds, values of lift–drag ratio would be greatest, but beyond certain limits aerodynamic quality cannot meet the larger increase in power required for flight. Although small birds tend to be less efficient flyers, the large difference between available and required power in small birds requires further explanation. However, as indicated, there are no metabolic data to support this contention.

A similar approach to this problem is given by Kokshaiski (1970), who calculated from data in the literature an aphid-to-pigeon equation for power output of active flight

$$P = 84.5\ W^{1.015}\ \text{cal/hour} = 0.29\ W^{1.015}\ \text{ml}\ O_2/\text{minute} \quad (3)$$

[body weight (W) in gm]. These calculated results are theoretically explained by change in lift–drag ratio with weight. From the increase in flight speed with weight

$$v = 13.8\ W^{0.288}\ \text{km/hr} \quad (4)$$

derived by Kokshaiski (1970), it can be calculated that power required for flight (= mass × speed) is proportional to $W^{1.29}$ and therefore increases even more with body weight than that assumed by Pennycuick (1968b) and others. Hence, from the comparison of power actually expended with the expression for power required, it follows that a flight-quality factor (including both efficiency and Re number) should increase with body weight according to

$$K = \text{mass} \times \text{speed/power expended} = 0.38\ W^{0.278}. \quad (5)$$

The assumptions of Kokshaiski (1970) are only partly supported by other workers. A low efficiency was shown for small artificial wings (von Holst, 1943) and it was calculated for flight of several small insects as 0.5–2.0% (Sotavalta, 1952). The compilation of many data for oxygen consumption in flying insects (Sotavalta and Laulajainen, 1961), however, showed an increase with body weight proportional to about $W^{1.15}$ to $W^{1.20}$. Apparently it is not valid to calculate a common insect–bird regression.

It has been reasoned (Wilkie, 1959; Pennycuick, 1968b) that, because the slope of the power equation for flight [Eq. (3)] is much steeper than the equation for metabolism under resting conditions, there must be an upper limit to the size for flying animals (about 10 kg) owing to the marked increase in power required to fly with increasing weight. We do not concur with Wilkie's (1959) assumption that maximum steady-state mechanical power output corresponds to 4.8 times basal metabolism and that, due to the efficiency factor of 25%, total energy production would be 4 times higher (altogether 20 times basal). Likewise, the equation for flight power, $P = 0.00233W^{1.167}$ (power output in horsepower, weight in kg) cannot be considered valid, because in birds with a body weight less than 4.8 kg, the power required to fly would be less than basal metabolism.

The high rate of oxygen consumed in flight is demonstrated by comparison with maximal oxygen consumption in mammals. During cold exposure, heat production is increased up to 4–8 times basal metabolism in bats and rodents at temperatures down to 0°C (Herreid and Schmidt-Nielsen, 1966; Segrem and Hart, 1967b; Stones and Wiebers, 1967; McNab, 1969; O'Farrell and Studier, 1970), and similar data are reported for hummingbirds and small passeriformes (Lasiewski, 1963a; Lasiewski et al., 1964, 1967; Lasiewski and Lasiewski, 1967; Calder, 1964; Mugaas and Templeton, 1970; Hissa and Palokangas, 1970). All these data are clearly lower by 0.3–0.5 than those in flight in birds when computed to $W^{0.7}$.

Another way to increase metabolism is by exercise on treadmills (additionally combined with cold exposure). In short exhausting runs of 1–2 minutes duration in mice, golden hamsters, rats and guinea pigs (body weight 30–900 gm), the maximal oxygen uptake is given by

$$\dot{V}_{O_2} = 0.436W^{0.73} \quad \text{ml/minute (body weight in gm)} \qquad (6)$$

(Pasquis et al., 1970) which is a constant rate of 6–7 times standard metabolism (Kleiber, 1947). The highest $\dot{V}_{O_2}$ in longer runs up to 40–60 minutes in mice, hamsters, and rats (Hart, 1950; Jansky, 1959a,b; Hart and Jansky, 1963) was similar to that of Pasquis et al. (1970), with a maximal value for the mouse of 0.4 ml O_2/gm minutes, and a regression for "maximal" metabolism of

$$\dot{V}_{O_2} = 0.06W^{0.73} \quad \text{ml/minute} \qquad (7)$$

It is not clear whether or not colinearity exists with much larger animals as proposed by Jansky (1965) or if dog, man, and horse, which show larger increases (Astrand, 1952; Brody, 1945; Cerretelli *et al.*, 1964a; Chatonnet and Minaire, 1966; Flandrois *et al.*, 1962; Young *et al.*, 1959), should be separated. For the latter, Pasquis *et al.* (1970) calculated maximal oxygen consumption as

$$\dot{V}_{O_2} = 0.654W^{0.79} \text{ ml/minute} \tag{8}$$

which is a mean increase of 17–19× basal, compared with standard metabolism given by Kleiber (1947) with

$$\dot{V}_{O_2} = 0.057W^{0.75} \text{ ml/minute} \tag{9}$$

In any case, in the range considered here for flying birds, the highest metabolic values obtained in short tests in exercising mammals do not reach values obtained in birds in longer flights, which are almost twice as great for a given body weight. The only value resembling bird flight metabolism is found in a flying bat, *Phyllostomus hastatus* (body weight 70–110 gm) with 27.5 ml O_2/gm/hour (Thomas and Suthers, 1970). The oxygen consumption in flying animals is higher than that in any other form of energy continuously expended (Weis-Fogh, 1961; Oehme, 1968b).

IV. Circulation

A. Relation of Heart Rate to Heart Weight and Body Weight

The increased oxygen uptake in flight compared with rest is necessarily combined with increased oxygen transport in the circulatory system. The equality between oxygen transport by respiration and circulation is given by the equation

$$\dot{V}_{O_2} = QD = RV_sD = RO \tag{10}$$

where $\dot{V}_{O_2}$ = oxygen uptake per unit time,
Q = cardiac output,
D = fractional difference in oxygen content between arterial and mixed venous blood,
R = heart rate,
V_s = stroke volume,
O = oxygen pulse.

5. PHYSIOLOGY AND ENERGETICS OF FLIGHT

The only circulatory data so far available in flying birds are measurements of heart rate. Combined with oxygen consumption, the determination of oxygen pulse is possible, and from comparison with resting and working mammals, estimates of minimal values for stroke volume and cardiac output are obtained. It should be realized that the values for V_S and Q, as commonly used, stand for either ventricle. The actual performance of the heart is twice the calculated amount.

Heart-rate measurements have been obtained from electrocardiograms in flying birds by Wick (1964) and Hart and Roy (1966) in pigeons, by Berger *et al.* (1970b) in nine species ranging in size from waxbill to pheasant, and by Aulie (1971a) in the Budgerigar and seedeater (*Sporophila* sp.). Blood-pressure measurements of flying Great Black-backed Gulls (*Larus marinus*)(Eliassen, 1963a,b) are not included in the calculations since the methods were not unequivocal and the results differed greatly from those of all other species.

The heart rate in flying birds (R_F) obtained in 12 species is related to body weight according to the equation

$$R_F = 24.1 W^{-0.146} \qquad (11)$$

(rate in beats per second, body weight in grams, see Fig. 5). The 95% confidence limits are very small, being only ±0.6 per second for the mean size of 48 gm and mean rate of 13.7 per second.

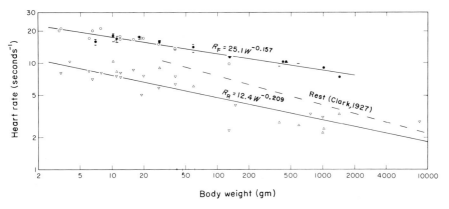

FIG. 5. Heart rates of birds during different activities: —, flight (range); ●, postflight (mean); ○, highest resting rates; △ and ▽, resting rates. From Berger *et al.* (1970b), using data of various authors.

Since heart rates in flight are often masked by myograms of wing muscles, it is fortunate that the rates obtained immediately after landing (post-flight, or PF) do not differ significantly from those in flight

(Berger et al., 1970b). The regression obtained from such data in thirteen species is

$$R_{PF} = 24.8W^{-0.152} \text{ per second} \tag{12}$$

Maximal heart rates under nonflight conditions can reach flight values (Berger et al., 1970b; Aulie, 1971a). Such data of Odum (1945), Lasiewski (1964), Lasiewski et al. (1964, 1967), when combined with postflight data (1–2 seconds) of Berger et al. (1970b) reveal

$$R_{max} = 25.1W^{-0.157} \text{ per sec (28 species)} \tag{13}$$

There is no significant difference between the above equations. The relations differ significantly in slope from all other heart rate–body weight relationships obtained in resting animals.

The early studies of heart rate in resting birds (R_R) (Clark, 1927) were apparently conducted in partially active birds since the data are variable and too high. Using selected data from the literature (Odum, 1941, 1945; and others) as well as more recent data, the equations

$$R_R = 12.7 W^{-0.23} \text{ per second} \quad \text{(Calder 1968)} \tag{14}$$

$$R_R = 12.4 W^{-0.21} \text{ per second} \quad \text{(Berger et al., 1970b)} \tag{15}$$

were obtained (difference not significant). Judging from the regression lines, the heart rates in flight are about 2.4 times those at rest in small birds (10–20 gm), and three times as great in large birds (500–1000 gm). The largest observed differences were found in the pigeon, with a mean flight rate of 3.5 times resting rate (Hart and Roy, 1966) and in the Black Duck, with a maximal of about four times (Berger et al., 1970b).

Although body weight is usually used as the basis for comparison of heart rates among species, we can expect that heart weight would serve equally or even better as a reference, since, for theoretical reasons, the frequency of pumps is related to their size when maximal efficiency is achieved (von Miller, 1965). From nine species (Berger et al., 1970b; Aulie, 1971a), heart rate during flight can be related to heart weight (H) according to the regression:

$$R_F = 12.5H^{-0.174} \text{ per second (heart weight in gm)} \tag{16}$$

The combination of Eqs. (11) and (21) yields

$$R_F = 12.7H^{-0.162} \text{ per second} \tag{17}$$

It appears likely that there may be a difference between the resting heart rates of birds and mammals at a given body weight. While the compilation of Clark (1927) included high "resting" rates in birds resulting from activity, which are about 15% higher than those for mammals, the data for mammals are indeed in agreement with those of Stahl (1967):

$$R_R = 22.6 W^{-0.25} \text{ per second (body weight in gm)} \tag{18}$$

and with those of Holt et al. (1968)

$$R_R = 22.2 W^{-0.25} \text{ per second} \tag{19}$$

Hudson (pp. 72–73) in Leitner and Nelson (1967) found that resting heart rates in sixteen species of small mammals can be described by

$$R_R = 13.6 W^{-0.25} \text{ per second} \tag{20}$$

which is similar to that in resting birds and lower than that predicted by Eqs. (18) and (19). Similarly, Berger (1964) found that heart rates of ten species of mammals (weight up to 1 kg) under undisturbed conditions were clearly lower than predicted by Eq. (18). For birds, we have assumed that the data summarized by Calder (1968) represent normal resting conditions [Eq. (14)] and conclude that resting heart rates of mammals ranging in size up to about 1 kg are 50–70% higher than those for birds of the same weight. In other words, birds have resting heart rates that are similar to those in mammals that are 6–10 times heavier. However, if it is considered that the heart weight of birds is about equal to that of mammals twice as large, and that the rates computed by Stahl (1967) for mammals do not represent truly resting conditions, it is apparent that there may be very little difference between resting heart rates of birds and mammals when related to heart weight.

Heart rates during flight in mammals have been reported for only two species. In the Big Brown Bat (*Eptesicus fuscus pallidus*) weighing 17.7–25.7 gm, resting heart rates of 7.5 per second, and heart rates of 17.0 ± 0.4 per second (SE) during short flights of 2–4 seconds were observed (Studier and Howell, 1969). In *Phyllostomus hastatus* (70–110 gm), heart rate increased from 7.0 per second to 12–13 per second (Thomas and Suthers, 1970). Both rates correspond very well to heart rates of flying birds; Eq. (11) predicts 15.4 and 12.5 per second, respectively, for the two species. Heart rate and oxygen consumption in flight, as well as heart size in bats (Hesse, 1921) show agreement with birds rather than with other mammals during exercise. However, during heavy exercise, the heart rates of nonflying mammals can approach those of birds during flight, as judged by data of Bailie *et al.* (1961), Segrem and Hart (1967a,b), Gollnick and Ianuzzo (1968), and Spector (1956).

B. Heart Weight and Body Weight in Birds and Mammals

Various studies have been conducted on the relation between heart weight and body weight in birds and mammals (Hesse, 1921; Clark, 1927; Crile and Quiring, 1940; Rensch, 1948; Hartmann, 1955, 1961), but a full treatment is beyond the scope of this review. However, it is considered important to point out certain differences between mammalian and bird hearts using the most reliable but selected data. Hesse (1921) in an extensive review provided many data but felt that a single regression for all birds masked certain comparative information. It is true that some families are clearly different in regard to heart size, and the accidental selection of certain species could influence the results. However, since it is useful to demonstrate a difference between birds and mammals, we have calculated least-squares regressions, using the means of all species in which three or more animals were analyzed, and obtained for birds:

$$H_B = 0.0196 W^{0.899} \tag{21}$$

($n = 78$, body and heart weights in gm, 95% confidence limits of the exponent is ± 0.089) and for mammals

$$H_M = 0.00894 W^{0.935} \tag{22}$$

($n = 30$, body and heart weights in gm, with 95% confidence limits of the exponent of ± 0.073). Also, Clark (1927) found that in large homeotherms the relative heart weight is small, and that birds and mammals differ in heart size, the equations for birds being

$$H_B = 0.02 W^{0.9} \tag{23}$$

and for mammals

$$H_M = 0.014 W^{0.9} \tag{24}$$

On the average, birds have hearts 1.4–2 times as large as mammals of

5. PHYSIOLOGY AND ENERGETICS OF FLIGHT

the same weight, but confidence limits are overlapping. The data of Hartmann (1961) for birds treated in the same way as Hesse's gave

$$H_B = 0.0198 W^{0.834} \tag{25}$$

($n = 416$, body and heart weights in gm, 95% confidence limits of the exponent is ± 0.017). In this relationship, many species of hummingbirds appear clearly to influence the slope of the regression, and without this family we obtained

$$H_B = 0.0163 W^{0.874} \tag{26}$$

($n = 380$, body and heart weights in gm, 95% confidence limit of the exponent is ± 0.018). For hummingbirds alone, the regression was

$$H = 0.0244 W^{0.937} \tag{27}$$

($n = 36$, 95% confidence limit of the exponent is ± 0.051).

Holt et al. (1968) recently measured heart weights in mammals, and demonstrated an increase with $W^{1.1}$. We do not compare these data with other mammals and birds, since they are mainly from domestic animals of fairly large size compared with those in question here. It was suggested by Hartmann (1955) and Brush (1966, using selected data of Hartmann), that the relationship of heart weight to body weight in birds up to about 100–200 gm differs from that of larger birds. Errors in calculations by Brush and the new data of Hartmann in our opinion negate this assumption.

C. OXYGEN PULSE AND MINIMUM CARDIAC OUTPUT

Determination of the amount of oxygen uptake or oxygen transport per heart beat is possible for some birds under flight conditions. We have summarized the available data (Table IV) and have added some results from other birds and mammals in which high metabolic rates were achieved by cold exposure or exercise.

Oxygen pulse in flight is markedly increased compared with that during rest. From the combination of Eqs. (2) and (11), the oxygen pulse is obtained as

$$O = \dot{V}_{O_2}/R_F = 0.70 W^{0.87} \; \mu l \; O_2/beat \tag{28}$$

All available data for flight are clearly higher than those for birds and

TABLE IV

HEART RATE, OXYGEN PULSE, MINIMAL CARDIAC OUTPUT, AND STROKE VOLUMES DURING FLIGHT AND OTHER ACTIVITIES IN BIRDS AND MAMMALS

Species	Body weight (gm)	Heart rate seconds^{-1}	Heart rate Increase vs. rest	Oxygen pulse μl/beat	Oxygen pulse Per heart weight (μl/beat/gm)	Oxygen pulse Increase vs. rest	Cardiac output, Q_{min} Per heart weight (ml/gm/minute)	Cardiac output, Q_{min} Per body weight (ml/gm/minute)	Stroke volume per heart weight (ml/gm)	References[a]
Bird flight										
Melopsittacus undulatus	35	15.5	2.2	14	29	5.8	130	1.8	0.14	a,b,c,d
Coccothraustes vespertinus	60	14.1	2.4	40	—	6.1	—	2.8	—	e
Columba livia	380	9.3	3.4	140	29	2.5	79	1.0	0.14	d,f,g
Larus delawarensis	420	10.2	2.9	110	31	3.8	91	0.78	0.15	c,d,e
Anas rubripes	1030	9.0	3.7	440	34	3.1	91	1.1	0.17	e
Bird cold exposure										
Calypte costae	3.0	15	1.8	0.97	15	3.5	67	1.5	0.045	h
Phalaenoptilus nuttallii	40	8.3	2.3	4.8	—	1.8	—	0.30	—	i
Colius striatus	50	8.0	2.0	6.3	—	2.5	—	0.30	—	r

Mammal flight								
Phyllostomus hastatus	90	12.5	—	55	—	—	2.3	—	k
Mammal cold exposure								
Peromyscus leucopus	25	13	2.0	6.9	—	2.2	1.1	—	l
Pteropus scapulatus	362	4.1	1.5	45	—	1.9	0.15	—	m
Mammal exercise								
Peromyscus leucopus	25	12.5	1.9	6.1	—	1.9	0.92	—	l
Rattus norvegicus (wild)	243	8.2	1.9	53	55	1.5	0.54	0.28	n
Rattus norvegicus (dom.)	345	9.1	1.5	51	49	**2.7**	0.40	0.24	n
Dog	25000	4.7	2.5	7800	—	3.5	0.44	—	o
Man	70000	3	3	16700	56	3.7	0.21	0.28	p

*a*References: (a) Tucker, 1968b; (b) Aulie, 1971a; (c) Lasiewski and Dawson, 1967; (d) Hartmann, 1961; (e) Berger *et al.*, 1970b; (f) LeFebvre, 1964; (g) Hart and Roy, 1966; (h) Lasiewski, 1964; (i) Bartholomew *et al.*, 1962; (k) Thomas and Suthers, 1970; (l) Segrem and Hart, 1967a,b; (m) Bartholomew *et al.*, 1964; (n) Segrem and Anderson, 1968; (o) Cerretelli *et al.*, 1964a,b; (p) Asmussen and Nielsen, 1952; (r) Bartholomew and Trost, 1970.

mammals during cold exposure or exercise (Table IV). If the oxygen pulse is related to heart weight, using regression equations for $\dot{V}_{O_2}$ (2), heart rate (11) and heart weight (21), near independency of oxygen pulse on body weight can be established:

$$O/H = \dot{V}_{O_2}/R_F H = 35.6 W^{-0.033} \; \mu\text{l/gm-beat}. \tag{29}$$

Owing to the relatively high heart weight in birds, the value of O/H for birds in flight (35 μl/gm-beat) is in the range or even lower than that found in exercising rats, dogs, and men (Segrem and Andersen, 1968; Cerretelli et al., 1964b). However, many data on mammals are uncertain owing to absence of reliable data on heart weights.

The increase of oxygen pulse from rest to flight can be considered to be remarkably great, particularly in small birds. This increase depends on body weight, attaining about six times resting in small birds and about three to four times resting in larger species. In pigeons, with their large difference between resting and flight heart rate, the oxygen pulse in flight is only 2.5 times resting pulse. During cold exposure in mammals and birds, we determined from data in the literature increases in rates varying between 1.2 and 3.5 times resting in birds and between 1.5 and 3.7 times resting in mammals.

The large increase in oxygen pulse during flight requires increase in stroke volume and/or arteriovenous (A-V) oxygen difference. At present there is no information on the extent to which both parameters can increase. However, it is known that blood pressure in birds is generally higher than in mammals (Groebbels, 1932; Sturkie, 1970), and that the venous return is probably supported by coordinated action between wing and heart muscles (Aulie, 1971a). From this it can be inferred that stroke volume could increase more during flight than is known from exercise in mammals (Asmussen, 1965; Andersen, 1968), but experimental evidence is lacking.

It follows from Eq. (10) that cardiac output and stroke volume can be determined if A-V difference in oxygen content (D) of the blood is known. Since there are no determinations of A-V differences in flight, only minimal figures can be calculated assuming the maximal A-V difference of 0.20. Using Eqs. (2) and (11) for oxygen consumption and heart rate we obtain for minimum cardiac output (Q) the equation

$$Q_{\min} = 5\dot{V}_{O_2} = 5W^{0.72} \; \text{ml/minute} \tag{30}$$

and for minimum stroke volume

$$V_{S\,\min} = Q_{\min}/R_F = 3.5 W^{0.87} \; \mu\text{l} \tag{31}$$

As shown for the oxygen pulse [Eq. (29)], the minimal stroke volume is almost constant when related to heart size [combined with Eq. (21)]

$$V_{S\ min}/H = 0.18W^{-0.033} \quad \text{ml/gm heart weight} \tag{32}$$

Minimal estimates of cardiac output and V_S (Table IV) involve uncertain assumptions and are hardly comparable with those for resting values or any other activity. It is also recognized that since these equations are based on relatively few species, they have relatively low predictive reliability.

V. Temperature Regulation

During rest, heat production is determined by the ambient temperature, showing an increase below the thermoneutral range. The lower critical temperature ranges for many species lie between 20° and 35°C, and lower values are found in large species (for summarized data, see King and Farner, 1961, 1964; Kendeigh, 1969a). During flight, heat production is necessarily increased in association with the increased power output. Thus, heat production is apparently not primarily stimulated by thermoregulatory demands, but becomes a byproduct of activity, contrary to that during rest (at least at moderate and higher temperatures). To illustrate these relations, the relatively few data on body temperature measurements, heat production assumptions, and heat flow and heat loss measurements in flying birds are discussed below.

A. Body Temperature

As found in resting birds, the temperatures of different parts of the body can vary appreciably, and therefore deep-body, subcutaneous, and cutaneous temperatures of different parts have to be considered separately (Table V).

In a comprehensive study of 327 species, Wetmore (1921) determined deep-body temperatures of a few active birds. The birds were shot and the temperature reading (cloaca or esophagus) was taken immediately.

Deep-body temperatures measured during or immediately after flight show an increase compared with rest or preflight in *Pachyptila, Phaethon, Columba,* and *Amazilia* (Table V). The figures given vary roughly between 1° and 2°C above resting. Judging from thoracic temperatures measured in pigeons (Hart and Roy, 1967), the postflight cooling rate of the body is highest at low ambient temperatures,

TABLE V
TEMPERATURES IN BIRDS FLYING OR AT FLIGHTLIKE CONDITIONS

Species	T_a (°C)	Point of measurement	T_b (°C)	Flight −rest	Temperature measured	Reference
Pachyptila turtur	—	Cloaca	41.5±0.21	1.6	Postflight	Farner (1956)
Puffinus tenuirostris	—	Cloaca	39.9±0.97	2.0	Postflight	Farner and Serventy (1959)
Phaethon rubricauda	26–29	Esophagus near stomach, or stomach	40.9	1.9	Postflight	Howell and Bartholomew (1962)
Phaethon rubricauda	26–29	Foot web	36.7	4.7	Postflight	Howell and Bartholomew (1962)
Columba livia	4–30	Thoracic	44.5±0.2	1.79±0.15	Flight > 2 minutes	Hart and Roy (1967)
Columba livia	4–30	Subcutaneous over pectoral muscle	43.0±0.3	0.93±0.25	Flight > 2 minutes	Hart and Roy (1967)
Columba livia	25–29	Cloaca	43.0	2.0	Descending flight (3°)	Aulie (1971b)
Columba livia	13–26	Esophagus	41.5–43.8	—	Postflight (shot)	Pearson (1964)
Amazilia fimbriata	0	Cloaca	40.4	—	Postflight < 1 minute	Berger and Hart (1972)
Amazilia fimbriata	10	Cloaca	40.9	—	Postflight < 1 minute	Berger and Hart (1972)
Amazilia fimbriata	20	Cloaca	41.7	—	Postflight < 1 minute	Berger and Hart (1972)
Amazilia fimbriata	30,35	Cloaca	42.5	≥0.7–1.4	Postflight < 1 minute	Berger and Hart (1972)
Larus marinus	−7 to −14	Cloaca	42–43	—	Restrained wind tunnel gliding	Eliassen (1962)

5. PHYSIOLOGY AND ENERGETICS OF FLIGHT

Species		Site			Condition	Reference
Larus marinus	15–19	Cloaca	42–43	—	Restrained wind tunnel gliding	Eliassen (1962)
Larus marinus	−7 to −14	Thoracic muscle	39.5–41	—	Restrained wind tunnel gliding	Eliassen (1962)
Larus marinus	15–19	Thoracic muscle	39.5–41	—	Restrained wind tunnel gliding	Eliassen (1962)
Larus marinus	−7 to −14	Skin of ventral thorax	36–37.5	—	Restrained wind tunnel gliding	Eliassen (1962)
Larus marinus	15–19	Skin of ventral thorax	36–37.5	—	Restrained wind tunnel gliding	Eliassen (1962)
Larus marinus	−7 to −14	Skin of wing	26–27.5	−5	Restrained wind tunnel gliding	Eliassen (1962)
Larus marinus	15 to 19	Skin of wing	31.5–32.5	0	Restrained wind tunnel gliding	Eliassen (1962)
Larus marinus	−7 to −14	Web	9–12.5	—	Restrained wind tunnel gliding	Eliassen (1962)
Larus marinus	15 to 19	Web	25–26	—	Restrained wind tunnel gliding	Eliassen (1962)
Anas rubripes	−26	Nasal tissue	4.9	—	Short flights	Berger et al. (1971)
Anas rubripes	−7	Nasal tissue	14.0	—	Short flights	Berger et al. (1971)
Anas rubripes	4	Nasal tissue	19.3	—	Short flights	Berger et al. (1971)
Anas rubripes	19	Nasal tissue	26.5	—	Short flights	Berger et al. (1971)
Tyto alba	—	Cloaca	38.8–40.6	—	Postflight (shot)	Wetmore (1921)
Tachycineta thalassina	—	Interthoracic (esophagus)	39.9–41.6	—	Postflight (shot)	Wetmore (1921)
Progne subis	—	Interthoracic (esophagus)	40.4–40.6	—	Postflight (shot)	Wetmore (1921)
Melopsittacus undulatus	25–29	Cloaca	42.0	1.0	Wind tunnel, horizontal flight	Aulie (1971b)

and therefore the reported postflight data reflect true flight temperatures only when taken within seconds after landing. In the hummingbirds, at least at high ambient temperatures (30° and 35°C), body temperatures are higher than in rest, since the birds started panting after landing and body temperatures fell within a few minutes by 0.7–1.4°C. These postflight temperatures are also higher than those found in other resting hummingbirds (Lasiewski, 1964).

The increase in thoracic temperatures during flight in pigeons (Hart and Roy, 1967) lasted about 1.5–3 minutes (with the highest rate attaining about 1°C per minute). After this time a plateau was reached. "No correlation was observed between rise in deep body temperature and air temperature." In contrast, air temperatures had a marked effect on the subcutaneous temperatures in flight. "At low air temperatures (4–14°C) subcutaneous temperatures during flight tended to fall below resting levels, and to rise above resting levels at ambient temperatures above 10–15°C. The greatest rise (up to 2.0°C) occurred at high air temperatures" (22°–30°C).

In wind-tunnel experiments (Aulie, 1971b) at 25°–29°C, pigeons were not able to regulate body temperature to a constant level when flying horizontally, or at 1° or 2° descent. Only at 3° descending flight, when flight work was apparently considerably reduced, a body temperature of 43.0°C (~2°C above resting) could be maintained during about 10 minutes of flight. In the same study, Budgerigars kept a constant body temperature of 42.1°C (about 1°C above resting) during level flight.

In tests with gulls in a wind tunnel, Eliassen (1962) did not mention any change in thoracic and pectoral cutaneous temperatures, but his conditions do not reflect active flight. The remaining data show dependency on ambient temperature and that the dependency is stronger as the site where temperature was taken is distant from the core.

Torpidity in resting birds has been demonstrated for some species of the Apodidae, Trochilidae, Caprimulgidae (McAtee, 1947; Bartholomew et al., 1957) and Coliidae (Bartholomew and Trost, 1970). Apparently in sustained flight, body temperatures (T_b) are high, but there are also reports that birds are able to fly with fairly low temperatures. Poorwills (Phalaenoptilus nuttallii) and White-throated Swifts (Aeronautes saxatalis) can fly with a T_b between 34.0° and 37.5°C (Miller, 1950; Marshall, 1955; Bartholomew et al., 1957; Howell and Bartholomew, 1959). The Glittering-throated Emerald (Amazilia fimbriata), showed forward flight after torpidity at a body temperature (cloacal) of 36.7°C and was able to hover at about 39°C (Berger and Hart, 1972). In hummingbirds, according to Morrison (1962) "flight was not possible below 36°C and even at 38°C the birds appeared handicapped." Recently, Austin and Bradley (1969) reported that in the Poorwill, after torpidity, wing flutter begins at T_b between 19.0° and 26.5°C (mean 22.4°), and that good flight with some gain of altidude was achieved with T_b between 27.4° and 30.8°C (mean 28.5°). These are the lowest values for flying birds. In flying bats, body temperatures between 20° and 22°C (*Pipistrellus hesperus*) and 24.3°C (*Plecotus townsendi*) were measured (Bradley and O'Farrell, 1969; Reeder and Cowles, 1951).

For comparison, we mention two instances of body temperatures in flying poikilotherms (insects). In the Monarch Butterfly (*Danaus plexippus*) (Kammer, 1970), thoracic temperatures increased in "fixed" flight, but it was apparently not regulated to a constant level at different ambient temperatures. The wing rate was clearly dependent on thoracic temperatures. On the other hand, the moth *Manduca sexta* (Sphingidae) was able to fly only if thoracic temperatures were 38°C or higher (Heinrich, 1970). These species have relatively small wings relative to body weight and consequently a high wing rate. We can see correspondence with flying birds and mammals in the fact that low body temperature in flight can be tolerated if the wing loading and therefore wing rate are low (bat, goatsucker). Those species with higher wing loading and relatively high wing rate (swift, hummingbird) may require higher body temperatures for flight.

B. Heat Production and Efficiency

A complete conversion of metabolism into useful work is impossible. Part of the metabolic rate (M) appears as heat production in the body (H_p), part is used as work output E_f to keep the bird in the air. This latter part is finally converted into heat after the air, accelerated by the wings, has become still. The heat production H_p may lead after take-off to an increase in body temperature T_b and to an increase in total heat loss H_l. In longer flights, after an equilibrium is reached, T_b is assumed to reach a constant value, and $H_p = H_l$. The symbols are used here for power and energy (amount of work) since in quotients the time is dropped.

In horizontal flight, part of the external energy E_f is converted to useful work W_f and part is lost by friction and turbulence of air (E_t) at the body surface. The energetic relations are described by the equation

$$M_{hor} = E_f + H_p = W_f + E_t + H_p \tag{33}$$

In ascending flight, the potential energy E_p of the bird increases with gain of height according to the equation

$$M_{asc} = H_{p\ asc} + W_f + E_t + E_p \qquad (34)$$

In downward flight, as in any downward movement, potential energy cannot be transformed into chemical energy of the muscle, but catabolic processes are reduced.

Other than $\dot{V}_{O_2}$ and E_p, none of the parameters have been measured in flying birds. However, various calculations provide approximations in the energy relations that are quite inexact but at least indicate the order of magnitude of the flight efficiency.

The gross (or total) efficiency E_f/M and net (or partial) efficiency of total work $E_f/M_f = E_f/(M - M_b)$ are assumed in flying birds to be 0.20–0.25 (Tucker, 1968b; Pennycuick, 1968b) as in working mammals (Brody, 1945; Kleiber, 1961). The metabolism of flight muscles is assumed to be $M_f = M_{tot}$ minus resting metabolism. During flight, however, it is very likely that the metabolism of other organs M_b is also increased. Comparison of calculated E_f (Pennycuick, 1968b) for the pigeon with measured M (LeFebvre, 1964) yield the high value of 0.34 for gross efficiency and 0.38 for net efficiency (using the lowest value of flight power). (If correction for the difference in body weight is made, we obtain 0.33 and 0.37, respectively.) If the power at the flight speed of 58 km/hour assumed by LeFebvre for the pigeon is taken for comparison, efficiencies of 0.41 and 0.47 are obtained, which seem too high. For hovering hummingbirds calculations of E_f (Hertel, 1963; Pennycuick, 1968b) compared with measurements of V_{O_2} (Lasiewski, 1963b; Berger and Hart, 1972) yield a gross efficiency of 0.07–0.09, and a net efficiency of 0.08–0.11 (using data from Lasiewski, 1963b; for resting metabolism). In these physiological quotients, all external work E_f (including the work to overcome drag, i.e., profile and parasitic power, Pennycuick, 1968b) is considered, unlike that for mice or horses on treadmills, where no air resistance arises.

In extensive studies on the Budgerigar flying in a wind tunnel, all efficiencies were found to be dependent on flight speeds (Fig. 6) (Tucker, 1968b). Total efficiency estimates (E_p/M), however, take only change in potential energy into account, and this is an unknown fraction of total flight work performed. Therefore, the figures for total efficiency are fairly low (maximum for ascending 5%, for descending 11%). The partial efficiencies for ascending $E_p/(M_{asc} - M_{hor})$ and descending flight $E_p/(M_{desc} - M_{hor})$ describe the relation of change in height to the difference in metabolism compared to that in level flight. Highest values over 0.50 are found in ascending flight at slow speeds.

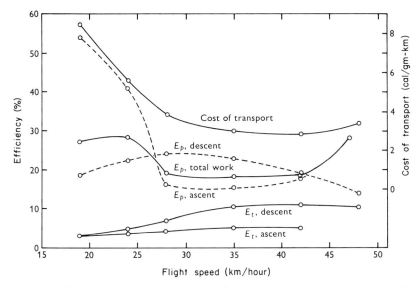

FIG. 6. Efficiencies for change in potential energy in the Budgerigar in ascending and descending flight (5°) at various flight speeds. From Tucker (1968b).

At 27 and 43 km/hour these efficiencies are the same whether the bird is ascending or descending. Tucker (1968b) points out that "at either of these speeds a budgerigar flying at an angle of 5° could ascend to and descend from a given altitude and use no more energy than if it spent the same time in level flight at these speeds."

The aerodynamic efficiency W_f/E_f describes the relation between induced work and total work done, i.e., $(1-W_f)/E_f$ gives losses from drag and friction in relation to total work. The efficiency W_f/E_f was assumed in hovering hummingbirds to be 0.80 as in helicopters (Hertel, 1963), and from calculations of Pennycuick (1968b), a value of 0.65 was obtained.

In summary, there exist estimates on efficiencies in flying birds, but no comprehensive data are available from which heat production can be calculated. At present, heat production can be given only by approximation (about 0.75–0.80 M as in working mammals; Brody, 1945; Kleiber, 1961; Asmussen, 1965), and this has been assumed in the following section.

C. Heat Loss

The heat produced during flight in the body is lost by two routes — through the skin and by expired air via the respiratory tract. In both routes, loss by evaporation of water and dry heat loss must be considered. Loss of liquid water (fecal) does not serve any function in temperature regulation. The problem of nonrespiratory heat loss is treated by data on heat transfer through the skin (conduction). It can be assumed that convective heat loss to the air accounts for nearly all of the dermal dry heat loss. Radiative loss as well as absorption of solar radiation in flight are probably insignificant when compared to total heat production especially when the feathers are vented.

From total heat loss by the flying bird, only that from the respiratory part has been determined directly in some experiments (Table VI). Evaporative heat loss of respiratory air is calculated from water loss assuming that 0.58 kcal/gm of evaporated water is lost. Convective heat loss of expired air is usually considered to be negligible due to low heat capacity of the air. At lower temperatures, however, it should be taken into account. In the Black Duck at T_a of +19°C, it amounts to 24% of respiratory heat loss or 4.3% of total heat produced. At −16°C it is 56% of respiratory heat loss and 11% of total heat production. The reduction of nose tissue temperature in the cold, while not changing total respiratory loss (about 19% of heat production), markedly reduces the portion lost through evaporation of water.

As in rest, evaporative water loss in flight and consequently heat loss by respired air becomes more important at high temperatures. In the Budgerigar, it amounts to 47% of heat production in flights at 37°C (Tucker, 1968b), and in hovering hummingbirds, maximal figures are about 40% at 35°C (Berger and Hart, 1972). Although data are not available, we consider it unlikely that the total heat produced can be dissipated to the environment entirely by respiratory evaporation in active flight, unless flight work is greatly reduced (e.g., by gliding).

It is unknown to what extent respiratory heat loss in flight can be regulated or if it has a more or less fixed value determined only by external factors (temperature, pressure, humidity). A regulation could be accomplished by change of expired air temperature independently of T_a, by change in minute volume or percentage of water vapor saturation, but no information is available on the subject. The data from hummingbirds and Budgerigars at T_a of 35° and 37°C suggest that at high temperatures, respiratory water loss can be regulated to some extent. However, this would be a small portion of the total heat loss.

From the foregoing, it follows that in the species mentioned, at low and medium temperature, 80% or more of the heat production in flight is lost through the skin. Evaporation through the skin was usually thought to be insignificant since birds lack sweat glands. Recent investigations of Smith (1969) in pigeons demonstrate the contrary. Evaporation through the skin during rest can exceed respira-

TABLE VI
RESPIRATORY HEAT LOSS IN FLYING BIRDS

	T_a (°C)	Respiratory heat loss (kcal/hour)			Heat produced (kcal/hour)	Respiratory heat loss (% of heat production)		Reference
		Evaporation	Convection	Total		Evaporation	Total	
Amazilia fimbriata	0	0.09			1.1	8.6		Berger and Hart (1972)
	10	0.11			1.0	11		
	20	0.15			1.0	15		
	30	0.23			0.9	25		
	35	0.35			0.9	39	39	
Melopsittacus undulatus	18–24	0.41			2.8	15		Tucker (1968b)
	29–31	0.50			2.8	18		
	36–37	1.30			2.8	47	47	
Columba livia	—	—		2.2	18	—	12	LeFebvre (1964)
Anas rubripes[a]	−16	4.5	5.8	10.3	—	8.5	19	Berger et al. (1971)
	−7	5.6	4.9	10.5	—	11	20	
	7	7.2	3.5	10.7	—	14	19	
	19	7.2	2.3	9.5	53	14	17	

[a] Heat production in *Anas rubripes* was not measured at all temperatures, but assumed to be constant.

tory loss by 2–4-fold at a T_b of 43°C, with a strong dependency on T_b and on the speed of the air movement over the feathers. With this result, evaporation through the skin during flight cannot be ignored, but no measurements have been made.

Dry heat loss through the skin is considered to account for the largest portion of the total heat loss. Dry heat loss was partially estimated in flying pigeons by implanting heat-flow disks subcutaneously (Hart and Roy, 1967). The values obtained in flight over the pectoral muscles range between 20 and 40 cal/cm²-hour (Fig. 7) with

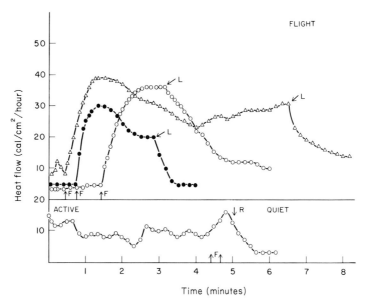

FIG. 7. Heat flow between pectoral muscle and skin before, during, and after flight in pigeons. The arrows F and L mark take-off and landing. In lower section is a record of a nonflying bird when other birds were present and when quiet after other birds were removed (R). From Hart and Roy (1967).

an average of 28.3. Within 1 minute after take-off, usually a maximum of 30–40 cal/cm²-hour was reached, and frequently it tended to fall thereafter. This is the range that would be expected if 16 kcal/hour (see Table VI) were lost equally over the whole skin. The average increase from rest to flight was 6–7-fold, which is of the same magnitude as the increase in heat production (LeFebvre, 1964). This agreement suggests "that the entire body surface and not just the wings may be important for convective heat transfer during flight" (Hart and Roy, 1967). Notwithstanding that, the heat flow is not likely to be equally distributed, and the undersurface of the wings (Eliassen, 1962) and feet (Howell and Bartholomew, 1962) could play a special role, particularly at extreme temperatures. In flying bats the wings appear to be important for regulating body temperature (Kluger and Heath, 1970).

Insulative properties of the body have been characterized by conductance values that have been calculated in quite diverse ways, beyond the scope of this review. In comparisons made in this section, (1) dry and wet heat loss are not separated owing to lack of measurements and different dependencies on temperature, and (2) conductance values are related to an area (Kleiber, 1970). Since both respiratory and skin heat loss are involved, we cannot determine appropriate surface areas but relate total heat production to $W^{2/3}$.

The data for flying birds, calculated as

$$C = \frac{0.8 M}{W^{2/3}(T_b - T_a)} \quad \frac{\text{cal}}{\text{gm}^{2/3} \text{ minute } °C}$$

show a strong dependency on temperature, contrary to that for rest (Table VII). These values must be considered with caution since C as calculated may not be independent of body weight. Nevertheless,

TABLE VII
THERMAL CONDUCTANCE DURING FLIGHT (C_f) AND AT REST (C_r)

Species (reference)	T_a (°C)	ΔT (°C)	Heat produced (cal/minute)	$C_f{}^a$	$C_r{}^b$	C_f/C_r
Amazilia fimbriata	0	40.5	18.1	0.14	0.050	2.8
(Berger and Hart,	10	31.1	17.2	0.17	0.050	3.4
1972)	20	21.8	16.2	0.23	0.050	4.6
	30	12.4	15.4	0.39	0.050	7.8
	35	7.7	14.9	0.62^e		
Melopsittacus undulatus	20	22	49.2	0.21	0.036	5.8
(Tucker, 1968b)	30	13	49.2	0.35	0.036	9.7
	37	7	49.2	0.57^c		
Columba livia (LeFebvre, 1964; Hart and Roy, 1967)	20^d	24.5	293	0.23	0.024	9.6

[a] Data calculated according to the equation $C_f = 0.8M/W^{2/3}(T_b - T_a)$ cal/gm$^{2/3}$ °C-minute.
[b] Resting conductance calculated from equation of Lasiewski et al. (1967).
[c] 13% of heat production stored.
[d] This temperature is assumed. The actual value was not given.
[e] Heat storage not considered.

flight data of the three species seem quite constant at a given temperature and show an increase with temperature. For comparison, resting conductance values are derived from the equation given by Lasiewski et al. (1967), which is similar to that derived by Herreid and Kessel (1967). Because these resting data decrease with increasing weight when related either to body weight or $W^{2/3}$, the largest increase from rest to flight is found in the largest species.

It has been repeatedly demonstrated in many resting birds that heat production (chemical thermoregulation) abruptly increases at a critical temperature below a thermal neutral range. Since no such abrupt change in oxygen consumption was found either in flying Budgerigars between 20° and 37°C (Tucker, 1968b) or in hovering hummingbirds between 0° and 35°C (Berger and Hart, 1972), physical thermoregulation by changed insulation can be assumed to occur over the entire temperature range.

The question why the conductance can be still further decreased during flight at temperatures lower than during rest leads to a comparison of absolute values. Conductance values in flight are much higher than in rest (see Table VII), and even the lowest figures found at medium and low ambient temperatures are 3–4 times those of conductance during rest. If the values equivalent to birds at rest could be reached in flight, thermobalance in hummingbirds would be hypothetically possible at lower temperatures than would ever be found in the natural environment.

The variable conductance over a wide range of temperature and relative constancy of heat loss observed in certain species suggests new mechanisms for controlling heat loss in flight that are not available to resting birds. The mechanisms considered to regulate heat loss, vasomotor control of blood supply to the skin and pteromotor control of plumage insulation, may be maximally effective under conditions of high metabolism in a moving air stream during flight. While forced convection in a model system resulted in only a moderate increase in heat loss (Hart and Roy, 1967), a slight venting of the feathers could have a marked influence, as suggested by these authors. Vasomotor regulation of blood supply to thinly feathered surfaces of wings (Eliassen, 1962) or feet may be similarly effective in an air stream.

Evidence for relative constancy of heat production in flight at different temperatures observed by Tucker (1968b) and Berger and Hart (1972) must be confirmed by further studies. Flight metabolism can be considered theoretically independent of temperature but efficiency could be variable. Evidence for the assumption of independence of flight metabolism and ambient temperature is also based on the following: (1) metabolism in the observed tests is probably close to maximal and therefore would increase very little in the cold, (2) shivering, which is the only known source of regulatory thermogenesis in the cold (West, 1965), is completely suppressed by flight (Hart, 1960). Thus, heat production associated with flight would be expected to substitute for that due to cold (see, also, review by Dawson and Hudson, 1970).

VI. Water Loss

Water is lost in flying birds by at least two routes: by evaporation from the respiratory system and by the feces and urine. A third possibility, evaporation from the skin, has been observed in several species (Bernstein, 1969, 1971a,b; Lasiewski et al., 1971; Smith, 1969; Smith and Suthers, 1969), and this route of evaporation may be important in flight, because in resting pigeons this was found to be dependent on air flow rate over the skin (Smith, 1969). Measurements available on four species for respiratory and total water loss (Table VIII) suggest that birds continuously lose water during flight.

The experimental findings show that water loss and net weight loss depend on ambient temperature, with the highest rates occurring at the highest temperatures as found in resting birds. Correlation with relative humidity of the air, as found for resting birds (Salt, 1964; Lasiewski et al., 1966) has not been demonstrated in flying birds. The water lost is in part compensated by production of metabolic water. Since energy expenditure and, therefore, water production is assumed to change little with temperature, the net loss of water (difference between absolute water loss and water production) increases with increasing temperatures. Respiratory net water loss via respiration alone can exceed 350 or 400% over water production at ambient temperatures of 35°–37°C (Tucker, 1968b; Berger and Hart, 1972), with a net loss of 5–8% of body weight per hour. Such losses would certainly limit extended flights at high temperatures. Even at 20°C net losses range between 0.4 and 2% of body weight per hour. These are much greater than values calculated by Keskpaik (1968), which are in the range of resting birds owing to unclear calculations of respiratory volumes.

An interesting comparison with the few flight data is given by the experiments of Taylor et al. (1971) on Rheas (*Rhea americana*) running at high speeds in which oxygen consumption increased to a level similar to that in flight. Highest rates of respiratory water loss were associated with high external heat load (high T_a) or high internal heat

TABLE VIII
RESPIRATORY AND TOTAL WATER LOSS IN FLYING BIRDS[a]

Species	T_a (°C)	Metabolic water (gm/hour)	Respiratory water loss (gm/hour)	Respiratory water loss (% W/hour)	Respiratory net water loss (% metabolic water)	Respiratory net water loss (% W/hour)	Total net water loss (% metabolic water)	Total net water loss (% W/hour)
Amazilia fimbriata	0	0.16	0.16	2.9	0	0	—	—
	10	0.15	0.19	3.5	25	0.7	—	—
	20	0.14	0.25	4.5	70	1.9	—	—
	30	0.14	0.40	7.3	190	4.6	—	—
	35	0.13	0.60	11	350	8.2	—	—
Melopsittacus undulatus	18–20	0.45	0.71	2.0	56	0.7	89	1.1
	29–31	0.45	0.87	2.5	93	1.2	124	1.6
	36–37	0.45	2.24	6.4	400	5.1	430	5.5
Columba livia	20[b]	2.6	3.8[c]	1.0[c]	—	—	46	0.32
Anas rubripes[d]	−16	—	7.7	0.75	−5	0	—	—
	−7	—	9.6	0.93	18	0.1	—	—
	7	—	12	1.2	48	0.4	—	—
	19	8.1	12	1.2	48	0.4	—	—

[a] Data sources same as Table VII.
[b] Assumed temperature.
[c] Estimates are total weight losses.
[d] Metabolic water production in *Anas rubripes* assumed constant at all temperatures.

load (high speed). While during rest more than 100% of heat production could be lost by respiratory evaporation, during fast running only about 30% was lost (0.9% of body weight/hour).

Birds during rest can easily sustain a water loss of 15% of body weight (Bartholomew and Cade, 1956; Bartholomew and Dawson, 1954), and also in migrational birds dehydration can occur (Salomonsen, 1969). The importance of water economy for migrating birds pointed out by Yapp (1956, 1962) suggests that the potential flight duration or range as derived from fat reserves (Fig. 9) could be shortened by excessive water loss. We do not know of studies on migrating birds in which dehydration was found while there were still reserves of fat, from which it can be inferred that on long-distance migration, water loss is reduced to a minimum by physiological or ecological adaptation. However, statistically valid measurements of both water and fat loss are required under actual migratory conditions to settle this important issue.

In resting birds under normal temperature conditions (20–30°C), evaporative water loss usually exceeds production of metabolic water with a progressively increasing difference with decrease in body weight, the loss being five times greater than production in the smallest birds (Bartholomew and Cade, 1963; Crawford and Lasiewski, 1968). However, at low temperatures water loss is reduced but production of metabolic water increases, the effect being greatest in small species. Data for the Common Cardinal (*Cardinalis cardinalis*) (Dawson, 1958), Black-rumped Waxbill (*Estrilda troglodytes*)(Lasiewski *et al.*, 1964), and Red-breasted Nuthatch (*Sitta canadensis*) (Mugaas and Templeton, 1970) demonstrate water homeostasis at ambient temperatures of 14°–18°C. Larger birds, however, such as the Ostrich (*Struthio camelus*)(Crawford and Schmidt-Nielsen, 1967), pigeon, and Greater Roadrunner (*Geococcyx californianus*) (Calder and Schmidt-Nielsen, 1967), Speckled Mousebird (*Colius striatus*) (Bartholomew and Trost, 1970), Spotted Screech-Owl (*Otus trichopsis*) (Ligon, 1969), had at 14°–18°C a 50–80% higher water loss per unit heat production than the smaller birds mentioned above. At low temperatures, loss of water is apparently of minor importance.

The limited data on water loss and water production in relation to energy expenditure in flight (Fig. 8) are in the range of resting data. If respiratory water loss is considered to be the largest part of water and weight loss, we can estimate from four species a mean figure of 0.17–0.21 gm water lost per kcal of flight energy expenditure at 20°C (= 4.8–5.8 kcal/gm water). The figure for caloric density (or energy equivalent) of weight loss for swallows (Keskpaik, 1968) of 4.80 and

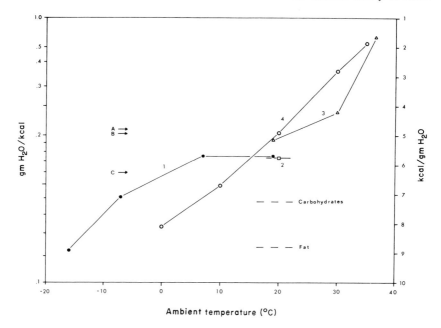

FIG. 8. Relation between total energy expenditure and water loss during flight in relation to ambient temperature. (1) Black Duck (*Anas rubripes*) respiratory water loss (Berger et al., 1971); (2) Pigeon (*Columba livia*) total water loss (LeFebvre, 1964); (3) Budgerigar (*Melopsittacus undulatus*) evaporative water loss (Tucker, 1968b); (4) Glittering-throated Emerald (*Amazilia fimbriata*) respiratory water loss (Berger and Hart, 1972). Under these experimental conditions water homeostasis in flight is reached only at low ambient temperatures. In addition, values for energy equivalent of weight loss measured in rest and assumed for flight are given: (A) Barn Swallow (*Hirundo rustica*) and (B) House Martin (*Delichon urbica*) (Keskpaik, 1968), (C) three species of finches (Dolnik and Blyumental, 1967).

4.95 kcal/gm water is within this range, the value of 6.28 kcal/gm which was measured in resting finches and assumed for flight (Dolnik and Blyumental, 1967) is somewhat higher than those in Fig. 8. Energy equivalents in the range of resting water homeostasis (about 9 kcal/gm water) can be expected only at low temperatures. The agreement of these quotients with those for resting birds should not be confused with the absolute values for the rate at which water is produced and lost, which are much higher in flight than at rest.

The few data on weight loss in flying bats are still higher than those reported above. Three species of desert bats (*Eptesicus fuscus, Antrozus pallidus, Leptonycteris sanborni*) had rates of water loss of 3% of body weight per hour or more at 45–55% relative humidity and 23°C ambient temperature (Carpenter, 1969). In a fish-eating bat (*Pizonyx vivesi*), water loss in flight was still higher (Carpenter, 1968).

VII. Energy Turnover in Migratory Flights

A. MATERIAL BALANCE STUDIES

Studies on energy expenditure were divided by Farner (1970) into three groups: (1) material balance studies, (2) measurements of oxygen consumption in wind tunnel or free flights, and (3) oxygen consumption during hovering in a closed system. Problems under (2) and (3) have been discussed in Section III; material balance studies have been reported by Dolnik et al. (1963), Nisbet (1963, 1967), Raveling and LeFebvre (1967), Hussell (1969), Farner (1970), Lyuleeva (1970), and others. Further calculations have been made by Brown (1961), Oehme (1963, 1965, 1968a, 1970), Pennycuick (1968a,b, 1969), and Pennycuick (Volume V of this treatise).

It is known from many species that in long migratory flights birds may undergo tremendous changes in both weight and body constituents (Salomonsen, 1969). As judged from several studies, it is apparent that the highest weight loss in a species is associated with the highest amount of flight work performed. Two types of material balance studies can be distinguished: fat loss and total weight loss studies.

Fat loss and carbon dioxide production were studied in pigeons by LeFebvre (1964), and by both measurements a flight energy of 22 kcal/hour (= 0.057 kcal/gm-hour) was obtained. In migrating Chaffinches with known flight speed and time interval, fat loss of 0.365 gm/hour (= 3.5 kcal/hour = 0.155 kcal/gm-hour) was found in two samples with different specimens (Dolnik et al., 1963). In the Myrtle Warbler (*Dendroica coronata*) (Hussell and Caldwell, 1972), the fat loss was different in adult and immatures (0.8 and 1.4% of body weight per hour equal 0.07 and 0.13 kcal/gm-hour). In this study, it is assumed that difference in time of arrival at the light-house reflects difference in flight time (same starting time assumed for all migrants arriving at different times). When compared with standard metabolism (Lasiewski and Dawson, 1967), the increase was 14-fold in the pigeon and 10-fold in the Chaffinch. When related to existence metabolism (Kendeigh, 1970) the flight metabolism of various species varied from 2–6-fold increase (Dolnik and Blyumental, 1964).

Weight loss in flight can be related to the material balance as follows:

$$\text{Change in body weight} = \text{weight of food taken} + \text{weight of inspired water} + \text{weight of oxygen consumed} - \text{weight of expired water} - \text{weight of carbon dioxide produced} - \text{weight of feces} \quad (35)$$

The change in body weight can be positive in natural environments over a certain time in all birds that are feeding in flight (e.g., swallows, swifts, goatsuckers). In most species, however, especially in ground-feeding birds and night migrants, there is no food intake and a continuous decrease in weight can be followed. The change of weight due to gaseous exchange (oxygen consumed minus carbon dioxide produced) appears to be of minor importance. Actually, the difference is zero at RQ of 0.723, and at a higher RQ there is a weight loss (0.153 gm per liter of oxygen consumed at RQ of 0.8). In the Black Duck, this would amount to 0.21% of body weight per hour ($T_a = 19°C$, RQ assumed 0.78) compared to 1.2% of body weight per hour of respiratory water loss. In the hummingbird *Amazilia fimbriata*, in addition to respiratory water loss of 4.5% of body weight per hour (20°C, RQ = 0.77) the gaseous weight loss is about 0.4% of body weight per hour (i.e., 8% of respiratory weight loss is caused by the carbon dioxide–oxygen difference).

Since evaporative water loss accounts for such a large portion of the weight loss in flight (see Section VI), the concept of water homeostasis in flight, i.e., no net loss (Odum *et al.*, 1964) should be reexamined. Because water loss is highly dependent on ambient temperature (Fig. 8) while total energy output appears not to be, it follows that at medium and especially at high temperatures the water loss and also weight loss may greatly exceed the production of metabolic water. Therefore, we do not agree that the measurement of total weight loss is an adequate measure for energy turnover. To reduce somewhat the uncertainties, Keskpaik (1968) and Dolnik and Blyumental (1967) have measured the caloric equivalent of the weight loss at rest and assume its application to flight, but over a large temperature range this cannot be accepted as constant.

In spite of these various errors and assumptions it seems rather remarkable that estimates of energy expenditure in migrational flight from weight-loss calculations (Nisbet *et al.*, 1963; Pearson, 1964; Raveling and LeFebvre, 1967; Dolnik and Blyumental, 1967; Hussell, 1969) agree quite well with estimates derived from oxygen consumption measurements in flight (Hart and Berger, 1972). On the other hand, estimates for Barn Swallows (*Hirundo rustica*) and House Martins (*Delichon urbica*) (Lyuleeva, 1970) were only about half those derived from oxygen consumption. The low weight loss in martins and swifts might be associated with their gliding habits and possibly with feeding during flight. With the present uncertainties in material balance studies, any agreements with flight energy estimates based on oxygen consumption may be fortuitous since (1) actual flight energy and fat loss may be much less than that derived

5. PHYSIOLOGY AND ENERGETICS OF FLIGHT

from measurements of metabolism during flight, and (2) the fat loss may be less than the measured weight loss owing to a large loss in water not derived from fat metabolism.

B. FLIGHT DURATION AND RANGE

Energy expenditures and available energy can be combined to calculate the possible flight duration and range. These are hypothetical figures since the true range is subject to great modification by environmental factors. Various estimates have already been given by Odum et al. (1961), Pennycuick (1969), and Tucker (1971) using different assumptions.

Energy expenditure in relation to body weight can be estimated from long migratory flights. Physiological measurements indicate that maximal continuous power for flight can be given by Eq. (2), transformed to

$$P = 0.29W^{0.72} \text{ kcal/hour } (W = \text{body weight in gm}) \qquad (36)$$

which applies for nearly maximal energy expenditure (P). It is likely that in several species energy expenditure in long flights is considerably less than given by this equation. Under experimental conditions usually continuous flapping was observed whereas many birds during migration show quite regularly an alternating pattern of flapping phases and pauses (quantitative data given by Bruderer, 1971).

Fat is considered to be the most important and concentrated energy reserve for migrational flight (George and Berger, 1966), and the amount of available fat determines the potential duration of nonstop flight, when no food is consumed during flight. In small birds (up to about 50 gm total weight), especially in long-distance migrants, fat content up to 50% of total body weight has been found, and 30–40% is given as the mean value for several species (Odum and Connell, 1956; Odum, 1958; Odum et al., 1961; Caldwell et al., 1963; Child, 1969). The potential flight duration is given by available energy E and energy expenditure P [Eq. (36)] according to

$$t = E/P = 53W^{0.28}f/(2-f)^{0.72} \text{ hours} \qquad \text{(Fig. 9)} \qquad (37)$$

where W = initial body weight and f = initial fractional fat content. With insignificant error, this equation can be simplified to

$$t = 36fW^{0.28} \text{ hours} \qquad (38)$$

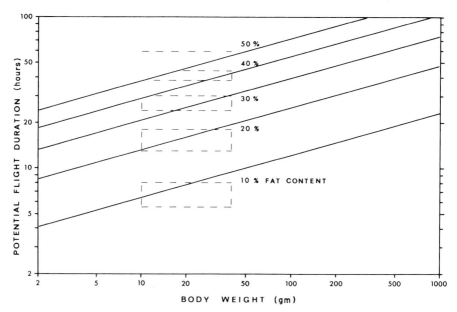

FIG. 9. Potential flight duration in relation to fat content and body weight, according to Eq. (37). Rectangles and dashed line include ranges given by Odum et al. (1961).

For a 4 gm bird with 40% fat (hummingbird) a total time in the air of 21 hours would be possible, and for a 30 gm bird with 40% fat, 37 hours. For the well known long-distance migrant, the Lesser Golden Plover (*Pluvialis dominica*) (weight 150 gm, 25% fat; Johnston and McFarlane, 1967), a flight time of 37 hours has been estimated. The potential flight range is determined as

$$R = EV/P = tV \qquad (V = \text{speed}) \qquad (39)$$

On the average, flight-speed increases with increasing body weight. For a rough calculation for small birds body weight can be neglected and 50 km/hr assumed as a reasonable mean value (Odum et al., 1961). From this, the flight range for a 40-gm bird with 40% fat is predicted as 1900 km. Such a value appears not unrealistic (maximal expenditure assumed) if balance of body water is reached. An even greater range could be predicted for lower energy expenditures.

The predictions (Fig. 9) are in agreement with the estimates of flight ranges for 10–40-gm birds by Odum et al. (1961). The range taken from this curve is about 1000 km if 25% of total body weight is usable fat, and 3000 km for 50% fat. The equation for flight range given by Pennycuick (1969) is derived from calculations of flight

5. PHYSIOLOGY AND ENERGETICS OF FLIGHT

power, and with the difference between initial weight W_i and final weight W_f being considered combustible fat, the range in kilometers is

$$R = 820 \ln (W_i/W_f)(L/D)_{eff} \qquad (40)$$

[$(L/D)_{eff}$ = effective lift-drag ratio]. This equation predicts in general for small birds larger ranges and for larger birds smaller ranges than Eq. (38), which is derived from empirical data. Using data of flying insects and birds, Tucker (1970) calculated the cost of transport for flight [see Eq. (42)], and from this equation flight range in kilometers was estimated (Tucker, 1971) as

$$R = 74.6 \lambda m_0^{0.227} \qquad (41)$$

where λ = percent mass loss and m_0 = initial body mass in kilograms. For a hummingbird of 3.7 gm, 50% fat or mass loss, $L/D = 4.1$, speed = 9 m/second, the potential flight distances would be 920 km [Eq. (38) and (39)], 2300 km [Eq. (40)], and 1050 km [Eq. (41)], respectively. The flight range in shorebirds (Charadriidae, Scolopacidae) was estimated by McNeil (1969) from assumed flight metabolism (12 times basal, Raveling and LeFebvre, 1967), observed flight speed and estimated fat content. For 12 species, the longest distances range between about 1500 and 2500 km.

The distances that are overcome by flying animals are greater than any migrations on land. Annual migration distances in birds can reach up to 40,000 km (Arctic Tern, *Sterna paradisaea*) and 6000–10,000 km are covered by many small species. If energy expenditure is compared with distance covered, flying animals are much more economical than walking or running mammals which utilize 10–15 times more energy for the same distance (Tucker, 1969). The equation for the flight data given by Tucker (1970) for cost of transport (C) in insects and birds is (body weight, W, in kg)

$$\begin{aligned} C &= 1.25 W^{-0.227} \quad \text{kcal/kg-km} \\ &= 1.25 W^{0.773} \quad \text{kcal/km} \end{aligned} \qquad (42)$$

This cost of transport might be lower only in swimming animals.

From theoretical considerations (Pennycuick, 1969), it is understood that the speed of minimal cost of transport (= maximum range speed) is higher than the speed at which lowest energy is expended (= minimum power speed). For migration, the former would be important. In the Budgerigar (Tucker, 1968b), the lowest oxygen consumption in level flight was observed at flight speed of 35 km/hour (Fig. 4) while at 42 km/hour the cost of transport was lowest (Fig. 6).

C. TEMPERATURE AND ALTITUDE

The potential flight duration and time is influenced by external factors and may be rarely attained under the less-than-ideal conditions assumed above. Favorable winds (tail winds and up-currents) and unfavorable winds have direct effects on flight range, while duration can be shortened by high temperatures. It is concluded from Table VIII and Fig. 9 that birds flying at normal or high temperatures lose more water than is available from metabolic processes. It is unknown for most species to what extent such dehydration of the fat-free mass can be tolerated in flight, but for long flights it is evident that water loss must be reduced to a minimum. This is possible by flying at low temperatures.

There is considerable evidence of high-altitude flight in some birds. Radar measurements, with limited accuracy at high altitudes, have indicated heights up to 6000 m in some migrations (Lack, 1960; Nisbet, 1963; Nisbet and Drury, 1967), and flights over mountains have been reported at similar altitudes (Aymar, 1935; Meinertzhagen, 1955; Besson, 1969). Hunt (1954) noted birds flying on Mount Everest at 7940 m and Manville (1963) reported collision of an aircraft with a Mallard at 6400 m. Studies of Himalayan birds showed that at heights of at least 5700 m above sea level migrations of several bird species occur (Hingston, 1925; Diesselhorst, 1968; Martens, 1971).

High-altitude flight would definitely be favorable for water homeostasis, especially when desert areas are crossed. Temperatures of 0°–10°C are reached at altitudes below 3000 m in summer at all latitudes, so that considerable water economy might be expected for flight even at these moderate altitudes. Small passerine birds such as wagtails, pipits, warblers, and flycatchers are trans-Sahara migrants (Moreau, 1961), and long flights near the ground would be impossible owing to dehydration of the bird and reduced convective heat loss.

The other possibility to avoid high temperatures during migration are flights during the night. Analyzing the migrational pattern of birds crossing the Alps, Dorka (1966) demonstrated that long-distance migrants and the early migrants in the fall are predominantly night migrants, and day migrants fly predominantly at dawn and later in the fall. This was found in general, but also in closely related species; e.g., the European Turtle-Dove (*Streptopelia turtur*), is a night and a trans-Sahara migrant leaving in August and September, while the

Wood Pigeon (*Columba palumbus*), and Stock Dove (*C. oenas*), were observed to cross the Alps mainly during daytime in October and migrate only into Mediterranean countries. Long-distance migrants flying during the daytime rely on external energy, such as food (swallows) or upcurrents (storks). Due to limited space, we cannot go further into details of this interesting field of ecological adaptation of flight behavior to comply with physiological demands and limitations.

The advantages of high-altitude flight for water economy and temperature regulation are obvious, but the decreasing partial pressure of oxygen may reduce the capability for oxygen uptake. Not only soaring, but also active flight has been observed at altitudes up to 9000 m (Swan, 1970) where barometric pressure P_B is 230 torr and partial pressure of oxygen P_{O_2} is 48 torr. At $P_B = 380$ torr ($=1/2 P_{B_n}$, $P_{O_2} = 79$ torr, 5,500 m altitude) species of several families were found flying (see summary by Groebbels, 1932, and migrational observations of Martens, 1971). At such heights the alveolar P_{O_2} even in resting birds (at sea level about 100 Torr) is markedly decreased. In sparrows at 6100 m altitude, Tucker (1968a) determined $P_{A_{O_2}}$ as 28 Torr and calculated arterial blood to be only 24% saturated, if equilibrated at that pressure.

The tolerance of low barometric pressures has been examined by various investigators since the time of Bert (1878) by measuring the survival time of mammals and birds placed in decompression chambers, but this technique gives no information on the comparative ability to undergo strenuous exertion at reduced pressure. Physiological information for birds in flight seems to be limited to observations by Tucker (1968a), who found that Budgerigars were unable to fly for more than a few seconds at simulated altitudes of 6100 m without artificial oxygen supply. This suggests severe physiological limitation to high altitude flight in this species.

Comparison of the effect of reduction of inspired oxygen on oxygen consumption during exercise in mammals has invariably shown a reduction with declining P_{O_2}. Comparisons made by Segrem and Hart (1967a) showed that the White-footed Mouse (*Peromyscus leucopus*) and man had similar characteristics and that results with Mallards (J. S. Hart, unpublished data) exercising at $\dot{V}_{O_2}$ of four times resting, were not notably different from the mammalian data (Fig. 10). In other words there is nothing to suggest that birds when not flying are superior to mammals in regard to their ability to consume oxygen at low pressures during strenuous exertion. While the ability to run in a treadmill in ducks at altitudes equivalent to 6100 m seemed little

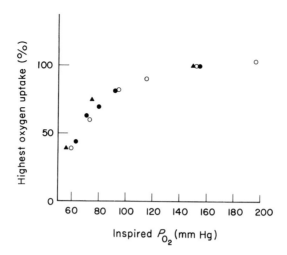

FIG. 10. Percentage reduction in highest observed oxygen uptake during treadmill or bicycle exercise with reduction in inspired P_{O_2}. For man (●) and White-footed Mouse (*Peromyscus leucopus*) (○) (Segrem and Hart, 1967a), near-maximal oxygen uptake was attained at sea level values of P_{O_2}. For Mallards (*Anas platyrhynchos*) (▲), the exercise gave only four times resting levels or about one-third of the $\dot{V}_{O_2}$ found during flight (J. S. Hart, unpublished).

reduced compared with sea level, it must be remembered that flight metabolism reaches about three times the level of metabolism attained in these tests (Berger *et al.*, 1970b). It is obvious that study of similar capabilities of birds accustomed to high-altitude flight would be required to clarify this important problem.

Calculations by Pennycuick (1969) show that the speed to achieve maximum range (and equally power) is proportional to (air density)$^{-1/2}$. As can be seen from Eq. (39), the range would not be changed. At an air density half that of sea level both speed and oxygen consumption would increase to $\sqrt{2}$ times the sea-level value. By means of radar measurements (Bruderer, 1971), the increase of speed with height was shown and the increase of about 10% per 1000 m height (up to 3000 m in height) is greater than the predicted value. This was explained by the assumption that faster flying birds would prefer greater altitudes.

The increasing disadvantage for energy turnover at high altitudes is demonstrated by these estimates. It may well be that at heights up to about 3000 m the negative effect on oxygen uptake and transport is minimal, but thermoregulation and water economy might be favored, leading to an optimal height for long-distance migration.

REFERENCES

Andersen, K. L. (1968). The cardiovascular system in exercise. In "Exercise Physiology" (H. B. Falls, ed.), pp. 79–128. Academic Press, New York.

Aschoff, J., and Pohl, H. (1970). Der Ruheumsatz von Vögeln als Funktion der Tageszeit und der Körpergrösse. J. Ornithol. 111, 38–47.

Asmussen, E. (1965). Muscular exercise. In "Handbook of Physiology" (Amer. Physiol. Soc., J. Field, ed.), Sect. 3, Vol. II, pp. 939–978. Williams & Wilkins, Baltimore, Maryland.

Asmussen, E., and Nielsen, M. (1952). The cardiac output in rest and work determined simultaneously by the acetylene and the dye injection method. Acta Physiol. Scand. 27, 217–230.

Astrand, P. O. (1952). "Experimental Studies of Physical Working Capacity in Relation to Sex and Age." Munksgaard, Copenhagen.

Aulie, A. (1971a). Co-ordination between the activity of the heart and the flight muscles during flight in small birds. Comp. Biochem. Physiol. A 38, 91–97.

Aulie, A. (1971b). Body temperatures in pigeons and budgerigars during sustained flight. Comp. Biochem. Physiol. A. 39, 173–176.

Aulie, A. (1972). Co-ordination between the activity of the heart and the pectoral muscles during flight in the pigeon. Comp. Biochem. Physiol. 41A, 43–48.

Austin, G. T., and Bradley, W. G. (1969). Additional responses of the Poor-will to low temperatures. Auk 86, 717–725.

Aymar, G. C. (1935). "Bird Flight." Dodd, Mead, New York.

Baer, M. (1896). Beiträge zur Kenntnis der Anatomie und Physiologie der Atemwerkzeuge bei den Vögeln. Z. Wiss. Zool. 61, 420–498.

Bailie, M. D., Robinson, S., Rostorfer, H. H., and Newton, J. L. (1961). Effects of exercise on heart output of the dog. J. Appl. Physiol. 16, 107–111.

Bartholomew, G. A., and Cade, T. J. (1956). Water consumption of House Finches. Condor 58, 406–412.

Bartholomew, G. A., and Cade, T. J. (1963). The water economy of land birds. Auk 80, 504–539.

Bartholomew, G. A., and Dawson, W. R. (1954). Body temperature and water requirements in the Mourning Dove, Zenaidura macroura marginella. Ecology 35, 181–187.

Bartholomew, G. A., and Trost, C. H. (1970). Temperature regulation in the Speckled Mousebird, Colius striatus. Condor 72, 141–146.

Bartholomew, G. A., Howell, T. R., and Cade, T. J. (1957). Torpidity in the White-throated Swift, Anna Hummingbird, and Poor-will. Condor 59, 145–155.

Bartholomew, G. A., Hudson, J. W., and Howell, T. R. (1962). Body temperature, oxygen consumption, evaporative water loss, and heart rate in the Poor-will. Condor 64, 117–125.

Bartholomew, G. A., Leitner, P., and Nelson, J. E. (1964). Body temperature, oxygen consumption, and heart rate in three species of Australian flying foxes. Physiol. Zool. 37, 179–198.

Berger, M. (1964). Untersuchungen über die Reaktionsgeschwindigkeit von Warmblütern bei kurzen optischen und akustischen Reizen. Zool. Jahrb., Abt. Allg. Zool. Physiol. Tiere 70, 513–538.

Berger, M. (1974). Energiewechsel von Kolibris beim Schwirrflug unter Höhenbedingungen. J. Ornithol. 115 (in press).

Berger, M., and Hart, J. S. (1972). Die Atmung beim Kolibri *Amazilia fimbriata* während des Schwirrfluges bei verschiedenen Umgebungstemperaturen. *J. Comp. Physiol.* **81**, 363-380.

Berger, M., Roy, O. Z., and Hart, J. S. (1970a). The co-ordination between respiration and wing beats in birds. *Z. Vergl. Physiol.* **66**, 190-200.

Berger, M., Hart, J. S., and Roy, O. Z. (1970b). Respiration, oxygen consumption and heart rate in some birds during rest and flight. *Z. Vergl. Physiol.* **66**, 201-214.

Berger, M., Hart, J. S., and Roy, O. Z. (1971). Respiratory water and heat loss of the Black Duck during flight at different ambient temperatures. *Can. J. Zool.* **49**, 767-774.

Bernstein, M. H. (1969). Cutaneous and respiratory evaporation in Painted Quail, *Excalfactoria chinensis*. *Amer. Zool.* **9**, 1009.

Bernstein, M. H. (1971a). Cutaneous and respiratory evaporation in the Painted Quail, *Excalfactoria chinensis*, during ontogeny of thermoregulation. *Comp. Biochem. Physiol.* A **38**, 611-617.

Bernstein, M. H. (1971b). Cutaneous water loss in small birds. *Condor* **73**, 468-469.

Bert, P. (1878). "La pression barométrique." Masson, Paris.

Besson, J. (1969). Migration de bergeronettes printantières, *Motacilla flava* à très haute altitude. *Oiseaux* **30**, 23.

Blyumental, T. I., and Dolnik, V. R. (1970). Body weight, wing length, fat depots and flight in birds. *Akad. Nauk SSSR, Zool. J.* **49**, 1069-1072 (in Russian).

Bradley, W. G., and O'Farrell, M. J. (1969). Temperature relationships of the Western Pipistrelle (*Pipistrellus hesperus*). In "Physiological Systems in Semiarid Environments" (C. C. Hoff and M. L. Riedesel, eds.), pp. 85-96. Univ. of New Mexico Press, Albuquerque.

Brody, S. (1945). "Bioenergetics and Growth." Van Nostrand-Reinhold, Princeton, New Jersey.

Brown, R. H. J. (1961). The power requirements of birds in flight. *Symp. Zool. Soc. London* **5**, 95-99.

Bruderer, B. (1971). Radarbeobachtungen über den Frühlingszug im Schweizerischen Mittelland. *Orn. Beob.* **68**, 89-158.

Brush, A. H. (1966). Avian heart size and cardiovascular performance. *Auk* **83**, 266-273.

Calder, W. A. (1964). Gaseous metabolism and water relations of the Zebra Finch, *Taeniopygia castanotis*. *Physiol. Zool.* **37**, 400-413.

Calder, W. A. (1968). Respiratory and heart rates of birds at rest. *Condor* **70**, 358-365.

Calder, W. A., and Schmidt-Nielsen, K. (1966). Evaporative cooling and respiratory alkalosis in the pigeon. *Proc. Nat. Acad. Sci. U.S.* **55**, 750-756.

Calder, W. A., and Schmidt-Nielsen, K. (1967). Temperature regulation and evaporation in the pigeon and the roadrunner. *Amer. J. Physiol.* **213**, 883-889.

Caldwell, L. D., Odum, E. P., and Marshall, S. G. (1963). Comparison of fat levels in migrating birds killed at a central Michigan and Florida Gulf coast television tower. *Wilson Bull.* **75**, 428-434.

Carpenter, R. E. (1968). Salt and water metabolism in the marine fish-eating bat, *Pizonyx vivesi*. *Comp. Biochem. Physiol.* **24**, 951-961.

Carpenter, R. E. (1969). Structure and function of the kidney and the water balance of desert bats. *Physiol. Zool.* **42**, 288-302.

Cerretelli, P., Piiper, J., Mangili, F., and Ricci, B. (1964a). Aerobic and anaerobic metabolism in exercising dogs. *J. Appl. Physiol.* **19**, 25-28.

Cerretelli, P., Piiper, J., Mangili, F., Cuttica, F., and Ricci, B. (1964b). Circulation in exercising dogs. *J. Appl. Physiol.* **19**, 29-32.

Chatonnet, J., and Minaire, Y. (1966). Comparison of energy expenditure during exercise and cold exposure in the dog. *Fed. Proc., Fed. Amer. Soc. Exp. Biol.* **25**, 1348–1350.

Child, G. I. (1969). A study of nonfat weights in migrating Swainson's Thrushes (*Hylocichla ustulata*). *Auk* **86**, 327–338.

Clark, A. J. (1927). "Comparative Physiology of the Heart." Macmillan, New York.

Crawford, E. C., and Lasiewski, R. C. (1968). Oxygen consumption and respiratory evaporation of the Emu and Rhea. *Condor* **70**, 333–339.

Crawford, E. C., and Schmidt-Nielsen, K. (1967). Temperature regulation and evaporative cooling in the Ostrich. *Amer. J. Physiol.* **212**, 347–353.

Crile, G., and Quiring, D. P. (1940). A record of the body weight and certain organ and gland weights of 3690 animals. *Ohio J. Sci.* **40**, 219–259.

Dawson, W. R. (1958). Relation of O_2 consumption and evaporative water loss to temperature in the cardinal. *Physiol. Zool.* **31**, 37–48.

Dawson, W. R., and Hudson, J. W. (1970). Birds. *In* "Comparative Physiology of Thermoregulation" (G. C. Whittow, ed.), Vol. 1, pp. 223–310. Academic Press, New York.

Dawson, W. R., and Schmidt-Nielsen, K. (1964). Terrestrial animals in dry heat: Desert birds. *In* "Handbook of Physiology" (D. B. Dill, ed.), Sect. 4, pp. 481–492. Amer. Physiol. Soc., Washington, D.C.

Dejours, P., Kellog, R. H., and Puce, N. (1963). Regulation of respiration and heart rate responses in exercise during altitude acclimatization. *J. Appl. Physiol.* **18**, 10–18.

Diesselhorst, G. (1968). Beiträge zur Ökologie der Vögel Zentral- und Ost-Nepals. *In* "Khumbu Himal" (W. Hellmich, ed.), Vol. 2, Wagner, Innsbruck and Munich.

Dolnik, V. R. (1969a). Bioenergetics of the flying bird. *Zh. Obshch. Biol.* **30**, 273–291 (in Russian).

Dolnik, V. R. (1969b). Theoretical limits of energy expenditure in flying birds. *In* "Ornithology in USSR" (Materials 5th All-Union Ornithol. Cong.), Vol. 2, pp. 198–202. Ashkhabad. (in Russian).

Dolnik, V. R. (1970). Fat metabolism and bird migration. *Coll. Int. Cent. Nat. Rech. Sci.* **172**, 351–364.

Dolnik, V. R. (1971). Energetics of bird migration. *In* "Zoology of Vertebrates, 1970, Ornithol. Problems" (L. P. Poznanin, ed.), pp. 52–81. Moskwa (in Russian).

Dolnik, V. R., and Blyumental, T. I. (1964). Bioenergetics of migration in birds. *Usp. Sovrem. Biol.* **58**, 280–301 (in Russian).

Dolnik, V. R., and Blyumental, T. I. (1967). Autumnal premigratory and migratory periods in the Chaffinch (*Fringilla coelebs coelebs*) and some other temperate-zone passerine birds. *Condor* **69**, 435–468.

Dolnik, V. R., Gavrilov, V. M., and Ezerskas, L. J. (1963). Energy losses of small birds to natural migrational flight. *In* "Tezisy Dokladov Pyatoi Pribaltiiskoi Ornitologicheskoi Konferentsii," pp. 65–67. Akad. Nauk, Inst. Zool. Bot. (in Russian).

Dorka, V. (1966). Das jahres- und tageszeitliche Zugmuster von Kurz- und Langstreckenziehern nach Beobachtungen auf den Alpenpässen Cou/Bretolet (Wallis). *Ornithol. Beob.* **63**, 165–223.

Drummond, G. I. (1967). Muscle metabolism. *Fortschr. Zool.* **18**, 359–429.

Drummond, G. I., and Black, E. C. (1960). Comparative physiology: Fuel of muscle metabolism. *Annu. Rev. Physiol.* **22**, 169–190.

Duncker, H.-R. (1968a). Das Lungen-Luftsacksystem der Vögel. Ein Beitrag zur funktionellen Anatomie des Respirationsorgans. Habilitation Paper. Med. Fak., Univ. Hamburg.

Duncker, H.-R. (1968b). Der Bronchialbaum der Vogellunge. *Anat. Anz.* **121**, 287–292.
Duncker, H.-R. (1971). The lung air sac system in birds. *Ergebn. Anat. Entwichlungsgesch.* **45** (No. 6).
Eliassen, E. (1962). Skin temperatures of sea gulls exposed to air currents of high speed and low temperatures. *Arbok Univ. Bergen, Mat.-Naturvitensk. Ser.* **17**.
Eliassen, E. (1963a). Preliminary results from new methods of investigating the physiology of birds during flight. *Ibis* **105**, 234–237.
Eliassen, E. (1963b). Telemetric registering of physiological data in birds in normal flight. *In* "Biotelemetry" (L. E. Slater, ed.), pp. 257–265. New York.
Epting, R. J., and Casey, T. M. (1973). Power output and wing disc loading in hovering hummingbirds. *Am. Naturalist* **107**, 761–765.
Farner, D. S. (1956). Body temperature of the Fairy Prion (*Pachyptila turtur*) in flight and at rest. *J. Appl. Physiol.* **8**, 546–548.
Farner, D. S. (1970). Some glimpses of comparative avian physiology. *Fed. Proc., Fed. Amer. Soc. Exp. Biol.* **29**, 1649–1663.
Farner, D. S., and Serventy, D. L. (1959). Body temperature and ontogeny in the Slender-billed Shearwater. *Condor* **61**, 426–433.
Flandrois, R., Puccinelli, R., Houdas, Y., and Lefrançois, R. (1962). Comparaison des consommations d'oxygène maximales mesurées et théoriques d'une population française. *J. Physiol. (Paris)* **54**, 337–338.
Fogden, M. P. L. (1972). Premigratory dehydration in the Reed Warbler *Acrocephalus scirpaceus* and water as a factor limiting migratory range. *Ibis* **114**, 548–552.
Fraenkel, G. (1934). Der Atmungsmechanismus der Vögel während des Fluges. *Biol. Zentralbl.* **54**, 96–101.
George, J. C., and Berger, A. J. (1966). "Avian Myology." Academic Press, New York.
Gollnick, P. D., and Ianuzzo, C. D. (1968). Colonic temperature response of the rat during exercise. *J. Appl. Physiol.* **24**, 747.
Greenewalt, C. H. (1960). The wings of insects and birds as mechanical oscillators. *Proc. Amer. Phil. Soc.* **104**, 605–611.
Greenewalt, C. H. (1962). Dimensional relationships for flying animals. *Smithson. Misc. Collect.* **144**, No. 2.
Groebbels, F. (1929). *Tagung Deut. Physiol. Ges.*, Kiel.
Groebbels, F. (1932). "Der Vogel," Vol. 1. Borntraeger, Berlin.
Hainsworth, F. R., and Wolf, L. L. (1969). Resting, torpid, and flight metabolism of the hummingbird *Eulampis jugularis*. *Amer. Zool.* **9**, 1100–1101.
Hainsworth, F. R., and Wolf, L. L. (1972a). Power for hovering flight in relation to body size in hummingbirds. *Amer. Naturalist* **106**, 589–596.
Hainsworth, F. R., and Wolf, L. L. (1972b). Crop volume, nectar concentration and hummingbird energetics. *Comp. Biochem. Physiol.* **42A**, 359–366.
Hart, J. S. (1950). Interrelations of daily metabolic cycle, activity, and environmental temperature of mice. *Can. J. Res., Sect. D* **28**, 293–307.
Hart, J. S. (1960). The problem of equivalence of specific dynamic action: exercise thermogenesis and cold thermogenesis. *Cold Injury, Trans. 6th Conf. Josiah Macy, Jr. Found., 1958* pp. 271–302.
Hart, J. S., and Berger, M. (1972). Energetics, water economy and temperature regulation during flight. *Proc. Int. Ornithol. Congr., 15th, 1970* pp. 189–199.
Hart, J. S., and Jansky, L. (1963). Thermogenesis due to exercise and cold in warm and cold acclimated rats. *Can. J. Biochem. Physiol.* **41**, 629–634.
Hart, J. S., and Roy, O. Z. (1966). Respiratory and cardiac responses to flight in pigeons. *Physiol. Zool.* **39**, 291–305.

Hart, J. S., and Roy, O. Z. (1967). Temperature regulation during flight in pigeons. *Amer. J. Physiol.* **213**, 1311–1316.

Hartmann, F. A. (1955). Heart weight in birds. *Condor* **57**, 221–238.

Hartmann, F. A. (1961). Locomotor mechanisms of birds. *Smithson. Misc. Collect.* **143**, No. 1.

Heinrich, B. (1970). Thoracic temperature stabilization by blood circulation in a free-flying moth. *Science* **168**, 580–582.

Hemmingsen, A. M. (1960). Energy metabolism as related to body size and respiratory surfaces, and its evolution. *Rep. Steno Hosp., Copenhagen* **9**, Part 2.

Herreid, C. F., and Kessel, B. (1967). Thermal conductance in birds and mammals. *Comp. Biochem. Physiol.* **21**, 405–414.

Herreid, C. F., and Schmidt-Nielsen, K. (1966). Oxygen consumption, temperature, and water loss in bats from different environments. *Amer. J. Physiol.* **211**, 1108–1112.

Hertel, H. (1963). "Struktur, Form, Bewegung." Krausskopf, Mainz.

Hesse, R. (1921). Das Herzgewicht der Wirbeltiere. *Zool. Jahrb., Abt. Allg. Zool. Physiol. Tiere* **38**, 242–364.

Hingston, R. W. G. (1925). Animal life at high altitudes. *Geogr. J.* **65**, 185–198.

Hissa, R., and Palokangas, R. (1970). Thermoregulation in the Titmouse (*Parus major* L.). *Comp. Biochem. Physiol.* **33**, 941–953.

Holt, J. P., Rhode, E. A., and Kines, H. (1968). Ventricular volumes and body weight in mammals. *Amer. J. Physiol.* **215**, 704–715.

Howell, T. R., and Bartholomew, G. A. (1959). Further experiments on torpidity in the Poor-will. *Condor* **61**, 180–185.

Howell, T. R., and Bartholomew, G. A. (1962). Temperature regulation in the Red-tailed Tropic Bird and the Red-footed Booby. *Condor* **64**, 6–18.

Hunt, J. (1954). "The Conquest of Everest." Dutton, New York.

Hussell, D. J. T. (1969). Weight loss of birds during nocturnal migration. *Auk* **86**, 75–83.

Hussell, D. J. T., and Caldwell, L. D. (1972). Flight metabolism of the Myrtle Warbler (*Dendroica coronata*) during nocturnal migration. *Proc. Int. Ornithol. Congr., 15th 1970* Abstract, pp. 121–122.

Jansky, L. (1959a). Oxygen consumption in white mice during physical exercise. *Physiol. Bohemoslov.* **8**, 464–471.

Jansky, L. (1959b). Einfluss der Arbeit und der niedrigen Temperaturen auf den Sauerstoffverbrauch des Goldhamsters. *Acta Soc. Zool. Bohemoslov.* **23**, 266–274.

Jansky, L. (1965). Adaptability of heat production mechanisms in homeotherms. *Acta Univ. Carol., Biol.* **1**, 1–91.

Johnston, D. W. (1968). Body characteristics of Palm Warblers following an overwater flight. *Auk* **85**, 13–18.

Johnston, D. W. (1973). Cytological and chemical adaptations of fat deposition in migratory birds. *Condor* **75**, 108–113.

Johnston, D. W., and McFarlane, R. W. (1967). Migration and bioenergetics of flight in the Pacific Golden Plover. *Condor* **69**, 156–168.

Kammer, A. E. (1970). Thoracic temperature, shivering, and flight in the Monarch Butterfly, *Danaus plexippus* (L.). *Z. Vergl. Physiol.* **68**, 334–344.

Kendeigh, S. C. (1969a). Tolerance of cold and Bergmann's rule. *Auk* **86**, 13–25.

Kendeigh, S. C. (1969b). Energy responses of birds to their thermal environments. *Wilson Bull.* **81**, 441–449.

Kendeigh, S. C. (1970). Energy requirements for existence in relation to size of bird. *Condor* **72**, 60–65.

Kendeigh, S. C., Kontogiannis, J. E., Mazac, A., and Roth, R. R. (1969). Environmental regulation of food intake by birds. *Comp. Biochem. Physiol.* **31**, 941–957.

Keskpaik, J. (1968). Heat production and heat loss of swallows and martins during flight. *Izv. Akad. Nauk. Est. SSR, Biol.* **17**, No. 2, 179–191 (in Russian).

Keskpaik, J., and Lyuleeva, D. (1968). Temporary hypothermia in swallows. *Soobshch. Pribalt. Kom. Izuch. Migrat. Ptits.* **5**, 122–145 (in Russian).

Keskpaik, J., and Horma, P. (1972a). Heart rate during the birds flight. *Eesti NSV Teaduste Akad. Toimetised 21, Biol.* **1**, 78–85 (in Russian).

Keskpaik, J., and Horma, P. (1972b). Body temperature and heart rate of flying Sea-Gulls (*Larus m. marinus* L.). *Eesti NSV Teaduste Akad. Toimetised 21, Biol.* **2**, 109–116 (in Russian).

King, J. R., and Farner, D. S. (1961). Energy metabolism, thermoregulation and body temperature. *In* "Biology and Comparative Physiology of Birds" (A. J. Marshall, ed.), Vol. 2, pp. 215–288. Academic Press, New York.

King, J. R., and Farner, D. S. (1964). Terrestrial animals in humid heat: Birds. *In* "Handbook of Physiology" (D. B. Dill, ed.), Sect. 4, pp. 603–624. Amer. Physiol. Soc., Washington, D.C.

Kleiber, M. (1947). Body size and metabolic rate. *Physiol. Rev.* **27**, 511–541.

Kleiber, M. (1961). "The Fire of Life." Wiley, New York.

Kleiber, M. (1970). Conductivity, conductance and transfer constant for animal heat. *Fed. Proc., Fed. Amer. Soc. Exp. Biol.* **29**, 660 (abstr.).

Kluger, M. J., and Heath, J. E. (1970). Vasomotion in the bat wing: A thermoregulatory response to internal heating. *Comp. Biochem. Physiol.* **32**, 219–226.

Kokshaiski, N. V. (1970). Flight energetics of insects and birds. *Zh. Obshch. Biol.* **31**, No. 5, 527–549 (in Russian).

Lack, D. (1960). The height of bird migration. *Brit. Birds* **53**, 5–10.

Lasiewski, R. C. (1962). The energetics of migrating hummingbirds. *Condor* **64**, 324.

Lasiewski, R. C. (1963a). The energetic cost of small size in hummingbirds. *Proc. Int. Ornithol. Congr., 13th, 1962* pp. 1095–1103.

Lasiewski, R. C. (1963b). Oxygen consumption of torpid, resting, active, and flying hummingbirds. *Physiol. Zool.* **36**, 122–140.

Lasiewski, R. C. (1964). Body temperatures, heart and breathing rate, and evaporative water loss in hummingbirds. *Physiol. Zool.* **37**, 212–223.

Lasiewski, R. C., and Calder, W. A. (1971). A preliminary allometric analysis of respiratory variables in resting birds. *Resp. Physiol.* **11**, 152–166.

Lasiewski, R. C., and Dawson, W. R. (1967). A re-examination of the relation between standard metabolic rate and body weight in birds. *Condor* **69**, 13–23.

Lasiewski, R. C., and Lasiewski, R. J. (1967). Physiological responses of the Blue-throated and Rivoli's Hummingbird. *Auk* **84**, 34–48.

Lasiewski, R. C., Hubbard, S. H., and Moberly, W. R. (1964). Energetic relationships of a very small passerine bird. *Condor* **66**, 212–220.

Lasiewski, R. C., Acosta, A. L., and Bernstein, M. H. (1966). Evaporative water loss in birds. II. A modified method for determination by direct weighing. *Comp. Biochem. Physiol.* **19**, 459–470.

Lasiewski, R. C., Weathers, W. W., and Bernstein, M. H. (1967). Physiological responses of the Giant Hummingbird, *Patagona gigas*. *Comp. Biochem. Physiol.* **23**, 797–813.

Lasiewski, R. C., Bernstein, M. H., and Ohmart, R. D. (1971). Cutaneous water loss in the Roadrunner and Poor-will. *Condor* **73**, 470–472.

LeFebvre, E. A. (1964). The use of D_2O^{18} for measuring energy metabolism in *Columba livia* at rest and in flight. *Auk* **81**, 403–416.

Leitner, P., and Nelson, J. E. (1967). Body temperature, oxygen consumption and heart rate in the Australian False Vampire Bat, *Macroderma gigas*. *Comp. Biochem. Physiol.* **21**, 65–74.

Lifson, N., Gordon, G. B., and McClintock, R. (1955). Measurement of total carbon dioxide production by means of D_2O^{18}. *J. Appl. Physiol.* **7**, 704–710.

Ligon, J. D. (1969). Some aspects of temperature relations in small owls. *Auk* **86**, 458–472.

Lilienthal, O. (1889). "Der Vogelflug als Grundlage der Fliegekunst." Gaertner, Berlin.

Lord, R. D., Bellrose, F. C., and Cochran, W. W. (1962). Radiotelemetry of the respiration of a flying duck. *Science* **137**, 39–40.

Lyuleeva, D. S. (1970). Flight energy of swallows and swifts. *Dokl. Biol. Sci.* **190**, No. 6, 1467–1469.

McAtee, W. L. (1947). Torpidity in birds. *Am. Midl. Natur.* **38**, 191–206.

McGahan, J. (1973). Flapping flight of the Andean Condor in nature. *J. Exp. Biol.* **58**, 239–253.

McNab, B. K. (1969). The economics of temperature regulation in neotropical bats. *Comp. Biochem. Physiol.* **31**, 227–268.

McNeil, R. (1969). La détermination du contenu lipidique et de la capacité de vol chez quelques espèces d'oiseaux de rivage (Charadriidae et Scolopacidae). *Can. J. Zool.* **47**, 525–536.

Manville, R. H. (1963). Altitude record for mallard duck. *Wilson Bull.* **75**, 92.

Marey, E. J. (1890). "Le vol des oiseaux." Masson, Paris.

Marshall, J. T. (1955). Hibernation in captive goatsuckers. *Condor* **57**, 129–134.

Martens, J. (1971). Zur Kenntnis des Vogelzuges im nepalischen Himalaya. *Vogelwarte* **26**, 113–128.

Meinertzhagen, R. (1955). The speed and altitude of bird flight (with notes on other animals). *Ibis* **97**, 81–117.

Miller, A. H. (1950). Temperatures of Poor-wills in the summer season. *Condor* **52**, 41–42.

Moreau, R. E. (1961). Problems of Mediterranean-Saharan migration. *Ibis* **103a**, 373–427 and 580–618.

Moreau, R. E., and Dolp, R. M. (1970). Fat, water, weights and wing-lengths of autumn migrants in transit on the northwest coast of Egypt. *Ibis* **112**, 209–228.

Morrison, P. R. (1962). Modifications of body temperature by activity in Brazilian hummingbirds. *Condor* **64**, 315–323.

Mugaas, J. N., and Templeton, J. R. (1970). Thermoregulation in the Red-breasted Nuthatch *(Sitta canadensis)*. *Condor* **72**, 125–132.

Nisbet, I. C. T. (1963). Weight loss during migration. Part II. Review of other estimates. *Bird-Banding* **34**, 139–159.

Nisbet, I. C. T. (1967). Aerodynamic theories of flight versus physiological theories. *Bird-Banding* **38**, 306–308.

Nisbet, I. C. T., and Drury, W. H. (1967). Orientation of spring migrants studied by radar. *Bird-Banding* **38**, 173–186.

Nisbet, I. C. T., Drury, W. H., and Baird, J. (1963). Weight loss during migration. I. Deposition and consumption of fat by the Blackpoll Warbler *Dendroica striata*. *Bird-Banding* **34**, 107–138.

Odum, E. P. (1941). Variations in the heart rate of birds: A study in physiological ecology. *Ecol. Monogr.* **11**, 229–326.

Odum, E. P. (1945). The heart rate of small birds. *Science* **101**, 153–154.

Odum, E. P. (1958). The fat deposition picture in the White-throated Sparrow in comparison with that in long-range migrants. *Bird-Banding* **29**, 105–108.

Odum, E. P., and Connell, C. E. (1956). Lipid levels in migrating birds. *Science* **123**, 892–894.

Odum, E. P., Connell, C. E., and Stoddard, H. L. (1961). Flight energy and estimated flight ranges of some migratory birds. *Auk* **78**, 515-527.
Odum, E. P., Rogers, D. T., and Hicks, D. L. (1964). Homeostasis of the nonfat components of migrating birds. *Science* **143**, 1037-1039.
Oehme, H. (1963). Flug und Flügel von Star und Amsel. *Biol. Zentralbl.* **82**, 413-454 and 569-587.
Oehme, H. (1965). Über den Kraftflug grosser Vögel, *Beitr. Vogelkunde* **11**, 1-31.
Oehme, H. (1968a). Der Flug des Mauerseglers (*Apus apus*). *Biol. Zentralbl.* **87**, 287-311.
Oehme, H. (1968b). Physiologische und morphologische Aspekte der Muskelleistung fliegender Tiere. *Biol. Rundsch.* **6**, 203-210.
Oehme, H. (1970). Der Rüttelflug des Gartenrotschwanzes (*Phoenicurus phoenicurus*). *Beitr. Vogelkunde* **15**, 417-433.
O'Farrell, M. J., and Studier, E. H. (1970). Fall metabolism in relation to ambient temperatures in three species of *Myotis*. *Comp. Biochem. Physiol.* **35**, 697-703.
Pasquis, P., Lacaisse, A., and Dejours, P. (1970). Maximal oxygen uptake in four species of small mammals. *Resp. Physiol.* **9**, 298-309.
Pearson, O. P. (1950). The metabolism of hummingbirds. *Condor* **52**, 145-152.
Pearson, O. P. (1964). Metabolism and heat loss during flight in pigeons. *Condor* **66**, 182-185.
Pennycuick, C. J. (1960). Gliding flight of the Fulmar Petrel. *J. Exp. Biol.* **37**, 330-338.
Pennycuick, C. J. (1968a). A wind-tunnel study of gliding flight in the pigeon *Columba livia*. *J. Exp. Biol.* **49**, 509-526.
Pennycuick, C. J. (1968b). Power requirements for horizontal flight in the pigeon *Columba livia*. *J. Exp. Biol.* **49**, 527-555.
Pennycuick, C. J. (1969). The mechanics of bird migration. *Ibis* **111**, 525-556.
Pennycuick, C. J. (1972). Soaring behaviour and performance of some East African birds, observed from a motor-glider. *Ibis* **114**, 178-218.
Piiper, J., Drees, F., and Scheid, P. (1970). Gas exchange in the domestic fowl during spontaneous breathing and artificial ventilation. *Resp. Physiol.* **9**, 234-245.
Portmann, A. (1950). Les organes de la circulation sanguine. In "Traité de Zoologie" (P. P. Grassé, ed.), Vol. 15, pp. 243-256. Masson, Paris.
Prechtl, J. J. (1846). "Untersuchungen über den Flug der Vögel." Gerolds Verlag. Wien.
Raspet, A. (1960). Biophysics of bird flight. *Science* **132**, 191-200.
Raveling, D. G., and LeFebvre, E. A. (1967). Energy metabolism and theoretical flight range of birds. *Bird-Banding* **38**, 97-113.
Reeder, W. G., and Cowles, R. B. (1951). Aspects of thermoregulation in bats (Chiroptera). *J. Mammal.* **32**, 389-403.
Rensch, B. (1948). Organproportionen und Körpergrösse bei Vögeln und Säugetieren. *Zool. Jahrb., Abt. Allg. Zool. Physiol. Tiere* **61**, 337-412.
Roy, O. Z., and Hart, J. S. (1966). A multi-channel transmitter for the physiological study of birds in flight. *Med. Biol. Eng.* **4**, 457-466.
Rüppell, G. (1969). Flugstudien an felsbewohnenden Vögeln (*Fulmarus glacialis* L. und *Uria lomvia* L.) mit Hilfe kinematographischer Methoden. *Res. Film* **6**, 445-457.
Salomonsen, F. (1969). "Vogelzug." Bayerischer Landwirtschaftsverlag, Munich.
Salt, G. W. (1964). Respiratory evaporation in birds. *Biol. Rev. Cambridge. Phil. Soc.* **39**, 113-136.

Salt, G. W., and Zeuthen, E. (1960). The respiratory system. *In* "Biology and Comparative Physiology of Birds" (A. J. Marshall, ed.), Vol. 1, pp. 363–409. Academic Press, New York.

Scharnke, H. (1934). Die Bedeutung der Luftsäcke für die Atmung der Vögel. *Ergeb. Biol.* **10**, 177–206.

Scharnke, H. (1938). Experimentelle Beiträge zur Kenntnis der Vogelatmung. *Z. Vergl. Physiol.* **25**, 548–583.

Schmidt-Nielsen, K. (1972). Locomotion: energy cost of swimming, flying, and running. *Science* **177**, 222–228.

Schmidt-Nielsen, K., Hainsworth, F. R., and Murrish, D. E. (1970). Counter-current heat exchange in the respiratory passages: Effect on water and heat balance. *Resp. Physiol.* **9**, 263–276.

Segrem, N. P., and Andersen, K. L. (1968). Unpublished data (cited in Andersen, 1968, pp. 79–151).

Segrem, N. P., and Hart, J. S. (1967a). Oxygen supply and performance in *Peromyscus*. Metabolic and circulatory responses to exercise. *Can. J. Physiol. Pharmacol.* **45**, 531–541.

Segrem, N. P., and Hart, J. S. (1967b). Oxygen supply and performance in *Peromyscus*. Comparison of exercise with cold exposure. *Can. J. Physiol. Pharmacol.* **45**, 543–549.

Smith, R. M. (1969). Cardiovascular, respiratory, temperature and evaporative water loss responses of pigeons to varying degrees of heat stress. Ph.D. Thesis, Dept. of Anatomy and Physiology, Indiana University, Bloomington.

Smith, R. M. (1972). Circulation, respiratory volumes and temperature regulation of the pigeon in dry and humid heat. *Comp. Biochem. Physiol.* **43A**, 477–490.

Smith, R. M., and Suthers, R. (1969). Cutaneous water loss as a significant contribution to temperature regulation in heat stressed pigeons. *Physiologist* **12**, 358.

Sotavalta, O. (1952). The essential factor regulating the wing-stroke frequency of insects in wing mutilation and loading experiments and in experiments at subatmospheric pressure. *Ann. Zool. Soc. "Vanamo"* **15**, 1–67.

Sotavalta, O., and Laulajainen, E. (1961). On the sugar consumption of the drone fly (*Eristalis tenax* L.) in flight experiments. *Ann. Acad. Sci. Fenn., Ser. A4* **53**, 1–25.

Spector, W. S., ed. (1956). "Handbook of Biological Data," p. 279. Saunders, Philadelphia.

Stahl, W. R. (1967). Scaling of respiratory variables in mammals. *J. Appl. Physiol.* **22**, 453–460.

Stolpe, M., and Zimmer, K. (1959). Atmungs- und Luftsacksystem, Luftzirkulation. *In* "Naturgeschichte der Vögel" (R. Berndt and W. Meise, eds.), Vol. I, pp. 134–141. Franckh'sche Verlagshandlung, Stuttgart.

Stones, R. C., and Wiebers, J. E. (1967). Temperature regulation in the little brown bat, *Myotis lucifugus*. *In* "Mammalian Hibernation" (K. C. Fisher *et al.*, eds.), Vol. III, pp. 97–109. Oliver & Boyd, London.

Studier, E. H., and Howell, D. J. (1969). Heart rate of female big brown bats in flight. *J. Mammal.* **50**, 842–845.

Studier, E. H., and O'Farrell, M. J. (1972). Biology of *Myotis thysanodes* and *M. lucifugus* (Chiroptera: Vespertilionidae). I. Thermoregulation. *Comp. Biochem. Physiol.* **41A**, 567–595.

Sturkie, P. D. (1970). Circulation in aves. *Fed. Proc., Fed. Amer. Soc. Exp. Biol.* **29**, 1674–1679.

Suthers, R. A., Thomas, S. P., and Suthers, B. J. (1972). Respiration, wing-beat and ultrasonic pulse emission in an echo-locating bat. *J. Exp. Biol.* **56**, 37–48.

Swan, L. W. (1970). Goose of the Himalayas. Nat. Hist., Dec. 1970, 68–75.

Taylor, C. R., Dmi'el, R., Fedak, M., and Schmidt-Nielsen, K. (1971). Energetic cost of running and heat balance in a large bird, the rhea. *Amer. J. Physiol.* **221**, 597–661.

Teal, J. M. (1969). Direct measurement of CO_2 production during flight in small birds. *Zoologica (New York)* **54**, 17–23.

Thomas, S. P., and Suthers, R. A. (1970). Oxygen consumption and physiological responses during flight in an echolocating bat. *Fed. Proc., Fed. Amer. Soc. Exp. Biol.* **29**, 265 (abstr.).

Thomas, S. P., and Suthers, R. A. (1972). The physiology and energetics of bat flight. *J. Exp. Biol.* **57**, 317–335.

Tomlinson, J. T. (1963). Breathing of birds in flight. *Condor* **65**, 514–516.

Tomlinson, J. T., and McKinnon, R. S. (1957). Pigeon wing-beats synchronized with breathing. *Condor* **59**, 401.

Tucker, V. A. (1966). Oxygen consumption of a flying bird. *Science* **154**, 150–151.

Tucker, V. A. (1968a). Respiratory physiology of House Sparrows in relation to high-altitude flight. *J. Exp. Biol.* **48**, 55–66.

Tucker, V. A. (1968b). Respiratory exchange and evaporative water loss in the flying Budgerigar. *J. Exp. Biol.* **48**, 67–87.

Tucker, V. A. (1969). The energetics of bird flight. *Sci. Amer.* **220**, No. 5, 70–78.

Tucker, V. A. (1970). Energetic cost of locomotion in animals. *Comp. Biochem. Physiol.* **34**, 841–846.

Tucker, V. A. (1971). Flight energetics in birds. *Amer. Zool.* **11**, 115–124.

Tucker, V. A. (1972). Respiration during flight in birds. *Resp. Physiol.* **14**, 75–82.

Tucker, V. A. (1972). Metabolism during flight in the Laughing Gull, *Larus atricilla*. *Am. J. Physiol.* **222**, 237–245.

Tucker, V. A. (1973). Bird metabolism during flight: Evaluation of a theory. *J. Exp. Biol.* **58**, 689–709.

Tucker, V. A., and Parrott, G. C. (1970). Aerodynamics of gliding flight in a falcon and other birds. *J. Exp. Biol.* **52**, 345–367.

Utter, J. M., and LeFebvre, E. A. (1970). Energy expenditure for free flight by the Purple Martin (*Progne subis*). *Comp. Biochem. Physiol.* **35**, 713–719.

von Helmholtz, H. (1874). Ein Theorem über geometrisch ähnliche Bewegungen flüssiger Körper, nebst Anwendung auf das Problem, Luftballons zu lenken. *Monatsber. Kgl. Preuss. Akad. Wiss. Berlin* (26.6.1873), pp. 501–514.

von Holst, E. (1939). Die relative Koordination als Phänomen und als Methode zentralnervöser Funktionsanalyse. *Ergeb. Physiol., Biol. Chem. Exp. Pharmakol.* **42**, 228–306.

von Holst, E. (1943). Untersuchungen über Flugbiophysik. I. Messungen zur Aerodynamik kleiner schwingender Flächen. *Biol. Zentralbl.* **63**, 289–317.

von Lucanus, F. (1929). "Zugvögel und Vogelzug." Springer-Verlag, Berlin.

von Miller, R. (1965). "Lexikon der Energietechnik und Kraftmaschinen," Vol. 6 of Lueger Lexikon der Technik. Deut. Verlagsanstalt, Stuttgart.

Weis-Fogh, T. (1961). Power in flapping flight. *In* "The Cell and the Organism" (J. S. Ramsay and V. B. Wigglesworth, eds.), pp. 283–300. Cambridge Univ. Press, London and New York.

Weis-Fogh, T. (1972). Energetics of hovering flight in hummingbirds and in *Drosophila*. *J. Exp. Biol.* **56**, 79–104.

Weis-Fogh, T. (1973). Quick estimates of flight fitness in hovering animals, including novel mechanisms for lift production. *J. Exp. Biol.* **59**, 169–230.

West. G. C. (1965). Shivering and heat production in wild birds. *Physiol. Zool.* **38,** 111–120.

Wetmore, A. (1921). A study of the body temperature of birds. *Smithson. Misc. Collect.* **72,** No. 12, 1–52.

Wick, H. (1964). Elektromyographische Untersuchungen an der fliegenden Taube. *Helv. Physiol. Acta* **22,** C88–C90.

Wilkie, D. R. (1959). The work output of animals: Flight by birds and by man-power. *Nature (London)* **183,** 1515–1516.

Wolf, L. L., and Hainsworth, F. R. (1971). Time and energy budgets of territorial hummingbirds. *Ecology* **52,** 980–988.

Yapp, W. B. (1956). Two physiological considerations in bird migration. *Wilson Bull.* **68,** 312–319.

Yapp, W. B. (1962). Some physiological limitations on migration. *Ibis* **104,** 86–89.

Young, D. R., Mosher, R., Erve, P., and Spector, H. (1959). Energy metabolism and gas exchange during treadmill running in dogs. *J. Appl. Physiol.* **14,** 834–838.

Zeuthen, E. (1942). The ventilation of the respiratory tract in birds. *Kgl. Dan. Vidensk. Selsk., Biol. Medd.* **17,** No. 1.

Zimmer, K. (1935). Beiträge zur Mechanik der Atmung bei den Vögeln in Stand und Flug. *Zoologica (Stuttgart)* **33,** No. 88.

AUTHOR INDEX

Numbers in italics refer to the pages on which the complete references are listed.

A

Aakhus, T., 14, 46, 59
Abel, W., 37, 46
Ábrahám, A., 14, 15, 30, 31, 32, 35, 37, 41, 44, 46, 47
Abrams, R. M., 261, 400
Acosta, A. L., 273, 305, 330, 333, 404, 455, 472
Adams, T., 262, 393
Åkerman, B., 356, 393
Akester, A. R., 23, 24, 31, 32, 33, 38, 47
Akester, B., 23, 31, 47
Alder, L. P., 372, 394
Alexander, G., 368, 394
Alexander, R. McN., 138, 251
Allen, D. J., 357, 358, 368, 369, 370, 394
Allen, T. E., 262, 395
Allee, W. C., 301, 401
Altman, P., 281, 394
Ambache, N., 45, 47
Ames, P., 135, 137, 251
Andersen, A. E., 13, 47, 65
Andersen, H. T., 14, 47
Andersen, K. L., 441, 442, 467, 475
Andersson, B., 356, 393
Andres, K. H., 13, 47
Ariëns Kappers, J., 90, 114
Arimoto, K., 32, 33, 47
Armas, A. E., 386, 413
Armstrong, E. A., 383, 394
Aschoff, J., 270, 271, 281, 321, 346, 360, 365, 366, 394, 412, 431, 467
Aschraft, D. W., 34, 47
Ash, R. W., 39, 47
Ashhurst, D. E., 9, 47

Asmussen, E., 441, 442, 449, 467
Astrand, P. O., 434, 467
Augee, M. L., 350, 394
Aulie, A., 314, 394, 435, 436, 441, 442, 444, 445, 446, 467
Austin, G. T., 346, 394, 447, 467
Axelrod, J., 42, 43, 62, 80, 85, 86, 109, 115, 118
Aymar, G. C., 464, 467
Azzena, G. B., 12, 47, 63

B

Baer, M., 416, 467
Baher, E., 319, 320, 405
Bailie, M. D., 438, 467
Baird, J., 460, 473
Baker, P. T., 381, 394
Bakken, G. S., 263, 394
Balda, R. P., 227, 252, 383, 394
Ballantyne, B., 39, 44, 47, 55
Barfield, R. J., 82, 115
Barfuss, D. W., 82, 114
Bargmann, W., 89, 114
Barker, D., 10, 11, 12, 47
Barnett, G. R., 27, 53
Barnett, L. B., 375, 378, 380, 394
Barnikol, A., 2, 47, 129, 134, 135, 251, 256
Bartholomew, G. A., 301, 302, 305, 323, 329, 333, 338, 341, 342, 344, 345, 346, 347, 351, 353, 371, 373, 387, 394, 397, 401, 404, 410, 441, 444, 446, 453, 457, 467, 471
Bartlett, A. L., 34, 47
Bartuňkova, R., 317, 328, 402
Batekhina, N. K., 37, 48
Bateman, G. C., 383, 394

479

Baumel, J. J., 20, 48
Baumgarten, H. G., 23, 40, 48
Baur, M., 22, 48
Becker, W. C., 85, 89, 115
Beecher, W. J., 135, 187, 251
Bell, C., 23, 32, 33, 48
Bellrose, F. C., 417, 473
Bellairs, A. D'A., 123, 132, 251
Benedeczky, I., 41, 48
Benedict, F. G., 350, 380, 395, 408
Bennett, H. S., 9, 48
Bennett, T., 13, 14, 16, 21, 22, 23, 24, 25, 27, 28, 29, 30, 31, 32, 33, 35, 36, 38, 40, 41, 42, 43, 45, 46, 48, 49, 51
Bentley, P. J., 330, 395, 409
Benzinger, T. H., 355, 395
Berger, A. J., 9, 55, 123, 134, 135, 137, 188, 189, 251, 254, 429, 461, 470
Berger, M., 312, 313, 314, 395, 417, 419, 420, 421, 422, 423, 424, 425, 426, 427, 428, 429, 435, 436, 437, 441, 444, 445, 446, 448, 450, 451, 453, 454, 455, 458, 460, 466, 467, 468, 470
Berman, A., 311, 312, 395
Bernard, C., 264, 395
Bernstein, M. H., 273, 274, 275, 305, 327, 329, 330, 333, 350, 351, 395, 404, 405, 433, 436, 454, 455, 468, 472
Bert, P., 465, 468
Berthold, H., 388, 395
Berthold, P., 388, 395
Besson, J., 464, 468
Best, A. T., 345, 402
Bidder, 27, 49
Binkley, S., 86, 88, 114, 346, 395
Birkebak, R. C., 267, 287, 288, 290, 291, 292, 294, 302, 307, 308, 395, 397
Bischoff, M. B., 41, 49, 92, 114
Biswal, G., 40, 41, 49
Blaber, L. C., 22, 49
Black, E. C., 429, 469
Blem, C. R., 388, 389, 395
Bligh, J., 262, 354, 359, 395
Blyumental, T. I., 458, 459, 460, 469
Boas, J. E. V., 132, 134, 158, 251
Bock, W. J., 126, 127, 128, 129, 130, 131, 132, 133, 134, 135, 137, 138, 139, 140, 145, 146, 147, 149, 150, 151, 153, 154, 155, 156, 157, 160, 162, 163, 164, 165, 176, 178, 181, 182, 183, 184, 187, 188, 190, 192, 193, 194, 195, 197, 198, 199, 202, 204, 205, 206, 212, 219, 222, 224, 226, 227, 229, 230, 231, 232, 233, 234, 236, 237, 238, 239, 240, 241, 242, 243, 244, 246, 247, 251, 252, 254, 257
Boeke, J., 44, 49
Boesiger, B., 10, 49
Bogdanov, R. Z., 36, 49
Böker, H., 226, 253
Bolton, T. B., 32, 33, 49, 50
Bonsdorff, E. J., 2, 50
Bonting, S. L., 39, 50
Booser, J., 350, 396
Borbely, E., 385, 412
Bortolami, R., 4, 12, 13, 38, 50, 63
Boswall, J., 373, 395
Botezat, E., 10, 32, 50
Bowman, R. I., 151, 253
Bowman, W. C., 32, 33, 34, 35, 50, 368, 395
Bowsher, D. R., 14, 60
Boyd, J. E., 42, 43, 62, 85, 86, 109, 115, 281, 395
Bradey, O. C., 135, 253
Bradley, W. G., 346, 394, 447, 467, 468
Brauer, A., 41, 50
Breipohl, W., 97, 115
Bremer, J. L., 20, 50
Breneman, W. R., 83, 117
Brenner, F. J., 373, 383, 395
Bretz, W. L., 327, 329, 333, 338, 339, 409
Brinkman, R., 7, 50
Brodkorb, P., 260, 264, 396
Brody, S., 269, 270, 396, 434, 448, 449, 468
Bron, A., 31, 71
Brooks, W. S., 382, 384, 396
Brown, R. H. J., 459, 468
Browne, M. J., 20, 28, 50
Bruderer, B., 461, 466, 468
Brumleve, S. J., 321, 398
Brush, A. H., 439, 468
Bryden, M. M., 381, 396
Bubien-Waluszewska, A., 2, 34, 50
Buchmüller, K., 290, 396
Buckley, G. A., 43, 50
Budd, S. M., 345, 346, 396
Budgell, P., 370, 373, 396
Budgell, P. W., 320, 365, 405
Bühler, P., 130, 138, 253

Burger, R. E., 2, 14, 33, 50, 54, 339, 379, 380, 398, 405
Burn, J. H., 16, 50
Burnstock, G., 24, 43, 46, 48
Burton, A. C., 267, 396
Burton, P. J. K., 127, 227, 230, 253, 255
Busse, P., 383, 396
Butler, P. J., 14, 15, 50, 51
Butt, E. M., 34, 35, 57

C

Cabanac, M., 261, 396
Cade, T. J., 302, 313, 330, 344, 346, 347, 351, 353, 373, 383, 394, 396, 400, 446, 457, 467
Cahn, R. D., 191, 255
Calder, W. A., 270, 272, 276, 278, 286, 302, 305, 320, 327, 329, 330, 333, 334, 335, 337, 338, 339, 340, 342, 349, 350, 371, 388, 396, 397, 401, 404, 422, 425, 428, 433, 437, 457, 468, 472
Caldwell, F. T., Jr., 261, 400
Caldwell, L. D., 459, 461, 468, 471
Calislar, T., 40, 57
Callingham, B. A., 23, 51
Cambier, E., 9, 51
Cameron, J. L., 385, 402
Campbell, B., 26, 27, 67
Campbell, G., 15, 16, 25, 51
Campbell, H. S., 45, 51
Canady, M. R., 39, 50
Canfield, S. P., 190, 192, 253
Cannon, P., 381, 397
Cantino, D., 28, 29, 30, 35, 51
Capanna, E., 13, 43, 51
Caravaggio, L. L., 39, 50
Cardinali, D. P., 83, 85, 114, 117
Carlson, L. D., 317, 397
Carpenter, F. W., 2, 25, 44, 51
Carpenter, R. E., 458, 468
Cass, R., 23, 51
Cassuto, Y., 317, 378, 397
Cerretelli, P., 434, 441, 442, 468
Chaffee, R. R. J., 262, 317, 374, 378, 397
Chambers, A. B., 305, 397
Chamberlain, F. W., 135, 253
Chang, H-Y., 2, 51
Chatonnet, J., 434, 469
Cheke, R. A., 281, 346, 397
Chen, T. Y., 14, 55, 71

Child, G. I., 461, 469
Chinoy, N. J., 10, 51
Chowdhary, D. S., 14, 51
Civitelli, M. V., 13, 43, 51
Clark, A. J., 436, 437, 438, 469
Clark, W. M., 264, 397
Cloudsley-Thompson, J. L., 383, 397
Clouse, M. E., 45, 51
Cobb, J. L. S., 13, 22, 23, 24, 25, 28, 29, 30, 32, 33, 35, 42, 43, 45, 46, 48, 49, 51
Cochran, W. W., 417, 473
Cohen, D. H., 2, 14, 17, 19, 31, 32, 51, 52, 62
Cohn, J. E., 15, 52, 327, 329, 338, 339, 409
Collin, J.-P., 92, 99, 101, 107, 114, 115
Connell, C. E., 461, 462, 463, 473, 474
Consiglio, M., 44, 52
Constantinescu, G., 37, 67
Cook, R. D., 14, 33, 52
Cordier, A., 37, 52
Cords, E., 2, 17, 52
Costa, M., 23, 28, 29, 52
Cottle, M. K. W., 39, 52
Cotton, T. E., 385, 402
Coulombe, N. H., 280, 281, 342, 397
Coulouma, P., 2, 28, 30, 35, 52
Coupland, R. E., 24, 41, 52
Couvreur, E., 2, 20, 32, 33, 34, 52
Cowie, A. F., 34, 52, 60
Cowles, R. B., 447, 474
Cracraft, J., 129, 131, 133, 134, 138, 159, 164, 237, 253
Crawford, E. C., Jr., 302, 307, 319, 322, 327, 333, 335, 337, 342, 394, 397, 457, 469
Cremers, C. J., 308, 395
Crile, G., 438, 469
Crosby, E. C., 2, 7, 13, 15, 17, 19, 59
Csoknya, M., 30, 34, 35, 37, 52
Cuello, A. C., 41, 52
Cuello, A. E., 83, 114
Cunningham, A., 22, 54
Cuthbert, A. W., 22, 49
Cuttica, F., 441, 442, 468
Cuvier, G., 17, 20, 53

D

Dahl, E., 40, 53
D'Albora, H., 11, 55

Dale, H. E., 40, 67
Dale, H. H., 16, 53
Daniels, F., Jr., 381, *394*
Davids, J. A. G., 134, *253*
Davis, E. A., Jr., 377, *397*
Davis, L. B., Jr., 292, *397*
Davis, L. E., 40, 67
Davydov, A., 368, *403*
Dawson, D. C., 85, 111, *117*
Dawson, T. J., 261, 321, *397, 409*
Dawson, W. R., 263, 266, 270, 271, 272, 273, 274, 276, 278, 280, 281, 282, 283, 302, 305, 316, 323, 324, 327, 329, 338, 242, 345, 346, 347, 350, 354, 377, 380, 383, 386, 387, 388, *397, 398, 404, 409,* 431, 433, 436, 441, 453, 455, 457, 459, *469, 472*
Day, R., 295, 321, *398, 411*
de Anda, G., 11, *53*, 67
DeJong, A. A., 307, 308, *398*
Dejours, P., 429, 433, 434, 464, *469, 474*
de Kock, L. L., 14, *53*
de Lisi, L., 39, *53*
De Lorenzo, A. J., 26, 27, *53*
De Lucchi, G., 43, *53*
de Meyer, R., 14, 31, 32, *53*
Demment, M., 373, *395*
Dempster, W. T., 146, 194, *253*
Den Hartog, J. P., 138, *253*
de Pérez Bedés, G. D., 85, *117*
D'Erchia, F., 44, *53*
Deshazer, J. A., 386, *398*
Desole, C., 12, *47*
Devries, A. L., 283, *398*
DeWitt, L., 11, *58*
Diesselhorst, G., 464, *469*
Djojosugito, A. M., 15, *53*
Dmi'el, R., 261, 322, 326, *411*, 431, 455, *476*
Dodt, E., 81, 90, *115*
Dogiel, J., 44, *53*
Dolezel, S., 23, 32, *53*
Dollander, A., 24, *57*
Dolnik, V. R., 375, 376, 377, 380, 388, *398, 399, 427, 428, 429, 432, 458, 459, 460, 469*
Donham, R. S., 82, *115*
Dorka, V., 464, *469*
Dorminey, R. W., 386, *413*
Dorward, P. K., 12, 13, *53*

Dow, D. D., 383, *398*
Doyon, M., 34, 36, *53, 54*
Drachman, D. B., 162, *256*
Drebin, M., 317, 378, *397*
Drees, F., 428, *474*
Drennan, M. R., 31, *54*
Drent, R., 275, 276, 281, 286, 298, *398*
Drummond, G. I., 429, *469*
Drury, W. H., 460, 464, *473*
Drury, L. N., 301, 322, *410*
DuBois, E. F., 264, *398*
Dudley, J., 43, *54*
Duncker, H.-R., 421, *469, 470*
Durfee, W. K., 14, 15, *54*
Dyachkova, L. N., 26, 27, *57*

E

Eakin, R. M., 90, 92, *115, 117*
Ealey, E. H. M., 350, *394*
Eberth, C. J., 33, *54*
Ederstrom, H. E., 321, *398*
Edwards, G. D., 134, *254, 255*
Ehinger, B., 26, 45, *54*
Eldred, E., 11, *63*
El Halawani, M. E.-S., 379, 380, *398*
Eliassen, E., 14, *54*, 313, 321, *398*, 435, 444, 445, 446, 454, *470*
Elliot, J., 82, 85, *116*
Elliot, T. R., 36, 37, *54*
Ellis, L. C., 82, *114*
Emlen, J. M., 382, *398*
Emmert, A. F., 17, 20, *54*
Enemar, A., 23, 28, 30, 31, 32, 37, 38, *54*
Engels, W. L., 133, 135, 137, *253*
Enger, P. S., 316, 349, *411*
Eränkö, L., 22, *54*
Eränkö, O., 22, 41, *54*
Eriksson, K., 384, *398*
Erve, P., 429, 434, *477*
Essex, H. E., 40, *62*
Evans, H. E., 2, 7, *54*, 135, 136, *253*
Evans, P. R., 384, *398*
Everett, S. D., 23, 24, 30, 34, 35, 37, 38, 50, *54*
Ezerskas, L. J., 427, 428, 429, *469*

F

Fabricius, E., 356, *393*
Falck, B., 23, 28, 30, 31, 32, 37, 38, *43, 54*

Fänge, R., 39, 43, 54
Farner, D. S., 82, 92, 101, 103, 105, *116*, 269, 270, 274, 278, 279, 281, 305, 308, 311, 312, 317, 320, 323, 333, 341, 360, 381, 384, 387, 390, 393, *399*, *403*, *412*, 431, 443, 444, 459, *470*, *472*
Favre, M., 11, 68
Fedak, M., 261, 322, 326, *411*, 431, 455, *476*
Fedde, M. R., 2, 14, 33, *54*, 67, 326, *399*
Feigl, E., 14, 55
Feldberg, W. A., 356, 358, *399*
Feng, T. P., 9, 11, 55, *74*, 213, *253*
Ferrando, G., 40, 55
Fiedler, W., 128, 134, *253*
Finch, V. C., 282, *399*
Fisher, C. D., 338, 341, *397*
Fisher, E., 10, 55
Fisher, H. I., 123, 127, 129, 130, 134, 135, 137, *253*, *254*
Fitzgerald, T. C., 135, *253*
Flandrois, R., 434, *470*
Foà, C., 83, *115*
Folk, G. E., Jr., 345, *400*
Folkow, B., 14, 15, 23, 32, *53*, 55
Follet, B. K., 360, *410*
Fourie, S., 133, *254*
Fourman, J., 39, 44, *47*, 55
Fowler, P. W., 319, *412*
Fraenkel, G., 416, 422, *470*
Frankel, H., 386, *412*
Frazier, A., 372, 383, *399*
Freedman, S. L., 20, 21, 39, 40, 41, *55*, 70
Freeman, B. M., 357, 358, 367, 368, 369, *399*
French, N. R., 372, *399*
Fry, F. E. J., 283, *399*
Fu, S. K., 14, *55*, *71*
Fujie, E., 41, 55
Fujita, T., 43, *60*
Funk, E. R. R., 366, 367, *413*
Fürbringer, M., 123, *254*
Furuta, F., 355, 356, 370, *405*
Fusari, R., 41, 55
Fuxe, K., 14, 23, 32, 55

G

Gadow, H., 2, 6, 17, 20, 33, *55*, 123, *254*
Gagge, A. P., 264, 267, 354, *399*, *400*
Ganfini, C., 40, 41, 55

Gans, C., 176, 182, 183, 184, *254*
Gardasson, A., 382, *399*
Garg, K. N., 357, 368, 369, *394*
Gaston, S., 85, 86, 87, 88, *115*
Gates, D. M., 260, 263, 287, 288, 289, 290, 291, 292, 294, 298, 299, 301, 306, 308, 381, *394*, *399*, *407*
Gavrilov, V. M., 377, 380, *399*, 427, 428, 429, *469*
Geberg, A., 32, 44, 55
Gebhart, B., 287, 294, 377, *399*
Geiger, R., 282, 289, 302, *399*
Gelineo, S., 272, 377, 384, 385, *399*
George, J. C., 9, 10, *51*, 55, 57, 123, 133, 134, 135, 137, 188, 189, *254*, 373, *400*, 429, 461, *470*
Germino, N. I., 11, 55
Gessaman, J. A., 301, 322, 345, *400*
Giacomini, E., 41, 55
Gibbs, O. S., 38, 55
Gilbert, A. B., 15, 32, 40, 56
Gilbreath, J. C., 386, 387, *400*
Ginsborg, B. L., 9, 10, 11, 56, 192, *254*
Gold, A., 330, *396*
Goldsmith, R., 281, *400*
Goller, H., 17, 56
Gollnick, P. D., 438, *470*
Goloube, D. M., 20, 41, 56
Goodman, D. C., 129, 130, 134, 135, *253*, *254*, 372, *400*
Goormaghtigh, N., 20, 41, 56
Gordon, G. B., 428, *473*
Gordon, M. S., 283, *400*
Goslow, G. E., Jr., 138, 165, *254*
Gossrau, R., 32, *56*
Gould, R. P., 43, 58
Govyrin, V. A., 23, 31, *56*
Graham, J. D. P., 14, 34, *56*
Gray, J. C., 40, 56, 142, 197, *254*
Greenewalt, C. H., 291, 297, *400*, 419, 421, *470*
Greenwald, L., 313, *400*
Grewe, F. J., 2, 39, *56*
Grignon, G., 24, 57
Gringer, I., 9, 57, 189, *254*
Groebbels, F., 416, 421, 422, 442, 465, *470*
Gruenhagen, A., 45, 57
Grunden, L. R., 357, *400*
Guedenet, J. C., 24, 57

Gunther, P. G., 9, 61

H

Hachmeister, U., 43, 57
Hainsworth, F. R., 331, 332, 333, 343, 344, 346, 348, 349, 371, 372, 400, 408, 409, 413, 426, 427, 428, 470, 475
Håkanson, R., 23, 28, 30, 31, 32, 37, 38, 54
Håkansson, C. H., 39, 57
Hall, L., 269, 402
Hall, V. E., 369, 403
Halpern, 15, 57
Hama, K., 22, 26, 58, 71
Hamilton, T. H., 393, 400
Hamilton, W. J., III, 293, 307, 400
Hammel, H. T., 261, 290, 292, 321, 354, 369, 396, 400, 409
Hammer, W., 82, 115
Hammond, W. S., 17, 20, 28, 41, 57, 74
Hamori, J., 26, 27, 57
Hanzlik, P. J., 34, 35, 57
Hardy, J. D., 261, 264, 267, 290, 302, 303, 354, 396, 399, 400
Harms, R. H., 382, 386, 387, 407
Harrison, P. C., 85, 89, 115
Hart, J. S., 287, 311, 312, 313, 314, 316, 317, 323, 366, 367, 374, 375, 377, 378, 379, 380, 385, 395, 400, 401, 413, 417, 419, 420, 421, 422, 423, 424, 425, 426, 427, 428, 429, 433, 435, 436, 438, 441, 443, 444, 445, 446, 448, 450, 451, 452, 453, 454, 455, 458, 460, 465, 466, 468, 470, 471, 474, 475
Hartmann, F. A., 41, 60, 316, 401, 438, 439, 441, 471
Hartung, R., 297, 401
Hartwig, H-G., 111, 116
Harvey, E. B., 135, 254
Hassa, O., 40, 57
Hassan, T., 24, 34, 35, 36, 47, 57
Hatch, D. E., 330, 401
Hatier, R., 24, 57
Hawkins, N. M., 39, 50
Heath, J. E., 344, 345, 351, 352, 355, 359, 366, 401, 406, 453, 472
Hebb, C., 22, 60
Hedlund, L., 105, 115
Heinrich, B., 351, 353, 401, 447, 471
Hellman, B., 43, 54
Hemingway, A., 367, 401

Hemmingsen, A. M., 272, 401, 431, 471
Hennig, W., 236, 254
Hensel, H., 359, 401
Henshaw, R. E., 349, 401
Hensley, M. M., 371, 372, 401
Heppner, F., 291, 292, 293, 307, 400, 401
Herrath, E., 28, 30, 35, 52
Herreid, C. F. II., 274, 275, 277, 319, 401, 411, 433, 454, 471
Hertel, H., 419, 448, 449, 471
Hess, A., 9, 10, 26, 27, 44, 57, 58
Hesse, R., 301, 401, 437, 438, 471
Heusner, A., 261, 403
Hicks, D. L., 460, 474
Hiestand, W. A., 362, 408
Hikida, R. S., 127, 190, 192, 252, 254
Hill, A. V., 206, 213, 214, 217, 254
Hillerman, J. P., 386, 401
Himms-Hagen, J., 369, 401
Hinds, D. S., 321, 338, 342, 388, 401, 409
Hingston, R. W. G., 464, 471
Hirakow, R., 32, 58
Hirsch, E. F., 31, 58
His, W., Jr., 28, 58
Hissa, R., 317, 377, 380, 401, 433, 471
Hník, P., 58
Ho, W. Y., 13, 73
Hochachka, P. W., 283, 399
Hock, R., 262, 264, 267, 273, 278, 387, 388, 406, 409
Hodges, R. D., 43, 58
Hodos, W., 2, 59
Hofer, H., 129, 134, 135, 254
Hoff, H. E., 14, 62
Höhn, E. O., 58
Hollands, K. G., 386, 412
Hollenberg, N. K., 14, 15, 58
Holstein, A. F., 23, 40, 48
Holt, J. P., 437, 439, 471
Holtzmann, H., 44, 58
Homma, K., 42, 43, 62, 85, 115
Höhn, E. O., 40, 58
Hopwood, D., 24, 41, 52
Hoshi, T., 41, 58
Houdas, Y., 434, 470
Howard, H., 132, 254
Howell, D. J., 437, 475
Howell, T. R., 305, 329, 342, 344, 346, 347, 351, 353, 373, 394, 401, 441, 444, 446, 453, 467, 471

Hsieh, T. M., 2, 3, 17, 19, 20, 21, 28, 31, 33, 34, 35, 38, 39, 40, 41, 42, 44, 58
Huang, S. K., 9, 74
Hubbard, S. H., 274, *404*, 433, 436, *472*
Huber, G. C., 2, 7, 11, 13, 15, 17, 19, 27, 58, 59
Huber, J. F., 7, 19, 20, 58
Hudson, G. E., 130, 134, 137, 161, *254*, *255*
Hudson, J. W., 263, 278, 281, 305, 316, 323, 327, 329, 342, 345, 346, 347, 350, 354, 387, 388, 389, *394*, *398*, *402*, *406*, 441, 455, *467*, *469*
Huggins, A., 298, *402*
Huik, P., 11, 58
Hulbert, A. J., 261, 397
Hull, W. E., 335, *402*
Hunt, J., 464, *471*
Hussell, D. J. T., 459, 460, *471*
Huston, T. M., 385, *402*
Hutchinson, J. C. D., 386, *402*
Hutt, F. B., 269, *402*
Huxley, J. S., 345, *402*

I

Ianuzzo, C. D., 438, *470*
Ignarro, L. J., 23, 58
Ihnen, K., 34, 58
Imaizumi, M., 22, 58
Irving, L., 262, 264, 267, 273, 278, 281, 298, 321, 346, 373, 383, 384, 387, 388, 390, *402*, *409*
Isomura, G., 2, 44, 73
Iwanow, J. F., 30, 34, 35, 36, 37, 58, 60

J

Jackson, D-C., 321, *409*
Jaeger, E. C., 344, 345, 347, 353, *402*
Jain, P. D., 31, 59
James, F. C., 390, 391, *402*
Jansky, L., 317, 328, *402*, 433, 434, *470*, *471*
Jaquet, M., 2, 59
Jegorow, J., 33, 43, 44, 45, 59
Jenden, D., 26, 27, 67
Jenkin, C. R., 123, 132, *251*
Jessen, C., 321, *412*
Jirmanovà, I., 11, 58, 75
Johansen, K., 14, 32, 46, 59, 68, 321, 365, *402*

Johnson, A. T., 360, 361, *410*
Johnson, E. A., 22, 69
Johnson, F., 262, 264, 267, 273, *409*
Johnson, H. McC., 373, *402*
Johnson, J. S., 39, 59
Johnson, R. E., 384, *402*
Johnston, D. W., 318, 369, *402*, 462, *471*
Johnston, P. H., 41, 67
Johnston, R. F., 391, *402*
Jollie, M. T., 183, *255*
Jones, D. R., 14, 15, *51*, 59, 321, 365, *402*
Jones, R. E., 298, *403*
Jordan, K. A., 386, *398*
Jowett, P., 22, *64*
Jungherr, E. L., 2, 7, 59
Jürgens, H., 32, 59

K

Kadono, H., 15, 59
Kagawa, K., 38, 59
Kahl, M. P., Jr., 321, 330, 331, *403*
Kaiser, H. E., 135, *254*
Kaiser, L., 7, 59
Kamar, G. A. R., 386, *403*
Kammer, A. E., 447, *471*
Kampe, G., 335, 337, 397
Kanematsu, S., 355, 356, *403*
Kano, M., 24, 41, 59
Kanwisher, J., 327, 329, 338, 339, *409*
Kaplan, N. O., 191, *255*
Kappers, C. U. A., 2, 7, 13, 15, 17, 19, 41, 59
Karten, H. J., 2, 59
Kasa, P., 22, *60*
Kashkin, V., 316, *403*
Kato, M., 85, *116*
Kato, Y., 355, 356, *403*
Kattus, A. H., 369, *403*
Kaupp, B. F., 2, *60*, 135, *255*
Kavanau, J. L., 371, *403*
Kayser, C., 261, *403*
Kazumoto, F., 36, *60*
Kear, J., 227, *255*
Keatinge, W. R., 381, 397
Keatts, H., 85, *115*
Keegan, J. J., 292, *399*
Kelso, L., 383, *411*
Kellog, R. H., 429, 464, *469*
Kelly, D. E., 80, 99, *115*, *118*

Kendeigh, S. C., 272, 285, 286, 298, 373, 393, *403*, 431, 443, 459, *471*
Kennedy, R. J., 372, *403*
Keskpaik, J., 368, 377, 380, 399, *403*, 428, 455, 457, 458, 460, *472*
Kessel, B., 274, 275, *401*, 454, *471*
Khalifa, M. A. S., 386, *403*
Kibyakov, A. V., 36, *49*
Kii, M., 355, 356, *403*
Kilham, L., 372, 383, *403*
Kimzey, S. L., 388, 389, *402*
King, A. S., 14, 33, 34, *52*, *60*
King, B., 372, *403*
King, J. R., 263, 269, 270, 274, 276, 278, 279, 281, 286, 305, 308, 311, 317, 320, 323, 333, 341, 372, 381, 383, 384, 387, 390, 393, *396*, *397*, *403*, 431, 443, *472*
Kines, H., 437, 439, *471*
Kirschstein, H., 41, *66*, 82, 85, 92, 99, 101, 103, 105, 107, *116*
Kitchell, R. L., 2, 14, 15, 33, *54*, *60*, 359, *403*
Kleiber, M., 263, 270, 272, 275, 287, 294, 313, 315, 380, *403*, *404*, 433, 434, 448, 449, 453, *472*
Klemm, R. D., 130, 134, 255
Klugger, M. J., 453, *472*
Kluth, E., 86, 88, 114, 346, *495*
Kluyver, H. N., 383, *404*
Knorr, O. A., 372, 383, *404*
Knouff, R. A., 41, *60*
Ko, R.-C., 386, 387, *400*
Kobayashi, H., 82, 92, 101, 103, 105, *115*, *116*, *117*
Kobayashi, S., 14, 43, *60*
Koch, J. C., 144, 255
Kočkova, J., 317, 328, *402*
Koering, H.-L., 26, *60*
Kokshaiski, N. V., 432, *472*
Kolossow, N. G., 15, 27, 30, 34, 35, 36, 37, *60*
Komarek, V., 40, *60*
Kondratjew, N. S., 31, *60*, *61*
Koppanyi, T., 45, *61*
Kose, W., 14, *61*
Koskimies, J., 347, 388, 389, *404*
Kostinowitsch, L. I., 20, *61*
Kovach, A. G. B., 15, *53*
Krabbe, K. H., 90, *115*
Kracht, J., 43, *57*

Krammer, E., 10, 44, 45, 75
Krog, J., 14, 15, 39, 43, *52*, *54*, 68, 281, 298, 321, 346, *402*
Krogis, A., 13, *61*
Krohn, A., 44, *61*
Krüger, P., 9, *61*, 127, 188, 190, 255
Kruse, H., 43, *57*
Kuhne, W., 10, *61*
Külbs, F., 31, *61*
Kummer, B., 144, 151, 155, 156, 163, 252, 255
Kuntz, A., 13, 28, 29, 37, 41, *61*
Kura, N., 41, *61*

L

Lacaisse, A., 433, 434, *474*
Lack, D., 226, 255, 464, *472*
Lackey, R. W., 355, *409*
Laffont, M., 2, 17, 32, *61*
Lahti, L., 388, 389, *404*
Lake, P. E., 40, *56*, *61*
Lakjer, T., 134, 137, 255
Lane, K. B., 82, *117*
Landmesser, L., 27, 44, *61*
Langendorff, O., 44, *61*, *62*
Langley, J. N., 15, 16, 20, 27, 28, 33, 43, 45, *62*
Lanzillotti, P. J., 134, *254*, 255
Laruelle, L., 32, 33, *62*
Lasiewski, R. C., 260, 265, 270, 271, 272, 273, 274, 275, 276, 278, 293, 302, 305, 307, 324, 326, 327, 329, 330, 331, 333, 337, 338, 339, 342, 343, 345, 346, 347, 350, 351, 352, 353, 364, 372, *394*, *397*, *404*, *405*, *407*, *409*, 425, 426, 427, 428, 431, 433, 436, 441, 446, 448, 453, 454, 455, 457, 459, *469*, *472*
Lasiewski, R. J., 343, 346, 350, 351, 352, 353, *404*, 433, *472*
Lauber, J. K., 42, 43, *62*, 85, 86, 109, *115*
Laulajainen, E., 433, *475*
Lebedinsky, N. G., 134, 255
Lee, D. H. K., 382, *408*
Lee, P., 327, 329, 330, *405*
Lee, R. C., 350, *395*
Lee, S. Y., 190, 191, 255
Le Febvre, E. A., 308, 312, 314, 390, *395*, *405*, *411*, 424, 427, 428, 429, 441, 448, 451, 452, 453, 459, 460, 463, *472*, *474*, *476*

Lefrançois, R., 434, *470*
Legait, E., 32, 33, 42, *62*
Legait, H., 42, *62*
Leiber, A., 129, 133, 135, *255*
Leighton, A. T., Jr., 275, *405*
Leitner, P., 437, 441, *467, 472*
Lenke, M., 43, *57*
Leontieva, G. R., 23, 31, 32, 56, *62*
Lepkovsky, S., 2, *74*, 355, 356, 370, *405*
Levi-Montalcini, R., 19, 20, *62*
Levins, R., 265, *405*
Levinson, I. J., 138, *255*
Lierse, W., 109, *115*
Lifson, N., 428, *473*
Ligon, J. D., 346, 372, *405*, 457, *473*
Lilienthal, O., 416, *473*
Lin, Y. C., 23, *62*, 379, *405*
Lindsley, J. G., 339, *405*
Lissak, K., 41, *48*
List, R. J., 305, *405*
Liu, H. C., 38, *62*
Löhrl, H., 372, 383, *405*
Long, E. C., 335, *402*
Lord, R. D., 417, *473*
Lowry, W. P., 288, 289, 302, *405*
Lucas, A. M., 2, *62*
Lukashin, V. G., 15, *62*
Lull, H. W., 288, 289, *408*
Lustick, S., 291, 292, 293, 307, 372, *405*
Lwin, S., 43, *50*
Lyuleeva, D., 457, *472*
Lyuleeva, D. S., 459, 460, *473*

M

McAtee, W. L., 446, *473*
McClintock, R., 428, *473*
McCrady, J. D., 14, *62*
MacDonald, E., 40, *62*
MacDonald, R. L., 2, 17, 19, 31, 32, 52, *62*
McEvey, A., 128, 132, 133, 234, *252*
McFarland, D. J., 265, 319, 320, *405*
McFarland, L. Z., 42, 43, *62*, 360, 361, 362, *413*
McFarlane, R. W., 462, *471*
Mackay, B., 9, 10, 11, *56*
Mackenzie, G. M., 42, *74*
Mackenzie, I., 14, *62*
McKinnon, R. S., 422, *476*
MacLean, G. L., 330, *396*
McLelland, J., 14, 33, 60, *62*

McMahon, T., 272, *405*
McMillan, J., 86, *115*
MacMillen, R. E., 261, 281, 338, *405*
McNab, B. K., 286, 393, *405*, 433, *473*
McNeil, R., 463, *473*
Maggi, G., 20, *75*
Magnien, L., 2, 17, *62, 63*
Maier, A., 11, *63*
Makita, T., 42, *63*
Malcus, B., 39, *57*
Malinovský, L., 2, 13, 20, 30, 33, 35, *63, 67*
Malmfors, T., 23, 24, 25, 27, 28, 29, 30, 32, 33, 35, 38, 40, 41, 42, 43, 45, 46, *48, 49*
Mandell, A. J., 357, *410*
Manger Cats-Kuenen, C. S. W., 129, 151, 161, *255*
Mangili, F., 434, 441, 442, *468*
Mangold, E., 15, 30, 35, 36, *63*
Mann, F. C., 40, *62*
Mann, S. P., 22, 23, 30, 31, 32, 37, 38, 47, *54, 60*
Manni, E., 12, *63*
Manville, R. H., 464, *473*
Marage, R., 2, 17, 20, 27, 28, 32, 33, *63*
Marder, J., 292, *405*
Marey, E. J., 416, 422, *473*
Marley, E., 356, 357, 358, 368, 369, 370, 394, 400, *406*
Marshall, J. T., 446, *473*
Marshall, M., 2, *63*
Marshall, S. G., 461, *468*
Martens, J., 464, 465, *473*
Martin, A. H., 7, 26, *50, 63*
Martin, E. W., 381, *406*
Marwan, F., 40, *60, 63*
Marwitt, R., 26, 27, 44, *63*
Matsushita, M., 19, 20, *64*
Mauger, H. M., Jr., 39, *64*
May, B., 364, *408*
Mayhew, W. W., 317, 378, *397*
Mayr, E., 240, 241, 242, 243, 255, 390, 393, *406*
Mayr, R., 44, *64*
Medway, Lord., 371, *406*
Mehrotra, P. N., 40, *64*
Meinertzhagen, R., 464, *473*
Mejsnar, J., 317, 328, *402*
Melkicn, A., 44, *64*
Meller, K., 97, *115*

Menaker, M., 82, 85, 86, 87, 88, *114*, *115*, *116*, 346, *395*
Meng, M. S., 384, *413*
Mikami, S.-I., 92, *116*, 360, *412*
Miller, A. H., 129, 255, 446, *473*
Miller, D. S., 377, *406*
Miller, W. DeW., 237, 241, 243, 247, 252
Mills, S. H., 355, 359, 366, *406*
Milokhin, A. A., 15, 35, *64*
Minaire, Y., 434, *469*
Minard, D., 323, *406*
Misch, M. S., 273, *406*
Mitchell, D., 290, *406*
Mitchell, G. A. G., 16, 20, *64*
Mivart, St. G., 133, 255, 256
Miyagawa, K., 32, 33, *47*
Moberly, W. R., 274, *404*, 433, 436, *472*
Moen, A. N., 322, *406*, *411*
Moller, W., 135, 256
Mollier, G., 135, 159, 256
Molony, V., 14, *60*
Moore, A. D., 319, *406*
Moreau, R. E., 464, *473*
Morioka, H., 126, 130, 133, 138, 157, 252
Morita, Y., 41, 42, 66, 85, 92, 99, 101, 103, *116*
Morony, J., 127, 134, 160, 252
Morowitz, H. J., 282, *406*
Morozov, E. K., 14, *64*
Morrison, P. R., 272, 278, *406*, 446, *473*
Mortimer, M. F., 14, *60*
Morton, M. L., 293, 372, **406**
Mosher, R., 429, 434, **477**
Moss, R., 382, *399*
Moyanahan, E. J., 22, *64*
Mudge, G. P., 129, 135, 256
Mugaas, J. N., 380, **406**, 433, 457, **473**
Müller, J., 41, *64*
Muratori, G., 14, 33, *64*, *65*
Murray, P. D. F., 162, 224, 256
Murrish, D. E., 273, 331, 332, 333, 366, *406*, *409*, 428, *475*

N

Nafstad, P. H. J., 13, *47*, *65*
Nafstad, P. H. J., 13, *65*
Nakayama, S., 34, *65*
Nakazato, Y., 24, 25, 34, 35, 36, *65*, 69
Nalbandov, A. V., 40, 55, 105, *115*
Necker, R., 358, 359, 364, *406*, *408*

Nelson, J. E., 261, *405*, 437, 441, *467*, *472*
Neumann, R., 278, *406*
Newton, I., 381, 384, *406*, *407*
Newton, J. L., 438, *467*
Nielsen, M., 441, *467*
Niestand, W. A., 14, *58*
Nilsson, N. J., 14, 55
Nisbet, I. C. T., 459, 460, 464, *473*
Nishida, S., 41, 42, 45, *63*, *65*, 69
Nolan, V., Jr., 372, 383, *399*
Nolf, P., 2, 28, 30, 31, 35, 36, 37, 38, 65
Nonidez, J. F., 14, 43, 65
Norris, K. S., 292, 293, *407*
Nott, M. W., 368, *395*
Nye, P., 297, *407*

O

Odake, G., 23, 68
Odum, E. P., 315, 316, **368**, **407**, 436, 460, 461, 462, 463, *468*, **473**, *474*
Oehme, H., 26, 44, 45, 65, 419, 434, 459, *474*
O'Farrell, M. J., 433, 447, *468*, *474*
Ogilvie, D. M., 369, *407*
Ohashi, H., 23, 34, 35, 65
Ohga, A., 23, 24, 25, 34, 35, 36, 65, 69
Ohmart, R. D., 293, 307, 327, 329, 345, 352, 372, *405*, *407*, 455, *472*
Ohno, K., 15, 59
Oishi, T., 85, *116*
Okada, T., 15, 59
Okamura, C., 30, 33, 35, 66
Oksche, A., 41, 42, 66, 81, 82, 85, 89, 90, 92, 95, 99, 101, 103, 105, 107, 111, 115, *116*, 360, *412*
Olech, B., 383, *396*
Oliver, R., 319, *412*
Oribe, T., 39, 40, 66
Øritsland, N. A., 292, *407*
Orr, Y., 298, 371, *407*
Ostmann, O. W., 13, 43, 66
Ossorio, N., 23, *70*
Otsuka, N., 23, 31, 66
Owman, C., 23, 54, 113, *116*

P

Page, S. G., 9, 66
Palme, F., 14, 66
Palmieri, G., 4, 11, 12, 13, 38, *47*, *50*, 66, 73

Palokangas, R., 317, 377, 380, *401*, 433, *471*
Pasquis, P., 433, 434, *474*
Pastăa, E., 20, 42, *66*
Pastăa, Z., 20, 42, *66*
Paton, D. N., 31, 32, *66*
Paul, E., 111, *116*
Pauwels, F., 144, 151, 157, 163, 223, *256*
Pearce, J. W., 39, *47*, *52*
Pearson, O. P., 261, 344, 346, 350, 351, 373, *407*, *426*, 444, 460, *474*
Peck, J. I., 7, *66*
Peiponen, V. A., 346, *407*
Pellegrino de Iraldi, A., 13, *66*, *68*
Pennycuick, C. J., 430, 431, 432, 433, 448, 449, 459, 461, 462, 463, 466, *474*
Pera, L., 19, 20, 28, 29, 36, *66*, *67*
Persons, J. N., 382, 386, 387, *407*
Peterson, D. F., 14, *54*, *67*
Peterson, R. A., 13, 43, *72*
Petry, G., 43, *73*
Peyton, L. C., 384, *402*
Pfuhl, W., 183, *256*
Piiper, J., 428, 434, 441, 442, *468*, *474*
Pilar, G., 10, 26, 27, 44, *58*, *61*, *63*, *67*
Pintea, V., 37, *67*
Pisskunoff, N. N., 31, *67*
Pitelka, F. A., 275, 381, *407*
Pitts, L. H., 2, 17, 31, 32, *51*, *52*
Poczopko, P., 369, *407*
Pohl, H., 270, 271, 272, 281, 313, 321, 346, 377, *394*, *407*, 431, *467*
Polacek, P., 13, *67*
Poorvin, D., 23, *70*
Popa, F., 20, *67*
Popa, G., 20, *67*
Popper, K. R., 234, *256*
Porte, A., 43, *70*
Porter, K. R., 9, *48*
Porter, W. P., 260, 273, 287, 290, 291, 292, 294, 301, 308, 381, *407*
Portman, A., 416, *474*
Poustilnik, E., 20, *67*
Power, D. M., 391, *407*
Prange, H. D., 261, 330, *408*
Prasad, A., 40, *67*
Prechtl, J. J., 416, *474*
Proctor, R. C., 270, *396*
Puce, N., 429, 464, *469*
Puccinelli, R., 434, *470*
Puppi, A., 41, *48*
Purves, M. J., 14, 15, *59*

Q

Quay, W. B., 41, 67, 89, 92, 101, 103, *117*
Quilliam, T. A., 13, *67*
Quiring, D. P., 438, *469*

R

Rabl, H., 41, *67*
Radostina, T. N., 30, 35, 37, *58*
Radu, C. M., 37, *67*
Raikow, R. J., 130, 134, *256*
Ralph, C. L., 82, 83, 84, 85, 111, *117*
Ramón y Cajal, S., 7, 20, *67*
Ramos, J., 371, *403*
Randall, W. C., 14, *58*, 362, 367, *408*
Raper, C., 32, *50*
Rapp, G. M., 264, 267, *399*
Raspet, A., 430, *474*
Rau, S. A., 41, *67*
Rautenberg, W., 364, 367, 378, *408*
Raveling, D. G., 390, *405*, 459, 460, 463, *474*
Ray, P. J., 14, *67*
Rebollo, M. A., 11, *53*, *67*
Rechardt, L., 22, *54*
Reeder, W. G., 447, *474*
Regaud, C., 11, *68*
Reifsnyder, W. E., 288, 289, *408*
Reite, O. B., 14, 32, 39, 43, *54*, *59*, *68*
Remak, R., 28, *68*
Rensch, B., 438, *474*
Renzoni, A., 90, 92, 101, 103, *117*
Reshetnikov, A. B., 35, *64*
Reumont, M., 32, 33, *62*
Rhode, E. A., 437, 439, *471*
Ricci, B., 434, 441, 442, *468*
Richards, L. P., 133, 181, 227, *256*
Richards, S. A., 14, 15, *68*
Richards, S. E., 339, 356, 360, 361, 362, 363, 365, 366, *408*
Ricklefs, R. E., 371, 372, *408*
Riddle, O., 380, *408*
Rigdon, R. H., 45, *51*
Rijke, A. M., 297, *408*
Ringer, R. K., 13, 43, *66*, *72*
Rising, J. D., 388, *408*

Roberts, J. C., 262, 374, *397*
Roberts, R., 82, 85, *116*
Robertson, J. I., 14, *62*
Robinson, K. W., 382, *408*
Robinson, M., 39, *54*
Robinson, S., 438, *467*
Rochas, F., 2, 17, 44, *68*
Rodbard, S., 261, *409*
Rodriguez-Perez, A. P., 13, 66, *68*
Rogers, D. T., 460, *474*
Rogers, F. T., 355, *409*
Romanoff, A. L., 316, *409*
Romijn, C., 311, *400*
Rosenberg, L. E., 135, *254*
Rosenzweig, M. L., 393, *409*
Rosner, J. M., 83, 85, *114*, *117*
Ross, G., 369, *403*
Ross, I. J., 386, *413*
Rossi, O., 22, *68*
Rostorfer, H. H., 438, *467*
Roy, O. Z., 312, 313, 314, *395*, *401*, 417, 419, 420, 421, 422, 423, 424, 425, 426, 427, 428, 435, 436, 441, 443, 444, 445, 446, 451, 452, 453, 454, 458, 466, *468*, *470*, *471*, *474*
Rouget, C., 9, *68*
Rüdeberg, C., 90, 92, 113, *116*
Rüppell, G., 419, *474*
Rutschke, E., 277, *409*

S

Sabussow, G. H., 27, 30, 34, 36, 37, *60*
Saglam, M., 11, *68*
Sakakibara, Y., 85, *115*
Salt, G. W., 416, 455, 457, *474*, *475*
Sano, Y., 23, *68*
Sato, H., 24, 25, 34, 35, 36, *68*, *69*
Sato, M., 24, 36, *65*
Saxena, B. B., 388, *409*
Saxod, R., 13, *69*
Sayler, A., 83, *117*
Schadé, J. P., 90, *114*
Scharnke, H., 428, *475*
Schatau, O., 13, 43, *69*
Scheid, P., 428, *474*
Schildmacher, H., 128, 138, *256*
Schmekel, L., 28, *74*
Schmidt, K. P., 301, *401*
Schmidt-Nielsen, K., 39, *54*, *69*, 261, 270, 276, 281, 282, 283, 302, 305, 319, 320, 321, 322, 323, 327, 329, 330, 331, 332, 333, 334, 335, 337, 338, 339, 340, 342, 371, 373, *395*, *397*, *398*, *405*, *408*, *409*, 425, 428, 431, 433, 455, 457, *468*, *469*, *471*, *475*, *476*
Schnall, A. M., 2, 14, 17, 31, 32, *52*
Schoener, T. W., 272, *409*
Scholander, P. F., 262, 264, 267, 273, 278, 384, 387, 388, 390, 393, *409*
Schoones, J., 133, *256*
Schrodt, G. R., 9, *73*
Schumacher, G. H., 183, *256*
Schüz, E., 384, *409*
Schwalbe, G., 44, *69*
Scott, N. R., 360, 361, *410*
Sears, M., 45, *65*
Segrem, N. P., 433, 438, 441, 442, 465, 466, *475*
Seibert, H. C., 384, *410*
Selander, R. K., 391, *402*
Selby, S. M., 305, *412*
Selenka, E., 123, *254*
Sellers, W. D., 282, 289, *410*
Serventy, D. L., 345, *410*, 444, *470*
Sethi, N., 22, *64*
Seto, H., 26, 44, *69*
Seymour, R. S., 268, 328, 329, *404*, *410*
Shannon, S., 15, *52*
Sharp, P. J., 360, *410*
Shear, C. R., 133, 134, 135, 137, 138, 160, 183, *252*
Shellabarger, C. J., 83, *117*
Shideman, F. E., 23, *58*
Shilov, I. A., 263, *410*
Shimada, M., 38, *69*
Shioda, T., 41, 42, *63*, *69*
Shvalev, V. N., 38, *69*
Siegel, H. S., 275, 301, 322, *405*, *410*
Siegel, P. B., 275, *405*
Silver, A., 10, 39, *47*, *69*
Simon, E., 321, *412*
Simonetta, B., 2, 3, *69*
Sinha, M. P., 14, *69*, 355, 365, *410*
Siple, P. A., 275, *411*
Sisson, S., 135, *256*
Sivaram, S., 41, *69*
Sjöstrand, N. O., 23, 41, *69*
Sklenska, A., 13, 67, *69*
Sladen, W. J. L., 281, *395*, *400*
Slater, C. R., 9, *66*

Slonaker, J. R., 44, 69
Slonim, A. D., 379, *410*
Smith, G. C., 380, *408*
Smith, J. L., 45, *51*
Smith, M. L., 2, 69
Smith, S. W., 99, *115*
Smith, R. M., 327, 329, 333, 335, *410*, 450, 455, *475*
Smyth, M., 373, *410*
Snapir, N., 311, 312, 355, 356, 370, *395*, *405*
Solomonsen, F., 457, 459, *474*
Soloviera, I. A., 15, 69
Sommer, J. R., 22, 69
Sonada, T., 355, 356, *403*
Sonnenschein, R. R., 14, 23, 32, 55
Sopyev, O., 371, *410*
Sotavalta, O., 432, 433, *475*
Southwick, E. A., 280, 384, *410*
Spector, H., 429, 434, *477*
Spector, W. S., 438, *475*
Spellerberg, I. F., 298, *410*
Spooner, C. R., 357, *410*
Spring, L. W., 130, 131, 138, 181, 222, *256*
Ssinelhikow, R., 31, 69
Staderini, R., 2, *70*
Stahl, W. R., 265, 272, 278, 286, *410*, 425, 437, *475*
Stalsberg, H., 83, *117*
Stammer, A., 11, 13, 14, 31, 41, 43, 44, *47*, *70*
Starck, D., 129, 134, 135, *256*
Staudacher, E. V., 19, *70*
Steen, I. B., 302, 321, 362, *411*
Steen, J., 316, 345, 346, 349, *410*, *411*
Stein, G., Jr., 366, *413*
Steinbacher, G., 247, *256*
Stephenson, J. D., 356, 357, 358, 370, *406*
Stettenheim, P. R., 2, *62*
Stevens, D., 322, *411*
Stiemens, M. J., 2, 20, 36, *70*
Stoddard, H. L., 461, 462, 463, *474*
Stoeckel, M. E., 43, *70*
Stolpe, M., 133, 159, 164, 257, 416, *475*
Stolwijk, J. A. J., 354, *400*
Stonehouse, B., 275, 276, 281, 286, *398*
Stone, W. B., 313, *400*
Stones, R. C., 433, *475*
Storer, R. W., 133, 134, 257, 264, *411*
Streeter, G. L., 7, *70*

Ström, L., 15, *60*, 359, *403*
Strunk, T. H., 263, *411*
Stübel, H., 32, 36, *70*
Studier, E. H., 433, 437, *474*, *475*
Studnička, F. K., 89, 103, *117*
Sturkie, P. D., 23, 31, 32, 39, 40, 45, *55*, 62, *70*, *72*, 366, 379, *405*, *413*, 442, *475*
Suggs, G. W., 386, *398*
Sulkava, S., 373, 383, *411*
Sun, K. H., 45, *61*
Suthers, B. J., 421, *476*
Suthers, R., 329, 455, *410*, *475*
Suthers, R. A., 421, 434, 437, 441, *476*
Svensson, L., 356, *393*
Swan, J., 20, 28, *70*
Swan, L. W., 465, *476*
Sykes, A. H., 15, 40, 68, *70*, 386, *402*
Sy, M., 133, 159, 164, *257*
Szantroch, Z., 28, 31, *70*
Szepsenwol, J., 26, 31, *70*, *71*
Szentágothai, J., *70*

T

Takahashi, K., 26, *71*
Takewaki, T., 38, *59*
Takino, M., 14, 32, 33, *71*
Taxi, J., 26, *71*
Taylor, C. R., 261, 322, 326, 333, *409*, *411*, 431, 455, *476*
Taylor, L. W., 40, *62*
Tcheng, K.-T., 14, 55, *71*
Teager, C. W., 372, *411*
Teal, J. M., 312, 426, 429, *411*, *476*
Tello, J. F., 11, 12, 20, 28, 37, *71*
Templeton, J. R., 380, *406*, 433, 457, *473*
Terni, T., 2, 19, 20, 27, 42, 43, *71*, *72*
Terzuolo, C., 26, 44, *72*
Tetzlaff, M. J., 13, 43, 66, *72*
Thébault, V., 2, 17, 20, 28, 33, 34, 36, *72*
Thomas, S. P., 421, 434, 437, 441, *476*
Thompson, A. L., 315, *411*
Thompson, H. J., 345, 347, *404*
Thornthwaite, C. W., 302, 303, 304, *411*
Tiedemann, F., 17, 20, *72*
Tiegs, O. W., 10, *72*
Tigyi, A., 41, *48*
Tilney, F., 89, *117*
Tixier-Durivault, A., 2, 17, 31, *72*
Tobin, C. A., 330, *396*
Tomisawa, M., 23, 31, 66

Tomlinson, J. T., 417, 422, *476*
Tordoff, H. B., 273, 383, *398*
Traciuc, E., 40, *72*
Tracy, C. R., 263, *411*
Tramezzani, J. H., 83, *114*
Trewartha, G. T., 282, *399*
Trost, C. H., 281, 338, 345, *394*, *405*, 441, 446, 457, *467*
Tucker, V. A., 261, 312, 313, 314, 315, 321, 322, 326, 350, *411*, 417, 422, 423, 424, 425, 426, 427, 428, 429, 430, 441, 448, 449, 450, 451, 453, 454, 455, 458, 461, 463, 464, 465, *476*
Tummons, J. L., 23, 31, 32, 45, *62*, *72*
Turček, F. J., 269, 383, *411*

U

Uchida, S., 20, *72*
Ueck, A., 90, 92, *116*
Ueck, M., 41, *72*, 81, 90, 96, 97, 99, 101, 103, 105, 107, 109, 113, *115*, *116*, *117*
Underwood, H., 82, 85, *116*
Urra, M., 44, *72*
Utter, J. M., 312, *411*, 427, 428, 429, *476*
Uvnäs, B., 14, 15, *58*

V

Vallbona, C., 14, *62*
van Campenhout, E., 14, 20, 28, 33, 37, 40, *72*, *73*
van den Akker, L. M., 7, *73*
vanden Berge, J. C., 130, *257*
van der Linden, P., 15, *73*
Vander Wall, S. B., 227, *252*
van Dilla, M., 295, *411*
van Gehuchten, A., 7, 20, *73*
van Kampen, M., 329, 365, 366, *411*
van Tienhoven, A., 360, 361, *410*
Vaughan, P. C., 10, *67*
Vaupel-von Harnack, M., 41, 42, *66*, 89, 92, 95, 99, 101, 103, *116*
Vegetti, A., 4, 12, 13, *50*, *73*
Veghte, J. H., 273, 277, 281, 319, 377, 380, 381, *411*
Vignal, W., 31, *73*
Vitums, A., 360, *412*
Volkmann, 27, *49*
Volkov, N. I., 373, 383, *412*
von Golenhofen, K., 24, 43, *73*
von Harnack, M., 99, *116*

von Helmholtz, H., 431, *476*
von Holst, E., 421, 432, *476*
von Lenhossék, M., 7, 20, 25, 44, *73*
von Lucanus, F., 416, *476*
von Miller, R., 436, *476*
von Saalfeld, E., 355, 365, *412*
von Saint Paul, U., 360, 365, 366, *412*
von Wahlert, G., 139, 140, 145, 162, 212, 224, 226, 240, 241, 242, *252*, *257*
Vreugdenhil, E. L., 311, *409*
Vyklický, L., 11, 58, *75*

W

Wald, G., 289, *412*
Wales, E. E., Jr., 372, 383, *403*
Walker, S. M., 9, *73*
Wallgren, H., 381, 384, 385, 388, *412*
Walters, V., 262, 264, 267, 273, 278, 387, 388, *409*
Walther, O. E., 321, *412*
Wang, C. C., 13, *73*
Warham, J., 281, *412*
Warren, J. W., 345, 352, *412*
Warren, L. F., 89, *117*
Watanabe, T., 2, 3, 6, 17, 33, 34, 44, *73*
Watmough, D. J., 319, *412*
Watzka, M., 14, 43, *73*, *74*
Weakly, J. N., 26, 27, 44, *58*, *63*
Weast, R. C., 305, *412*
Weathers, W. W., 274, 275, 335, 350, 351, *404*, *412*, 433, 436, 454, *472*
Webb, C. S., 345, *402*
Webb, M., 2, 44, *74*
Weber, A., 20, *74*
Weber, E. H., 17, 20, 28, *74*
Wechsler, W., 28, *74*
Weidner, V. R., 292, *399*
Weis-Fogh, T., 434, *476*, *477*
Weiss, H. S., 385, 386, *412*
Wekstein, D. R., 358, 367, 368, 369, *412*
Weller, M. W., 371, *412*
Welty, J. C., 373, 383, *412*
West, G. C., 316, 318, 366, 367, 375, 377, 378, 381, 382, 384, 386, *402*, *412*, *413*, 455, *477*
Wetmore, A., 269, *413*, 443, 444, *476*
Wheat, S. D., 355, *409*
Whittow, G. C., 311, 354, 366, 386, *413*
Wick, H., 422, 435, *477*
Wiebers, J. E., 433, *475*

Wight, P. A. L., 42, *74*
Wilcox, C. J., 386, *413*
Wilkie, D. R., 431, 432, *477*
Williams, D., 368, *394*
Wilson, E. O., 234, *257*
Wilson, F. E., 82, *115*
Wilson, H. R., 382, 386, 387, *407, 413*
Wilson, W. O., 42, 43, 62, 360, 361, 362, 379, 380, 386, *398, 401, 413*
Winchester, C. F., 378, *413*
Winkel, K., 388, *413*
Winters, W. D., 357, *410*
Wirtz, B., 14, *74*
Wohlschlag, D. E., 283, *398*
Wolf, L. L., 343, 344, 346, 348, 349, 426, 427, *400, 413, 470*
Wolfe, D. E., 103, *118*
Wolfson, A., 83, *117*
Wu, W. Y., 11, *55*
Wurtman, R. J., 80, *118*

Y

Yagasaki, O., 38, *59*
Yamasaki, I., 32, 44, *74*
Yanagiya, I., 38, *59*
Yang, F. Y., 11, *55*
Yapp, W. B., 457, *477*

Yarbrough, C. G., 286, *413*
Yasuda, M., 2, 3, 7, 8, 44, *73, 74*
Yeh, Y., 9, *74*
Yntema, C. L., 17, 20, 28, 41, *57, 74*
Yonce, L. R., 14, 15, 32, *53, 55*
Yonezawa, T., 23, *68*
Young, D. R., 429, 434, *477*
Yousef, M. K., 360, 361, 362, *413*
Yousuf, N., 31, *74*

Z

Zar, J. H., 261, 270, 387, *413*
Zeglinski, N., 44, 45, *75*
Zeisberger, E., 317, 328, *402*
Zeitzschmann, O., 44, *75*
Zelená, J., 11, *58, 75*
Zemanek, R., 13, *63*
Zenker, W., 10, 44, 45, *64, 75*
Zeuthen, E., 416, 428, *475, 477*
Zimmer, K., 416, 421, 422, *475, 477*
Zimmerman, J. L., 384, *413*
Žlábek, K., 23, 32, *53*
Zolman, J. F., 358, 367, 368, 369, *412*
Zonov, G. B., 373, 383, *413*
Zorzoli, G. C., 20, *75*
Zotterman, Y., 15, *60*, 359, *403*
Zusi, R. L., 130, 133, 134, *257*

Index to Bird Names

A

Acanthis, 373, 383
  flammea, 349, 373, 376, 383, 384
  hornemanni, 384
Aeronautes saxatalis, 346, 446
Agelaius phoeniceus, 391
Aix sponsa, 420
Amazilia, 443
  fimbriata, 417, 424, 427, 429, 444, 446, 451, 453, 456, 458, 460
Ammoperdix heyi, 292
Anas
  platyrhynchos, 13, 297, 417, 419, 465, 466
  rubripes, 297, 314, 419, 420, 424, 425, 427, 428, 436, 440, 445, 450, 451, 456, 458, 460
Ani, Smooth-billed, *see Crotophaga ani*
Anseriformes, 270, 333
Apodiformes, 270
Apus apus, 347
Archilochus, 352
  alexandri, 347, 353
Ardea cinerea, 302, 321
Auriparus flaviceps, 371

B

Blackbird
  European, *see Turdus merula*
  Red-winged, *see Agelaius phoeniceus*
Blackcap, *see Sylvia atricapilla*
Bluebird, Eastern, *see Sialia sialis*
Booby, Masked, *see Sula dactylatra*
Branta canadensis, 290, 291, 390
Bubo, 231
Budgerigar, *see Melopsittacus undulatus*
Bunting, Snow, *see Plectrophenax nivalis*

C

Calypte
  anna, 344, 346, 349, 372
  costae, 347, 427, 440
Campylorhynchus brunneicapillus, 331, 332, 371, 372
Caprimulgus europaeus, 346
Cardinal, Common, *see Cardinalis cardinalis*
Cardinalis cardinalis, 273, 290, 301, 338, 342, 380, 386, 457
Carduelis
  chloris, 108, 109, 349
  spinus, 376
Carpodacus
  erythrinus, 376
  mexicanus, 82, 338
Cathartes aura, 330, 344, 345, 351–353
Cathartidae, 333
Certhia brachydactyla, 372
Chaffinch, *see Fringilla coelebs*
Cheramoeca leucosternum, 345
Chickadee, Black-capped, *see Parus atricapillus*
Chicken, 83, 88, 105, 109, 356–358, 360–364, 366–370, 378, 379, 382, 385, 386
Chordeiles minor, 342, 351, 371
Ciconia ciconia, 416
Ciconiiformes, 270
Cinclus mexicanus, 273, 366
Clark's Nutcracker, *see Nucifraga columbiana*
Coccothraustes vespertinus, 273, 317, 367, 374, 378, 380, 419, 420, 424, 425, 427, 428, 440
Cockerels, White Leghorn, 105
Colies, *see Colius*

494

INDEX TO BIRD NAMES

Colinus virginianus, 290
Colius, 345
   striatus, 440, 457
Columba, 443
   albitorques, 373
   livia, 89, 99, 109, 276, 327, 338, 342, 365, 379, 380, 420, 424, 427, 440, 444, 451, 453, 456, 458
   oenas, 465
   palumbus, 465
Columbiformes, 90, 270, 333
Coragyps atratus, 331, 430
Corvus, 146, 147, 153, 155, 156, 212, 231, 416
   brachyrhynchos, 417, 419–421
   corax, 280
   corone, 421
   corone cornix, 419
   ruficollis, 292
Cotingids, 160
Coturnix
   chinensis, 327, 338
   coturnix, 420
Cowbird, Brown-headed, *see Molothrus ater*
Crotophaga ani, 345, 352
Crow, *see Corvus*
   Carrion, *see Corvus corone*
   Common, *see Corvus brachyrhynchos*
   Hooded, *see Corvus corone cornix*
Cuculidae, 247
Cyanocitta
   cristata, 273
   stelleri, 292, 380, 381
Cygnus olor, 40

D

Delichon urbica, 428, 458, 460
Dendrocopos
   major, 11
   pubescens, 390, 391
   syriacus, 11
Dendroica coronata, 459
Diomedea, 373
Dipper, American, *see Cinclus mexicanus*
Dove, 336, 337
   European Turtle-, *see Streptopelia turtur*
   Inca, *see Scardafella inca*
   Mourning, *see Zenaida macroura*

   Ring, 320, 370
   Stock, *see Columba oenas*
Duck, 109
   Black, *see Anas rubripes*
   domestic, 83, 89, 95, 332, 364
   Mallard, *see Anas platyrhynchos*
   New Zealand Blue, 227

E

Emberiza
   citrinella, 384, 385
   hortulana, 376, 385
Emerald, Glittering-throated, *see Amazilia fimbriata*
Estrilda troglodytes, 417, 420, 457
Eugenes, 352
   fulgens, 346, 347, 349, 353
Eulampis jugularis, 347–349, 427
Euplectes hordaceus, 388
Eurostopodus guttatus, 338

F

Falco jugger, 430
Falcon, Laggar, *see Falco jugger*
Falconiformes, 270
Finch, 149, 150, 225, 247
   House, *see Carpodacus mexicanus*
   rosy-, *see Leucosticte*
   Zebra, *see Poephila guttata*
Flycatchers, tyrannid, 10
Fowl, domestic, 7, 89, 97, 275
Fringilla
   coelebs, 313, 376, 416, 427, 428, 459
   montifringilla, 349, 376
Frogmouth, Little Papuan, 338
Fulmar
   Giant, *see Macronectes giganteus*
   Northern, *see Fulmarus glacialis*
Fulmarus glacialis, 430

G

Galliformes, 270, 333
Gallus, 133
   gallus, 105
Geococcyx californianus, 307, 327, 337, 338, 342, 345, 352, 371, 372, 457
Goose
   Canadian, *see Branta canadensis*
   domestic, 369
Grackle, Common, *see Quiscalus quiscula*

Greenfinch, *see Carduelis chloris*
Grosbeak, Evening, *see Coccothraustes vespertinus*
Gull
  Black-headed, *see Larus ridibundus*
  Glaucous-winged, *see Larus glaucescens*
  Great Black-backed, *see Larus marinus*
  Herring, *see Larus argentatus*
  Laughing, *see Larus atricilla*

H

Hawk, 149
Heron, Gray, *see Ardea cinerea*
Hirundo rustica, 428, 458, 460
Hummingbird, 230, 247, 277, 345
  Anna's, *see Calypte anna*
  Black-chinned, *see Archilochus alexandri*
  Broad-tailed, *see Selasphorus platycercus*
  Calliope, *see Stellula calliope*
  Costa's, *see Calypte costae*

J

Jay
  Blue, *see Cyanocitta cristata*
  Gray, *see Perisoreus canadensis*
Junco hyemalis, 373
Junglefowl, Red, *see Gallus gallus*

L

Lagopus, 311
  lagopus, 290, 377, 380, 384
  mutus, 382
Lampornis, 352
  clemenciae, 353
Larus, 389
  argentatus, 419
  atricilla, 312, 313, 427, 429
  delawarensis, 420, 424, 427, 428, 440
  glaucescens, 290, 291, 321
  marinus, 302, 321, 362, 435, 444, 445
  ridibundus, 368
Leucosticte, 383
  arctoa atrata, 372
  arctoa tephrocotis, 372
Lophortyx gambelii, 371
Loxia, 384
Loxops coccinea, 227

M

Macronectes giganteus, 365
Magpie, Black-billed, *see Pica pica*
Manorina, 127, 162
Martin
  House, *see Delichon urbica*
  Purple, *see Progne subis*
Meadowlark, Eastern, *see Sturnella magna*
Melanerpes uropygialis, 372
Meliphagidae, 127, 162, 236, 248
Melithreptus, 122, 127, 157
Melopsittacus undulatus, 7, 135, 136, 313, 314, 321, 326, 327, 332, 417, 422, 424, 427–429, 435, 440, 445, 446, 448–451, 453, 456, 458, 465
Micrathene whitneyi, 372
Moho, 236, 248
Molothrus ater, 292, 307, 372
Montifringilla, 127
Mousebird, Speckled, *see Colius striatus*
Murres, *see Uria*
Mycteria americana, 321, 330
Myiarchus
  cinerascens, 372
  tyrannulus, 372

N

Nectarinia, 346
Nectariniidae, 236, 248
Nighthawk, Common, *see Chordeiles minor*
Nightjar
  European, **see *Caprimulgus europaeus***
  Spotted, **see *Eurostopodus guttatus***
Nucifraga, 384
  columbiana, 227
Nuthatch, *see Sitta*
Nyctea scandiaca, 301, 306–308, 322, 345

O

Oedistoma, 248
Oreotrochilus estella, 373
Ostrich, *see Struthio camelus*
Otus trichopsis, 457
Owl, 128, 138, 149, 247
  Burrowing, *see Speotyto cunicularia*
  Snowy, *see Nyctea scandiaca*
  Spotted Screech-, *see Otus trichopsis*

P

Pachyptila, 443
    turtur, 444
Panterpe insignis, 346, 347, 349
Parids, *see Parus*
Parrot, 129, 135, 136, 149, 225
Partridge, Desert, *see Ammoperdix heyi*
Parus, 383
    atricapillus, 277, 315, 345, 346, 373, 417, 420
    major, 318, 346, 349, 380
    montanus, 373, 383
Passer, 122, 127, 183, 211
    domesticus, 82, 85–89, 93, 94, 96–103, 105–107, 109–111, 113, 332, 349, 355, 359, 367, 373, 375–381, 389, 391, 431
    melanurus, 330
    montanus, 349
Passeriformes, 113, 228, 238, 271, 333
Patagona, 352
Pelecanidae, 333
Pelecanus erythrorhynchos, 297, 334
Pelican, White, *see Pelecanus erythrorhynchos*
Penguin, 281
    Adélie, *see Pygoscelis adeliae*
    Humboldt, *see Spheniscus humboldti*
Perisoreus canadensis, 273, 277, 380, 381
Petronia, 127
Phaethon, 443
    rubricauda, 444
Phalacrocoracidae, 333
Phalaenoptilus, 352
    nuttallii, 327, 329, 338, 342, 345–347, 353, 440, 446
Phasianus colchicus, 316, 321, 368, 420
Pheasant, Ring-necked, *see Phasianus colchicus*
Pica, 231
    pica, 101
Pici, 247
Piciformes, 270
Pigeon, 7, 101, 105, 111, 112, 189, 192, 277, 314, 316, 317, 332, 335–337, 355, 358, 364, 366, 374, 378
    domestic, *see Columba livia*
    Wood, *see Columba palumbus*
Plectrophenax nivalis, 273
Ploceidae, 127

Ploceus cucullatus, 327
Plover, 162
    Lesser Golden, *see Pluvialis dominica*
Pluvialis dominica, 462
Podargus ocellatus, 338
Poephila guttata, 301, 307, 327, 329, 330, 338, 342
Polioptila melanura, 371
Poorwill, *see Phalaenoptilus nuttallii*
Procellariiformes, 281
Progne subis, 290, 312, 427, 428, 445
Promerops, 236, 248
Psittaci, 247
Ptarmigan
    Rock, *see Lagopus mutus*
    White-tailed, 277
    Willow, *see Lagopus lagopus*
Pterocles namaqua, 330
Ptiloprora, 127
Puffinus tenuirostris, 444
Pygoscelis adeliae, 281
Pyrrhula pyrrhula, 379, 381
Pyrrhuloxia sinuata, 338, 442

Q

Quail
    Bobwhite, *see Colinus virginianus*
    Japanese, 83, 85, 105, 135, 357, 360, 362, 364, 368, 370
Quiscalus quiscula, 290, 317, 367

R

Ratites, 281
Raven, 277
    Brown-necked, *see Corvus ruficollis*
    Common, *see Corvus corax*
Redpoll, *see Acanthis*
Rhea, Common, *see Rhea americana*
Rhea americana, 322, 326, 431, 455
Roadrunner, Greater, *see Geococcyx californianus*

S

Salpinctes obsoletus, 373
Sandgrouse, Namaqua, *see Pterocles namaqua*
Scardafella inca, 338
Selasphorus platycercus, 350
Sialia sialis, 372, 383

Sitta, 383
   canadensis, 380, 457
   carolinensis, 372
   pygmaea, 372, 383
Sparrow
   Cape, *see Passer melanurus*
   Harris's, *see Zonotrichia querula*
   House, *see Passer domesticus*
   Tree, *see Spizella arborea*
   White-crowned, *see Zonotrichia leucophrys gambelii*
   White-throated, *see Zonotrichia albicollis*
Speotyto cunicularia, 342
Spheniscus humboldti, 281
Spizella arborea, 375, 381, 382
Sporophila, 435
Starling, European, *see Sturnus vulgaris*
Stellula calliope, 349, 372
Sterna paradisaea, 463
Stork, White, *see Ciconia ciconia*
Streptopelia
   "*risoria*," 373
   turtur, 464
Strigiformes, 270
Struthio camelus, 319, 322, 327, 329, 338, 339, 342, 457
Sturnella magna, 290
Sturnus vulgaris, 290, 373, 374, 380, 383, 419
Sula dactylatra, 301, 371, 373
Sunbirds, *see Nectarinia*
Swallow, 345
   Violet-green, *see Tachycineta thalassina*
   White-backed, *see Cheramoeca leucosternum*
Swan, Mute, *see Cygnus olor*
Swift, 247
   European, *see Apus apus*
   White-throated, *see Aeronautes saxatalis*
Sylvia atricapilla, 388
Sylvia borin, 388

T

Tachycineta thalassina, 345, 445
Tern, Arctic, *see Sterna paradisaea*

Thrush, Hermit, 332
Tit
   Great, *see Parus major*
   Willow, *see Parus montanus*
Toxorhamphus, 236
Troglodytes aedon, 298, 315
Trogons, 247
Turdus merula, 419
Turnices, 132
Tyto, 231
   alba, 445

U

Uria
   aalge, 131
   lomvia, 131, 419

V

Vireo, Red-eyed, *see Vireo olivaceus*
Vireo olivaceus, 290
Vulture
   Black, *see Coragyps atratus*
   Turkey, *see Cathartes aura*

W

Warbler
   Garden, *see Sylvia borin*
   Myrtle, *see Dendroica coronata*
Waxbill, Black-rumped, *see Estrilda troglodytes*
Woodpecker, 129, 135, 148, 149, 222
   Downy, *see Dendrocopos pubescens*
Wood-Stork, American, *see Mycteria americana*
Wren
   Cactus, *see Campylorhynchus brunneicapillus*
   Carolina, 332
   House, *see Troglodytes aedon*

Z

Zenaida macroura, 290
Zonotrichia
   albicollis, 86
   leucophrys, 372
   leucophrys gambelii, 82, 86, 101, 104, 105, 280, 307, 308, 384
   querula, 82

Subject Index

A

Abducent nerve, 4, 18
Absorptivity, 288
Accessory nerves, 6, 18
Acclimation, 374–388
Acclimatization, 374–388
 behavioral, 382–384
 insulative, 380, 381
 metabolic, 375–380
 nutritional, 381, 382
Acoustic nerve, 5, 18
Adaptation, physiological, 223–226
Adductor longus muscle, 8
Adductor magnus muscle, 8
Adrenal gland, 41
Adrenal plexus, 21
Adrenaline, 23, 357, 368
Albedo, 288, 289, 292
Alkalosis, respiratory, 339
Allen's Rule, 389, 393
Allometric analysis, 264, 265
Aminergic nerve fibers, 97
Ampulla, 5
Anterior commissure, 355, 356
Anterior mesenteric plexus, 21
Aortic plexus, 21, 38
Articular pads, 157
Articulations, 133, 162–164, see also specific types
Atrioventricular node, 31
Auerbach's plexus, 25, 30, 37
Auricular nerve, 19
Autonomic ganglia, 25–31
Autonomic innervation, 31–45
 cardiovascular, 31, 32
 digestive, 34–38
 endocrine, 41–43

excretory, 38, 39
eye, 44, 45
genital, 39–41
integumentary, 43, 44
pineal, 105–107
respiratory, 33, 34
Axillary nerve, 8

B

Basal metabolic rate, 269–272, 377, 380, 385, 387–389
Basitemporal articulation, 126
Bergmann's Rule, 389, 393
Biceps femoris, 8
Blood vessels, 32, 33
Body temperature, 86, 88, 279–284, 345, 443–447
 daily cycle, 281
 evaluation of, 282–284
 phylogenetic variation, 281, 282
Boundary layer, 295
Brown fat, 318
Burrowing, 382

C

Calorigenesis
 control of, 366–369
 nonshivering, 367–369, 378, 379
 shivering, 378, 379
Calorigenic effect, 311, 312
Carbon dioxide production, 429
Cardiac nerve, 31
Cardiac output, 439–443
Cardiovascular afferent nerves, 14, 15
Carotid body, 14
Cartilage, 157, 158
Caudofemoralis muscle, 8
Celiac plexus, 21

499

Cephalic carotid nerve, 19
Ceratomandibular muscle, 6
Cholinesterase, 22
Chorda tympani, 19
Ciliary ganglion, 17, 18, 25–27, 44
Circadian rhythmicity, 86–89
Cloaca, 36–38
Coccygeal nerves, 8
Cochlea, 5
Cochlear nerve, 5
Cold stress, 309
Comb, 366
Conductance, 388, 389
Conduction, 294–298
Conductivity coefficient, 295
Convection, 298–302, 322
  forced, 313
Coracobrachialis dorsalis muscle, 8
Countercurrent exchange, 331, 333
Cranial nerve, 1–7
Cranial pectoral nerve, 7
"Critical gradient," 278, 279
Critical temperature, 272, 273, 276–279
  lower, 267, 272, 377, 380
  upper, 267
Crop, 34, 35, 384
Cucullaris muscle, 6
Cutaneous colli muscle, 6
Cutaneous mechanoreceptors, 13

D

Deep gluteal muscle, 8
Deep pectoral muscle, 7, 8
Deltoideus muscle, 8
Depressor mandibulae muscle, 221
Duodenum, 29, 30, 36

E

Ecogeographic rules, 389–393
Ectethmoid-mandibular articulation, 126, 127
Edinger–Westphal accessory nucleus, 3
Electromyography, 316–319
Emission coefficients, 293
Emissivity, 288, 290
Energy transfer
  conduction, 294–298
  convection, 298–302
  evaporation, 302–307, 314
  during flight, 312–315, 321
  thermal radiation, 287–294
Enteric plexus, 30, 31
Epiphysis cerebri, 90, 91
Esophageal diverticula, 384
Esophageal plexus, 34
Esophagus, 34, 35
Ethmoidal ganglion, 3, 17, 18
Evaporation, 302–307
  respiratory, 331–337
External obturator muscle, 8
Eye, 44, 45, see also specific parts

F

Facial nerve, 4, 5, 18
Feeding, heat increment of, 311, 312
Felderstruktur, 9, 188–192
Fibrillenstruktur, 9, 188–192
Flight
  circulation, 434–443
  efficiency, 447–449
  energetics, 415–466
  migration, 459–466
  physiology, 415–466
  range, 461–464
  respiration, 415–435
  thermoregulation, 443–455
  water loss, 455–459
Fourier's law, 294, 295

G

Gasserian (semilunar) ganglion, 4, 18
Gastrointestinal afferent nerves, 15
Gemellus muscle, 8
Geniculate ganglion, 18, 19
Genital plexus, 21
Gizzard, 17, 35, 36
Glossopharyngeal nerve, 5, 18
Gular fluttering, 333–337

H

Habenular commissure, 103
Harderian gland, 19, 44
"Hatching" muscle, 127
Heart, 31, 32, see also Cardiac output
  rate, 434–438
  weight, 434–439
Heat, endogenous, source of, 310
Heat conductance, 453, 454
Heat exchange, countercurrent, 321

SUBJECT INDEX 501

Heat loss, 449–455
  center, 355
Heat maintenance center, 355
Heat production, 447–449
  of muscular exercise, 312–315
  nonshivering, 310, 316
  shivering, 310, 315–319
Heat storage, 322–326
Heat stress, 309
Heat transfer
  coefficients, 273–276, 340–342
  dry, 274
  evaporation, 274, 326–340
  maximum, 337, 338
Heliothermia, 344
Herbst corpuscles, 13, 128
Hibernation, 344
Homology, 229–234
Huddling, 383
Humidity, relative, 303–305
6-Hydroxydopamine, 24, 25
Hydroxyindole-O-methyltransferase, 82, 84, 85
5-Hydroxytryptamine, 356, 357
5-Hydroxytryptophan decarboxylase, 84
Hypogastric plexus, 21
Hypoglossal nerve, 6, 7, 18, 19
Hyoid apparatus, 222, 334
Hyperthermia, 268, 282, 322, 341, 355, 356, 386
  heat storage in, 322–326
Hypocapnia, 339
Hypothalamus, 355–359, 370
  temperature of, 360–364, 366
Hypothermia, 343–353, 355, 357, 369, 383, 387

I

Ileum, 37
Iliac muscle, 8
Innervation, see specific types
Insolation, 293
Insulation, 319
  plumage, 380, 381
  tissue, 380, 381, 389
Internal obturator muscle, 8
Internal pudic nerves, 8
Intestine, 36–38
Intrafusal muscle fibers, 11, 12
Iris, 10, 44

Ischiadic nerve, 8
Ischiadic plexus, 8

J

Jugular ganglion, 6, 18

K

Kidney, 38
Kirchoff's law, 288

L

Lagena, 5
Larynx, 17
Lateral geniculate nucleus, 3
Lateral rectus muscle, 4
Lateral tympanic muscle, 5
Lateral vastus muscle, 8
Latissimus dorsi muscle, 7
Ligaments, 133, 134, 158–162
  collagenous, 220, 221
  muscular, 221
Lung, 33, 34

M

M. branchiomandibularis, 222
M. complexus, 127
M. depressor mandibulae, 160, 161
M. latissimus dorsi anterior, 189, 191
M. latissimus dorsi posterior, 189
M. protractor quadrati et pterygoidei, 148
M. pseudotemporalis profundus, 210–212
M. pseudotemporalis superficialis, 161, 210–212, 220
M. pterygoideus, 146
M. serratus metapatagialis, 189, 191
Medial gluteal muscle, 8
Medial vastus muscle, 8
Median nerve, 8
Meissner's plexus, 30
Melatonin, 84
Membrane, nictitating, 4
α-Methylnoradrenaline, 357, 358, 368, 370
α-Methyltryptamine, 358, 370
Morphology
  adaptation, 141
  biological role, 140
  comparative, 228–250
  ecological, 226–228
  faculty, 140
  feature, 139

form, 139
function, 139
niche, 140
synerg, 140
Motor nerve endings, 10
  en grappe, 10, 12, 191
  en plaque, 10–12, 190
Muscle, *see also* specific muscles
  adaptation of, 206–220
  extraocular, 190
  fatigue, 189–190
  fiber types, 187–192
    red, 188
    tonus, 189–192
    twitch, 189–192
    white, 187–192
  isometric contraction, 213–220
  isotonic contraction, 213–220
  one-joint, 194–201
  parallel-fibered, 176–187
  special systems, 220–223
  striated, 9
  two-joint, 201–206
Muscle spindles, 11, 12
Muscular system, 134, 135
  physiology of, 164–169
Myoglobin, 189–191

N

Nasal-frontal hinge, 146–153, 162
Nasal gland, 39
Nerves, *see* specific nerves
Nervous system
  autonomic, 15–46
    adrenergic, 16
    cephalic division, 16–19
    cholinergic, 16
    enteric, 15, 16
    parasympathetic, 15, 16
    paravertebral trunk, 20, 21
    prevertebral ganglia, 20, 21
    spinal preganglionic division, 19, 20
    sympathetic, 15, 16
  peripheral, 1–15
Neuroeffector junctions, 22–25
  adrenergic, 23, 24
  cholinergic, 22, 23
Neurotransmitters, 356–358, 378
Nodose ganglion, 6
Noradrenaline, 23, 356–358

O

Obturator nerve, 8
Occipitomandibular ligament, 221
Oculomotor nerve, 3, 18
Oculomotor nucleus, 3, 4
Olfactory mucosa, 3
Olfactory nerve, 3
Optic nerve, 3
Optic tectum, 3
Osseous arch, 128
Ovary, 39, 40
Oviduct, 39, 40
Oxygen consumption, 426–429
Oxygen debt, 421, 429
Oxygen extraction, 428
Oxygen pulse, 434, 439–443

P

Pancreas, 43
Panting, 333–337, 356, 364, 365
  center, 355, 356
  threshold, 333
Parathyroid gland, 42, 43
Paravertebral ganglia, 19, 27, 28
Pectineus muscle, 8
Pelvic plexus, 21, 38
Peroneal nerve, 8
Petrosal ganglion, 6, 18
Petrosal nerve, 19
Pharynx, 17
Phenoxybenzamine, 358
Photoreception, 84–86
  extraretinal, 85, 86
Phylogeny, 236, 237
Pineal gland, 41, 42
Pineal organ, 79–113
  cell types, 90–101
  circadian rhythmicity and, 86–89
  embryology of, 90
  ependymal cells, 99
  follicles, 101
  functions of, 80–89
  nerve cells, 101–107
  photoreception and, 84–86
  reproduction and, 82–84
  secretory apparatus, 107–109
  structure of, 89–109
  vasculature, 109
Pinealocytes, 90–99

SUBJECT INDEX

Pinnate muscles, 181–187
Plumage
  insulation, 274, 276
  weight, 269, 380, 381, 388
Pneumotaxic center, 356
Poikilothermia, 344
Posterior commissure, 103
Preadaptations, 238–245
Preglossale, 127
Prevertebral ganglia, 19, 28
Propranolol, 358, 368, 369
Proventriculus, 17, 35, 36
Pterygopalatine nerve, 19
Ptiloerection, 341
Ptilomotion, 319, 320, 364, 365, 381
Pulmonary plexus, 33
Pyramidalis muscle, 4

Q

Q_{10} effect, 268, 343, 346, 368, 369
Quadrate muscle, 4
Quadratus femoris muscle, 8

R

Radial nerve, 8
Rectum, 29, 36–38
Rectus femoris muscle, 8
Reflectivity, 288, 290
Remak's nerve, 28, 29, 37
Renal plexus, 21
Reproduction, 82–84
Reserpine, 107, 109
Respiration
  frequency, 422–426
  tidal volume, 422–426
  ventilation, 422–426
Respiratory afferent nerves, 14
Respiratory quotient, 429
Roosting, 383

S

Sacculus, 5
Salivary glands, 5, 6, 34
Sarcomere, 169–174
Sartorius muscle, 8
Scapularis muscle, 8
Scholander model, 263, 267–269, 274, 276, 278, 283–286, 349
Schwann cells, 26
Semimembranosus muscle, 8

Semitendinosus muscle, 8
Sensory corpuscles, 12, 13
Serotonin, 84
Serratus muscle, 7
Shivering, 315–319, 364. 366–369
Sinoatrial node, 31
Skeletomuscular system, 119–257
  morphology of
    descriptive, 124–137
    functional, 137–223
Skeleton, 132, 133, 144–157
  free-body diagrams, 146–151, 193–210
  maximum-minimum principle, 145
Somatic afferent innervation, 11–13
Somatic efferent innervation, 9–11
Sphenopalatine ganglia, 17–19
Spinal ganglion, 7
Spinal nerves, 7, 8
  number of, 7
Spinatus muscle, 8
Splanchnic nerve, 21
Splenius capitis muscle, 127
Stefan–Boltzmann equation, 288, 293
Submandibular ganglia, 17, 18
Superficial gluteal muscle, 8
Superior cervical ganglion, 6, 17, 18, 42, 85, 105
Supracarial paraganglion, 14
Supracoracoid muscle, 7
Surface area, 275
Surface/volume ratio, 272, 295

T

Taste buds, 15
Taxonomy, 245–248
Tegmentum, 111
Temperature
  body, *see* Body temperature
  critical, *see* Critical temperature
  preferred, 369, 370
  regulation, 443–455
  tolerance, 374, 375
Temporolacrimal nerve, 19
Tendon organs, 12
Tendons, 158–162
  ossified, 161, 162
Tensor fasciae latae muscle, 8
Teres major muscle, 8
Teres minor muscle, 8
Testis, 40, 41

Thermal radiation
  emission of, 288–293
  net, 293, 294
Thermoceptors, 354, 358–367
Thermoneutral range, 267, 269
Thermoneutrality, 309
Thermoregulation, 354–374
  behavioral, 369–374
  central neural components, 354–358
  effector response, 365–374
  evolution, 386–393
  neurotransmitters, 356–358
  physiological, 354–369
  role of appendages, 365, 366
  sensory input, 358–365
  vasomotor effects, 365, 366
Thoracic nerve, 7
Thymus gland, 42
Thyroid gland, 42
Tibial nerve, 8
Tongue muscles, 5
Torpor, 281, 344
Trabeculae, 152–154
Tractus pinealis, 90, 91
Trapezius muscle, 6
Triceps brachii, 8
Trigeminal nerve, 4, 18
Trochlear nerve, 4, 18
Tympanic nerve, 19
Tyramine, 379

U

Ulnar nerve, 8
Ultimobranchial body, 43
Ureter, 38
Urohidrosis, 330, 331
Utriculus, 5

V

Vagal nucleus, 17
Vagus nerve, 6, 18, 19
Vapor-pressure deficit, 303
Vaporization, latent heat of, 303
Vas deferens, 40, 41
Vasomotor nerves, 32, 33
Vestibular ganglion, 5
Vestibular nerve, 5
Vidian nerve, 3
Visceral afferent innervation, 13–15
Visceral pleura, 33

W

Water
  loss, 455–458
    cutaneous, 329–331
    metabolic, 323–325
Wattles, 366
Wien's law, 288
Wing beat, 416–421
  body weight and, 419–421
  respiration and, 416–419
Wing-loading index, 419, 431